The Patterned Peatlands
of Minnesota

*This book is sponsored in part by
the Peat Program of the Division of Minerals
and
the Natural Heritage and Nongame Wildlife programs
of the Division of Fish and Wildlife,
Minnesota Department of Natural Resources
and by
the Hamilton P. Traub University Press Fund*

The Patterned Peatlands
of Minnesota

*H. E. Wright, Jr., Barbara A. Coffin,
and Norman E. Aaseng, editors*

University of Minnesota Press
Minneapolis • London

Library of Congress Cataloging-in-Publication Data

The patterned peatlands of Minnesota / H.E. Wright Jr., Barbara A.
   Coffin, and Norman E. Aaseng, editors.
       p.   cm.
     Includes index.
     ISBN 0-8166-1917-4
       1. Peatland ecology—Minnesota. 2. Peatlands—Minnesota.
     3. Human ecology—Minnesota. I. Wright, H. E. (Herbert Edgar),
     1917- . II. Coffin, Barbara. III. Aaseng, Norman E.
     QH105.M55P36 1992
     574.5'26325—dc20   91-36463
                                                                      CIP

Published by the University of Minnesota Press
2037 University Avenue Southeast, Minneapolis, MN 55414
Printed in the United States of America on acid-free paper

The University of Minnesota is an
equal-opportunity educator and employer

Contents

Glossary

Aapa mire: a peatland characterized by well-developed strings and flarks.

Anastomosing: branching and rejoining in a network.

Autecological: the ecology of individual species.

Bog: a peatland characterized by acidic waters with low concentrations of inorganic salts and a poor flora of acid-loving species. Bogs (in contrast to fens) are generally believed to be nourished solely from atmospheric precipitation (ombrotrophic).

Boreal: a northern forest region dominated by spruce and other northern conifers.

Carr: a shrub-covered fen.

Ericaceous: belonging to the plant family *Ericaceae*, including blueberry, heather, and other shrubs common on the peatlands.

Fen: a peatland characterized by weakly acidic or circumneutral waters with higher concentrations of inorganic salts than those found in bogs. Fens (in contrast to bogs) are nourished by waters that transport inorganic salts from mineral soil (minerotrophic).

Fennoscandia: Norway, Denmark, Sweden, and Finland.

Flark: a linear pool or hollow transverse to water flow in a water track.

Forb: an herb other than grass or sedge.

Mesic: a moderately moist habitat.

Minerotrophic: nourished by ions in waters previously in contact with mineral sediment or rock.

Oligotrophic: relatively poor in nutrients (in contrast to eutrophic).

Ombrotrophic: nourished only by the ions contained in atmospheric precipitation.

Paludification: expansion of peatland into forest or prairie.

Patterned peatland: a peatland marked by distinct patterns of vegetation adjusted to the seepage of water that carries nutrients essential to the growth of the plants forming the patterns.

Peatland: a waterlogged terrestrial ecosystem in which a layer of organic matter (peat) accumulates for a thickness of at least 30 cm (12 inches) because of perennial saturation with water.

Raised bog: a bog shaped like a dome and therefore not accessible to adjacent minerotrophic waters.

Riparian: located along the bank of a stream.

Spring fen: a fen localized by emergence of mineral-rich groundwater.

String: a linear peat ridge separating flarks in a water track.

Swamp: a tree-covered fen.

Water table: the top of water-saturated ground. The surface at which the fluid pressure in the pores of a porous medium is exactly equal to atmospheric pressure.

Water track: a path of concentrated water flow between raised bogs or from a mineral source.

Watershed: the area draining to a stream, lake, or peatland. Technically, the outer limit of such an area.

Foreword

Why are the pristine patterns of vegetation on Minnesota's peatlands worthy of protection? This book provides in-depth knowledge that will help thoughtful persons answer that question for themselves. But a short answer is that these peatlands exhibit one of the most remarkable displays on our planet of the complex adjustment of living organisms to environment. And while the research has only begun, the understanding of these relationships has already contributed greatly to the natural and environmental sciences that must lead us all to come to terms with the reality of our earth's life-support system.

The gradual accumulation of knowledge about Minnesota's patterned peatlands over the past century is a fascinating story in the history of the science of ecology. The parallel growth of an appreciation of the complexity of peatland ecosystems and of the importance of preserving at least the best representative examples of each peatland type is an equally fascinating chapter in the growth of environmental awareness in our nation.

Minnesota's vast northern peatlands were first noted in the scientific literature in some detail in Warren Upham's classic monograph on Glacial Lake Agassiz (Upham 1895). The next episode in expanding our knowledge of the patterned peatlands came as a result of the monumental but ill-fated effort to drain the peatlands for agriculture between about 1905 and 1922. This project produced the first topographic map of the vast Red Lake peatland, even though subsequent drainage work ignored known engineering principles and located ditches on section lines a mile or more apart without regard to actual peatland slopes (U.S. Geological Survey 1907-1908). (More than 50 years later this map became a vital aid to me in understanding the vegetation patterns of this fascinating peatland.) Other legacies of the drainage era were the work on peat types, peat depth, peat stratigraphy, and other factors by Soper (1919) and on the fertility and agricultural potential of the peatlands by Alway (1920).

During the drainage era, scientists working in the emerging field of plant ecology were struggling toward an understanding of the origin and development of peatlands. First Transeau (1903) and later Cooper (1913), working on Isle Royale out of the University of Minnesota, observed the centripetal rings of sedges, shrubs, and trees around bog lakes and postulated the classic lake-filling bog succession. Dachnowski's (1912) original drawings of this process were so vivid and widely reproduced in textbooks that in North America lake filling was the only process of peatland development recognized until mid-century.

But Dachnowski (1924, 1925) later became the first serious American student of the stratigraphic record of peatland development as preserved in the plant remains of the peat deposits themselves. It was his descriptions of the puzzling stratigraphy of certain Lake Agassiz peatland profiles (Dachnowski 1925) that led me to question the applicability of the lake-filling succession to Minnesota's vast glacial lake plains and to pursue the stratigraphic record myself. Auer (1928) suggested that paludification ("swamping") was a possibility in Minnesota, and the early topographic maps and some of the peat stratigraphy discussed by Soper reinforced that possibility. So did the work of Rigg (1940).

Just as World War II was commencing, one of the truly seminal insights in the history of science came to fruition in the now-famous paper of Raymond L. Lindeman (1942), based on his doctoral studies at Cedar Creek Bog Lake while he was at the University of Minnesota. Lindeman did not live to see his paper published or to receive the worldwide recognition his work ultimately earned, but much of modern ecosystem theory — particularly that dealing with trophic levels, food chains, energy flows, and nutrient cycling — originated in his studies of this then-obscure Minnesota bog lake (Cook 1977). The lesson in this tragic anecdote is that one never knows what keys to the life-support systems of the earth may be hiding in a peat bog or bog lake. What it takes to unlock the storehouse is a keen and creative young mind with a sense of wonder and curiosity — and a lot of hard work! Ray Lindeman had what it took, and he found the key but did not live to see the full storehouse he had opened.

After the war a British bog ecologist, Verona M. Conway, made an extensive study of the bogs of north-central Minnesota (Conway 1949), but her study areas did not include the huge peatlands on the abandoned lake beds of Glacial Lakes Agassiz, Aitkin, or Upham. Thus she did not encounter the patterned peatlands or significant evidence of paludification. Her sites seem mostly to have been more or less typical examples of the lake-filling process, which is the dominant mode of bog development on the glacial moraines and outwash plains of her study region.

About this same time, ecology professor Donald B. Lawrence of the University of Minnesota showed geology professor Herbert E. Wright, Jr., some aerial photographs of the Red Lake peatland that had been taken by a Naval Reserve pilot on training flights over the area. By then, high-altitude vertical air photos of the region taken by the U.S. Department of Agriculture were also available. The striking vegetation patterns caught the attention of both men. Wright, a pilot in World War II, was familiar with air photos. He recognized the potential scientific significance of the patterns but was puzzled by their nature and possible mode of origin. The two organized a field trip to the peatland, setting out a series of stakes in what we now call a patterned fen to test the theory that the peat ridges were caused by some sort of soil creep, possibly triggered by frost action. After a year or two the stakes were remeasured, but no significant movement was detected.

In 1955 I began my doctoral research in the Lake Agassiz peatlands. Fortunately for me, I took several courses in glacial geology and geomorphology from H. E. Wright, who became a valued member of my doctoral committee. At first my research focused on factors such as peat depth, soil acidity, temperatures, water movement, and peat types, which were thought to influence the growth rate of black spruce on peatlands more or less directly. This was the interest of my agency and the rationale for the support of my study. But I soon began to realize that my study site at Lindford Bog simply did not fit the existing North American ecological literature on peatland development. The peatland was clearly not a filled lake. It had distinct slopes, which I mapped after contract surveying, and it had certain, albeit faint, vegetation patterns related to surface slopes and water movement paths.

While I was conducting these studies, H. E. Wright often gave helpful advice, but he kept plying me with questions about the Red Lake peatland patterns. At first I dismissed his interest by noting that the patterns were probably irrelevant to black spruce site quality, and in any case the problems of their nature and origin were too nebulous to deal with in the time I had left. But finally I realized that elucidating the origin and nature of the vast Lake Agassiz peatlands might just be the key to understanding forest sites on peatlands, as well as a badly needed step forward in peatlands ecology. With only a year or two remaining for my research, I undertook a general survey of the Lake Agassiz peatlands, being careful to include field studies of the most puzzling examples of H. E. Wright's patterned terrain. It became one of the most fascinating episodes in my entire research career!

I was fortunate to have several key tools lacked by earlier bog scientists. The most critical were air photos, access to aircraft for reconnaissance flights, topographic maps (although they were crude at that time), and the published work of many who had come before me, notably the papers of Soper (1919), Dachnowski (1924, 1925), and Auer (1920, 1928). Then, as I worked in the field, a torrent of highly relevant new peatland papers came to my attention. The most of these were the studies of Gorham (1956, 1957) detailing the role of the chemical properties of peatland waters in bog and fen development; the work of Sjörs (1948, 1950, 1959, 1961) in Sweden and Canada on water chemistry and vegetation patterns; and the work of Drury (1956) in Alaska on patterned peatlands and bog processes. As the work progressed, I had several key discussions with H. E. Wright. In the end we both reached the conclusion that the Lake Agassiz peatland patterns were striking examples of a whole family of patterned peatland types that occur in varied form around the entire boreal region of the globe (Heinselman 1963). I owe a great debt to Herbert Wright for his constant encouragement and insights, for without them I surely would never have recognized the significance of the patterned peatlands.

While I was completing my research, Herbert Wright established the Limnological Research Center at the University of Minnesota. Soon a series of talented graduate students and visiting scientists began probing the mysteries of the patterned peatlands. The studies of vegetation and pollen profiles in the Myrtle Lake peatland and elsewhere by Janssen (1968, 1984), and in the Red Lake peatland by Hofstetter (1969) and Griffin (1975, 1977) are especially notable. While this work was under way, I carried out a new study in the Myrtle Lake peatland to test some of my theories about the history of Myrtle Lake and the development of the peatland (Heinselman 1970).

My work in the patterned peatlands made me acutely aware that pulpwood logging and Christmas-tree harvesting, as well as my own and others' use of muskeg tractors, were having an impact on the fragile ecosystems of these remarkable landscapes. The much greater impact of possible drainage, road construction, or peat mining was also clear. I discussed my concerns with several of my colleagues in the U.S. Forest Service and found strong support for a research natural area in Eugene Jamrock and Clarence Buckman, both of whom had worked in the Littlefork area and were supportive of my research in the Myrtle Lake peatland. In 1962, with the support of regional forester Art Keenan, Division of Forestry director Edward Lawson, and commissioner Clarence Prout, the Minnesota Department of Conservation designated the Myrtle Lake peatland as a research natural area, calling it the Lake Agassiz Peatlands Natural Area. In 1964 the National Park Service designated the same area as a Registered National Natural Landmark. In 1974 the Park Service also designated the Red Lake peatland as a National Natural Landmark, but its official registration was never accomplished because of local opposition. Thus began the first efforts to protect some of Minnesota's most distinctive patterned peatlands.

Then in 1975 came the request of the Minnesota Gas Company (Minnegasco) to lease 300,000 acres for peatland mining in the Red Lake peatland. This major threat to Minnesota's most diverse and outstanding patterned peatland never materialized, but it resulted in the Minnesota Peat Program, which spawned the many new studies reported in this volume. The threat also stimulated interest in designating certain peatlands for protection against future disturbance; this designation was accomplished in 1991, as recounted in the final chapter of this book. This brief historical overview is intended as a context for the fascinating new knowledge recorded here.

Miron L. Heinselman
May 1991

Literature Cited

Alway, F. J. 1920. Agricultural value and reclamation of Minnesota peat soils. University of Minnesota Agricultural Experiment Station Bulletin 188.

Auer, V. 1920. On the origin of the Strange in peat bogs. Acta Forestalia Fennica 12:23-145.

———. 1928. Some future problems of peat bog investigations in Canada. Commentations Forestales I.

Conway, V. M. 1949. The bogs of central Minnesota. Ecological Monographs 19:173-206.

Cook, R. E. 1977. Raymond Lindeman and the trophic-dynamic concept in ecology. Science 198:22-26

Cooper, W. S. 1913. The climax forest of Isle Royale, Lake Superior, and its development. Botanical Gazette 15:1-44, 115-140, 189-235.

Dachnowski, A. P. 1912. Peat deposits of Ohio. Ohio Geological Survey Bulletin 16.

———. 1924. The stratigraphic study of peat deposits. Soil Science 17:107-133.

———. 1925. Profiles of peatlands within limits of extinct Glacial Lakes Agassiz and Wisconsin. Botanical Gazette 80:345-366.

Drury, W. H. 1956. Bog flats and physiographic processes in the upper Kuskokwim River region, Alaska. Contributions of the Gray Herbarium, Harvard University 178.

Gorham, E. 1956. The ionic composition of some bog and fen waters in the English Lake District. Journal of Ecology 44:142-152.

———. 1957. The development of peatlands. Quarterly Review of Biology 32:145-166.

Griffin, K. O. 1975. Vegetation studies and modern pollen spectra from the Red Lake peatland, northern Minnesota. Ecology 56:531-546.

———. 1977. Paleoecological aspects of the Red Lake peatland, northern Minnesota. Canadian Journal of Botany 55:172-192.

Heinselman, M. L. 1963. Forest sites, bog processes, and peatland types in the Glacial Lake Agassiz region, Minnesota. Ecological Monographs 33:327-374.

———. 1970. Landscape evolution, peatland types, and the environment in the Lake Agassiz Peatlands Natural Area, Minnesota. Ecological Monographs 40:235-261.

Hofstetter, R. H. 1969. Floristic and ecological studies of wetlands in Minnesota. Ph.D. dissertation. University of Minnesota, Minneapolis, Minnesota, USA.

Janssen, C. R. 1968. Myrtle Lake: A late- and post-glacial pollen diagram from northern Minnesota. Canadian Journal of Botany 46:1398-1408.

———. 1984. Modern pollen assemblages and vegetation in the Myrtle Lake peatland, Minnesota. Ecological Monographs 54:213-252.

Lindeman, R. L. 1942. The trophic-dynamic aspect of ecology. Ecology 23:399-418.

Rigg, G. B. 1940. The development of Sphagnum bogs in North America. Botanical Review 6:666-693.

Sjörs, H. 1948. Mire vegetation in Bergslagen, Sweden. Acta Phytogeographica Suecica 21.

———. 1950. On the relation between vegetation and electrolytes in North Swedish mire waters. Oikos 2:241-258.

———. 1959. Bogs and fens in the Hudson Bay lowlands. Arctic 12:2-19.

———. 1961. Surface patterns in boreal peatland. Endeavour 20:217-224.

Soper, E. K. 1919. The peat deposits of Minnesota. University of Minnesota Geological Survey Bulletin 16.

Transeau, E. N. 1903. On the geographic distribution and ecological relations of the bog plant societies of northern North America. Botanical Gazette 36:401-420.

Upham, W. 1895. The Glacial Lake Agassiz. U.S. Geological Survey, Monograph 25.

U.S. Geological Survey. 1907-1908. Topographic map of the Ceded Lands, Red Lake Indian Reservation, Minnesota.

Preface

In a real sense we can attribute the beginnings of this volume to the energy crisis of 1974 and the proposal of Minnegasco in 1975 to lease 300,000 acres of the Glacial Lake Agassiz peatlands in order to mine peat as a supplemental energy source. The lease proposal prompted the state legislature to fund a Peat Program within the Minerals Division of the Department of Natural Resources. This division subsidized a number of research projects on the ecology, hydrology, and economic uses of peat, which helped spark renewed interest in understanding Minnesota's six million acres of peatland.

The editors and authors of this volume have an intimate understanding of certain aspects of the Minnesota patterned peatlands. H. E. Wright, a professor in the University of Minnesota geology department, was instrumental in stimulating the early research of M. L. Heinselman, C. R. Janssen, and others in the 1960s through his curiosity about the unusual patterns visible on aerial photos. Barbara Coffin was the staff ecologist and later director of the Peat Program of the Minnesota Department of Natural Resources. Norman Aaseng was the peatland ecologist for the Peat Program from 1979 to 1989.

Following the initiation of state-sponsored research in the late 1970s, a contract to Wright from the U.S. Department of Energy funded the continued study of the plant ecology and historical development of peatlands in the context of the climatic history of northern Minnesota. Additional research was undertaken as part of a major project funded by the National Science Foundation (Ecosystems Program), involving a transect of peatlands from Minnesota to Maritime Canada, headed by Professor Eville Gorham of the University of Minnesota. Paul Glaser and Jan Janssens, in particular, as research associates in the Limnological Research Center, were involved with the research on the vegeta-tion and vegetation history from the beginning. They have made major contributions to the understanding of peatland ecology and developmental history. C. R. Janssen, as occasional visiting scientist from the University of Utrecht in The Netherlands (where natural peatlands were long ago destroyed), independently studied the vegetation of the Myrtle Lake peatland and extended his research to contribute to this volume. Donald Siegel of the U.S. Geological Survey in St. Paul, Minnesota, a close associate of Paul Glaser, participated in joint projects to assess the role of groundwater in peatland development. K. F. Brooks of the University of Minnesota Department of Forest Resources headed the major hydrology contracts sponsored by the state's Peat Program. Kristine Bradof contributed revisions of two chapters of her M.S. thesis in the Department of Geology and Geophysics on the drainage ditches that had attracted hundreds of homesteaders, led the counties to bankruptcy, and indirectly resulted in the partial protection of the land from further exploitation by transfer of the abandoned homestead lands to the state.

The fauna section of this book includes detailed discussions of animal groups for which research and knowledge exists. Gerda Nordquist and Daryl Karns were two of the original investigators for the Peat Program, and their chapters on small mammals and on amphibians and reptiles are developed from their doctoral research in the University of Minnesota Department of Ecology and Behavioral Biology. The chapter on birds by Gerald Niemi and JoAnn Hanowski resulted from research sponsored by Northern States Power Company related to the construction of a power line across some of the largest peatlands. The chapter on large mammals by William E. Berg was recruited from wildlife research biologists of the Department of Natural Resources. No chapter on invertebrates is included, for very little information is available on this faunistic group, even though its role in the functioning of the peatland ecosystem is significant.

The two chapters on prehistoric and historic Native Americans of the peatland area were contributed by Mary Whelan and Melissa Meyer, who had completed related studies elsewhere in Minnesota as graduate students in the Departments of Anthropology and History respectively. The final chapters on economic applications and on protection were prepared by staff specialists Mary Keirstead and Norman Aaseng of the Peat Program and Robert Djupstrom of the Scientific and Natural Areas Program on the Minnesota Department of Natural Resources. Special note should be taken of the foreword by M. L. Heinselman, well-known advocate of wilderness protection in Minnesota and elsewhere, who initiated the first substantial research on the peatlands and set the stage for the subsequent research described in this volume.

The book therefore draws primarily on the experience of researchers and staff associated with the University of Minnesota and the Minnesota Department of Natural Resources. Generous financial support from the Minnesota Department of Natural Resources made production of this publication possible. The Nature Conservancy helped support graphic design efforts. Thanks are due to Rebne Karchefsky for drafting the sketches of landform development in chapters 2 and 3 and the maps, sketches, and stratigraphic diagrams in chapter 13 and to Don Luce for the graphics that begin each part of the book, for the illustrations of some of the peatland organisms, and for the transect of peatland communities used in several chapters. The Introduction and chapters 1-5, 12-14, and 17 are respectively contributions 416-425 of the Limnological Research Center, University of Minnesota.

Peatland research of the past decade has unraveled many secrets of an intricate ecosystem that has been little modified by human action since its development thousands of years ago. Our current understanding of the ecology, hydrology, and socioeconomic history of the Minnesota peatlands provides important perspectives for both the development and the preservation of peatlands wherever they may occur.

H. E. Wright, Jr.
Barbara A. Coffin
Norman E. Aaseng

Introduction

H. E. Wright, Jr.

The major vegetation formations of Minnesota (Fig. 1) reflect the climatic patterns, which themselves result from the frequency of different air masses (Fig. 2). Northern Minnesota is dominated by arctic air and

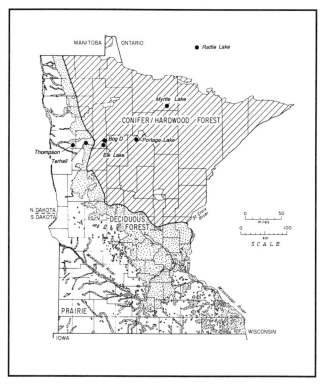

Fig. 1. Major vegetation formations of Minnesota, as existing before extensive agricultural development and forest clearance. The NW/SE trends reflect the regional climate — basically the decreasing temperature from south to north and the decreasing precipitation from northeast to southwest. The dots show the location of pollen-stratigraphic sites used to reconstruct the vegetation history (see Fig. 6).

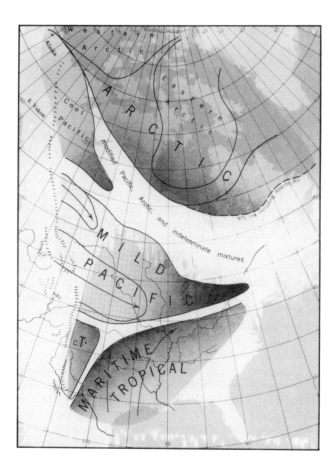

Fig. 2. Map showing by shading the areas dominated by the indicated air-mass types more than 50% of the time. From Bryson, R. A., and W. M. Wendland. 1967. In W. J. Mayer-Oakes, editor. Life, Land, and Water. University of Manitoba Press.

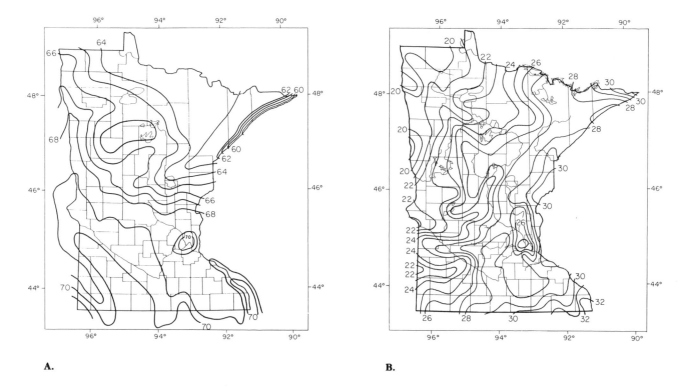

A.   **B.**

Fig. 3. A. Mean summer (June-August) temperatures in Minnesota in °F, illustrating the major latitudinal pattern. Note the warming effect of Lake Superior, as well as the slight cooling effect of the North Shore highland and of the moraine belt that lies east of the Red River Valley in northwestern Minnesota. Note also the heat island around the Minneapolis-St. Paul area, as well as the finger of warmer temperatures up the Mississippi Valley in the southeast. **B.** Mean annual precipitation in Minnesota in inches, showing the general increase from northwest to southeast, as a reflection of the primary moisture source in the southeast.

southern Minnesota by moist tropical air from the Gulf of Mexico and the Caribbean Sea. Thus in northeastern Minnesota the forest is covered largely by boreal conifers, which require relatively cool summers (Fig. 3). Deciduous forest extends in a band broadening to the southeast, reflecting the higher incidence of warm moist air. Prairie dominates in western Minnesota because of the frequent occurrence of summer drought under the influence of dry western air masses that have lost most of their moisture in crossing the western mountains.

The dense and productive upland forests of northern Minnesota produce great quantities of biomass — not equal to that of tropical rain forests, but still significant — and the forest cover protects the hillsides from erosion. Although the conifers (except tamarack) are evergreen and retain their needles a long time, they eventually die and contribute to the detritus on the forest floor, along with remains of deciduous trees, shrubs, and ground plants. There the plant detritus is slowly decomposed by microbes and fungi through various biochemical processes, and the products of decomposition provide the nutrients for new plant growth, assist in soil development, or escape to the atmosphere as gas or to streams and lakes in runoff. These slow processes can be temporarily accelerated

by forest fires, which can accomplish much the same breakdown of the organic detritus in an instant, and in the process of forest regeneration they serve essentially the same function as slow microbial decomposition. Thus over time under natural conditions the upland forests of northern Minnesota have endured, changing gradually in composition with the shifts in climate that have characterized the region since the retreat of the ice sheet more than 10,000 years ago.

Not so with the peat-filled lowlands of northern Minnesota, which belong to a different type of ecosystem. The basic control here is hydrological. Because lowlands are undrained or poorly drained, the rainwater and snowmelt do not move rapidly through the system, so the result is depletion of the oxygen that is necessary for many of the decompositional processes. The water stagnates, and only the least resistant plant tissues decompose. Wood and the more resistant stems, leaves, and fruits decompose very slowly. The resulting peat can accumulate as long as the water levels remain high.

Peatlands are of two topographic types, and both occur in northern Minnesota (Fig. 4). One is confined to depressions in the landscape, forming fringes on lakes or actually replacing lakes. A lake basin itself could have originated as an ice-block depression dur-

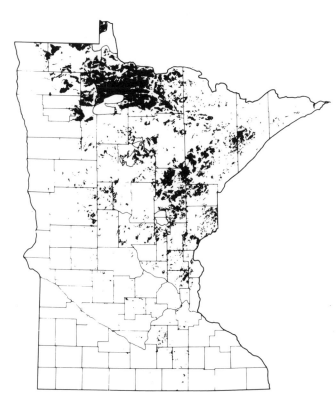

Fig. 4. Peatlands in Minnesota, including both topographic types. The major area north of Upper Red Lake, in the northwest, covers the floor of the eastern arm of Glacial Lake Agassiz.

ing wastage of the last ice sheet. Glacial ice typically contains a variable amount and distribution of rock fragments eroded from the substratum. When ice with a lot of rock debris melts, a hill results; when relatively clean ice melts, a depression results. Almost all of the lake basins in Minnesota originated in this way; only those in the rocky terrain of northeastern Minnesota, especially in the Boundary Waters Canoe Area, originated by differential erosion of relatively nonresistant or closely fractured rock. Original lake basins thus formed fill with water, and the water shallows as sediment is washed in from the surrounding hill slopes or is produced in the lake by growth of algae or littoral vegetation. Ultimately, shallow-water aquatic plants may grow over the entire sediment surface, usually starting at the margin and expanding toward the center. As long as water keeps coming into the basin the plant detritus will remain saturated with stagnant water and will accumulate. If the water level goes down for a period, the surface peat will dry out and partly decompose, and in a stratigraphic section it will appear as a dark band of greater density than the undecomposed peat above and below.

The other kind of peatland originates on flat or very gently sloping ground without the prior development of a lake. The main requirement again is poor drainage, caused primarily by the gentle slope but perhaps

also by the impermeability of the substratum, so that surface water will not drain underground. Of course a major factor in peatland development is a climate with adequate precipitation and summer temperatures that are not so high that evaporation greatly exceeds precipitation. In areas of very heavy rainfall, such as the western parts of Scotland and Ireland, peat can accumulate even on steep slopes, creating the blanket bogs that cover much of the landscape.

Apart from their relation to topography, peatlands can be classified in another important way. If the water flowing across the peatland surface originates on an adjacent upland, it will contain dissolved mineral ions that can provide the nutrients for growth of certain plants, and the vegetation may be quite distinctive. Such a peatland is called a fen. On the other hand, if the peatland surface is raised slightly above a level plain, the mineral-rich water will be diverted, and the peatland then receives its water and nutrients only from direct rainfall and snowmelt. Such a peatland is called a bog, and it is characterized by plants — like *Sphagnum* moss — that can grow under water conditions of relatively high acidity (low pH) and low levels of nutrients.

In northern Minnesota both of these types occur — those that are localized in depressions and those that blanket flat or gently sloping ground — and both of these types occur as fens or as bogs. By far the largest peatlands are those that developed on the plains of glacial lakes, especially Lake Agassiz. These peatlands did not originate as direct descendants of the glacial lakes, however, for the lake plains did not begin to accumulate peat until several thousand years after retreat of the ice sheet allowed the glacial lakes to drain. The reason for the delay was that at first the climate was not sufficiently wet and cool to assure the continuously high water table that is necessary to inhibit decomposition of accumulated plant material. We know this from pollen studies of the vegetation and climatic history in adjacent uplands. Peat formed only on the eastern part of the Lake Agassiz plain, for the portion in the Red River Valley and in adjacent North Dakota and southern Manitoba and Saskatchewan has never been wet enough in postglacial time.

The glacial uplands of northern Minnesota are commonly dotted with lakes, which contain in their organic sediments a continuous record of environmental conditions since the original ice masses melted out to form the depression. Pollen grains blown into the lake from the upland vegetation are preserved in the lake muds, which have a low oxygen content and are subject to little decomposition. The proportion of different types of pollen grains is a rough measure of the composition of the upland vegetation at a variable distance from the lake, depending on the size of the lake and the nature of the vegetation. If a continuous core of the lake sediments is taken, and if pollen samples are analyzed at regular intervals from bottom to top, it is possible to re-

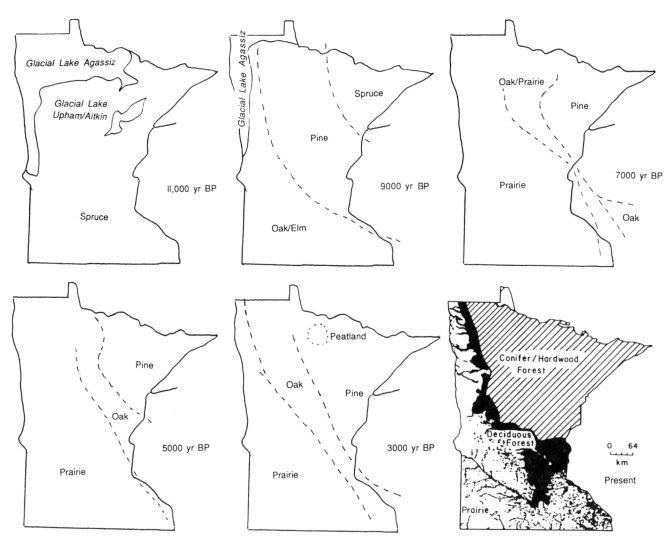

Fig. 5. Sequence of major vegetation types in Minnesota from 11,000 years ago to the present.

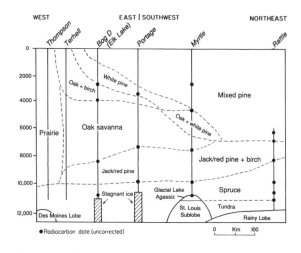

Fig. 6. Transect from west to east in northern Minnesota (see Fig. 1) to illustrate the vegetational history as inferred from pollen diagrams at the locations indicated.

construct the vegetational history and thus the climatic history of the area.

Results of such stratigraphic pollen analysis show that when the ice sheet retreated from northern Minnesota, the first vegetation was a treeless, tundralike formation, followed by a spruce forest that, except for its lack of jack pine, resembles today's boreal forest of central Canada (Figs. 5, 6). Jack pine or red pine immigrated into the area about 10,000 years ago, but it was soon replaced by oak and other hardwood trees of the deciduous forest. Open oak woodland or prairie then succeeded, reaching its full development 7,000 years ago and persisting for a few thousand years thereafter. This trend was a manifestation of a climate that gradually became warmer and drier, a climate that was not favorable for the accumulation of peat, even on flat plains where poor drainage might be expected. The plains of Glacial Lake Agassiz and other glacial lakes, which were drained 10,000 years ago, probably became wet meadows or even tree-covered swamps, but they were not wet enough to keep the water table at the surface, so decomposition kept pace with plant production, and no peat was formed. If any peat did accumulate, it was probably burned off by fires during dry intervals when the water table fell. Chapter 14 describes such a possibility for Myrtle Lake peatland, where a hiatus in the peat sequence is marked by a concentration of charcoal.

About 5,000 years ago the climatic trend reversed, and conditions became cooler and wetter. Pollen diagrams for upland sites show that the prairie or open oak woodland was heavily invaded by birch, and about 3,500 years ago white pine immigrated into north-central Minnesota from the east, to be followed about 1,000 years ago by red pine and jack pine. At the time, the peatlands began to form or expand. Topographic peatlands developed because many lakes had been shallowed by sedimentation, and during the dry periods their water levels may have been lower, allowing rooted aquatic plants to become established and keep pace with the rising lake level. But the blanket peats on the lake plains began to develop until this time, spreading from east to west as the climate became progressively cooler and wetter. Fens and bogs developed in different places, ultimately forming the intricate patterns we see today, as described in chapter 1.

Because the major glacial lake plains have provided the setting for the big peatlands of northern Minnesota, it is appropriate to describe how they came into existence and how they and the surrounding glacial landforms have affected peatland development.

All of the landforms of northern Minnesota owe their origin to the events of the last major glaciation. Several lobes of ice protruded from the main ice sheet, their form and direction of flow being determined primarily by the nature of the underlying topography. At the time of the glacial maximum about 20,000 years ago, these various ice lobes reached as far south as Iowa and Illinois, but as the main ice mass thinned, the lobes became less extensive and their flow direction was even more determined by the underlying topography (Fig. 7). The last ice lobe to affect the peatland area came from the northwest, carrying rock debris eroded from the Paleozoic limestone and dolomite of the Winnipeg lowlands. It formed the conspicuous hilly moraine south of Lower Red Lake as well as the long peninsula between the two Red Lakes.

When the ice first retreated from this area, a glacial lake was formed in its place. The first lake, called Lake Koochiching, drained eastward by the Savanna River northeast of Aitkin into the Mississippi River. It actually received inflow from the earliest phase of Lake Agassiz, which formed at the same time in the southern part of the Red River lowland. The connection was by way of a now-dry channel near Fosston, west of Bagley in Polk County. When the ice retreated farther, the drainage through the Fosston channel was reversed, and the two early lakes soon became confluent over a broad area — the Herman phase of Lake

Agassiz, dated at about 11,700 years ago. The outlet was southward down the Minnesota River Valley. This phase of the lake lasted long enough for beach ridges to form around its perimeter. Although the portion of the lake within the Red River lowland was a few hundred feet deep — deep enough for the deposition of silts and clays — the portion extending eastward over the Red Lakes lowland was too shallow for much deposition. In this region most of the original glacial-depositional landforms were smoothed out by wave and current erosion.

With further retreat of the ice into Canada, Lake Agassiz expanded northward, but at the same time its outlet to the Minnesota River Valley became intermittently eroded, resulting in a steplike development of outlet levels and thus of beach ridges. The portion of the lake occupying the Red River lowland thus became drained as early as about 11,000 years ago. Ultimately the rest of Lake Agassiz drained eastward when the ice retreated far enough to open an outlet to Lake Superior by way of Lake Nipigon in northwestern Ontario. Only a few lakes in Minnesota exist as remnants of Lake Agassiz, notably Upper and Lower Red Lake. Myrtle Lake is one that has persisted despite the thick accumulation of peat all around it, but smaller remnants elsewhere in the Myrtle Lake peatland were completely obliterated by peat growth (chapter 15).

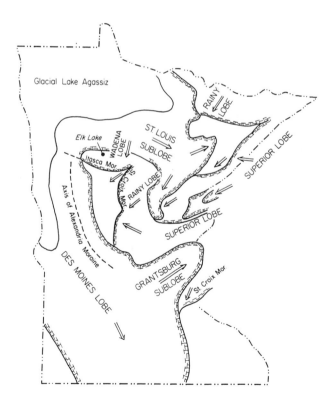

Fig. 7. Distribution of lobes of the Laurentide ice sheet in Minnesota during the last glaciation. The maximum extent of Glacial Lake Agassiz in Minnesota is also shown.

After Lake Agassiz drained, it was subject to erosion by tributaries of the Rainy River, especially the Big Fork and Little Fork rivers. This erosion has produced sharp valleys or gullies, while at the same time peat developed on the flat interfluves. These valleys have guided the development of peatland patterns, however, for they have provided foci for peatland drainage, leaving the intervening areas as sites for peat accumulation.

The great extent of the peatlands of northern Minnesota, then, can be attributed to the glacial history in combination with the climate. Extensive flatlands are necessary to begin with, and the plain of Glacial Lake Agassiz in particular provided the setting. Although Lake Agassiz covered the Red River Valley as well as extensive areas of southern Manitoba and Saskatchewan, the lake erosion and sedimentation did not smooth out the landscape in those areas as much as they did in Minnesota.

Northward from the Lake Agassiz peatlands of northern Minnesota and immediately adjacent northwestern Ontario, the peatlands are largely confined to topographic depressions, but some of the larger ones do contain the string patterns that are characteristic of the boreal forest region across Canada and Alaska as well as in the eastern hemisphere, particularly in Scandinavia and the Soviet Union. The climate throughout the boreal region has a distinctly positive moisture balance — substantially more annual precipitation than evaporation — and consequently the water table on gentle slopes is at or close to the surface, thus inhibiting decomposition of plant detritus.

The greatest area of peatland development in the boreal forest region of North America is the Hudson Bay lowlands, where the flat landscape owes its origin largely to marine deposits of an expanded Hudson Bay, formed as the ice sheet retreated into the bay about 7,000 years ago. There the peat accumulation has been continuous since the drainage of the plain, for the warm and dry period that characterized Lake Agassiz area in Minnesota between 9,000 and 5,000 years ago did not noticeably alter the moisture balance as far north as the Hudson Bay lowlands.

Because the most striking feature of the peatlands of northern Minnesota is the patterned vegetation, which is the subject of Part I, chapters 1 and 2 are devoted to a description of the patterns and the relation of the vegetation to the hydrology — the water flow and the water chemistry. Water is the key to the distribution of particular plant species in peatlands because it provides the nutrients required for plant growth: the plant roots do not reach into mineral soil the way they do in uplands. Water flow may control the oxygen availability for plant growth as well as the extent of microbial decomposition.

The major groups of animals in the peatlands — large and small mammals, birds, and amphibians and reptiles — are the subjects of Part II. Insects are not considered in the book, because of lack of data, although their most infamous representatives — black flies and mosquitoes — are certainly ubiquitous in northern Minnesota.

Because of the importance of water in affecting the nature of the peatland vegetation, hydrologic relations — both surface water and groundwater, as well as the effects of ditching on the flow of surface water — are the subject of Part III. The role of groundwater in peatland hydrology has usually been discounted as unimportant, but on extremely flat plains like the Red Lake peatland the groundwater flow systems can locally discharge mineral-rich water to the surface as a kind of spring and change a bog to a fen.

Part IV focuses on the historical development of the peatlands by the stratigraphic study of peat cores from two areas. Fossil fragments of mosses are especially useful in this respect, for species of *Sphagnum* in particular are highly sensitive to water chemistry and to water flow. Thus the moss content can reveal whether a sample of peat came from a bog or a fen, and the past degree of acidity (pH level) can be inferred from these fossils. If many cores are examined over an area of patterned fens and bogs, the history of pattern development can be reconstructed. Radiocarbon dating of the peat cores provides the chronology.

The final group of chapters is concerned with the use of the peatlands by people — first the Indians in their traditional lifestyle and in modern times, then the efforts to develop the area for agriculture by ditching and homesteading, and finally the recent evaluation of the peatlands for other possible purposes, including use as a source of fuel, for horticultural materials, for forest products, and for educational and scientific purposes. Much of the peatlands is completely undisturbed and represents a pristine piece of landscape in which the natural processes of a relatively simple ecosystem have developed over time with essentially no influence from either modern or ancient people — a wilderness in the truest sense.

PART I
Vegetation

CHAPTER 1

Peat Landforms

Paul H. Glaser

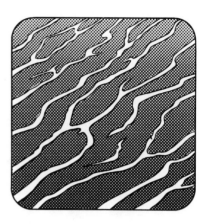

One of the most unusual landscapes in the United States is found on the glacial lake beds of northern Minnesota. Here the irregular mosaic of forest, meadow, and thicket on upland soils abruptly gives way to the symmetrical vegetation patterns on peatlands (Plates 1 and 2). The large uniform stands of vegetation in these peatlands often resemble the shape of geological landforms such as river channels, islands, and ripple marks. In places the patterns may also conjure the image of a fleet of ships steaming at sea. The "ships" have rounded bows and trailing wakes that interrupt the wavelike pattern of the ridges and troughs in the surrounding "ocean." The patterns seem to be in motion, with the ships gliding effortlessly over the waves. This spectacular landscape, however, is actually composed of plant communities that have developed orderly shapes in response to subtle gradients in the water chemistry and water movements.

The development of these vegetation patterns presents many intriguing questions. Natural vegetation usually exhibits some form of pattern, but this pattern is seldom expressed as regular geometric shapes with sharp boundaries. The similarity of the peatland patterns to fluvial landforms implies that they were molded by the erosive action of running water. However, the landscape is virtually flat, and water movements are generally imperceptible on the ground. The occurrence of these patterned peatlands, therefore, must indicate ecological processes that do not operate in upland terrestrial ecosystems.

Discovery and Early Investigation

Of all the ecosystems in Minnesota, the patterned peatlands were probably the last to be thoroughly investigated, because of their great size and inaccessible terrain. The striking vegetation patterns were not discovered until the 1940s, when the Soil Conservation and Stabilization Service produced the first comprehensive set of aerial photographs for the state. Naval Reserve pilots also observed the peatland patterns while training over the Red Lake peatland during World War II. One of these pilots took photographs, which he showed to his Minneapolis neighbor, Professor Donald B. Lawrence of the University of Minnesota. After the war Professor Lawrence organized a trip to investigate these peatland patterns with Professor H. E. Wright, Jr., of the Department of Geology. Working north of Waskish along Highway 71, they made preliminary measurements of the patterns and speculated on their origins. This field trip eventually laid the foundation for several generations of research by scholars working with the Limnological Research Center at the University of Minnesota.

In 1963 Heinselman published his pioneering work on the Glacial Lake Agassiz peatlands in which he interpreted the peatland patterns within the context of the established peatland literature from Europe. During the next phase, detailed vegetation work was conducted by Janssen (1967) and Heinselman (1970) in the Myrtle Lake peatland and by Hofstetter (1969) and Griffin (1975) in parts of the Red Lake peatland. Janssen (1968) and Griffin (1977) also produced the first pollen diagrams from these peatlands, documenting their development during the postglacial period. In this chapter I review the most recent phase of study, which combines the analysis of new types of remote sensing imagery with a regional field sampling program conducted by a helicopter.

Peat Landforms as Ecological Units

The vegetation patterns in boreal peatlands are commonly called peat landforms because of their visual similarity to geomorphic landforms. A peat landform

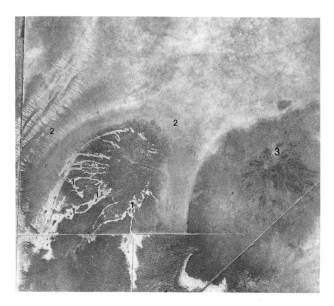

Fig. 1.1 Peat landforms in the Lost River peatland. Peat accumulation produced three types of peat landforms: (1) spring-fen channels on a spring-fen mound, (2) a water track, and (3) a raised bog. The surface elevations of the two peat mounds are nearly identical, although the spring-fen mound has developed over a rise in the mineral substratum, and the raised bog has developed over a depression (Almendinger *et al.* 1986, by permission, *Journal of Ecology;* Glaser *et al.* 1990).

new data on peatland development. Spatial transitions from one type of landform to another may represent important developmental trends, which can be tested by the stratigraphic analysis of peat cores (Glaser *et al.* 1981; Glaser 1987a, b). The physical dimensions of the landforms (length, width, and area) can also be measured and analyzed to infer the processes that formed them (Glaser 1987b).

Peat Landform Types

Three basic types of peat landform occur in northern Minnesota: raised bogs, water tracks, and spring-fen channels. These landform types are closely related to the principal types of peatland vegetation, which are discussed in more detail in the next chapter. Raised bogs have a raised profile in cross section, which isolates the bog surface from solute-rich runoff draining from the surrounding mineral uplands. In northern Minnesota raised bogs have a forested crest from which lines of spruce trees radiate downslope (Heinselman 1963, 1970). This pattern produces the characteristic radiating forest patterns that appear on aerial photographs (Fig. 1.2). On the lower bog flanks, the spruce forest gives way to nonforested lawns, which form an apron around the forested crest. In large peatlands, raised bogs have a streamlined margin where they are trimmed by fen vegetation. These streamlined forms

is characterized by (1) a distinctive shape in cross section and plan view, (2) a distinct vegetation assemblage, and (3) narrow ranges in water chemistry (Glaser *et al.* 1981; Glaser and Janssens 1986; Glaser 1987a). Certain types of landforms are also highly characteristic of discharge areas for groundwater (Tarnocai 1974; Glaser 1983a; Glaser *et al.* 1990).

Each type of landform has a characteristic Grossform (cross-sectional profile) and Kleinform (surface pattern) (Aario 1932; Paasio 1933; Glaser and Janssens 1986). A peat landform is therefore a three-dimensional feature that is composed of living vegetation growing on the surface of a massive deposit of organic matter (Fig. 1.1). These two components of a landform are highly correlated. The surface features visible on aerial photographs (plan view) are consistently oriented according to the prevailing slope. This relationship has been demonstrated by careful topographic leveling in Fennoscandia, eastern Canada, and Minnesota. The scales of these two components, however, are not comparable. Surface relief rarely exceeds 1 m/km in Minnesota and is often less than 50 cm/km. In plan view, however, a peat landform can cover an area up to 140 km^2 in Minnesota.

Each of the different landform types in Minnesota is associated with different ranges in water chemistry and different assemblages of species. The peat landforms therefore provide an important indicator of the peatland environment. The patterns also provide important

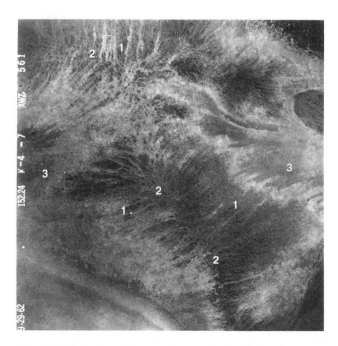

Fig. 1.2. A raised bog with radiating forest patterns. The bog is distinguished by lines of spruce trees (1), which radiate from the central bog crest (2). The lower flanks of the bog are fringed by a nonforested *Sphagnum* lawn (3). This aerial photograph of the Myrtle Lake peatland covers an area 2.5 km across.

Patterned fens may also contain fields of tree islands, which are oriented parallel to the prevailing slope (Fig. 1.4). These islands always contain rounded heads and tapering tails that stream downslope like the wake of a ship at sea.

The third type of peat landform is the spring fen, which is similar to water tracks with tree islands (Fig. 1.1; Plate 3). Spring fens consist of an anastomosing (branching and rejoining) network of nonforested channels that drain through a swamp forest and carry alkaline water. The forest is generally fragmented downslope into long streamlined fingers that split off into distinct tree islands.

Mire-Complex Types

Most of the larger peatlands in northern Minnesota contain more than one landform type and may best be described as mire complexes (*sensu* Cajander 1913). Mire complexes can be classified according to (1) their size and (2) the configuration of bog, fen, and mineral soil in a watershed. Most mire complexes in Minnesota are clearly separated from adjacent peatlands by mineral soil. The borders of these discontinuous complexes (types 1-7) are easier to delineate than the continuous complexes (types 8-11) of the Glacial Lake Agassiz region, in which the watershed divides are often covered by peat (Figs. 1.5, 1.6). The drainage divides in these continuous peatlands, however, may be determined by the vegetation patterns that are sensitively adjusted to the prevailing direction of water movement (color centerfold).

The simplest mire-complex type (type 1) consists

Fig. 1.3. Landsat MSS image of the western water track of the Red Lake peatland. This image (E-30042-16303-D) was recorded during spring breakup in 1978. The water track (*large white arrows*) drains eastward to a large raised bog complex (A), where it splits into two diverging branches. Surface drainage (*darker tones*) arises in narrow channels (*small dark arrows*) downslope from a beach ridge and flows around the central portion of the track, which is still covered with ice (*white tone*). Surface drainage on the adjacent bog (A) is also focused in internal water tracks (dark gray tones), in contrast to the surrounding snow-covered bog areas (*white*). The rectangular lines are drainage ditches spaced 1.6 or 3.2 km apart. Most of the image is covered by peatland except for the snow-covered beach ridge north of the western water track. The image covers an area approximately 64 km across. (See Plates 5-7 for comparison.)

simulate the shape of islands where the bog is completely surrounded by fen vegetation.

Water tracks (Sjörs 1948), in contrast, have a concave-to-flat profile in cross section and represent zones where runoff is channeled across the peat surface. Water tracks have the appearance of river channels on aerial photographs and are sharply delineated from the surrounding vegetation, which generally consists of swamp forests or raised bogs (Fig. 1.3). Three types of water track are found in northern Minnesota (Glaser 1987a). Featureless water tracks contain linear bands of trees and shrubs that are oriented parallel to the prevailing direction of flow. They usually consist of non-forested sedge meadows that are surrounded by wet swamp forests. Patterned fens (Aapamoor or Strangmoor of Fennoscandian authors) are water tracks that contain distinctive networks of peat ridges (strings) and pools (flarks) arranged perpendicular to the slope.

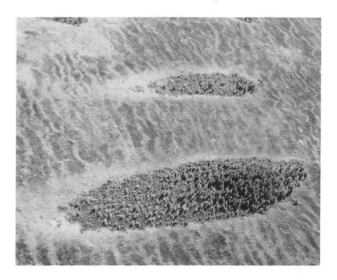

Fig. 1.4. Patterned fen with fields of tree islands. The streamlined tree islands have rounded heads and tapered tails. Notice the regular networks of pools (flarks) and peat ridges (strings) oriented perpendicular to the slope. (Oblique aerial photo by Paul Glaser south of the Lost River peatland, northwestern Minnesota.)

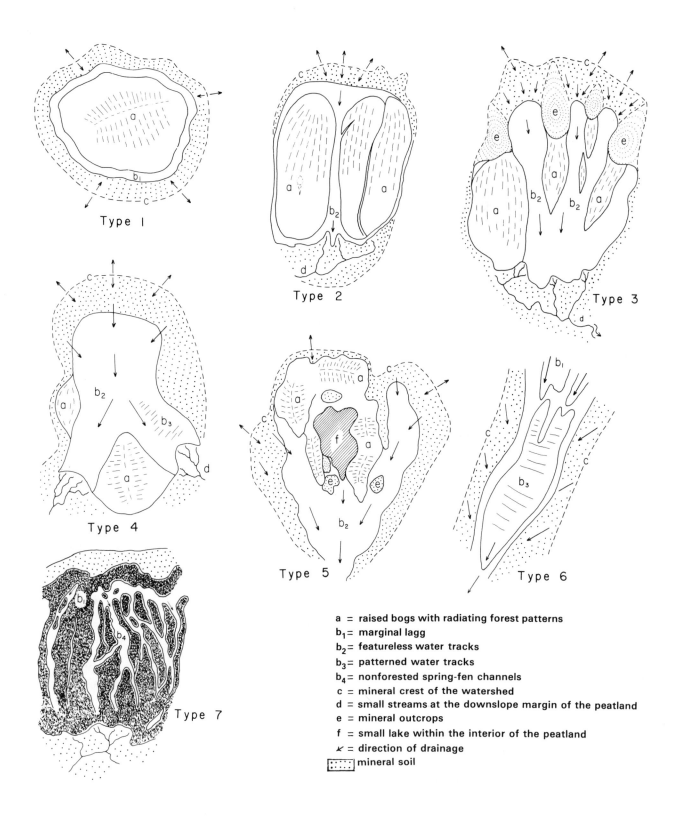

Type 1
Type 2
Type 3
Type 4
Type 5
Type 6
Type 7

a = raised bogs with radiating forest patterns
b_1= marginal lagg
b_2= featureless water tracks
b_3= patterned water tracks
b_4= nonforested spring-fen channels
c = mineral crest of the watershed
d = small streams at the downslope margin of the peatland
e = mineral outcrops
f = small lake within the interior of the peatland
↙ = direction of drainage
⬚ mineral soil

Fig. 1.5. Discontinuous mire-complex types in Minnesota. The types represent reference points along a continuous range of variation. In types 1-5 the area of fen relative to bog increases, culminating in the very large water tracks of type 4 (Glaser 1987a; from *USFWS Biological Report* 85 (7.14), drawn by Rebne P. Karchefsky). The arrows show the direction of water movement.

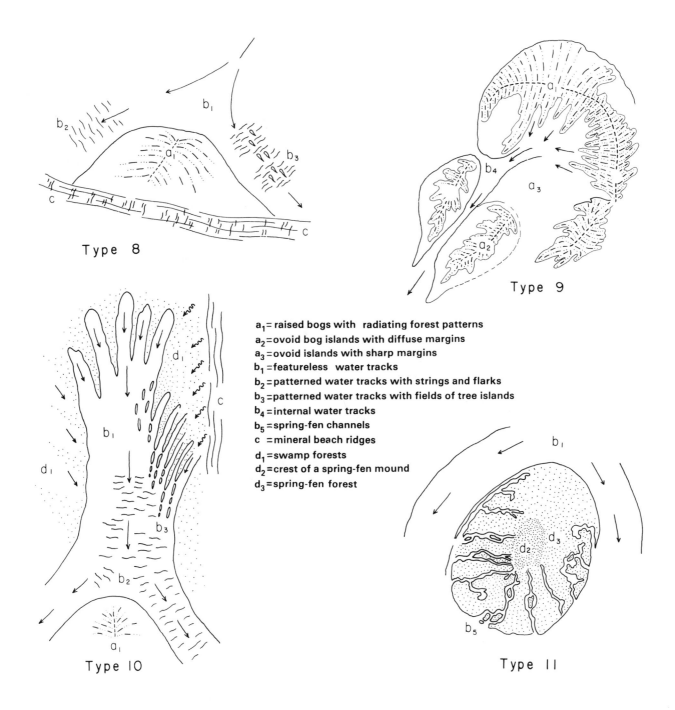

Type 8

Type 9

a₁ = raised bogs with radiating forest patterns
a₂ = ovoid bog islands with diffuse margins
a₃ = ovoid islands with sharp margins
b₁ = featureless water tracks
b₂ = patterned water tracks with strings and flarks
b₃ = patterned water tracks with fields of tree islands
b₄ = internal water tracks
b₅ = spring-fen channels
c = mineral beach ridges
d₁ = swamp forests
d₂ = crest of a spring-fen mound
d₃ = spring-fen forest

Type 10

Type 11

Fig. 1.6. Continuous mire-complex types in Minnesota (Glaser 1987a; from *USFWS Biological Report* 85 (7.14), drawn by Rebne P. Karchefsky).

of a single bog that almost completely fills the peat-land except for a narrow marginal lagg. These mire complexes are generally small (< 20 km²) and either straddle drainage divides or contain only a narrow strip of mineral soil at the crest of the watershed (Fig. 1.7; Plates 1 and 2, northeast quadrant). The bogs in these complexes may completely surround small lakes or outcrops of mineral soil. The larger bogs have pro-

nounced crests with conspicuous radiating patterns of forest.

In larger mire complexes (> 20 km²) the raised bogs are separated by water tracks of varying sizes. The water tracks originate in drainage channels at the mineral crest of the watershed and terminate in tributary streams at the downslope margin of the peatland. The bogs, in contrast, are located downslope from flow

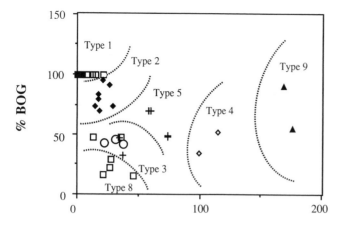

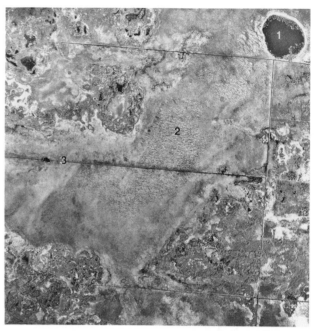

Fig. 1.7. Dimensions of selected mire-complex types in Minnesota. Mire-complex types may be categorized according to their proportion of bog to fen in relation to the total area of peatland.

Fig. 1.8. Patterned fen at Mud Lake (Luxemberg Peatland), northwestern Minnesota. Runoff drains from the surrounding uplands toward Mud Lake (1). The peatland contains string and flark patterns (2) oriented perpendicular to the slope and drainage ditches (3).

obstructions or develop over minor drainage divides. The majority of these mire complexes contain less than 50% bog and are smaller than 50 km² (Fig. 1.7). A few complexes, however, are huge, ranging up to 180 km². Despite these physiographic constraints, most bog and fen patterns in Minnesota can be assigned to a limited number of types.

The type 2 peatlands are nearly filled by several raised bogs that are separated by narrow water tracks (Plate 4). These peatlands are generally small (15-26 km²) and consist mostly of bog (> 75%). The bogs seldom have a conspicuous crest and generally have an eccentric cross section. This peatland type corresponds to watershed III at North Black River (Glaser 1983b).

The type 3 peatlands have larger water tracks that originate near the upslope margin of the peatland (Plate 5). Runoff from the adjacent uplands drains directly into these water tracks, which drain downslope toward tributary streams. These peatlands are generally small (20-37 km²), have nearly equal proportions of bog and fen (43:57), and have a wide zone of mineral soil at the watershed crest. The bogs are widely separated by the water tracks, which generally lack the string and flark patterns. This peatland type corresponds to watershed II at North Black River (Glaser 1983b).

The type 4 peatlands have one huge water track that arises from the upslope margin of the peatland and splits downslope into two branches around a large raised bog (Plate 5). Several raised bogs may also occupy the lateral margins of the water track. These peatlands are very large (about 100 km²) and range from 34% to 54% bog. This peatland type corresponds to the largest watersheds in the Myrtle Lake (Heinselman 1970) and North Black River (Glaser 1983b) peatlands.

The type 5 peatlands are distinguished by the con-

centration of raised bogs near the crest of the watershed, with broad water tracks downslope. One of these peatland types is small (37 km²), but the others are intermediate in size (58-79 km²), and the dimensions of several nearly completely overlap. Most of these peatlands have nearly equal proportions of bog and fen, but bogs cover over 70% of several type 5 complexes.

The remaining discontinuous mire complexes consist of patterned fens (type 6) and spring fens (type 7). These complexes lack raised bogs but differ greatly in size. The patterned fens range from the small fens in northeastern Minnesota (< 1 km²) to the huge patterned fens in the Glacial Lake Agassiz region (> 60 km²; Fig. 1.8). The spring fens, in contrast, are generally smaller (0.3-15 km²), although their watershed boundaries are somewhat arbitrary to define (Plate 3).

The continuous mire-complex types are restricted to the level terrain of the Glacial Lake Agassiz region, where peat nearly covers the regional landscape. The semi circular bogs of type 8 are situated along the linear beach ridges of Glacial Lake Agassiz, with their convex margin facing upslope (Plate 6). These bogs are associated with patterned water tracks, which split into two branches around the bog margin. These mire-complex types are fairly common and superficially resemble types 4 and 10. The type 8 peatlands, however, are smaller (< 36 km²) and are predominantly fen (> 50%).

The large bog complexes (type 9) of the Red Lake peatland have previously been described by Glaser *et*

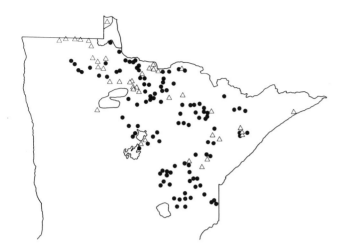

Fig. 1.9. Distribution of peat landforms in northern Minnesota. The peat landforms mapped are raised bogs (*circles*) and patterned fens (*triangles*).

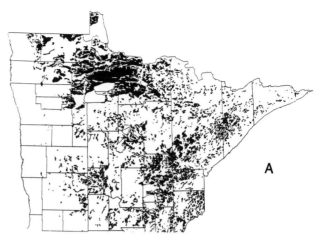

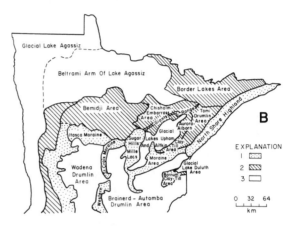

Fig. 1.10. Physiography and peatland distribution in northern Minnesota. The distribution of peatlands (**A**) is shown by the dark shading in relation to the major physiographic regions. The patterns in **B** represent (1) moraines and bedrock highlands, (2) till plains and drumlin areas, and (3) sand plains, lake plains, or large valleys. (From H. E. Wright, Jr., "Physiography of Minnesota," in *Geology of Minnesota: A Centennial Volume*, ed. P. K. Sims and G. B. Morey [Minnesota Geological Survey, 1972], by permission).

al. (1981). These complexes may cover more than 160 km² and consist largely of bog (55%-90%). The bogs are divided into streamlined lobes and islands by water tracks that arise near the bog crest (Plates 7 and 8). Several smaller bog complexes exhibit similar patterns but can be better placed in other peatland types on the basis of size.

The western water track (type 10) at Red Lake is also distinguished by its great size (> 160 km²) and the almost complete absence of bogs in the watershed (Fig. 1.3, Plate 9). Unlike the type 4 peatlands, the western water track is surrounded by a vast belt of swamp forests that grade downslope into linear fingers of forest and streamlined tree islands.

The Glacial Lake Agassiz region also contains several peat mounds that are dissected by spring-fen channels (type 11). These mounds are restricted to the Lost River peatland region and are surrounded by large areas of peatland (Fig. 1.1). The mounds range in size from 0.3 to 2.3 km².

Distribution of Patterned Peatlands in Minnesota

Patterned peatlands are restricted to the mixed-conifer hardwood forest region of northern Minnesota (Fig. 1.9). The peatland patterns are clustered on the glacial lake plains and Toimi Drumlin Area and are less common in the other physiographic units (Fig. 1.10). The very small peatlands in the extreme northeast are essentially featureless except for a single patterned fen near Grand Marais on the North Shore Highland (Glaser 1983c) and one near Ely in the Border Lakes area. Along their southern limit the peat landforms are less sharply defined.

Raised bogs are the most common peat landforms in northern Minnesota. They are evenly distributed across the region except in the northwest portion of Glacial Lake Agassiz, where they are scarce (Fig. 1.10). They mark the southern limits for patterned peatlands in Aitkin, Kanabec, and Pine counties near Mille Lacs Lake. A few bogs also occur beyond the prairie-forest border near Thief Lake in Marshall County. In these extreme locations, fire scars have obscured the radiating forest patterns, and the bog landforms must be identified by their streamlined margins.

Water tracks, in contrast, are concentrated in the Glacial Lake Agassiz region (Fig. 1.10). Here the largest and most highly developed water tracks occur, including the huge water tracks in the Red Lake, North Black River, and Myrtle Lake peatlands. Elsewhere, water tracks tend to be small and infrequent and usually lack string and flark patterns. The patterned fens in these smaller water tracks usually consist of oriented flarks without well-defined strings. Tree islands and spring-fen channels are not found outside the Glacial Lake Agassiz peatlands.

Regional and Local Controls on Pattern Development

The regional distribution of the mire-complex types in Minnesota provides an important indicator for the effects of different environmental factors on peatland development. Northern Minnesota is marked by a steep climatic gradient and by important changes in soils and physiography (chapter 3). The significance of the climatic gradient to pattern formation is indicated by the southern limit to patterned peatlands near Mille Lacs Lake. Raised bogs gradually disappear south of this zone, although peatlands remain abundant. Fire may contribute to the disappearance of raised bogs in this area because bogs in this transitional zone are highly altered by fire. Similarly, the bogs along the prairie-forest border in northwestern Minnesota are also altered by fire.

Perhaps one of the most surprising anomalies of the state's vegetation is the large expanse of patterned peatlands that straddle the prairie-forest border, where moisture is insufficient to support forest growth. Fire scars are very common in these peatlands and have greatly altered the peatland vegetation. Yet the peatlands exhibit no apparent effect of this decreasing moisture gradient. Some of the most highly developed patterns, for example, occur in the Red Lake peatland, which ends at the prairie-forest border.

The spread of peatlands in northern Minnesota and the origin of the striking landform patterns does not seem to be directly linked to specific climatic isopleths. The frequency of fire scars at the western and southern limits for peatland patterns may be related more to the construction of drainage ditches and roads than to the climatic gradient. Fire scars are common throughout the peatlands of northern Minnesota.

The distribution of peat landforms in Minnesota is strongly related to physiographic features. The most striking landform patterns are restricted to the flat plains of Glacial Lakes Agassiz, Upham, and Aitkin. The development of raised bogs and water tracks seems to be closely adjusted either to physiographic features that direct the path of runoff draining across a peatland or to tributary streams at their downslope margin. However, the topography of the peat surface does not always mirror the topography of the underlying mineral soil. In the Lost River peatland, two peat mounds have accumulated to similar heights, although one mound has developed over a rise in the mineral substrate, whereas the other is located over a depression (Almendinger *et al.* 1986; Fig. 1.1). The ridge and pool patterns in the large water tracks also have no relationship to the subsurface topography, which is typically flat in the areas surveyed. Thus physiographic features

may determine the configuration of bog and fen patterns in a peatland through their control on the path of runoff, but the three-dimensional shape of these peat landforms seems to develop independently of the underlying mineral soil. One of the striking anomalies in the state is the consistent recurrence of the same bog and fen landforms in different climatic and physiographic settings.

The distribution of peat landforms and soils in northern Minnesota are closely related. Mire complexes that are primarily composed of bog (types 1, 2, 5) are surrounded by clayey or loamy soils that have a relatively low hydraulic conductivity (Figs. 1.11, 1.12). Mire complexes that have large areas of fen (types 3, 4, 6, 7, 10), however, are consistently surrounded by sandy outwash or beach ridges. The largest water tracks in Minnesota, such as those at Red Lake and Myrtle Lake, all arise downslope from these sandy beach deposits.

Regional changes in soil types have a dramatic effect on the size and abundance of water tracks. In the northeastern portion of the Glacial Lake Agassiz region, raised bogs nearly cover the peatlands, which generally can be identified as type 1 or 2 mire complexes. This region is largely covered by relatively impermeable clayey lake sediments. An outlier of sandy outwash, however, is associated with the head of the large water track at North Black River. In the southeastern portion of the Lake Agassiz region, the bog complexes (types 3, 5, and 8) are almost completely surrounded by larger water tracks and featureless swamp forests. The peatlands in this zone are only locally interrupted by narrow sandy beach ridges.

To the east the beach ridges almost disappear underneath the Red Lake peatland, where large bog com-

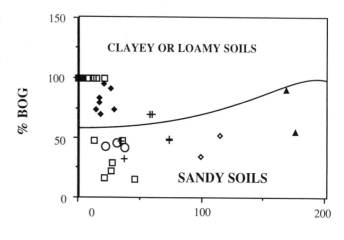

Fig. 1.11. The relationship of selected mire-complex types to soils in northern Minnesota. Peatlands dominated by bog landforms are generally located in areas of relatively impermeable clayey or loamy soils. Peatlands dominated by fens, however, are located in areas of permeable sands.

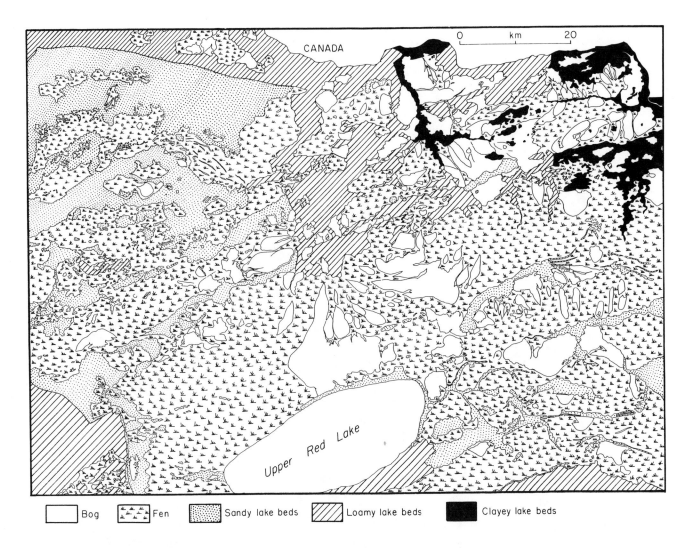

| | Bog | | | Fen | | Sandy lake beds | | Loamy lake beds | | Clayey lake beds |

Fig. 1.12. Classification of peatland types in the Glacial Lake Agassiz peatlands. This classification of the Landsat TM scene in the color centerfold distinguishes (1) fen vegetation, (2) bog vegetation, (3) exposures of sandy soil, (4) exposures of loamy soil, (5) exposures of clayey soil, and (6) areas with standing water. Most of the peatlands in this scene are fens, although the bog-dominated areas are primarily located in the northeast quadrant where the smaller peatlands are surrounded by relatively impermeable clayey or loamy soils. The large tracts of fen vegetation, in contrast, are located in areas with exposures of permeable sandy soils.

plexes have developed over drainage divides in the eastern and central watersheds. Here most of the landscape is covered by bog, but the bogs are finely divided by water tracks that arise from the lower bog flanks. The extremely large western watershed at Red Lake, however, is almost completely composed of fen vegetation, including the huge western water track. The edges of this watershed are marked by exposures of sandy outwash and beach ridges. This relationship is continued in the northwestern part of the Lake Agassiz peatlands. This area contains the largest patterned fens in Minnesota, which are surrounded by sandy outwash and beach deposits. Only a few bogs occur in this region despite the large expanse of peatlands.

This edaphic control on pattern development is difficult to explain without knowledge of the subsurface groundwater hydrology. The mineral soil throughout the Glacial Lake Agassiz peatlands is calcareous, although local changes occur in porosity and texture. These soils should weather to produce alkaline runoff or groundwater, which may restrict the spread of acidic bogs. Yet the greatest concentration of bogs in these peatlands occurs in the northeast quadrant (North Black River area), where the exposure of calcareous soils and hence the production of alkaline runoff is greatest (Fig. 1.12). In the southeast quadrant (Lost River area), however, mineral exposures are restricted to narrow beach ridges, yet the peatlands are predominantly fen. It is very unlikely that the volume of surface runoff draining from these narrow calcareous ridges would be sufficient to restrict the spread of bogs in the adjacent peatlands.

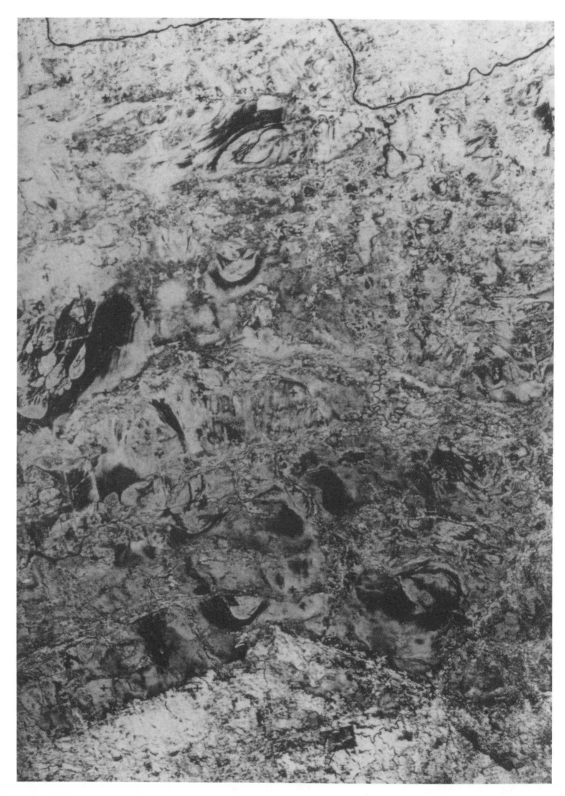

Fig. 1.13. Landsat MSS image of the Glacial Lake Agassiz peatlands during spring breakup of 1978. The white tones on this image are snow-covered bogs, swamp forests, and exposures of mineral soil. The darker tones are water tracks with flowing water. These water tracks flow around the streamlined margins of large bogs. Smaller "internal" water tracks are also located within the interior of the larger bogs. The image covers an area approximately 70 km across and is just east of the area shown in Fig. 1.3.

The regional distribution of bog and fen, however, is closely related to the texture of the adjacent soils. Peatlands are dominated by bogs where the adjacent clayey or loamy soils are relatively impermeable. Mire complexes, however, are dominated by fens where the adjacent mineral exposures consist of porous sandy outwash or beach ridges. These deposits are always found near the source of the largest water tracks or spring fens. The most likely explanation for this relationship is that the sandy deposits are conduits for alkaline groundwater discharging from the calcareous soils that underlie the peatlands.

Hydrology

Landsat imagery strongly supports the hypothesis that groundwater supplies the alkalinity found in the fen waters of northern Minnesota (Fig. 1.13). Imagery taken during spring breakup of 1978 indicates that the large water tracks are open and flowing while the surrounding bogs and mineral exposures are still frozen and covered with snow (Glaser 1987a, 1989). The heat required to melt these water tracks in advance of the surrounding areas most likely was supplied by groundwater upwelling from the mineral substrate. This interpretation is further supported by detailed analysis of Landsat imagery of the Red Lake peatland (Fig. 1.3). The "internal" water tracks that arise from the lower flanks of the raised bogs have thawed and are flowing in advance of the surrounding bog, which is still frozen and covered by snow. The most likely source of the heat that has melted the ice in these water tracks is groundwater discharging in the water tracks. The large western water track at Red Lake also seems to be derived from groundwater discharge.

The western water track at Red Lake exhibits a similar pattern. The water track is surrounded by sandy beach ridges and swamp forests, which were frozen and covered with snow when the images were recorded. Water is flowing, however, in a series of narrow channels that arise downslope from the beach ridges. The channels converge into the main part of the water track, which is still frozen in its center (Fig. 1.3).

Conclusion

The distribution of peat landforms in northern Minnesota indicates an important edaphic control on peatland development that may be associated with groundwater hydrology. The proportion of fen versus bog in a peatland is related to the exposure of sand deposits at the crest of these watersheds. These permeable sand layers may represent local discharge zones for alkaline groundwater that (1) check the advance of bog and (2) maintain fen vegetation in the water tracks. The morphology of the basic landform types (raised bog, water track, spring-fen channels), however, is remarkably uniform across northern Minnesota, indicating that their development is controlled by autogenic feedback systems. These processes can only be determined if the vegetation and water chemistry associated with these peatland patterns is adequately known.

Literature Cited

Aario, L. 1932. Pflanzentopographische und paläogeographische mooruntersuchungen in N-Satakunta. Fennia 55:1-179.

Almendinger, J. C., J. E Almendinger, and P. H. Glaser. 1986. Topographic fluctuations across a spring fen and raised bog in the Lost River peatland, northern Minnesota. Journal of Ecology 74:393-401.

Baker, D. G., and J. H. Strub, Jr. 1963a. Climate of Minnesota. Part 1. Probability of occurrence in the spring and fall of selected low temperatures. University of Minnesota Agricultural Experiment Station Technical Bulletin 243:1-40.

———— and J. H. Strub, Jr. 1963b. Climate of Minnesota. Part II. The agriculture and minimum temperature-free seasons. University of Minnesota Agricultural Experiment Station Technical Bulletin 255:1-31.

————, D. A. Haines, and J. H. Strub, Jr. 1967. Climate of Minnesota. Part V. Precipitation facts, normals, and extremes. University of Minnesota Agricultural Experiment Station Technical Bulletin 254:1-43.

————, W. W. Nelson, and E. L. Kuehnast, 1979. Climate of Minnesota. Part XXI. The hydrological cycle and soil water. University of Minnesota Agricultural Experiment Station Technical Bulletin 322:1-23.

Cajander, A. K. 1913. Studien über die moore Finnlands. Acta Forestalia Fennica 2:1-208.

Glaser, P. H. 1983a. *Eleocharis rostellata* and its relation to spring fens in Minnesota. Michigan Botanist 22:19-21.

————. 1983b. Vegetation patterns in the North Black River peatland, northern Minnesota. Canadian Journal of Botany 61:2085-2104.

————. 1983c. A patterned fen on the north shore of Lake Superior, Minnesota. Canadian Field-Naturalist 97:194-199.

————. 1987a. The patterned boreal peatlands of northern Minnesota: A community profile. U.S. Fish and Wildlife Service Biological Report 85(7):1-98.

————. 1987b. The development of streamlined bog islands in the continental interior of North America. Arctic and Alpine Research 19:402-413.

————. 1989. Detecting biotic and hydrogeochemical processes in large peat basins with Landsat TM imagery. Remote Sensing of Environment 28:109-119.

————, G. A. Wheeler, E. Gorham, and H. E. Wright, Jr. 1981. The patterned mires of the Red Lake peatland, northern Minnesota: Vegetation, water chemistry, and landforms. Journal of Ecology 69:575-599.

———— and J. A. Janssens. 1986. Raised bogs in eastern North America: transitions in surface patterns and stratigraphy. Canadian Journal of Botany 64:395-415.

————, J. A. Janssens, and D. I. Siegel. 1990. The response of vegetation to chemical and hydrological gradients in the Lost River peatland, northern Minnesota. Journal of Ecology 78:1021-1048.

Griffin, K. O. 1975. Vegetation studies and modern pollen spectra from the Red Lake Peatland, northern Minnesota. Ecology 56:172-192.

————. 1977. Paleoecological aspects of the Red Lake peatland, northern Minnesota. Canadian Journal of Botany 55:172-192.

Heinselman, M. L. 1963. Forest sites, bog processes, and peatland types in the Glacial Lake Agassiz region, Minnesota. Ecological Monographs 33:327-372.

———. 1970. Landscape evolution, peatland types, and the environment in the Lake Agassiz Peatland Natural Area, Minnesota. Ecological Monographs 40:235-261.

———. 1974. Interpretation of Francis J. Marschner's original vegetation of Minnesota. Text on the reverse of Minnesota map by F. J. Marschner, 1930. U.S. Department of Agriculture Forest Service, North Central Forest Experiment Station, St. Paul, Minnesota, USA.

Hofstetter, R. H. 1969. Floristic and ecologic studies of wetlands in Minnesota. Ph.D. thesis. University of Minnesota, Minneapolis, Minnesota, USA.

Janssen, C. R. 1967. A floristic study of forests and bog vegetation, northeastern Minnesota. Ecology 48:751-765.

———. 1968. Myrtle Lake: A late-and post-glacial pollen diagram from northern Minnesota. Canadian Journal of Botany 46:1397-1408.

Paasio, I. 1933. Über die vegetation der hochmoore Finlands. Acta Forestalia Fennica 39:1-210.

Sjörs, H. 1948. Myrvegetation i Bergslagen. Acta Phytogeographica Suecica 21:1-299.

Tarnocai, C. 1974. Peat landforms and associated vegetation in Manitoba. Pages 6-20 *in* J. H. Day, editor. Proceedings of the Canada Soil Survey Committee. Organic Soil Mapping Workshop, Winnipeg, Canada.

Wright, H. E., Jr. 1972. Physiography of Minnesota. Pages 515-560 *in* P. K. Sims and G. B. Morey, editors. Geology of Minnesota: A centennial volume. Minnesota Geological Survey, Minneapolis, Minnesota, USA.

CHAPTER 2

Vegetation and Water Chemistry

Paul H. Glaser

Peat landforms develop unique plant communities because the component species reside entirely within the biomass produced by the ecosystem. The living vegetation merely forms a cap to a massive peat deposit, which consists of the partially decomposed remains of vegetation. Decomposition is inhibited within this deposit by anoxia, which restricts the internal cycling of nutrients by microbial metabolism. The essential minerals for plant nutrition must therefore be transported into the peatland from an external source such as the atmosphere or surrounding mineral soil (Fig. 2.1). The supply of these essential mineral ions, however, is determined not only by hydrologic processes, which represent the main mechanism for transport, but also by the dynamics of peat accumulation, which determines the topography of any peat landform.

The formation of orderly peat landforms in boreal peatlands poses several important questions for vegetation science: Are these landforms characterized by different vegetation types? How uniform are the vegetation patterns across conspicuous climatic and edaphic gradients? Does the uniformity and regularity of these patterns imply stable plant communities that are buffered from environmental changes?

These questions can best be answered by a regional survey of peat landforms across northern Minnesota. Peatlands have spread over a wide range of soils and physiographic conditions in this region, and the steep climatic gradient has produced major changes in the vegetation of the uplands. Similar changes should be expected in the peatland vegetation if external environmental controls are paramount. However, if the peat landforms represent tightly integrated systems similar to the ideal of Clements, then the communities and landforms should remain uniform. Vegetation also

provides essential data for formulating developmental models that can be tested by stratigraphic or hydrologic methods.

Bog and Fen

Northern peatlands have traditionally been separated into bog and fen on the basis of their (1) peat landforms, (2) indicator species, (3) water chemistry, and (4) inferred hydrology (Fig. 2.1). Raised bogs can always be identified in Minnesota by their forested crest, acid waters (pH < 4.2; Ca concentration < 2 mg 1^{-1}), and absence of fen indicator species. Fens, in contrast, have concave landforms, less acid to alkaline waters (pH 4.2-7.2; Ca concentration 2-50 mg 1^{-1}), and at least one indicator species present. These contrasting characteristics are linked to the hydrological properties of bogs and fens. The acid waters of bogs are maintained by internal sources of H^+ ions and by the absence of any significant external source for a base. The H^+ ions may be generated internally by the cation-exchange capacity of *Sphagnum* (Clymo 1963, 1967; Clymo and Hayward 1982) or by the release of organic acids through decomposition of *Sphagnum* peat (Gorham *et al.* 1984). The most important base in surface waters or groundwaters, however, is bicarbonate, which is readily weathered from mineral soil by atmospheric precipitation (Drever 1982). Bicarbonate has a low concentration in bog waters because (1) the surface of the bog is raised above the flood level of runoff draining from the adjacent uplands, and (2) groundwater cannot move through the dense accumulation of bog peat. Thus bogs are believed to be *ombrotrophic*, or rain-nourished, in contrast to fens, which are *minerotrophic* and receive at least some water that has percolated through mineral soil.

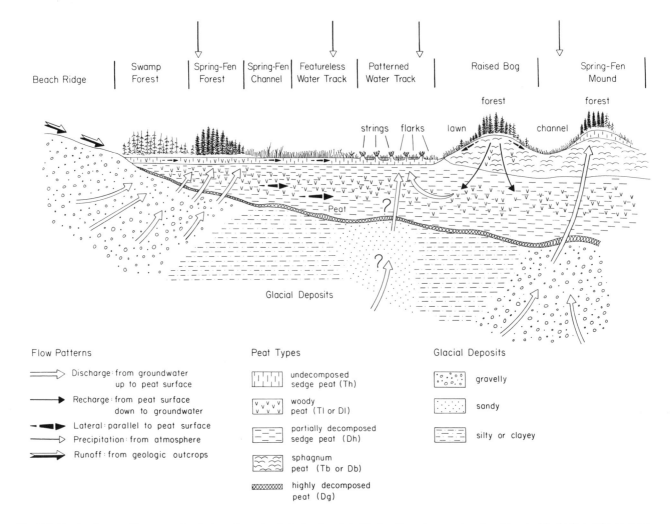

Fig. 2.1. Peatland types. The vegetation in boreal peatlands is closely related to landform type, water chemistry, and hydrology. The sources for water and mineral ions are indicated by the arrows. The alkalinity in the surface water decreases with increasing distance from the source of groundwater or runoff draining from mineral soil. This decline in alkalinity is associated with different species assemblages and different peat landforms, culminating in the development of a raised bog.

Raised Bogs

The ombrotrophic flora in Minnesota contains fewer than 20 species of vascular plants (Table 2.1; Fig. 2.2). All of these species can also be found in fens, so bogs are distinguished by the *absence* of fen indicator species. The two best-developed vegetation types (noda) on raised bogs are the *Carex trisperma-Vaccinium vitis-idaea* nodum in densely forested stands and the *Carex oligosperma* nodum on nonforested lawns (Glaser *et al.* 1981). Most of the bog flora occurs on both lawns and forested stands, but *C. trisperma, V. vitis-idaea, Gaultheria hispidula*, and *Smilacina trifolia* are generally restricted to forested stands, whereas *C. oligosperma* is most common on lawns.

The stands of radiating forest generally consist of small trees of *Picea mariana* (< 10 m tall) with occasional trees of *Larix laricina*. The trees are largest near the crest and gradually become smaller and more clumped downslope as the radiating lines of forest are more widely separated by nonforested bog drains (Plates 8A, 10A). The lawns are dominated by *C. oligosperma*, although small layered clumps of *Larix* and *Picea* are usually present (Fig. 2.3; Plate 8B). The ground layer consists of a continuous layer of *Sphagnum*, which forms easily compressible hummocks and ill-defined hollows.

In burned areas *C. oligosperma* becomes dominant, and the assemblage assumes the character of a nonforested lawn. Where the water table has been artificially lowered by drainage ditches, however, *C. oligosperma* does not colonize the burned area, which instead is dominated by a luxuriant growth of *Sphagnum* along with bog ericads, especially *Chamaedaphne calyculata*.

Sphagnum cuspidatum hollows (schlenke of north-

Table 2.1. Vascular plants associated with raised bog landforms in Minnesota.

	RADIATING BOG FOREST n=8		SCHLENKE n=2		NONFORESTED BOG n=8		SPHAGNUM LAWN POOR FEN n=10		INCIPIENT WATER TRACK n=2	
pH	3.7 - 4.1		4		3.8 - 4.5		3.7 - 4.6		4.2 - 4.4	
Ca (mg/l)	0.6 - 1.7		0.3 - 0.08		0.4 - 4.8		0.6 - 5.1		5.5 - 6.1	
K corr.	1 - 55		44 - 48		2 - 64		24 - 77		63 - 73	
	FREQ	COVER	FREQ	COVER	FREQ	COVER	FREQ	COVER	FREQ	COVER
Vaccinium vitis-idaea	0.38	1.00	-	-	-	-	-	-	-	-
Vaccinium myrtilloides	0.50	+	-	-	-	-	-	-	-	-
Gaultheria hispidula	0.63	1.00	-	-	-	-	0.13	+	-	-
Carex trisperma	0.75	3.00	-	-	0.13	+	-	-	-	-
Smilacina trifolia	0.63	1.00	-	-	0.13	+	0.40	1.00	-	-
Ledum groenlandicum	1.00	2.00	-	-	0.38	1.00	0.50	+	-	-
Picea mariana	1.00	5.00	-	-	0.75	1.00	0.70	2.00	-	-
Eriophorum spissum	1.00	2.00	0.50	+	0.88	1.00	0.60	1.00	-	-
Kalmia polifolia	1.00	2.00	0.50	+	0.88	1.00	1.00	2.00	1.00	1.00
Chamaedaphne calyculata										
Vaccinium oxycoccos	1.00	1.00	-	-	1.00	+	1.00	+	1.00	+
Carex pauciflora	0.63	1.00	-	-	0.50	+	1.00	1.00	0.50	+
Drosera rotundifolia	0.63	+	0.50	1.00	0.25	+	0.30	+	1.00	+
Carex paupercula	0.13	+	-	-	0.13	+	0.20	2.00	-	-
Andromeda glaucophylla	0.50	1.00	1.00	+	0.63	+	0.70	1.00	1.00	0.50
Sarracenia purpurea	0.50	+	0.50	+	0.63	+	0.80	+	1.00	+
Larix laricina	0.13	+	-	-	0.50	+	0.50	+	1.00	+
Carex oligosperma	0.13	1.00	1.00	1.00	1.00	3.00	1.00	3.00	0.50	+
Scheuchzeria palustris	-	-	1.00	2.00	0.25	+	5.00	+	1.00	1.00
Carex limosa	-	-	1.00	2.00	0.13	+	0.40	1.00	1.00	2.00
Carex rostrata	-	-	-	-	-	-	0.20	+	-	-
Carex chordorhizza	-	-	-	-	-	-	-	-	-	-
Carex canescens	-	-	-	-	-	-	0.10	+	-	-
Carex aquatilis	-	-	-	-	-	-	0.50	2.00	0.50	+
Potentilla palustris	-	-	-	-	-	-	0.10	+	-	-
Betula pumila var. glandulifera	-	-	-	-	-	-	0.10	+	-	-
Rhyncospora alba	-	-	1.00	4.00	-	-	-	-	1.00	3.00
Menyanthes trifoliata	-	-	-	-	-	-	-	-	1.00	1.00
Carex lasiocarpa	-	-	-	-	-	-	-	-	1.00	+

Note: For each species the frequency (proportion of plots occupied, e.g., 1 = all occupied) and cover (selective abundance) are given. The symbols for cover are as follows: (+) sparsely present; (1) plentiful but small cover value; (2) covering 1/20 to 1/4 of area; (3) any number of individuals covering 1/4 to 1/2 of area; (4) any number of individuals covering 1/2 of 3/4 of area; (5) covering more than 3/4 of area.

ern European authors) are rare in Minnesota, occurring only near the bog crests at North Black River, Myrtle Lake, and Sturgeon River (Plates 10B, 10C). The wetter moss carpet in these hollows contains a rare bog assemblage in Minnesota that includes *Carex limosa, Scheuchzeria palustris, Rhynchospora alba*, and *Utricularia cornuta*, These species are generally absent from most ombrotrophic sites in Minnesota, where the fluctuating water table may drop as much as 70-100 cm below the surface during a dry period.

With increasing pH and Ca concentration, a number of minerotrophic indicator species appear on the *Sphagnum* lawns, representing a subtle change from ombrotrophic bog to minerotrophic poor fen. *Carex aquatilis* is the most common of these poor-fen indicators, followed by *Carex rostrata, C. chordorrhiza,* and *C. lasiocarpa.* Except for these indicator species,

the poor-fen relevés are almost indistinguishable from those of ombrotrophic bog.

Water Tracks and Spring Fens

In Minnesota the fen vegetation is distinguished by (1) richer species assemblages, (2) the presence of minerotrophic fen indicators, and (3) the low representation of *Sphagnum* relative to the Amblystigeaceae mosses (Fig. 2.3; Table 2.2). The indicator species are related to different ranges in water chemistry, although the indicators are most common in nonforested stands. Each of the different landform types is distinguished by a characteristic species assemblage, which can be identified as noda. These noda are actually reference points along a continuous gradient of vegetation change.

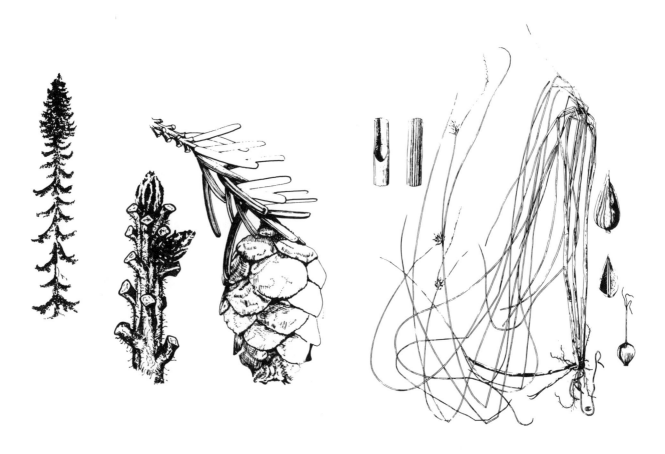

A. *Picea mariana*   **B.** *Carex oligosperma*

Fig. 2.2. Vascular plants that characterize the two-bog noda. **A.** *Picea mariana* (black spruce, A) forms a stunted forest at the crest of raised bogs in Minnesota. **B.** Downslope the forest is replaced by a *Sphagnum* lawn dominated by *Carex oligosperma* (sedge, B). [Reprinted from Gleason (1952) and Mackenzie (1940; H. C. Creutzburg, illustrator) with permission from The New York Botanical Garden.]

Spring-Fen Channels

(*Scirpus hudsonianus-Cladium mariscoides* nodum)

The spring-fen channels are characterized by the *Scirpus hudsonianus-Cladium mariscoides* nodum (Glaser *et al*. 1990). These nonforested channels are dominated by sedges, the most important of which are *Scirpus cespitosus, Cladium mariscoides, Carex lasiocarpa*, and *C. exilis* (Plate 10D). Also present are *Carex limosa, C. livida, Scirpus hudsonianus*, and *Rhynchospora alba*. The very alkaline waters (pH >6.8; Ca concentration >20 mg 1^{-1}) in these channels are associated with a number of extremely rich fen indicators including *Muhlenbergia glomerata, Cladium mariscoides, Parnassia palustris*, and *Thuja occidentalis*. A number of rare plants occur in these channels, which are discussed in chapter 5. The channels have standing water and scattered tussocks of sedges, which are slightly raised above the water level.

Flarks

(*Triglochin maritima-Drosera intermedia* nodum)

The flarks in northern Minnesota are variable with respect to their area and depth of standing water. The largest and deepest flarks are restricted to pristine water tracks in the Glacial Lake Agassiz region. Elsewhere the flarks are commonly small and shallow with little standing water. In the wettest locations *Triglochin maritima, Utricularia minor, Drosera intermedia, D. anglica*, and *D. linearis* are usually present (Glaser *et al.* 1981). The drier flarks usually lack these species. All flarks, however, are dominated by sedges including *Carex lasiocarpa, C. limosa, C. livida, C. chordorrhiza, Rhynchospora alba*, and *Menyanthes trifoliata* (Plate 11A, 11B). The relative abundance of these species changes with respect to water level and water chemistry. The flarks of pristine water tracks also contain several rare plants, which are restricted to fens in undisturbed locations.

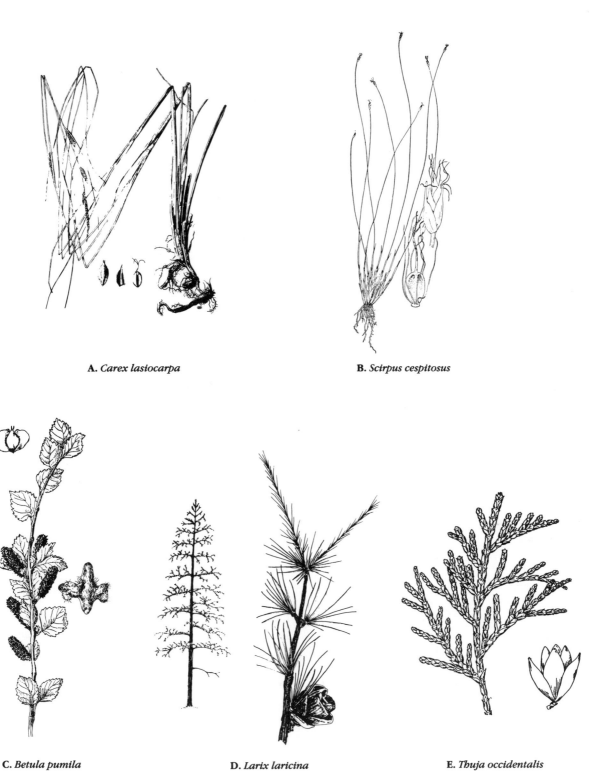

A. *Carex lasiocarpa*

B. *Scirpus cespitosus*

C. *Betula pumila*   **D.** *Larix laricina*   **E.** *Thuja occidentalis*

Fig. 2.3. Vascular plants that characterize the six fen noda. *Carex lasiocarpa* (sedge, **A**) is the typical dominant in the flarks but is locally replaced by other species such as *Scirpus cespitosus* (tussock bullrush, **B**), in the spring-fen channels. In patterned water tracks the transverse networks of strings is dominated by *Betula pumila* var. *glandulifera* (bog birch, **C**), whereas the tree islands and forested fingers are dominated by *Larix laricina* (tamarack, **D**). The spring-fen forests, however, are dominated by black spruce although they contain the extremely rich fen indicator *Thuja occidentalis* (northern white cedar, **E**). [Reprinted from Gleason (1952) and Mackenzie (1940; H. C. Creutzburg, illustrator) with permission from The New York Botanical Garden.]

Table 2.2. Vascular plants associated with fen landforms in northern Minnesota.

	SPRING-FEN CHANNELS (n=4)		FLARKS (n=26)		FEATURELESS WATER TRACKS (n=15)		STRINGS (n=7)		FOREST FINGERS (n=3)		TREE ISLANDS (n=4)		SPRING-FEN FORESTS (n=3)	
pH	6.6 - 7.6		4.8 - 7.4		4.2 - 7.1		6.2 - 7.2		4.8 - 6		6.4 - 7.2		6.8 - 7.2	
Ca concen. mg/l	55.9 - 98.5		2.0 - 56.5		1.5 - 30.4		2.1 - 65.8		1.5 - 11.1		13 - 65.9		30.7 - 45.6	
K corr.	44 - 64		22 - 149		26 - 181		20 - 129		42 - 129		20 - 128		21 - 35	
	FREQ.	COVER	FREQ.	COVER	FREQ.	COVER	FREQ.	COVER	FREQ.	COVER	FREQ.	COVER	FREQ.	COVER
Aster junciformis	0.75	+	-	-	-	-	-	+	-	-	-	-	-	-
Thuja occidentalis	1	+	0.11	+	-	-	0.14	2	-	-	0.75	+	-	-
Carex exilis	0.25	2	0.19	2	-	-	-	-	-	-	-	-	-	-
Cladium mariscoides	0.75	+	-	-	-	-	-	-	-	-	-	-	-	-
Parnassia palustris	1	+	0.15	+	-	-	0.14	+	-	-	-	-	-	-
Lobelia kalmii	1	+	0.26	+	-	-	0.43	+	-	-	-	-	-	-
Scirpus cespitosus	1	2	0.3	2	-	-	-	-	-	-	-	-	-	-
Drosera anglica	0.5	+	0.35	1	0.07	+	-	-	-	-	-	-	-	-
Drosera linearis	-	-	0.21	1	0.07	1	-	-	-	-	-	-	-	-
Utricularia minor	-	-	0.21	+	-	-	-	-	-	-	-	-	-	-
Triglochin maritima	0.5	-	0.41	+	0.13	+	0.14	+	-	-	0.25	+	-	-
Carex livida	0.5	+	0.9	1	0.2	+	0.14	+	-	-	-	-	-	-
Carex limosa	1	+	0.79	2	0.87	1	0.29	+	0.67	1	-	-	-	-
Menyanthes trifoliata	0.75	+	0.9	1	0.73	1	0.29	1	0.67	1	-	-	-	-
Rhyncospora alba	0.75	+	0.83	1	0.67	1	0.29	+	-	-	-	-	-	-
Utricularia intermedia	0.75	1	0.51	1	0.46	+	-	-	0.33	+	-	-	-	-
Eleocharis compressa	0.75	+	0.45	1	0.27	1	0.14	+	0.33	+	0.5	+	-	-
Equisetum fluviatile	-	-	0.52	1	0.4	1	0.43	+	-	-	0.33	+	-	-
Carex lasiocarpa	0.75	2	0.97	2	1	3	0.86	2	1	2	0.25	1	-	-
Betula pumila var. glandulifera	1	+	0.56	1	0.93	1	0.71	3	1	2	1	1	-	-
Salix pedicellaris var. hypoglauca	0.25	+	0.14	+	0.47	+	0.71	1	0.67	1	0.5	+	-	-
Andromeda glaucophylla	0.5	+	0.79	1	1	1	0.71	1	1	1	0.75	1	0.33	+
Drosera rotundifolia	0.25	+	0.34	+	0.53	+	0.43	+	0.66	+	0.75	+	0.66	+
Sarracenia purpurea	1	+	0.69	1	0.8	+	0.71	+	0.66	+	0.75	+	1	+
Carex chordorrhiza	-	-	0.82	1	0.93	1	0.71	3	1	1	0.5	1	0.33	+
Scheuchzeria palustris	0.25	+	0.82	1	0.53	1	0.53	+	0.33	+	-	-	-	-
Drosera intermedia	0.25	+	0.45	1	0.27	1	-	-	-	-	-	-	-	-
Kalmia polifolia	-	-	0.21	+	0.67	+	-	-	0.66	1	-	-	-	-
Chamaedaphne calyculata	-	-	0.41	1	0.8	1	0.29	1	1	2	0.75	1	0.66	+
Vaccinium oxycoccos	0.25	+	0.52	+	0.73	1	0.43	+	1	1	1	+	0.66	+
Larix laricina	0.75	1	0.28	1	0.47	1	0.56	1	1	4	1	4	-	-
Picea mariana	0.25	+	0.17	1	0.4	1	0.14	+	0.66	2	0.75	2	1	5
Smilacina trifolia	-	-	0.17	+	0.33	1	-	-	0.66	+	0.5	1	0.66	1
Carex paupercula	-	-	0.03	+	0.27	1	0.14	+	-	-	0.5	+	0.66	+
Ledum groenlandicum	-	-	0.07	+	0.27	+	0.14	+	1	+	1	2	1	3
Carex diandra	-	-	0.03	+	-	-	0.57	+	0.33	+	-	-	-	-
Thelypteris palustris var. pubescer	-	-	-	-	-	-	0.71	1	-	-	0.25	+	-	-
Viola pallens var. mackloskeyi	-	-	-	-	-	-	0.57	+	0.33	+	0.25	+	-	-
Bromus ciliatus														
Agrostis scabra	0.25	+	0.1	+	-	-	0.43	+	-	-	-	-	-	-
Potentilla palustris	-	-	0.1	+	0.13	+	0.57	+	0.33	+	1	1	-	-
Typha latifolia	0.75	1	-	-	-	-	0.29	1	0.33	2	0.75	1	-	-
Carex leptalea	-	-	0.03	+	-	-	0.57	+	0.33	+	0.5	+	1	+
Galium labradoricum	-	-	0.07	+	0.07	+	0.57	+	-	-	1	+	0.33	+
Carex tenuiflora	-	-	-	-	0.07	+	0.14	+	1	+	0.5	+	0.67	+
Carex trisperma	-	-	-	-	-	-	-	-	-	-	0.5	+	0.67	2
Lysimachia thrysiflora	-	-	-	-	-	-	-	-	0.33	2	0.75	+	-	-
Vaccinium vitis-idaea	-	-	-	-	-	-	-	-	-	-	0.25	+	1	1
Pyrola secunda var. obtusata	-	-	-	-	-	-	0.29	+	-	-	0.5	+	0.33	+
Rumex orbiculatus	-	-	-	-	-	-	-	-	-	-	0.75	+	0.33	+
Dryopteris cristata	-	-	-	-	-	-	-	-	-	-	0.5	+	0.33	+
Carex disperma	-	-	-	-	-	-	-	-	-	-	0.75	+	-	-
Caltha palustris	-	-	-	-	-	-	-	-	-	-	0.5	+	0.33	+
Cornus stolonifera	-	-	-	-	-	-	-	-	-	-	0.5	+	-	-
Carex gynocrates	-	-	-	-	-	-	-	-	-	-	0.25	1	-	-

Note: See Table 2.1 for cover symbols.

Table 2.2 (cont.). Vascular plants associated with fen landforms in northern Minnesota.

	SPRING-FEN CHANNELS (n=4)		FLARKS (n=26)		FEATURELESS WATER TRACKS (n=15)		STRINGS (n=7)		FOREST FINGERS (n=3)		TREE ISLANDS (n=4)		SPRING-FEN FORESTS (n=3)	
pH	6.6 - 7.6		4.8 - 7.4		4.2 - 7.1		6.2 - 7.2		4.8 - 6		6.4 - 7.2		6.8 - 7.2	
Ca concen. mg/l	55.9 - 98.5		2.0 - 56.5		1.5 - 30.4		2.1 - 65.8		1.5 - 11.1		13 - 65.9		30.7 - 45.6	
K corr.	44 - 64		22 - 149		26 - 181		20 - 129		42 - 129		20 - 128		21 - 35	
	FREQ.	COVER	FREQ.	COVER	FREQ.	COVER	FREQ.	COVER	FREQ.	COVER	FREQ.	COVER	FREQ.	COVER
Cornus canadensis	-	-	-	-	-	-	-	-	-	-	0.75	1	-	-
Gaultheria hispidula	-	-	-	-	0.07	+	-	-	-	-	0.5	+	0.33	+
Scirpus acutus	0.5	2	-	-	-	-	0.14	+	-	-	0.25	+	-	-
Habenaria clavellata	0.25	+	-	-	-	-	-	-	-	-	-	-	-	-
Scirpus hudsonianus	0.25	+	0.13	1	0.07	+	0.29	+	-	-	-	-	-	-
Potentilla fruticosa	0.5	+	-	-	-	-	0.14	+	-	-	-	-	-	-
Phragmites communis	0.25	+	0.07	2	-	-	0.29	2	-	-	-	-	-	-
Carex interior	0.5	+	0.03	+	0.07	+	0.29	+	-	-	0.5	+	-	-
Eleocharis rostellata	0.25	2	-	-	-	-	-	-	-	-	-	-	-	-
Carex cephalantha	0.25	1	0.03	+	-	-	0.14	+	-	-	-	-	-	-
Carex aquatilis	0.25	+	0.07	+	0.07	+	-	-	-	-	0.5	+	-	-
Muhlenbergia glomerata	0.25	+	0.1	+	0.07	+	0.42	+	-	-	0.25	+	-	-
Solidago uliginosa	-	-	0.07	+	-	-	0.14	+	-	-	-	-	-	-
Viola sp.	-	-	-	-	-	-	0.14	+	-	-	0.25	+	-	-
Epilobium leptophyllum	-	-	0.03	+	-	-	0.14	+	-	-	0.25	+	0.33	+
Utricularia cornuta	-	-	0.14	+	-	-	-	-	-	-	-	-	-	-
Eriophorum chamissonis	-	-	0.1	+	0.29	+	-	-	-	-	-	-	-	-
Pogonia ophioglossoides	-	-	0.24	+	0.07	+	0.14	+	-	-	-	-	-	-
Eriophorum gracile	-	-	0.13	+	0.13	+	-	-	0.33	+	-	-	-	-
Campanula aparinoides	-	-	0.1	+	-	-	0.43	+	-	-	0.5	+	-	-
Carex intumescens	-	-	-	-	-	-	-	-	0.33	1	-	-	-	-
Eriophorum chamissonis	-	-	0.07	+	-	-	0.29	+	-	-	-	-	-	-
Salix serissima	-	-	0.03	+	-	-	0.14	+	-	-	0.25	+	-	-
Eriophorum tenellum	-	-	0.21	+	0.2	+	-	-	-	-	0.25	+	-	-
Rhynchospora fusca	-	-	0.1	3	-	-	-	-	-	-	-	-	-	-
Xyris montana	-	-	0.1	1	-	-	-	-	-	-	-	-	-	-
Triadenum fraseri	-	-	0.07	+	-	-	-	-	-	-	-	-	-	-
Juncus canadensis	-	-	0.07	+	-	-	-	-	-	-	-	-	-	-
Graminaea	-	-	0.04	+	0.13	-	-	-	-	-	-	-	-	-
Carex pauciflora	-	-	0.04	1	-	-	0.33	+	-	-	-	-	-	-
Juncus stygius	-	-	0.07	+	0.2	+	-	-	-	-	-	-	-	-
Eriophorum sp.	-	-	0.04	+	0.07	+	0.14	+	-	-	-	-	-	-
Carex oligosperma	-	-	0.04	+	0.13	2	-	-	-	-	-	-	-	-
Lycopodium inundatum	-	-	0.04	+	-	-	-	-	-	-	-	-	-	-
Vaccinium macrocarpon	-	-	0.04	+	-	-	-	-	-	-	-	-	-	-
Myrica gale	-	-	-	-	0.07	+	0.14	1	-	-	-	-	-	-
Pedicularis lanceolata	-	-	-	-	0.13	+	-	-	-	-	-	-	-	-
Carex canescens	-	-	-	-	0.13	+	-	-	-	-	-	-	-	-
Lonicera villosa	-	-	-	-	-	-	0.14	+	-	-	0.25	+	-	-
Rhamnus alnifolia	-	-	-	-	-	-	0.29	+	-	-	0.25	1	-	-
Bromus ciliatus														
Umbelliferae	-	-	-	-	-	-	0.14	+	-	-	-	-	-	-
Rubus sp.	-	-	-	-	-	-	0.14	+	-	-	-	-	-	-
Rubus pubescens	-	-	-	-	-	-	0.14	+	-	-	-	-	-	-
Calla palustris	-	-	-	-	-	-	0.14	+	-	-	-	-	-	-
Rubus acaulis	-	-	-	-	-	-	0.14	+	-	-	-	-	-	-
Trientalis borealis	-	-	-	-	-	-	0.14	+	-	-	-	-	-	-
Eriophorum spissum	-	-	-	-	-	-	0.14	+	0.25	+	-	-	0.33	+
Lonicera oblongifolia	-	-	-	-	-	-	0.14	+	-	-	-	-	-	-
Carex rostrata	-	-	-	-	0.07	+	-	-	-	-	-	-	-	-
Iris versicolor	-	-	-	-	-	-	-	-	-	-	0.25	+	-	-
Vaccinium angustifolium	-	-	-	-	-	-	-	-	0.33	+	-	-	-	-
Salix	-	-	-	-	-	-	-	-	0.66	1	-	-	-	-
Salix candida	-	-	-	-	-	-	-	-	0.33	+	-	-	-	-

Strings
(*Carex cephalantha-Potentilla fruticosa* nodum)

Strings are very variable across northern Minnesota. They are best developed in water tracks that have been ditched, whereas they are barely perceptible above the water level in the most pristine water tracks. Strings are also impossible to define in certain areas with small and shallow flarks. Strings are generally dominated by *Betula pumila* var. *glandulifera*, *Potentilla fruticosa*, *Salix pedicellaris* var. *hypoglauca*, *Carex diandra*, *C. cephalantha*, *Thelypteris palustris* var. *pubescens*, and *Viola pallens* var. *mackloskeyi*. In the wettest water tracks, however, strings also have sedges such as *Carex lasiocarpa* and *C. chordorrhiza*.

Featureless water tracks
(*Carex limosa-C. lasiocarpa* nodum)

Most water tracks have nonforested lawns that lack oriented pools. These sedge lawns are dominated by *Carex lasiocarpa* and *Rhynchospora alba*, with *Carex limosa*, *C. chordorrhiza*, and *Betula pumila* var. *glandulifera*. These sedge lawns are very similar to the flark nodum but lack the more aquatic species such as *Drosera anglica*, *D. linearis*, *Utricularia minor*, and *Triglochin maritima*. Probably the most notable species absent from these featureless water tracks is *Carex livida*, which is virtually restricted to flarks and spring-fen channels in Minnesota. The featureless water tracks usually have standing water, but there may be a continuous carpet of mosses including various Sphagna.

Forested Fingers
(*Larix laricina-Carex chordorrhiza* nodum)

The margins of water tracks in the Glacial Lake Agassiz region are often fringed by fingers of forest that extend out into the track (Plate 11C) and are dominated by *Larix laricina* with *Picea mariana*. The understory of these stands is dominated by *Carex chordorrhiza*, *C. lasiocarpa*, *C. leptalea*, *Betula pumila* var. *glandulifera*, *Ledum groenlandicum*, and *Chamaedaphne calyculata*. The forest canopy in these stands is not continuous, and the forest floor consists of a continuous carpet of *Sphagnum*, with high hummocks around the base of the trees and low, moist depressions.

Tree Islands
(*Carex pseudocyperus-Aronia melanocarpa* nodum)

Several water tracks in the Glacial Lake Agassiz peatlands have small, streamlined tree "islands." These islands are dominated by *Larix laricina* associated with *Picea mariana* and occasionally *Abies balsamea*. The islands are floristically similar to the forested finger nodum but are generally much richer in species. *Carex pseudocyperus* and *Aronia melanocarpa* are consistently present, along with *Betula pumila* var. *glandulifera*, *Ledum groenlandicum*, *Vaccinium oxycoccos*, *Potentilla palustris*, *Galium labradoricum*, and *Lysimachia thrysiflora*. *Carex lasiocarpa* and *C. chordorrhiza*, however, are generally absent. The tree islands consist of moss-covered hummocks around the base of the trees and water-filled depressions (Plate 11D).

Spring-Fen Forest
(*Picea mariana-Carex gynocrates* nodum)

The spring-fen forests are dominated by *Picea mariana* but also contain *Larix laricina*, *Abies balsamea*, and *Thuja occidentalis*. *Carex gynocrates* is consistently present in the understory with *Smilacina trifolia*, *Carex paupercula*, *Ledum groenlandicum*, *Carex trisperma*, and *Vaccinium vitis-idaea*. The spring-fen forests usually have a nearly continuous canopy and a continuous carpet of moss on the substrate. Standing water is unusual in these stands, but the water table is usually close to the surface.

Water Chemistry

The water chemistry from the vegetation samples may be divided into two major classes on the basis of pH and calcium concentration. The ombrotrophic bog samples have a pH below 4.2 and calcium concentrations below 2 mg 1^{-1}, whereas the minerotrophic fen samples have values above this level.

The bog samples have very narrow ranges for pH (3.7-4.1) and calcium concentration (0.6-2.0 mg 1^{-1}) across northern Minnesota. A few bogs have unexpectedly high concentrations of Ca but otherwise seem to be ombrotrophic. These samples were taken from pits because standing water was not present at the surface, and they may represent complex exchanges with the subsurface peat. The water chemistry of the bog samples does not differentiate between the two types of vegetation landform on bogs, except that absorbance readings tend to be lower at the bog crest, indicating higher rates of flow.

Extremely poor fens (*sensu* Sjörs 1950) are distinguished by a pH of 3.8-5.0. In northern Minnesota these fens generally exhibit the most sensitive floristic response to small changes in water chemistry. A group of poor fens in Minnesota that have boglike vegetation is characterized by a pH of 4.1-4.6 and Ca concentrations of 2.2-5.5 mg 1^{-1}. These samples were taken from relevés containing one or more minerotrophic indicators (*sensu* Sjörs 1963, 1983; Glaser *et al.* 1981; Wheeler *et al.* 1983) with low cover values.

The other group of poor fens recognized by Sjörs (1950) has considerable overlap in pH and is characterized by Heinselman (1970) as weakly minerotrophic, with a pH of 4.3 to 5.8 and Ca concentrations of 3-10 mg 1^{-1}. The northern Minnesota samples that fall within this range were taken from relevés whose vege-

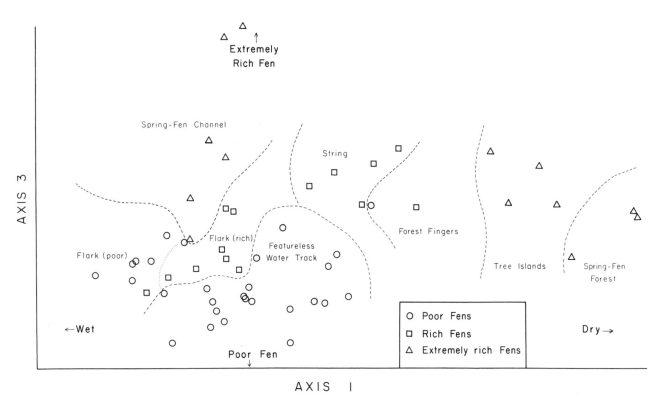

Fig. 2.4 Ordination of fen vegetation by detrended correspondence analysis. The fen vegetation is separated into groups that correspond to landform type and water chemistry. The first ordination axis corresponds most closely to a moisture gradient with the wetter stands on the left and drier stands on the right. The third ordination axis corresponds to the chemical gradient, with the poor-fen waters below and extremely rich fen waters above.

tation was very similar to that of the water tracks. These samples had a pH of 4.1-5.9 and Ca concentrations of 0.9-13.0 mg 1^{-1}. Many of the small patterned fens from northeastern Minnesota fall into this class, along with water tracks that have been partially drained by ditches.

The transitional rich fens of Sjörs (1950) or minerotrophic class of Heinselman (1970) are characterized by a pH of 5.8 to 7 and Ca concentrations of 10-25 mg 1^{-1}. The water samples from northern Minnesota that fall into this class have a pH of 5.9-6.8 and Ca concentrations of 10-32 mg 1^{-1}. Other samples, however, are transitional to the poor-fen class and are difficult to categorize with certainty.

Extremely rich fens, which Sjörs (1950) distinguishes as having a pH of 7 to about 8.5, are the only class of fen samples that are generally restricted to a single landform unit. Water samples from the spring fens have a pH of 6.8-7.4 and Ca concentrations of 20-45 mg 1^{-1}. Only a few samples from patterned fens in Minnesota have such high values. The chemistry of these fens seems to indicate groundwater discharge.

Classification of Peatlands

Boreal peatlands have been traditionally classified by northern Europeans on the basis of their landforms,

vegetation, water chemistry, and inferred hydrology. The fundamental assumption of this system is that the source of water and solutes in a peatland controls the course of peatland development. Thus raised bogs are distinguished from fens by their sole dependence on atmospheric precipitation (Aiton 1811; Dau 1823; Weber 1902), whereas fens derive at least some solutes from a mineral source.

In practice, however, the hydrology of a peatland is usually inferred indirectly from the chemistry of the surface waters and the topography of the peatland. Thunmark (1942) determined that in Sweden bog waters may be distinguished from fen waters by a pH lower than 4.0 and Ca concentration less than 1 mg 1^{-1}. A chemical classification system for peatlands that was closely correlated with the distribution of different indicator species was then developed (Whitting 1947, 1948, 1949; Sjörs 1946, 1948, 1950; Du Rietz 1949). This system has been successfully applied by Sjörs (1963) and Heinselman (1963, 1970) to North America, where many of the same indicator species respond to the chemical gradient in a similar fashion.

A parallel method of classifying peatlands that emphasized landform patterns and forest site types was developed in Finland (Cajander 1913). The landform patterns of a peatland were described according to

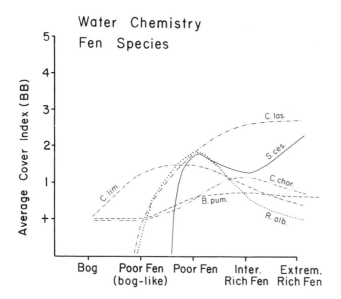

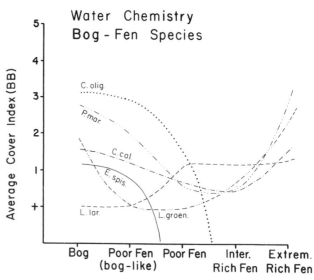

Fig. 2.5. Direct gradient analysis of the major vascular plant species along a chemical gradient. The average cover value (Braun-Blanquet index) of each species is plotted in relation to the water chemistry. The symbols for the Braun-Blanquet index are as follows: (+) sparsely present; (1) plentiful but low cover value; (2) very numerous, or covering at least 5% of the area; (3) very numerous, or covering 25-50% of the area; (4) any number of individuals covering 50-75% of the area; (5) covering more than 75% of the area. The ranges in water chemistry are bog (pH < 4.2; Ca concentration < 2 mg l^{-1}); poor fen: boglike (pH 4.1-4.6; Ca concentration 1.5-5.5 mg l^{-1}); poor fen: fenlike (pH 4.1-5.8; Ca concentration < 10 mg l^{-1}); intermediate rich fen (pH 5.8-6.7; Ca concentration 10-32 mg l^{-1}); extremely rich fen (pH > 6.7; Ca concentration > 30 mg l^{-1}).

Fig. 2.6. Direct gradient analysis of the major vascular plants in both bogs and fens. Dual peaks in abundance for various species may indicate that these species respond to a different environmental factor. See Fig. 2.5 for the average cover values and ranges in water chemistry.

their cross-sectional profile (Grossform) and surface relief (Kleinform) (Aario 1932, Paasio 1933). Regional surveys of peatlands in Finland determined a striking zonation of landform types related to the latitudinal gradient (Ruuhijärvi 1960, 1983; Eurola 1962; Eurola *et al.* 1984). These surveys combined the landform classification with a classification of the vegetation and water chemistry developed in Sweden, which provided a comprehensive regional classification for boreal peatlands.

Environmental Gradients

The vegetation in the patterned peatlands of Minnesota is closely related to landform type and water chemistry. This relationship is documented by detrended correspondence analysis (DCA), a multivariate method for determining patterns in an entire data set composed of many different variables (Hill 1979). The fen relevés, for example, are consistently grouped according to landform type and secondarily by water chemistry (Fig. 2.4). This close correspondence among the vegetation, landform, and water chemistry is surprising

when the regional spacing of the samples is considered and when many relevés from disturbed peatlands are included.

The DCA ordination indicates that two environmental gradients control the composition of the vegetation. Most of the variation in the data set is expressed along axis 1, which corresponds closely to the moisture gradient. The driest forested stands are located on the right side of the ordination, whereas the wettest flarks are located on the left. The influence of water chemistry is indicated by axis 3, with poor fens positioned at the bottom and richer fens toward the top of the axis.

The effect of water chemistry on the relative abundance of the dominant mire species is best illustrated by direct gradient analysis (*sensu* Whittaker 1967). The major fen dominants attain their peak in abundance within specific ranges of water chemistry with the exception of *Scirpus cespitosus*, which has separate peaks in the poor-fen and extremely rich fen range, and *Betula pumila* var. *glandulifera*, which has a fairly uniform distribution across the entire rich-fen range (Fig. 2.5). The direct gradient analysis of the major bog–fen species, however, demonstrates that many of these species are relatively insensitive to changes in water chemistry (Fig. 2.6). *Picea mariana* and bog ericads such as *Chamaedaphne calyculata* and *Ledum groenlandicum* have two peaks in abundance at the opposite ends of the chemical gradient. These species apparently respond to the moisture gradient and become dominant on the driest landforms irrespective of the water chemistry. *Larix*, however, is more common

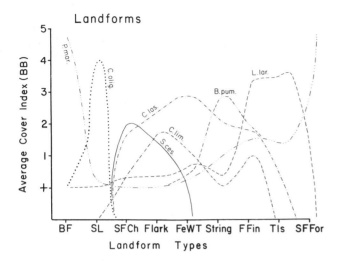

Fig. 2.7. Direct gradient analysis of the major vascular plant species with respect to landform type. Each species reaches its peak abundance on a different landform type. The landform types are bog forest (BF), *Sphagnum* lawn (SL), spring-fen channel (SF Ch), flark, featureless water track (Fe WT), string, forested finger (F Fin), tree island (T Is), and spring-fen forest (SF For). See Fig. 2.5 for average cover values.

in the fen range and is similar in behavior to *Betula pumila*.

Plant succession may also play an important role in determining the vegetation patterns in these peatlands. Two types of poor fen with similar ranges in water chemistry and water levels but very different types of vegetation occur in northern Minnesota. One type occurs on the *Sphagnum* lawns and is almost indistinguishable from ombrotrophic bog vegetation, whereas the other type is very similar to that found in the more minerotrophic flarks and featureless water tracks. The major division between these two types of poor fen actually separates the overall peatland vegetation into two contrasting types much better than the more subtle floristic changes that occur at a pH of 4.2 and Ca concentration of 2 mg 1^{-1}. These two types of poor fen seem to represent the opposing end products of development for bogs and fens.

The factor that best integrates the varying effect of water chemistry, moisture, and plant succession on the mire vegetation is the landform patterns. Direct gradient analysis indicates the degree to which different species attain dominance on different landform types, particularly the different types of poor fen (Fig. 2.7). Subtle changes in the form of these patterns also provide an important tool for interpreting the direction of plant succession and influence of the major environmental controls.

Conclusions

Peat landforms in northern Minnesota are associated with large uniform stands of vegetation. Each landform type is characterized by a distinctive species assemblage and a relatively narrow range in water chemistry. The bog vegetation is essentially invarient across the region, whereas a continuous series of species assemblages occurs in the water tracks and spring-fen channels. The underlying structure of the fen data set, however, corresponds most closely to landform type and also to water level and water chemistry. No apparent pattern in the data set can be related to the regional changes in climate, physiography, and soils that largely control vegetation patterns on the surrounding uplands. The process of peat accumulation seems to produce landforms with a relatively uniform set of environmental conditions. These landforms therefore seem to represent predictable steps along a common developmental pathway that is unique to these boreal peatlands.

Literature Cited

Aario, L. 1932. Pflanzentopographische und paläeogeographische mooruntersuchungen in N-Satakunta. Fennia 55:1-179.

Aiton, W. R. E. 1811. Treatise on the origin, qualities, and cultivation of moss-earth with directions for converting it into manure. Wilson and Paul, Air, U.K.

Cajander, A. K. 1913. Studien über die moore Finnlands. Acta Forestalia Fennica 2:1-208.

Clymo, R. S. 1963. Ion exchange in *Sphagnum* and its relation to bog ecology. Annals of Botany (London) N.S. 27:309-324.

———. 1967. Control of cation concentrations and in particular of pH in *Sphagnum*-dominated communities. Pages 273-284 *in* H. L. Golterman and R. S. Clymo, editors. Chemical environment in the aquatic habitat. North-Holland, Amsterdam, Netherlands.

——— and P. M. Hayward. 1982. The ecology of *Sphagnum*. Pages 229-289 *in* A. J. E. Smith, editor. Bryophyte ecology. Chapman & Hall, London, U.K.

Dau, H. C. 1823. Neus handbuch über den torf dessen natur, entstehung und Wiedererzeugung. J. C. Hinrichschen Buchhandlung, Leipzig, Germany.

Drever, J. I. 1982. The geochemistry of natural waters. Prentice-Hall, Englewood Cliffs, New Jersey, USA.

Du Rietz, G. E. 1949. Huvudenheter och huvudgränser i svensk myrvegetation. Summary: main units and main limits in Swedish mire vegetation. Svensk Botanisk Tidskrift 43:274-309.

Eurola, S. 1962. Über die regionale einteilung der südfinnischen moore. Annales Botanici Societatis Zoologicae Botanicae "Vanamo" 33:1-243.

———, S. Hicks, and E. Kaakinen. 1984. Key to Finnish mire types. Pages 11-117 *in* P. D. Moore, editor. European mires. Academic Press, New York, New York, USA.

Glaser, P. H., G. A. Wheeler, E. Gorham, and H. E. Wright, Jr. 1981. The patterned mires of the Red Lake peatland, northern Minnesota: Vegetation, water chemistry, and landforms. Journal of Ecology 69:575-599.

———, J. A. Janssens, and D. I. Siegel. 1990. The response of vegetation to chemical and hydrological gradients in the Lost River peatland, northern Minnesota. Journal of Ecology 78:1021-1048.

Gleason, H. A. 1952. The new Britton & Brown illustrated flora of the northeastern United States and adjacent Canada. Vols. 1, 2, 3. New York Botanical Garden, New York, New York, USA.

Gorham, E., S. J. Eisenreich, J. Ford, and M. V. Santlemann. 1984. The chemistry of bog water. Pages 339-363 *in* W. Strumm, editor. Chemical processes in lakes. Wiley, New York, New York, USA.

Heinselman, M. L. 1963. Forest sites, bog processes, and peatland types in the Glacial Lake Agassiz region, Minnesota. Ecological Monographs 33:327-372.

————. 1970. Landscape evolution, peatland types, and the environment in the Lake Agassiz Peatland Natural Area, Minnesota. Ecological Monographs 40:235-261.

Hill, M. O. 1979. DECORANA: A FORTRAN program for detrended correspondence analysis and reciprocal averaging. Cornell University, Ithaca, New York, USA.

Mackenzie, K. K. 1940. North American *Cariceae*. Vols. 1, 2. New York Botanical Garden, New York, New York, USA.

Paasio, I. 1933. Über die vegetation der hochmoore Finlands. Acta Forestalia Fennica 39:1-210.

Ruuhijärvi, R. 1960. Über die regionale Einteilung der Nordfinnischen Moore. Annales Botanici Societatis Zoologicae Botanicae "Vanamo" 31:1-360.

————. 1983. The Finnish mire types and their regional distribution. Pages 47-67 *in* A. J. P. Gore, editor. Mires: Swamp, bog, fen, and moor. Vol. 4B. Regional Studies. Elsevier, Amsterdam, Netherlands.

Sjörs, H. 1946. Myrvegetationen i övre långanomradet i Jämtland. Arkiv för Botanik 33:1-96 (English summary).

————. 1948. Myrvegetation i Bergslagen. Acta Phytogeographica Suecica 21:1-299.

————. 1950. On the relation between vegetation and electrolytes in North Swedish mire waters. Oikos 2:241-258.

————. 1963. Bogs and fens on Attawapiskat River, northern Ontario. Museum of Canada Bulletin, Contributions to Botany 186:45-133.

————. 1983. Mires of sweden. Pages 69-94 *in* A. J. P. Gore, editor. Mires: Swamp, bog, fen, and moor. Elsevier, Amsterdam, Netherlands.

Thunmark, S. 1942. Über rezente Eisenocker und ihre Mikroorganismengemeinschaften. Bulletin of the Geological Institutions of the University of Uppsala 29:1-285.

Weber, C. A. 1902. Vegetation und Entstehung des Hochmoors von Augstumal im Memeldetta mit vergleichenden Ausblicken auf andere Hochmoore der Erde. Verlagsbuchhandlung, Paul Parey, Berlin, Germany.

Wheeler, G. A., P. H. Glaser, E. Gorham, C. M. Wetmore, F. D. Bowers, and J. A. Janssens. 1983. Contributions to the flora of the Red Lake peatland with special attention to *Carex*. American Midland Naturalist 110:62-96.

Whitting, M. 1947. Kationbestämmningar i myrvatten. Botaniska Notiser 1947:287-304. (English summary).

————. 1948. Preliminärt meddelande om forsatta kationbestämmningar i myrvatten sommaren 1947. Svensk Botanisk Tidskrift 42:116-134. (English summary).

————. 1949. Kalciumhalten i nagra nordsvenska myrvatten. Svensk Botanisk Tidskrift 43:715-739.

Whittaker, R. H. 1967. Gradient analysis of vegetation. Biological Reviews 42:207-264.

CHAPTER 3

Ecological Development of Patterned Peatlands

Paul H. Glaser

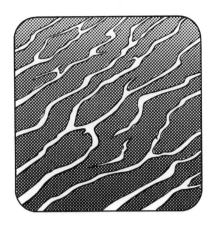

Peatlands have spread over a diverse landscape in northern Minnesota and cross important environmental thresholds such as the prairie-forest border. The regional changes in physiography, soils, and climate have produced important changes in the vegetation of mineral uplands. Similar changes should be expected in the peatland vegetation unless autogenic processes buffer these peatlands from external environmental factors.

The patterned peatlands, however, are remarkably uniform across Minnesota with respect to their landform patterns and associated vegetation assemblages and water chemistry. Raised bogs, for example, have the same radiating forest patterns from their southeastern limits on Glacial Lake Upham to their western limits on Glacial Lake Agassiz. The vegetation on these bogs is nearly identical, changing little in terms of species composition or species richness. Relevé plots of 100 or 400 m^2 consistently contain the same nine to thirteen species irrespective of their location or size. Even more surprising is the narrow range in the chemistry of the bog waters across this region despite the increase in calcareous dustfall toward the western prairies (Munger and Eisenreich 1983; Gorham and Tillton 1978).

The size and orientation of raised bogs within a peatland, however, are altered by local factors that control the hydrology and water chemistry. The patterned peatlands are apparently formed by simple feedback systems that develop between these factors and the growth of the major peat formers (Glaser 1987a, b). Transitions between different types of patterns indicate potential developmental trends that can be tested by stratigraphic and hydrologic analysis (Glaser *et al.* 1981). Drainage ditches have also locally altered the peat landforms, providing evidence for the hydrological and chemical controls on peatland development (Glaser 1987a).

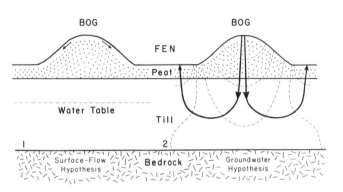

Fig. 3.1. Hydrological controls on bog development. According to the surface-flow hypothesis (1), all water movements in a peatland are restricted to the surface layer. A bog will spread outward until its growth is checked by the higher alkalinity in runoff draining from mineral uplands. The groundwater hypothesis (2: Siegel 1981, 1983), however, predicts that the outward growth of a bog will be blocked by groundwater discharging at the bog margin. According to this hypothesis, raised bogs contain groundwater mounds that will generate sufficient head to drive local flow cells. The water in these cells moves downward from the bog surface into the underlying calcareous till, where it picks up alkalinity and cations. The water then moves upward and discharges at the margin of the bog, producing a fen-water track.

Hydrology

The hydrology of peatlands is controlled by either surface runoff (Von Post and Granlund 1926; Ivanov 1981) or groundwater processes (Siegel 1981, 1983; Siegel and Glaser 1987; Fig. 3.1).

The surface-runoff hypothesis was originally proposed by Von Post and Granlund (1926) and was rigorously developed by peatland hydrologists in the Soviet Union (Ivanov 1981), Great Britain (Ingram 1983), and the United States (Boelter 1965, 1969, 1972; Boelter and Verry 1977). According to this hypothesis, water will flow only through the uppermost layer of a peatland, called the acrotelm. The peat in this layer is only slightly decomposed, and its large pores offer little resistance to water movement. The deeper layers of peat are decomposed, however, and water movement is restricted by the small size and low degree of connectivity of the pores. Water is so tightly held to the peat particles in this lower layer (catotelm) that flow is believed to be negligible.

The groundwater hypothesis, however, predicts that water will flow through any porous medium as long as there is sufficient hydraulic head to drive the flow. On the flat lake plains of northern Minnesota, this head gradient may be generated by groundwater mounds under the raised bogs. Computer simulations by Siegel (1981, 1983) show that the bogs may become recharge areas for groundwater, with water moving downward through the peat into the mineral substratum. The water will then pick up cations and alkalinity within the calcareous parent material before rising to discharge in a fen around the bog margins. The Siegel model therefore predicts an autogenic limit on the continued expansion of a bog margin, in contrast to the largely allogenic limit of the Von Post and Granlund model.

Surface-Runoff Controls

If water flows only through the uppermost layer of peat, the peatland waters can acquire alkalinity only from (1) precipitation, (2) chemical transformations within the peat, and (3) runoff from the surrounding mineral soil. As peat spreads over the landscape, these soils are lost as a potential source of cations and alkalinity. Fens will then develop only along drainage lines where runoff from mineral soil flows across a peatland. These waters must contain sufficient alkalinity to (1) provide the necessary requirements for the fen-indicator species and (2) block the spread of the bog-building Sphagna. Bogs are elevated above contact with these waters by their raised mounds of peat.

The volume and alkalinity of runoff draining onto a peatland can be estimated by the rational formula for runoff (Chow 1964):

$$Q = CIA$$

where

Q = peak discharge in $f^3 sec^{-1}$
C = runoff coefficient dependent on characteristics of the watershed
I = runoff intensity in hr^{-1}
A = drainage area in acres

Runoff will increase with area, whereas the alkalinity of runoff will change with different parent material. The surface-runoff model can therefore be tested by

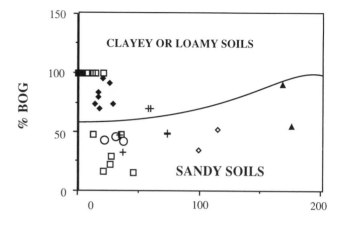

Fig. 3.2. The relationship between soils and bog development. The proportion of bog and fen vegetation in a peatland is related to the soil type in the surrounding uplands. Bogs predominate in smaller peatlands surrounded by relatively impermeable clayey or loamy soils. Fens cover more than 50% of peatlands surrounded by more permeable sandy soils, which may be discharge zones for groundwater.

measuring the bog-to-fen ratio in peatlands of different sizes. This model predicts that (1) small peatlands ($< 30 km^2$) with a *large* area of mineral soil in their catchment should be dominated by fens, (2) small peatlands ($< 30 km^2$) with a *small* area of mineral soil in their catchments should be dominated by bogs, and (3) large peatlands ($> 30 km^2$) with *little* mineral soil in their catchments should be dominated by bogs.

The first prediction is satisfied by small ($< 30 km^2$) discontinuous peatlands surrounded by clayey or loamy soils (Fig. 3.2). These peatlands are surrounded by a narrow strip of mineral soil, and the peatlands are largely covered by bogs. The appearance of small water tracks in the type 2 peatlands is also related to a slight increase in the drainage area of mineral soil in the catchment. The large fraction of bogs in these watersheds corresponds to the small area of mineral soil at the watershed crest.

However, the other peatland types do not fit the predictions of the surface-runoff model. The large water tracks in the type 3, 4, 5, 8, and 9 peatlands are not associated with a corresponding increase in mineral exposures at the watershed crest (chapter 1). The source of the alkalinity necessary to maintain these large fens is not apparent from surface drainage patterns. The exposure of sandy soil at the heads of these water tracks suggests that groundwater may be discharging through the porous sand deposits into the peatlands.

Groundwater Hypothesis

The large continuous peatlands of the Glacial Lake Agassiz region should be dominated by bogs, according to the surface-runoff hypothesis. As the mineral soil in this region is covered by peat, the alkalinity from precipitation should be insufficient to prevent the spread of the bog-building Sphagna. Only where exposures of mineral soil protrude through the peat would runoff carry sufficient alkalinity to maintain small areas of fen vegetation.

However, in the Glacial Lake Agassiz region, the majority of the peat surface consists of fen-water tracks and swamp forests despite the limited input of runoff from mineral uplands (color centerfold). Only in the northeast quarter (North Black River area) are most peatlands filled by bogs, despite the large exposures of mineral soil. Elsewhere the advance of bogs may by restricted by the discharge of alkaline groundwater from the underlying calcareous parent material.

The most obvious landform associated with groundwater discharge in the Lake Agassiz region is the spring fen. Without the continual discharge of groundwater, the exceptionally high pH and Ca concentration in the spring-fen channels should change to lower levels as the waters equilibrate with atmospheric CO_2. Spring fens (type 7) are always associated with sandy beach deposits in the Glacial Lake Agassiz region. The few spring-fen mounds (type 11) in Minnesota may be associated with sand deposits buried under the peat. A peat core taken from a channel in the Lost River area had coarse gravel and sand at its base (Glaser et al., 1990).

The discharge of groundwater may also play a role in the formation of the larger water tracks in the Glacial Lake Agassiz region that contain fields of tree islands. These water tracks are fed by systems of channels that drain through the surrounding swamp forest, which becomes progressively restricted to narrow forested fingers and islands as the channels expand and converge into a water track. The appearance of these nonforested channels is similar to the anastomosing (branching and rejoining) networks of the spring fens, indicating a similar mode of origin. The water-track channels also arise from the margins of narrow beach ridges that protrude through the peat.

The association of these channels with groundwater discharge is strengthened by an interpretation of Landsat imagery taken during spring breakup. An image from the Red Lake peatland (E-30042-16303-D; Fig. 3, chapter 1) indicates that water is flowing in the peripheral channels that feed into the western water track at a time when the surrounding peat surface is still frozen and covered with snow. The transport of heat to the surface by the upwelling groundwater should focus surface drainage in these discharge zones as the winter snowpack begins to melt in the spring. The gradual decline in minerotrophy in these water tracks from the marginal channels to the central patterned area also indicates that groundwater discharge is focused in the channels.

Another potential discharge zone in the Glacial Lake Agassiz peatlands occurs within the internal water tracks or seeps of the larger bogs (type 9). The source of minerotrophy in these water tracks is unclear because they are completely surrounded by ombrotrophic peat. Small rates of discharge from the underlying calcareous substrate may supply the alkalinity necessary to convert the bog waters into fen waters, but the water in these tracks is too dilute (pH < 6.5; Ca concentration < 6 mg 1^{-1}) to unequivocally indicate groundwater discharge (Siegel 1981, 1983). Alternatively, the alkalinity may be supplied by the release of a base from the peat by the enhanced decomposition along lines of flow (Glaser et al. 1981).

Landsat imagery taken during spring breakup on April 16, 1984 (E-50046-16311) indicates that surface drainage is focused in the internal water tracks in contrast to the surrounding snow-covered bogs. Groundwater discharge is the most likely source for the heat necessary to melt the ice in the tracks in advance of the surrounding ombrotrophic peat surface. The initiation of spring breakup in these tracks, however, may also be caused by (1) the thin snowpack that accumulates in the nonforested tracks because of their greater exposure to wind and (2) the collection of drainage that seeps from under the thicker snowpack farther upslope. Detailed hydrologic studies are therefore necessary to resolve this question, which is linked to the development of all large raised bogs from Minnesota to Hudson Bay.

All the large water tracks in Minnesota arise downslope from sandy beach, outwash, or moraine deposits. The huge western water track at Red Lake, for example, is fringed by beach ridges to the north and west, and the large water track at Myrtle Lake arises from a partially paludified beach ridge mapped by Eng (1980). The large water tracks at Wawina on Glacial Lake Upham and Sand Lake in Lake County also arise from the edge of sandy beaches or moraines. The close correlation between these permeable sand deposits and the high percentage of fen in the adjacent peatlands indicates that these permeable sand deposits may act as conduits for upwelling groundwater. This hypothesis, however, must be tested by hydrogeologic methods.

Chemical Transformations

Although hydrological processes determine the flux of water and solutes through a peatland, the chemistry of the mire waters is also controlled by chemical transformations within a peatland. The most striking effects of these transformations on the mire waters in northern Minnesota are (1) the decline in pH and Ca concentration from the margin of a peatland to the patterned interior, (2) the increasing acidity of peatlands dominated by Sphagnum, and (3) the downslope rise in pH and Ca concentration on the flanks of raised bogs.

The decline in the alkalinity of minerotrophic runoff within a peatland may be a product of (1) dilution (mixing with precipitation), (2) adsorption of cations onto peat by cation exchange, (3) uptake of ions by living plants, or (4) deposition as a precipitate. The decline in alkalinity is apparent in the large water tracks of the Red Lake peatland, but it can also be detected in very small peatlands, such as the Grand Marais fen in northeastern Minnesota (Glaser 1983a). The linear gradient in pH and Ca indicates that groundwater discharge is focused at the head and margins of these water tracks and does not occur within the central core of the track.

The acidity of raised bogs is maintained by either the cation-exchange mechanism of *Sphagnum* or the release of organic acids by decomposition of *Sphagnum* peat. The bog waters will therefore consume a certain quantity of base before the pH can rise. The buffering of these waters by humic acids probably accounts for the narrow range of water chemistry for bogs across northern Minnesota. The volume of alkaline water required to raise bog waters into the fen range should therefore be in excess of the amount calculated by Siegel (1981). However, the chemistry of the interstitial pore waters in the Lost River peatland indicates that the mixing of alkaline groundwater with acid bog waters can produce abrupt changes in the water chemistry (Siegel and Glaser 1987; Glaser *et al.*, 1990). These changes are also demonstrated by the discharge of minerotrophic water from drainage ditches onto bogs.

On large raised bogs (20 km²), however, there is no apparent source for the base required to raise the pH of the bog waters on the lower bog flanks (Glaser 1987a). Decomposition of *Sphagnum* should release organic acids that lower the pH rather than the reverse (Gorham *et al.* 1984). The release of a base by enhanced decomposition along lines of flow as suggested by Glaser *et al.* (1981) therefore seems unlikely. However, the alternate hypothesis also has problems. The necessary base could be supplied by groundwater discharging from the calcareous parent material that underlies these peatlands. But it is unlikely that this discharge would be focused near the crest of the raised bogs, which are often the highest points on the nearly flat landscape. This problem needs to be thoroughly explored because it is fundamental for understanding the development of all large raised bogs across the continental interior of North America (Glaser 1987a, b).

Vegetation Processes

Feedback Systems

Glaser (1987a, b) proposed a simple feedback model to explain the development of streamlined bogs and water tracks. This model is based on the contrasting characteristics of the major peat formers in bogs and fens. The major peat formers in bogs are Sphagna, which grow best in waters with a low pH (Clymo 1963, 1967; chapter 2). With increasing pH and Ca concentration, these Sphagna are replaced by sedges such as *Carex lasiocarpa*, the dominant peat former in the water tracks. *C. lasiocarpa* spreads by means of rhizomes

and forms a very permeable layer of peat composed of rhizomes, rootlets, and large sedge fragments. The porosity of this sedge peat is several orders of magnitude higher than that of undecomposed *Sphagnum* peat. As the degree of decomposition increases, the porosity of the *Sphagnum* peat falls faster than that of the sedge peat until the porosity of both peat types attain minimum value (Fig. 3.3).

The development of bogs and fens will then depend on the dynamics of runoff draining across a peatland. Rather than flowing across a peatland as sheet flow, runoff is channeled into water tracks first by topographic features at the mineral crest and secondarily by the accumulation of peat with different hydraulic properties (Glaser 1983a, 1987a, b). Greater flow rates in the water tracks promote the spread of *C. lasiocarpa* lawns, which lay down a permeable mat of rhizomes

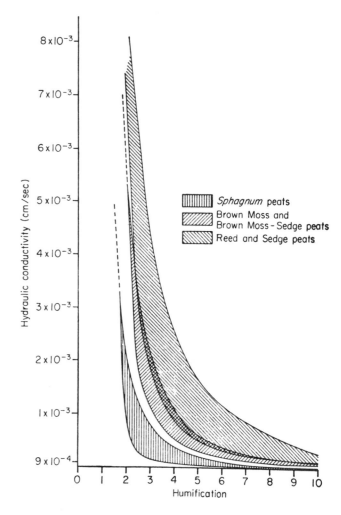

Fig. 3.3. Hydraulic conductivity of different peat types (Baden and Eggelsman 1963; Rycroft *et al.* 1975). For any given degree of decomposition (humification), *Sphagnum* peat has the lowest hydraulic conductivity (porosity) and *Carex* peat the highest. For the lowest grades of humification, the porosity of *Sphagnum* peat is several orders of magnitude less than that for *Carex* peat.

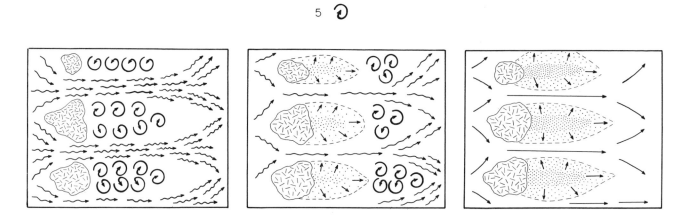

Fig. 3.4. Model for the development of primary bog islands (Glaser 1987a, b). The symbols represent (1) flow obstructions, (2) *Sphagnum* mat, (3) tortuous flow lines, (4) laminar flow lines, and (5) zones of sluggish flow. See text (Feedback Systems) for explanation. (Reproduced by permission of the Regents of the University of Colorado from *Arctic and Alpine Research* 19 [1987]).

and rootlets. *Sphagnum*, however, can become established downslope from flow obstructions, which split lines of flow and create zones with sluggishly flowing water downslope (Fig. 3.4). These zones will be acidified by the spreading *Sphagnum* mat by means of its cation-exchange capacity or release of organic acids. The *Sphagnum* mat will spread outward until its advance is blocked by higher alkalinity and flux of cations along the main drainage paths. A sharp boundary will then develop between the bog and fen vegetation.

Bog Complexes

The streamlined shape of the raised bogs appears to represent an equilibrium form that minimizes turbulent mixing of the alkaline waters in the water tracks (Glaser 1987a, b). The physical dimensions (length, width, and area) of the bog islands in Minnesota are highly correlated, and their average length-to-width ratio is approximately 3. Experiments with airfoils in wind tunnels and models in flumes indicate that landforms with this shape exhibit the minimum drag coefficient or resistence to a flowing fluid (Komar 1983, 1984). The most stable configuration of a bog island would therefore be a streamlined shape that (1) minimizes turbulent mixing around its edge and (2) permits a boundary layer of poor-fen water to develop around an island.

Bog islands are rare in the discontinuous peatlands

of Minnesota because of their small size (< 100 km^2) and numerous exposures of mineral soil. Bog islands, however, are common in the Red Lake peatland and occur on all large raised bogs in the continental interior of North America. The development of these islands appears to be different from that of the smaller discontinuous peatlands. The large bog complexes may be fragmented into streamlined lobes and islands by the origin of water tracks. These water tracks may arise by (1) chemical transformations of runoff draining from the bog crest as discussed above or (2) discharge of groundwater on the bog surface, which will convert the bog to fen (Glaser 1987a, b; Fig. 3.5). Sharp changes from bog to fen have been described from the peat stratigraphy of the Lost River peatland and Red Lake peatland (chapter 1). The discharge of water from drainage ditches may also create local water tracks on bogs, as at Wawina (Fig. 3.6).

Bog Margin

Despite their striking pattern on aerial photographs, only a slight elevational difference separates the surface of a water track from the margin of a raised bog. The bog margin is therefore sensitively adjusted to the position of the water table and expands or retreats in response to small changes in rates of flow or peat accumulation. An indication of these changes is provided by the landform patterns, which can best be interpreted

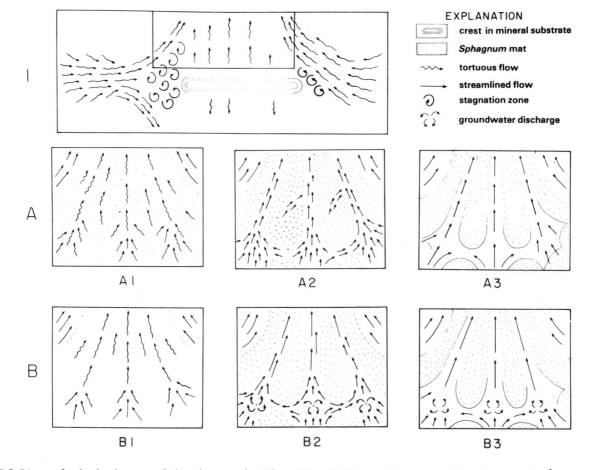

Fig. 3.5. Diagram for the development of a large bog complex (Glaser 1987a, b). The initial formation of a large bog (> 20 km²) is illustrated in step **1**. The origin of water tracks and streamlined islands within this bog can be explained by means of surface processes (step **A**) or the discharge of groundwater (step **B**). (Reproduced by permission of the Regents of the University of Colorado from *Arctic and Alpine Research* 19 [1987]).

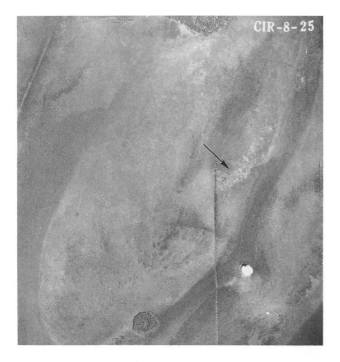

Fig. 3.6. Aerial photo of a bog island in the Wawina peatland in northeastern Minnesota. Minerotrophic water is discharging from a drainage ditch within the interior of the bog island, creating a zone of fen vegetation (*arrow*) downslope.

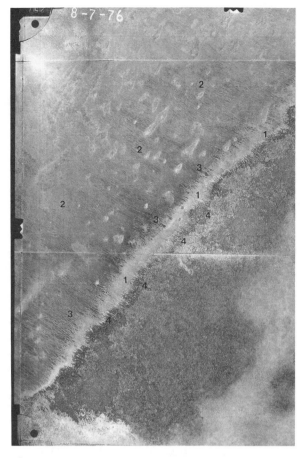

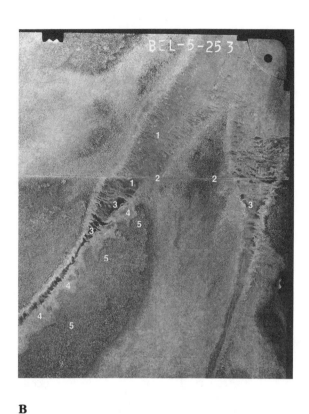

A   **B**

Fig. 3.7.A. Aerial photograph of an expanding bog margin. The position of the bog margin is maintained by the volume of runoff draining around its edges. The bog margin will spread outward in areas where the water table has been artificially lowered by drainage ditches. The marginal *Sphagnum* lawn (*1*) will spread into the adjacent water track (*2*), first over the strings (*3*) and then in the lower flarks. The trees then advance over the former strings, producing a sawtooth edge to the bog margin. This photo from the Red Lake peatland covers an area 2.5 km across. **B.** Aerial photograph of a retreating bog margin. The bog margin contracts in areas where the volume of runoff has increased. Water tracks (*1*) may become flooded as water discharges from a drainage ditch (*2*) dammed by beavers. The wide deep flarks (*3*) are aligned to wet portions of the *Sphagnum* lawn (*4*) that have deeply intruded into the surrounding bog forest (*5*). This aerial photograph from the Red Lake peatland covers an area 2.5 km across.

from areas that have been unambiguously altered by drainage ditches (Glaser 1987b).

In the Red Lake peatland the bog margin has a characteristic sawtooth margin in areas where the water table has been artificially lowered by drainage ditches (Glaser 1987a; Fig. 3.7a). Here a continuous carpet of *Sphagnum* has spread out from the bog margin onto the water track, colonizing first the strings and then coalescing over the wetter flarks. Trees have then spread out from the bog forest over the built-up strings, reinforcing the sawtooth pattern that can be seen from the air. An expanding bog margin will therefore create a broad nonforested *Sphagnum* apron, which retains the initial pattern of strings and flarks in the water tracks.

Local flooding from a dammed drainage ditch will produce a similar but less regular pattern along a bog margin. The best example of such a retreating mar-

gin occurs in the west-central watershed at Red Lake, where a beaver dam has backed up flow in a drainage ditch, causing it to discharge and flood several narrow water tracks (Fig. 3.7b; Plates 12B, 12C). Floodwaters from the water track have spilled over onto the bog margin, creating an irregular jagged edge. The marginal *Sphagnum* lawn is similarly altered as it intrudes deeply into the bog forest, producing a series of irregular forested lobes. Very similar patterns occur in pristine areas of the eastern watershed at Red Lake, where the water track also contains deep pools.

In featureless water tracks, a retreating bog margin is represented by a scalloped edge consisting of a series of disintegrating circular clones of *Sphagnum* (Plate 12B). Similar patterns have been observed on the Silver Flowe bogs in Scotland (Boatman 1983), where a series of circular hummocks and peninsulas of *Sphag-*

num occurs around the margin of coalescing pools. An expanding bog margin, in contrast, is marked by a smooth edge bordered by a nonforested *Sphagnum* lawn. Numerous small trees occur on these lawns, but a distinct border exists with the inner continuous bog forest. The third type of bog margin trimmed by these featureless water tracks has only a very narrow apron, with the trees growing close to the bog margin. This pattern seems to represent a stable margin, because the absence of vegetation patterns indicates recession or expansion.

Water Tracks

A water track develops within a zone where various lines of flow converge and produce distinctive vegetation patterns. The two most distinctive peat landforms in these water tracks are fields of tree islands entirely composed of peat and networks of transverse peat ridges and troughs called strings and flarks.

String-Flark Patterns

The linear networks of peat ridges and pools in circumboreal fens are perhaps the most uniform type of vegetation landform in the boreal zone. The uniformity of these vegetation patterns across such a broad region suggests a common mode of origin, but attempts to uncover a single mechanism have proven elusive. The large and controversial literature on the subject has been reviewed by Auer (1920), Washburn (1979), and Seppälä and Koutaniemi (1985), among others. Boatman (1983) and Glaser and Janssens (1986) also review the literature on pool formation as it relates to raised bogs.

Most explanations stress the formation of either strings or flarks by various physical and biotic processes. In northern Minnesota, flark formation seems to be the primary factor in the origin of patterned fens (Glaser 1983a, b). Flarks generally begin to appear on the downslope portion of featureless water tracks, where they are often aligned along one side of the track. The flarks in these situations are often well-defined pools separated by an otherwise featureless *Carex lasiocarpa* lawn. In the better-developed water tracks, the pools expand and deepen downslope, gradually restricting the *Carex* meadow into sinuous strips of firmer peat that are barely perceptible above the water surface.

The development of patterned fens therefore seems to be related to a progressive flooding of the peat surface, which is inundated in low spots parallel to the contour. This flooding process is produced by (1) the channeling of runoff, which converges from a wide area into a narrow water track, (2) the development of nearly level peat slopes as a result of peat accumulation, which inhibits surface drainage, and (3) the increasing resistance to infiltration produced by the accumulation of more-decomposed and less-porous peat

in the flarks (Fig. 3.8). The pools tend to form parallel to the contour interval because the nearly level slope produces a minimum force for downslope drainage. In patterned fens with tree islands, the flooding hypothesis is supported by the progressive paludification of the forested islands adjacent to the flarks.

The water chemistry in these patterned fens also supports the flooding hypothesis for flark formation. The downslope transition from marginal swamp forest to featureless water track to patterned fen is accompanied by a marked decline in the minerotrophy of the surface waters. In the western water track at Red Lake, for example, the pH declines from 6.5 at the featureless upslope portion of the track to 5.9 where isolated pools first begin to appear. On the lower portions of the track where the pools are deepest and the string-flark patterns most highly developed, the pH drops further to 5.7. The concentration of Ca shows a corresponding decline from 19.2 to 6.8 and 4.3 respectively at these same sites.

This decline is caused by four factors: (1) increasing isolation of the surface water from the source of mineral ions at the periphery of the peatland, (2) adsorption of ions to peat, (3) uptake of mineral nutrients by the absorptive organs of the plants, and (4) dilution by mixing with precipitation. The slower rates of surface drainage in the patterned areas would increase the proportion of mixing with rainwater and lower the minerotrophy of the surface waters. The large number of patterned fens in Minnesota with poor-fen waters therefore indicates a potentially important role of dilution with rainwater.

The flooding hypothesis is also supported by the vegetation changes produced by the drainage ditches (Glaser *et al.* 1981; Glaser 1987b; Fig. 3.9). Drainage ditches locally lower the water table (Boelter 1972) and thus permit more vigorous growth of trees and shrubs. This effect is demonstrated by the long internodes on *Larix* and bog birch growing in ditched areas and the signature of the vegetation on color infrared aerial photographs that indicate dense chlorophyll in the leaves (Hagen and Meyer 1979). Across a series of ditches, the drier strings and flarks become covered by dense stands of dwarf birch and tamarack until the characteristic landform patterns virtually disappear. Thus the peatland reverts to its original forested state in which tree growth is promoted by the better aeration of the surface layers of peat.

Alternate hypotheses for flark formation do not seem applicable for northern Minnesota. Permafrost is completely absent from the state and is very unlikely to have been present during the Holocene (Janssen 1968; Griffin 1977). Peat profiles from flarks contain no evidence for either truncation of the upper portion of the peat profile following "corrosive oxidation" (Sjörs 1946; Lundqvist 1951) or redeposition of peat following fissuring and solifluction (Pearson 1960, 1979; Pearsall 1966). Griffin's (1977) pollen diagrams from the western water track at Red Lake contain no ev-

PLAN VIEW

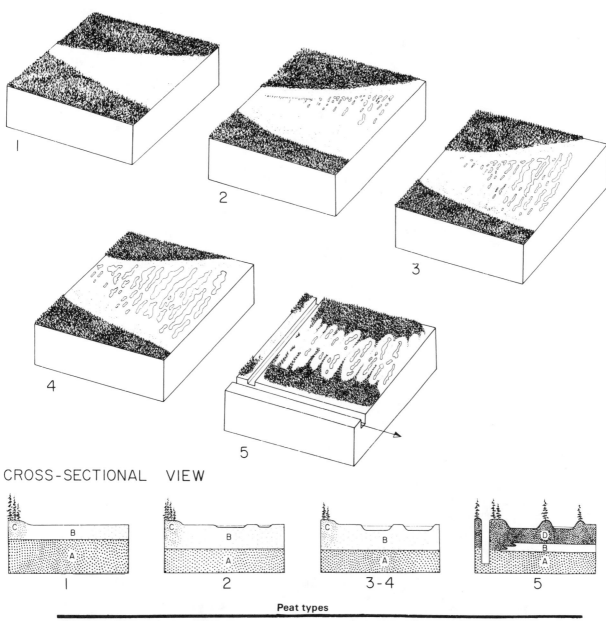

CROSS-SECTIONAL VIEW

Peat types

A highly decomposed wood peat
B porous sedge peat, particularly the rhizomatous network of *Carex lasiocarpa*

C humified wood peat under the swamp forest
D highly humified and compacted peat near the drainage ditches

Fig. 3.8. Development of patterned fen. A patterned fen begins as a featureless water track in a swamp forest (step *1*). The water track expands as drainage converges into the zone of very porous peat (*B*) that is deposited by the hydrophilous sedges growing in the water track. The water table rises in the track, and water ponds up at the surface, because peat accumulation produces nearly level slopes and runoff converges into the track (step *2*). The ponds expand parallel to the contour, restricting the sedge lawn into very narrow sinuous ridges (steps *3* and *4*). The progressive flooding of the track is checked, however, when drainage ditches lower the water table, permitting trees and shrubs to reinvade the water track (step *5*).

A

B

Fig. 3.9.A. A pristine patterned fen in the Myrtle Lake peatland. The role of the water table in maintaining a water track is indicated by a comparison of pristine and ditched peatlands. The characteristic strings and flarks in this pristine water track are dominated by *Carex lasiocarpa* and other sedges. (Oblique aerial photo by Paul Glaser, 1984.) **B.** The invasion of strings by shrubs. Where drainage ditches have lowered the water table, strings are generally covered by *Betula pumila* var. *glandulifera* and other shrubs. (Oblique aerial photo by Paul Glaser, 1984.)

C

D

C. The invasion of a water track by shrubs. Where drainage ditches have lowered the water table more, shrubs may spread from the strings into the adjacent flarks (Oblique aerial photo by Paul Glaser, Winter Road Lake peatland, 1984.) D. The invasion of a water track by trees. A still-deeper drawdown of the water table by drainage ditches permits trees to reinvade a water track. Notice the linear orientation of the trees, which have colonized strings in the water track. A further lowering of the water track will favor its conversion into a swamp forest. (Oblique aerial photo by Paul Glaser, Norris Camp peatland, 1984.)

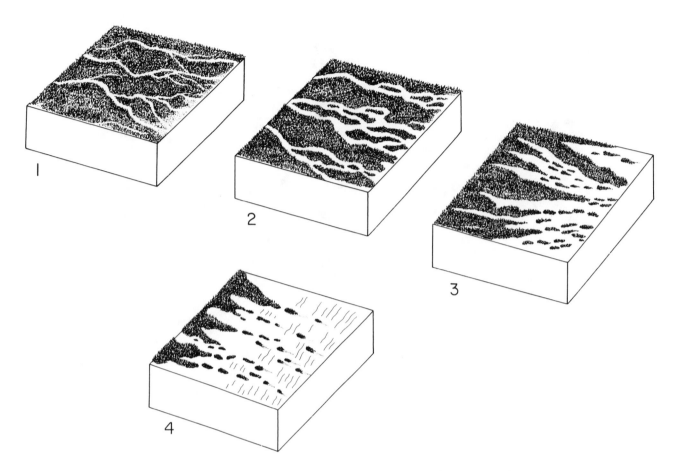

Fig. 3.10. The development of water tracks with tree islands. Runoff within a swamp forest is channeled into nonforested drains that fragment the swamp forest into linear forested fingers (step *1*). The forest cover in the drains becomes thinner downslope because of the locally rising water table. As the drains continue to expand, the forested fingers between the channels are constricted until clumps of forest are split off as discrete islands (steps *2* and *3*). The drains coalesce farther downslope, forming a patterned water track in which remnants of the original forest appear as tree islands (step *4*). See Plates 13A to 13D for examples of the various steps.

idence for these processes but instead indicate that assemblages characteristic of a patterned fen appear as much as 170 cm below the surface.

The collection of flotsam into windrows described by Drury (1956) is also unlikely to account for the string patterns in the Red Lake peatland. Hofstetter (1969) set up a screen barrier in the western water track to test this hypothesis in the mid 1960s as part of his doctoral study. In 1980 the screen fence was still intact but contained no debris.

Tree Islands

A few patterned fens in Minnesota contain fields of tree islands that grade into the surrounding belt of swamp forest (Plate 11C). The close link between the islands and the surrounding forest is indicated by a series of transitional landforms and by the striking alignment of the islands to forested fingers that protrude out onto the central water track (Glaser *et al.* 1981; Glaser

1987a). Teardrop tree islands are also conspicuously absent from most patterned fens in Minnesota, which are surrounded by only a narrow lagg or a number of raised bogs.

A general model of tree-island development may be constructed from the sequence of landforms that are consistently associated with these features in Minnesota (Fig. 3.10). The swamp forests that border these water tracks are dissected by a series of sinuous drainage paths that channel runoff onto a central track. The channels first appear as narrow zones of deteriorating forest, but downslope they are dominated by featureless sedge lawns (Plates 13A, 13B). The expansion and coalescence of these channels downslope restrict the forest to tapering fingers that fragment downslope to form detached islands (Plates 13A to 13D); Fig. 3.11A). The continual rise in the water table downslope restricts forest growth to a small circular head as most of the island becomes converted into a trailing tail of brush (Figure 3.11B). Thus channeled runoff across a

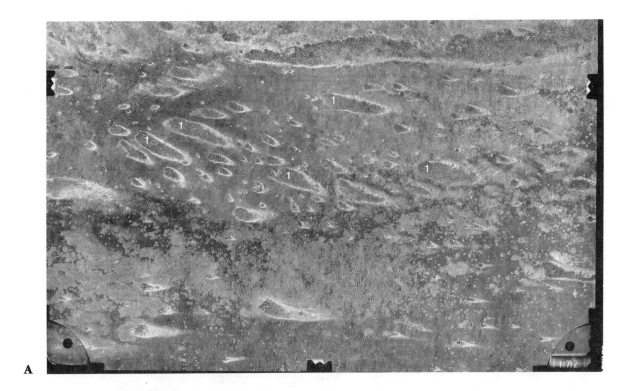

A

B

Fig. 3.11.A. Recently formed tree islands. These islands (*1*) are completely forested and are located near the forested margin of the water track. This aerial photograph from the western water track of the Red Lake peatland covers an area 2.5 km across. **B.** Partially paludified tree islands. These islands (*2*) have a small head of trees (*2a*) and a broad tapering tail of brush (*2b*). The direction of water movements is from left to right. Notice the effects of drainage ditches that have lowered the water table downslope and permitted the spread of shrubs. This aerial photograph from the western water track at Red Lake covers an area 2.5 km across. (Fig. 3.11.C on p. 40)

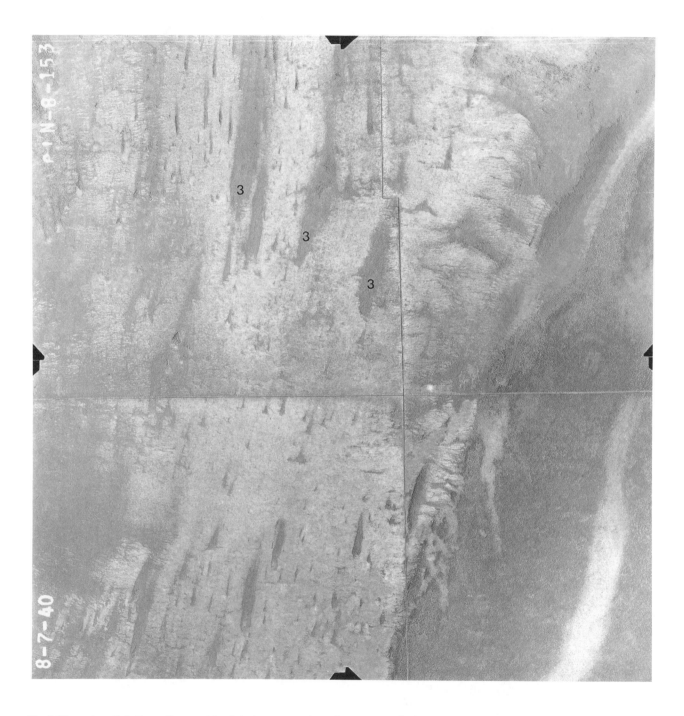

(Fig. 3.11 continued) **C.** Expanding tree island. At the extreme downslope margin of a water track, trees may expand over the entire island (*3*), which expands in size particularly downslope. This expansion is produced by a lowering of the water level caused by either drainage ditches or tributary streams eroding into the edge of the peatland. This aerial photograph from the western water track in the Red Lake peatland covers an area 2.5 km across.

swamp forest initiates a progressive flooding of a water track, gradually restricting the forest to smaller streamlined remnants. This sequence, however, is interrupted where drainage ditches have lowered the water table and have permitted the spread of trees and shrubs over formerly flooded surfaces (Fig. 3.11C).

A similar sequence occurs on patterned fens where echelons of islands grade into protruding fingers from a marginal swamp forest (Plate 11C). The fragmentation of these swamp forests into discrete fingers and ultimately detached islands is vividly illustrated by the spring-fen peatlands (types 7 and 11; Plate 3). The sedge channels in these peatlands contain buried wood layers with *Sphagnum* peat, indicating a previously more continuous cover of forest (Glaser 1983c; Glaser *et al.*, 1990). The channels are aligned with zones of deteriorating tree growth that seem to represent incipient channels developing toward the crest of the fen.

Less support for this hypothesis is provided by Griffin's (1977) stratigraphic survey of a tree island along the southern edge of the western water track at Red Lake. Griffin's cores indicate that this tree island arose at the same time as the adjacent water track. However, a shallow zone with *Larex* needles was found under the sedge peat of the water track. The *Carex diandra/ Carex aquatilis* peat underneath this horizon has no analog in the Red Lake peatland today.

Conclusions

The peat landforms represent an interesting anomaly in vegetation science. The dominant species of vascular plants vary continuously along environmental gradients and attain their peak abundance at different points along the chemical and moisture gradient. In this respect they conform to the continuum model of vegetation proposed by Whittaker (1957, 1975) and Curtis (1959). However, the major vegetation patterns in these patterned peatlands are spatially discontinuous and typically consist of large uniform stands. These sharp discontinuities are actually products of ecosystem development, because the pattern of peat accumulation alters the environmental gradient. The properties of a few major peat formers may therefore exert a controlling influence on the course of peatland development. The uniformity of the peat landform patterns can then be explained by simple feedback systems that control the development of these remarkable ecosystems.

Literature Cited

Auer, V. 1920. Über die Entstehung der strange auf den Torfmooren. Acta Forestalia Fennica 12:1-145.

Baden, W., and R. Eggelsman. 1963. Zur Durchlässigkeit der Moorböden. Z. Kulturtech. Flurbereinig. 4:226-254.

Boatman, D. J. 1983. The Silver Flowe National Nature Reserve, Galloway, Scotland. Journal of Biogeography 10:163-274.

Boelter, D. H. 1965. Hydraulic conductivity of peats. Soil Science 100:227-231.

———. 1969. Physical properties of peats related to degree of decomposition. Soil Science of America Proceedings 33:606-609.

———. 1972. Water table drawdown around an open ditch in organic soils. Journal of Hydrology 15:329-340.

——— and E. S. Verry. 1977. Peatland and water in the northern lake states. U.S. Department of Agriculture Forest Service General Technical Report NC-31:1-22.

Chow, V. T. 1964. Runoff. Pages 14-27 *in* V. T. Chow editor. Handbook of applied hydrology. McGraw-Hill, New York, New York, USA.

Curtis, J. T. 1959. The vegetation of Wisconsin. University of Wisconsin Press, Madison, Wisconsin, USA.

Drury, W. H., Jr. 1956. Bog flats and physiographic processes in the Upper Kuskokwim River Region, Alaska. Harvard University, Gray Herbarium Contribution 178:1-130.

Eng, M. T. 1979. An evaluation of the surficial geology and bog patterns of the Red Lake Bog (map). Minnesota Department of Natural Resources, Division of Minerals, St. Paul, Minnesota, USA.

———. 1980. Surficial geology, map of Koochiching County, Minnesota. Minnesota Department of Natural Resources, Division of Minerals, St. Paul, Minnesota, USA.

Glaser, P. H. 1983a. A patterned fen on the north shore of Lake Superior, Minnesota. Canadian Field-Naturalist 97:194-199.

———. 1983b. Vegetation patterns in the North Black River peatland, northern Minnesota. Canadian Journal of Botany 61:2085-2104.

———. 1983c. *Eleocharis rostellata* and its relation to spring fens in Minnesota. Michigan Botanist 22:19-21.

———. 1987a. The patterned boreal peatlands of northern Minnesota: A community profile. U.S. Fish and Wildlife Service Biological Report 85(7).

———. 1987b. The development of streamlined bog islands in the continental interior of North America. Arctic and Alpine Research 19:402-413.

———, G. A. Wheeler, E. Gorham, and H. E. Wright, Jr. 1981. The patterned mires of the Red Lake peatland, northern Minnesota: Vegetation, water chemistry, and landforms. Journal of Ecology 69:575-599.

——— and J. A. Janssens. 1986. Raised bogs in eastern North America: Transitions in surface patterns and stratigraphy. Canadian Journal of Botany 64:395-415.

———, J. A. Janssens, and D. I. Siegel. 1990. The response of vegetation to chemical and hydrological gradients in the Lost River peatland, northern Minnesota. Journal of Ecology 78:1021-1048.

Gorham, E., and D. L. Tilton. 1978. The mineral content of *Sphagnum fuscum* as affected by human settlement. Canadian Journal of Botany 56:2755-2759.

———, S. J. Eisenreich, J. Ford, and M. V. Santlemann. 1984. The chemistry of bog water. Pages 339-363 *in* W. Strumm, editor. Chemical processes in lakes. Wiley, New York, New York, USA.

Griffin, K. O. 1977. Paleoecological aspects of the Red Lake peatland, northern Minnesota. Canadian Journal of Botany 55:172-192.

Hagen, R., and M. Meyer. 1979. Vegetation analysis of the Red Lake peatlands by remote sensing methods. Minnesota Department of Natural Resources, St. Paul, Minnesota, USA.

Janssen, C. R. 1968. Myrtle Lake: A late- and post-glacial pollen diagram from northern Minnesota. Canadian Journal of Botany 46:1397-1408.

Hofstetter, R. H. 1969. Floristic and ecologic studies of wetlands in Minnesota. Ph.D. thesis. University of Minnesota, Minneapolis, Minnesota, USA.

Ingram, H. A. P. 1983. Hydrology. Pages 67-158 *in* A. J. P. Gore, editor. Mires: Swamp, bog, fen, and moor. Vol. 4A. General Studies. Elsevier, Amsterdam, Netherlands.

Ivanov, K. E. 1981. Water movement in mirelands. (Vodoobmen v Golotnykh landschaftkh.) A. Thomson and H. A. P. Ingram, translators. Academic Press, London, U.K.

Komar, P. D. 1983. Shapes of streamlined islands on Earth and Mars: Experiments and analyses of the minimum-drag form. Geology 11: 651-654.

———. 1984. The lemniscate loop-comparisons with the shapes of streamlined landforms. Journal of Geology 92:133-145.

Lundqvist, G. 1951. Beskriving till jordartskarta över Kopparbergs Län. Sveriges Geologiska Undersökning Ser. Ca. 21:1-213.

Munger, J. W., and S. J. Eisenreich. 1983. Continental-scale variations in precipitation chemistry. Environmental Science and Technology 17:32-42.

Pearsall, W. H. 1966. Two blanket bogs in Sutherland. Journal of Ecology 44:493-516.

Pearson, M. C. 1960. Muckle Moss, Northumberland. Journal of Ecology 48:647-666.

———. 1979. Patterns of pools in peatlands (with particular reference to a valley head mire in northern England). Acta Universitatis Oulensis Series A, Geologia 3:65-72.

Rycroft, D. W., D. J. A. Williams, and H. A. P. Ingram. 1975. The transmission of water through peat. I. Review. Journal of Ecology 63:535-556.

Seppälä, M., and L. Koutaniemi. 1985. Formation of a string and pool topography as expressed by morphology, stratigraphy, and current processes on a mire in Kuusamo, Finland. Boreas 14:287-309.

Siegel, D. I. 1981. Hydrogeologic setting of the Glacial Lake Agassiz peatlands, northern Minnesota. United States Geological Survey Water Resources Investigation 81-24.

———. 1983. Groundwater and the evolution of patterned mires, Glacial Lake Agassiz peatlands, northern Minnesota. Journal of Ecology 71:913-921.

——— and Glaser, P. H. 1987. Groundwater flow in a bog/fen complex, Lost River peatland, northern Minnesota. Journal of Ecology 75:743-754.

Sjörs, H. 1946. Myrvegetationen i övre Långanomradet i Jämtland. Arkiv fur Botanik 33:1-96 (English summary).

Von Post, L., and E. Granlund. 1926. Södra Sveriges torvtillgångar I. Sveriges Geologiska Undersökning, Ser. C335:1-127.

Washburn, A. L. 1979. Geocryology. Edward Arnold, London, U.K.

Whittaker, R. H. 1957. Recent evolution of ecological concepts in relation to the eastern forests of North America. American Journal of Botany 44:197-206.

———. 1975. Communities and ecosystems. 2nd edition. Macmillan, New York, New York, USA.

CHAPTER 4

Bryophytes

Jan A. Janssens

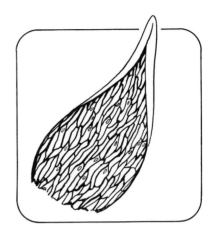

Bryophytes are primitive land plants and include mosses, bogmosses, liverworts, and hornworts, groups that are probably not very closely related to each other (Schofield 1985; Crandall-Stotler 1984; Mishler and Churchill 1984, 1985). They are mostly terrestrial, autotrophic (green) plants with two distinct alternating life forms: (1) a green gametophyte or sexual generation and (2) a partially parasitic sporophyte or asexual generation, *attached to the gametophore*, which is the mature part of the gametophyte. Bryophyte diversity, approximately 14,500 species in 700 genera worldwide, is higher than in any other group of terrestrial plants except angiosperms (Scagel *et al.* 1982). In addition, peatland landscapes dominated by bryophytes, and particularly *Sphagnum*, expanded during the Quaternary period, and preservation of *Sphagnum* remains in peat have resulted in the speculation that there is more carbon stored in *Sphagnum*, alive and dead, than in any other plant in the present biosphere (Clymo and Hayward 1982).

Mosses (Bryopsida) and bogmosses (Sphagnopsida) have been little studied in Minnesota, in contrast to the smaller group of liverworts (Hepaticopsida, Schuster 1977). The only preliminary checklist ever published dates from 1895 (Holzinger) and lists just two *Sphagnum* species and a total of 156 moss species (Janssens 1988a). Published floras that work perfectly well for this region are available, however (Crum and Anderson 1981; Crum 1983, 1984). A large herbarium, including more than 6,000 Minnesota collections dating from the end of the last century through the present day, is located at the University of Minnesota in St. Paul and with critical annotation could produce excellent baseline data for the Minnesota moss flora. Because of Minnesota's highly diverse biomes, about 400 moss species can be expected to have lived recently in the state, a number similar to that for the state of Wisconsin (Bowers and Freckmann 1979). This chapter is a

contribution to the effort in the form of a compilation of the author's moss collections made in Minnesota's peatlands, starting in 1981. Some of the results have been published earlier (Janssens 1988a; Janssens and Glaser (1984a and b, 1986).

The Moss Record of Minnesota's Peatlands

The Minnesota peatland sites visited by the author up to 1988 (165 ecotopes; see Janssens 1989) are mapped in Fig. 4.1. Collecting was started with the Red Lake peatland project during the summer of 1981 (Janssens and Glaser 1986).

All species collected by the author are listed alphabetically in Table 4.1. The nomenclature follows Ireland *et al.* (1987) for mosses and Isoviita (1966) for bog mosses. The total number of taxa (species and subspecific entities) identified in Table 4.1 is 164. The number of sites where each taxon has been collected is included in parentheses after species or subspecific epithet. The most commonly occurring bogmosses in more than half the sites are *Sphagnum magellanicum* and *S. angustifolium*. Common mosses are *Aulacomnium palustre, Bryum pseudotriquetrum, Campylium stellatum, Dicranum undulatum, and Polytrichum strictum*. Genera represented by a large number of species are *Brachythecium, Calliergon, Campylium, Dicranum, Drepanocladus, Plagiomnium*, and *Sphagnum*. Also included in Table 4.1 is the peatland-landform distribution for each taxon.

Peatlands are classified in two major groups, patterned and nonpatterned (chapter 1). Nonpatterned peatlands are usually landscape units small in areal extent and difficult to classify further along gradients of minerotrophy and hydrology. They have been subdivided here only as forested or open (nonforested). Patterned peatlands, however, can be subdivided into

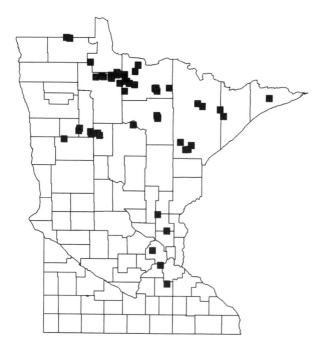

Fig. 4.1. Location of 165 Minnesota peatland sites where the author made bryophyte collections between 1981 and 1988.

three major groups: raised bogs, water tracks, and spring fens. Each of these can be further classified (see legend to Table 4.1). Each one of these entities is physiognomically distinct, and some are characterized by a particular position along the minerotrophic and hydrological gradients (chapter 2), the most significant environmental gradients controlling peatland vegetation and flora (Janssens and Glaser 1986; Janssens 1988b, 1989). Most common species, however, are quite tolerant of different minerotrophic and hydrological regimes and are thus found in most landforms. A more detailed ecological analysis is presented in the next section.

Figure 4.2 shows the individual distributions of the most common species in Minnesota (> 25 sites). Most of these species are found throughout the state where peatlands have been visited, except for the rich-fen species *Drepanocladus revolvens, Scorpidium scorpioides,* and *Tomenthypnum nitens* and the oligotrophic bogmoss *Sphagnum capillifolium*. These four species are distributed only among larger peatlands in the northern half of the state.

Ecology of the Mosses in Minnesota's Peatlands

The results of a detrended correspondence analysis (DCA: CANOCO program; ter Braak 1987; Jongman *et al.* 1987) of 170 bryophyte plots collected in Minnesota peatlands are illustrated in Fig. 4.3. Collection of plot data is explained in detail in Janssens and Glaser (1986) and in Janssens (1989). Position for each plot in the indirect gradient-analysis graph is indicated by one

of five symbols representing major landform groups. Forested bogs and poor fens are not differentiated, but three other landforms — forested fens, water tracks, and open bogs — are partly distinguished by their bryophyte composition (see also Janssens and Glaser 1986 for a cluster analysis of the Red Lake peatland bryophyte plots). The landform differentiation of this indirect gradient analysis forms the background for a stratigraphical trace presented in Fig. 13.3.

Figure 4.4 illustrates a species-environment direct-gradient ordination in the form of a biplot. The scores for 47 species and three environmental variables — pH, shade, and HMWT (height above the mean water table) — are derived by canonical correspondence analysis (CCA: Jongman *et al.* 1987; ter Braak 1987). All three environmental gradients were not measured at all 170 Minnesota plots: only 65 plots have been included in this ordination. The ranking of species along a particular environmental gradient can be inferred approximately by projecting species symbols perpendicular onto the environmental arrow. Fitted values on the positive side of an environment vector are higher than the mean; inferred values projecting on the negative side are lower. The angle between vectors reflects the correlation between gradients: very small angles indicate colinearity, angles approaching the perpendicular suggest no correlation, and opposite vectors indicate negative correlation. Arrow lengths indicate the relative importance of gradients in explaining differentiation among species. Distances between species points are approximations of the lengths of line segments joining species in multidimensional space, the axes of which are formed by species.

The most significant outliers among species are labeled in Fig. 4.4. It is clear from the right-angle relation between pH and HMWT that these two environmental gradients explain most biological variability among bryophyte communities and also that they are unrelated to each other in Minnesota's peatlands. Total variance accounted for by the two first axes of the species-environment biplot equals 80%. Figure 4.4 illustrates, for example, that both *Tomenthypnum nitens* and *Sphagnum wulfianum* are associated with highly minerotrophic habitats (pH gradient) but differentiate clearly along the hydrological gradient (HMWT). A similar species pair at the low pH extreme is *Sphagnum majus* and *S. fuscum*. Alternatively, some species pairs differentiate along the minerotrophic gradient but have similar response to water level (e.g., *Sphagnum capillifolium* versus *S. wulfianum*). Quantitative values for optima and tolerance of the most common peatland bryophytes to environmental gradients is presented in Table 12.2 Those results, rather than the qualitative illustration of the CCA, are used in the paleoenvironmental reconstruction of peatland development. The morphological structure of common bryophytes (Figs. 4.6, 4.7, 4.8, 4.12, 4.14, 4.15) and rare species (Figs. 4.5, 4.9, 4.10, 4.11, 4.13, 4.16) shows the high diversity among peatland mosses. The illustrations were drawn by the author.

Color Plates

Plates 1 and 2.
The Glacial Lake Agassiz peatlands. This Landsat TM image covers the heart of the Glacial Lake Agassiz peatlands including the Red Lake, North Black River, and Lost River peatlands. The raised bogs are distinguished by their radiating cover of spruce forest (brown tones), nonforested bog drains and lawns (tan), and streamlined margins trimmed by water tracks. The water tracks may contain bands of tamarack (light red) and shrubs oriented parallel to lines of flow, or consist of nonforested sedge lawns (green). Water tracks with string-flark patterns have wetter surfaces (dark green). The beach ridges that locally intersect these peatlands are distinguished by their pink tones and rough surfaces. This false color composite image is composed of TM bands 2, 3, and 4 (blue, green, and red) and was processed at the Goddard Space Flight Center under NASA contract #NAS 5-28740.

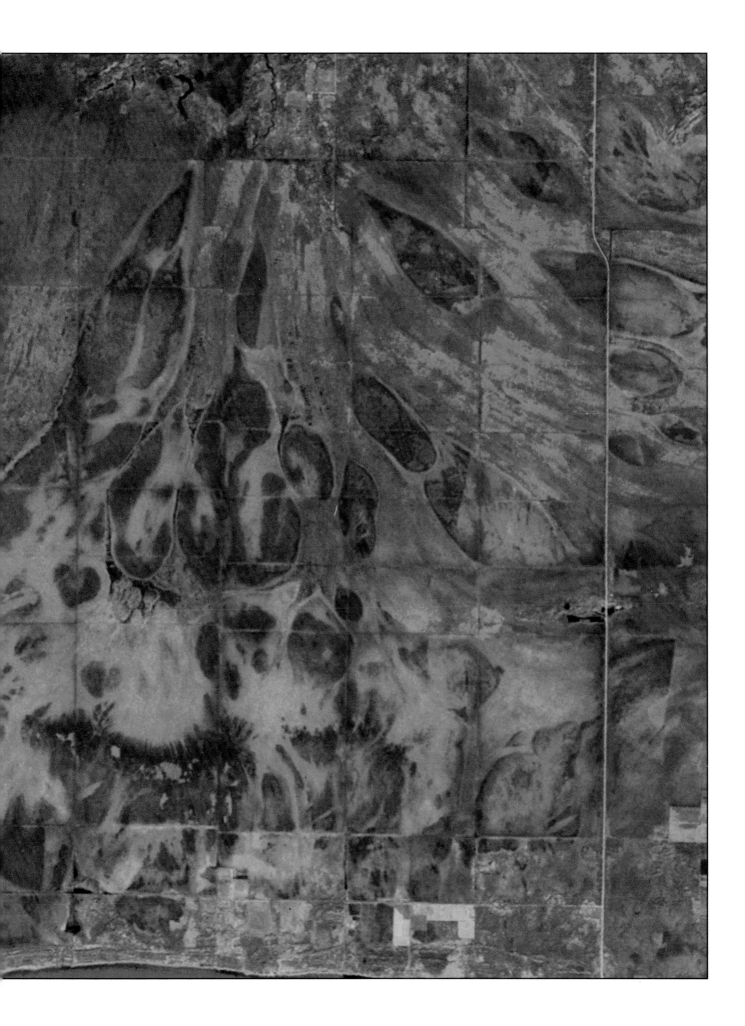

Plate 3. Spring-fen channels near Pine Creek. The peatland is characterized by an anastomosing network of channels (SFC) that drain southward through a swamp forest (SFF). The forest is fragmented downslope into long tapering fingers and discrete tree islands (TI). *(Color infrared aerial photo: USGS EROS Data Center)*

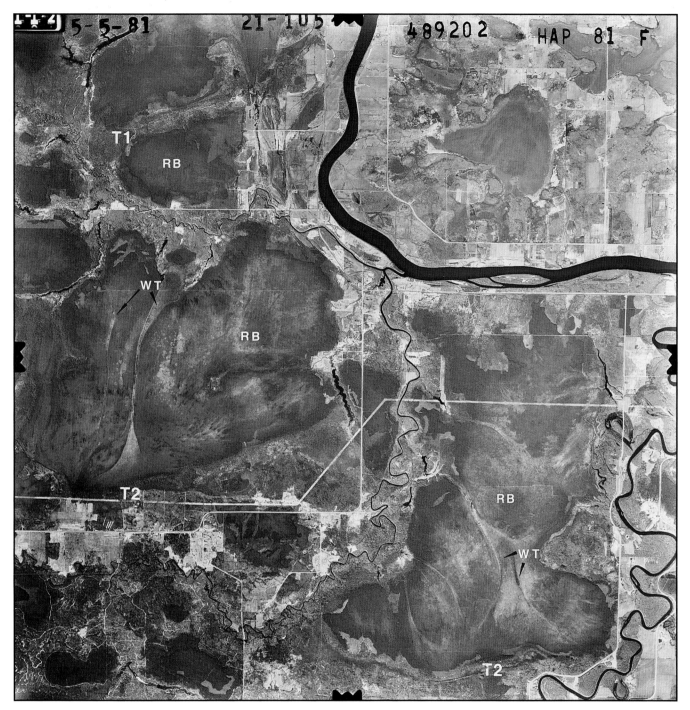

Plate 4. Peatlands (mire-complex types 1-2) in the North Black River area. The spruce and *Sphagnum* on the raised bogs (RB) produce a smooth red signature on the photos, in contrast to the coarser blue tones of the uplands, which at this time of year lack foliage. The smaller peatlands (type 1: T1) are filled by raised bogs, which have been partially burned or cut by loggers. The bogs in the larger peatlands (type 2: T2) are subdivided and streamlined by narrow water tracks (WT). This color infrared image covers an area approximately 12 km across. *(Color infrared aerial photo: USGS EROS Data Center)*

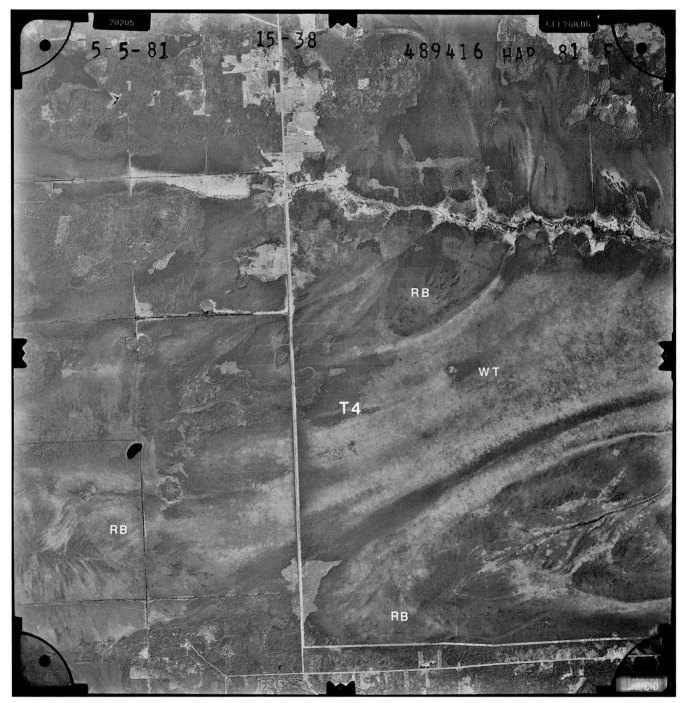

Plate 5. Peatlands (mire-complex types 3-4) in the North Black River area, northern Minnesota. Peatlands nearly cover this photo except for the fringe of uplands, the floodplain of the North Black River, and isolated mineral outcrops. The raised bog (RB) landforms are marked by red tones, in contrast to the blue tones and pink tones of the water tracks (WT). The type 3 peatland (T3: upper right corner) is equally divided into streamlined bogs and water tracks. The type 4 peatland (T4: center of photo) contains a huge water track and smaller bogs along its edges. Notice the small water tracks that arise from the interior of the large bog on the left. Each color infrared photo covers an area approximately 12 km across. *(Color infrared aerial photo: USGS EROS Data Center)*

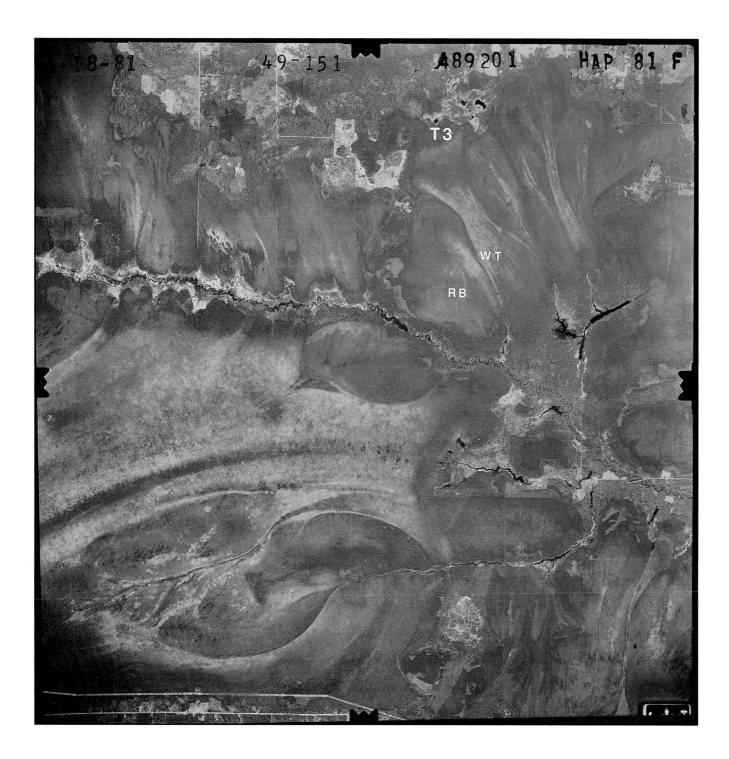

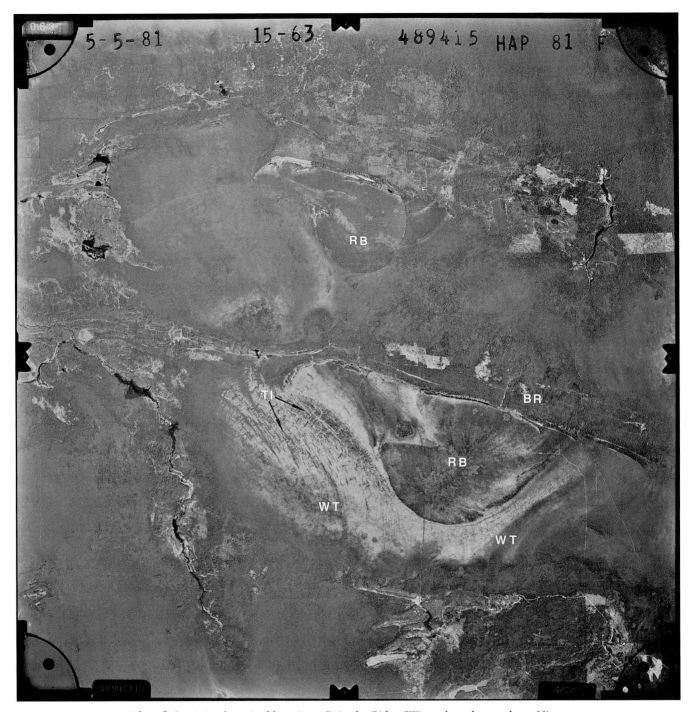

Plate 6. Semicircular raised bog (type 8) in the Ridge SW quadrangle, northern Minnesota. The large bogs (RB) have a convex margin facing upslope and resting on a linear beach ridge (BR) along its downslope margin. The water track (WT) arises from another beach ridge upslope and splits around the margins of the bog. The tree islands (TI) in the west branch of the water track are illustrated in Fig. 4. *(Color infrared aerial photo: USGS EROS Data Center)*

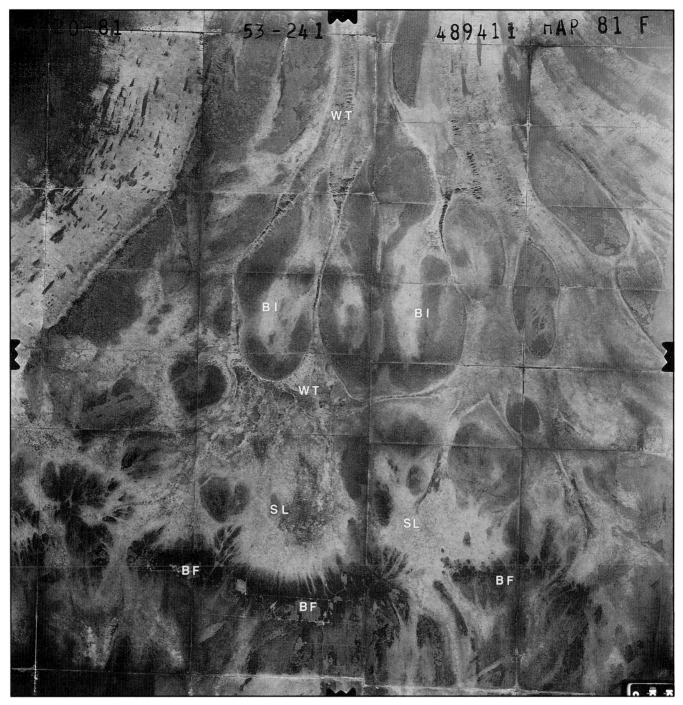

Plate 7. A large bog-complex in the Red Lake peatland. The crest of the bog is marked by radiating forest patterns (BF) in the lower portion of the photo, locally deforested by clearcuts. The forested crest (purple) gives way to a nonforested *Sphagnum* lawn (SL: pink) downslope on which internal water tracks (WT: blue-green) arise. The tracks expand downslope, fragmenting the lower bog flanks into large streamlined islands. The bog islands (BI) have horseshoe-shaped rings of forest (reddish) and nonforested interiors (pale pink). Water flows from the lower to the upper portion of the photo. See plate 8 for oblique aerial views. *(Color infrared aerial photo: USGS EROS Data Center)*

Plate 8. Oblique aerial photos of the large bog-complex in plate 7.

Plate 8a. The forested crest. Notice the radiating lines of forest and intervening nonforested bog drains. *(Photo by Paul Glaser)*

Plate 8b. *Sphagnum* lawn. The forested crest grades downslope into a nonforested lawn. *(Photo by Paul Glaser)*

Plate 8c. Internal water track with circular islands. Water tracks arise from the lower portion of the *Sphagnum* lawn, producing a zone with circular tree islands. *(Photo by Paul Glaser)*

Plate 8d. Streamlined bog islands. The water tracks convert the lower flanks of the bog into streamlined islands. The bog islands have an open ring of forest and nonforested interiors. *(Photo by Paul Glaser)*

Plate 9. The western water track of the Red Lake peatland. This huge water track (WT) arises from beach ridges along the periphery of the peatland and flows eastward for 36 km, where it splits into two branches around the margin of a large raised bog-complex (RB: illustrated in plate 7). The water track is surrounded by a swamp forest (SW: purple), which is locally interrupted by sinuous channels. The central core of the water track (blue) contains streamlined tree islands (TI: red) and networks of strings and flarks oriented perpendicular to the slope. Notice the changes in the water track downslope from the drainage ditches (rectangular grid), where the wet flarks (blue) are replaced by shrubs (red-pink) that have spread over the track. *(Color infrared aerial photo: USGS EROS Data Center)*

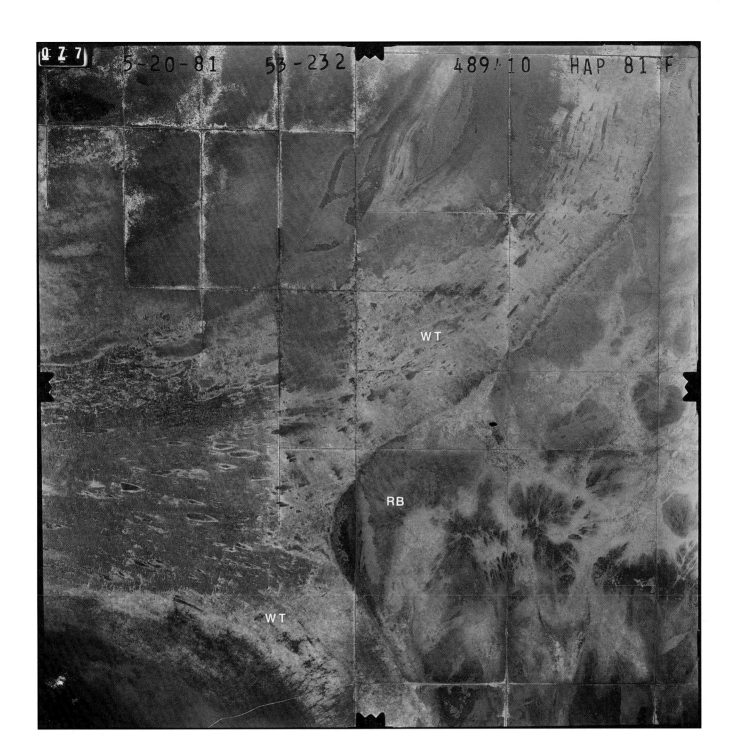

Plate 10. Vegetation types.

Plate 10a. A bog drain. The drain is dominated by a stand of *Carex oligosperma*. Notice the stunted clumps of *Picea mariana*, which are being overwhelmed by the rapidly accumulating peat surface. The surrounding bog forest is in the background. *(Photo by Paul Glaser, Red Lake peatland, 1978)*

Plate 10b. A bog drain with schlenke. A nonforested bog drain forms a V-shaped notch in the bog forest at the top of the photo. A series of wet hollows oriented perpendicular to the slope interrupts the *Carex oligosperma* lawn downslope. *(Oblique aerial photo by Paul Glaser, Sturgeon River peatland, 1984)*

Plate 10c. A schlenke. This hollow is dominated by *Rhyn-chospora alba, Utricularia cornuta,* and *Carex limosa.* The water table is near the surface of the *Sphagnum cuspidatum* mat. *(Photo by Paul Glaser, North Black River peatland, 1980)*

Plate 10d. A spring-fen channel. Notice the sharp margin between the nonforested channel and the surrounding spring-fen channel. *(Oblique aerial photo by Paul Glaser, Lost River peatland, 1982)*

Plate 11a. A flark. The typical flark assemblage is dominated by *Carex lasiocarpa* and other sedges. The bright yellow flowers are *Utricularia cornuta*. A tree island is in the background. *(Photo by Paul Glaser, Red Lake peatland, 1978)*

Plate 11b. A detailed view of a flark dominated by *Carex lasiocarpa*. Notice the standing water. *(Photo by Paul Glaser)*

Plate 11c. Forested fingers along the northern edge of the western water track at Red Lake. Notice how the fingers of forest grade into streamlined tree islands in the nonforested water track. *(Photo by Paul Glaser, Red Lake peatland, 1980)*

Plate 11d. A tree island. The tree islands are dominated by tamarack (*Larix laricina*) and occasional spruce (*Picea mariana*). Notice the substrate of dry moss-covered hummocks and water-filled depressions. The trees are *Larix laricina. (Photo by Paul Glaser, Red Lake peatland, 1978)*

Plate 12. Dynamic landform patterns in raised bogs.

Plate 12a. Expanding bog drain. This nonforested drain is expanding headward (toward the top of the photo) into the surrounding bog forest. Notice the transition from the nonforested center of the drain to the surrounding forest. The drain is fringed by a zone of stunted clumped spruce, which is being buried by a rapidly accumulating surface of *Sphagnum*. *(Photo by Paul Glaser, Red Lake peatland, 1984)*

Plate 12b. Retreating bog margin. A featureless water track has flooded the margin of a bog, which is fragmenting into circular clones. *(Photo by Paul Glaser, Red Lake peatland, 1980)*

Plate 12c. Retreating bog margin. The scalloped margin of the bog is dominated by *Sphagnum* and various sedges. The adjacent water track (dark tones to the right) have standing water. *(Photo by Paul Glaser, Red Lake peatland, 1980)*

Plate 13a. Initial stage in the formation of channels in swamp forests. The channels are dominated by stunted *Larix laricina. (Photo by Paul Glaser, Pine Creek peatland, 1984)*

Plate 13b. Later phase in channel development. The channels have expanded and are now dominated by sedges. The forest, in contrast, has been fragmented into tapering fingers that fragment into discrete tree islands downslope. *(Photo by Paul Glaser)*

Plate 13c-d. Final phase in channel development. The channels have now coalesced into a water track with isolated tree islands. Notice the fields of tree islands and string-flark patterns. *(Photos by Paul Glaser)*

Plate 14a. The *Abies* island in Myrtle Lake peatland, seen from the air looking southeast, in October 1976. Bedrock forms the core of the island and provides a source of mineral-rich water. Clearly visible is the mineral-rich swamp forest with tall black spruce and tamarack around the island and in the downstream track of mineral-rich water. October is the month in which peatland patterns are most beautifully displayed. A yellow patch of black ash occurs in the marginal swamp forest at the far side of the "head" of the island, reflecting extreme mineral-rich conditions in the substrate. White pine (*Pinus strobus*), Norway pine (*Pinus resinosa*), balsam fir (*Abies balsamea*), balsam poplar (*Populus balsamifera*), and bare rock are also visible. The two straight lines in the peatland this side of the island are the tracks of a bog vehicle that was used for transport on the ground. The clearest track was just a few weeks old at the time when this photograph was taken; the second track is three years old. It is very near these tracks that some of the studies on the recent vegetation and recent pollen deposition were carried out. The tracks show that the damage done by this kind of transporta-tion stays visible for a long time. Use of this mode of transportation should therefore be restricted as much as possible. Contorted faint lines in the peatland are probably animal tracks. *(Photo by C. R. Janssen)*

Plate 14b. A low-altitude aerial view of the ecological gradient between the cedar swamp forest (upper left) and the ombrotrophic spruce forest (lower left) at the northwestern side of Myrtle Lake (see plate 15). Yellow patches in the cedar swamp forest represent the fall (October 1976) colors of *Osmunda cinnamomea*. Note that the spruce forest reacts to the increasing mineral content of the substrate by growing taller before it is eliminated by competition from cedar, a plant better adapted to a substrate very rich in minerals. *(Photo by C. R. Janssen)*

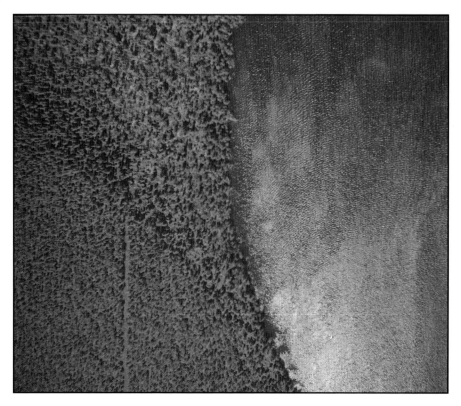

Plate 15. Color infrared aerial photograph of Myrtle Lake (lower right) and the southern part of the peatland. Photo overlaps northward with plate 16. Note the fan-shaped cedar swamp north of the lake and the ombrotrophic spruce forest on the east, west, and south. The contact between cedar and spruce forests is shown on Plate 14b. The water seeping northwestward from the cedar forest crosses a buried ridge (red tone) and there forms a broad water track with streamlined tree islands, the largest of which is the *Abies* island (Plate 14a). A string pattern is visible in the upper left corner of the plate. *(Color infrared aerial photo: USGS EROS Data Center)*

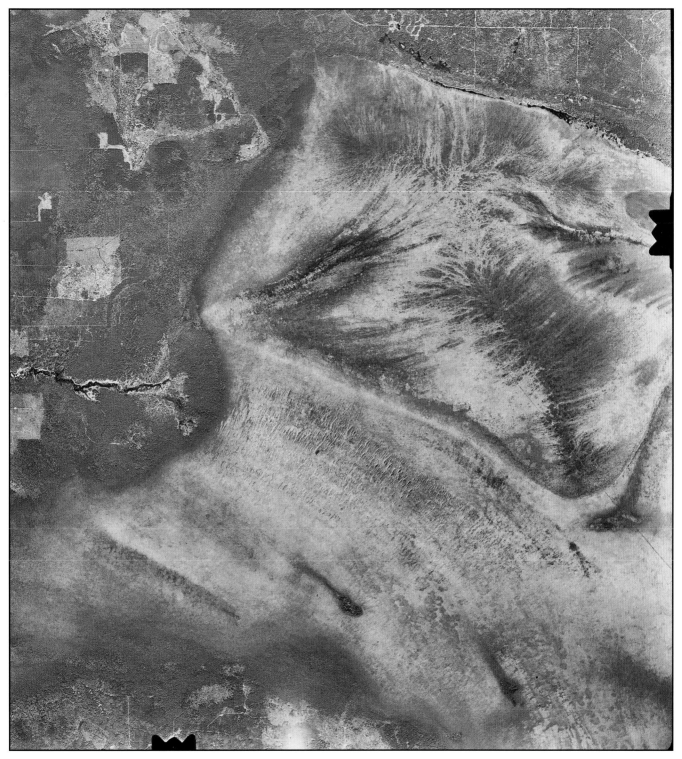

Plate 16. The raised-bog complex in the northern part of the Myrtle Lake peatland, with the *Alnus* island just to its south. The radial pattern on each of the two main components of the complex indicates the presence of a "crest" from which water drains. The broad water track with the string pattern (blue) is abruptly terminated to the northwest by a cedar swamp (red) at the head of a dissecting stream, and the strings at the sharp transition are marked by cedar trees. *(Color infrared aerial photo: USGS EROS Data Center)*

Table 4.1. Minnesota peatland moss flora and the landform[a] distribution of its species

Taxon	PRC	PRL	PWN	PWP	PWI	PSC	PSF	NFR	NNF
Amblystegium									
riparium (12)			●	●	●			●	●
serpens (7)							●		
var. *juratzkanum* (6)							●	●	●
varium (4)								●	●
Anomodon									
minor (3)								●	
Atrichum									
altecristatum (1)									●
Aulacomnium									
palustre (89)	●	●	●	●	●	●	●	●	●
Brachythecium									
calcareum (1)									●
campestre (9)				●			●	●	●
digastrum (1)									●
oedipodium (13)					●		●		●
oxycladon (1)							●		
plumosum (1)								●	
reflexum (2)								●	●
rivulare (7)					●		●	●	
rutabulum (2)								●	
salebrosum (17)				●	●		●	●	●
turgidum (5)						●	●	●	
velutinum (7)				●				●	●
Bryoerythrophyllum									
recurvirostrum (3)							●	●	
Bryum									
argenteum (1)				●					
lisae var.									
cuspidatum (3)							●		
pseudotriquetrum (43)			●	●	●	●	●		●
Callicladium									
haldanianum (39)	●	●		●	●	●	●	●	●
Calliergon									
aftonianum (3) Fig. 4.5							●		
cordifolium (14)					●		●	●	
giganteum (23)				●	●	●	●	●	●
richardsonii (7)								●	
stramineum (16) Fig. 4.6	●	●		●	●		●	●	●
trifarium (18) Fig. 4.7					●	●		●	●

[a]Patterned Peatlands
    Raised Bogs
        PRC = forested crest
        PRL = Sphagnum lawn
    Water Tracks
        PWN = nonpatterned
        PWP = patterned
        PWI = patterned with islands
    Spring Fens
        PSC = channels
        PSF = forest
Nonpatterned Peatlands
    NFR = forested
    NNF = open

Table 4.1 (cont.). Minnesota peatland moss flora and the landform distribution of its species

Taxon	PRC	PRL	PWN	PWP	PWI	PSC	PSF	NFR	NNF
Calliergonella									
cuspidata (20)				•	•	•	•	•	•
Campylium									
chrysophyllum (6)					•	•	•		•
hispidulum (13)					•	•	•	•	•
polygamum (24)		•		•	•	•	•	•	•
radicale (7)				•			•	•	
stellatum (50) Fig. 4.8			•	•	•	•	•	•	•
Catoscopium									
nigritum (3)						•			
Ceratodon									
purpureus (6)				•	•		•	•	
Cinclidium									
stygium (19) Fig. 4.9				•	•	•	•		
Climacium									
americanum (3)								•	
dendroides (18)					•		•	•	•
Conardia									
compacta (1)							•		
Cratoneuron									
filicinum (5)					•			•	
Dicranum									
flagellare (15)	•	•			•		•	•	
fragilifolium (1)								•	
fuscescens (1)								•	
montanum (8)	•				•		•	•	
ontariense (12)	•			•	•		•	•	
polysetum (27)	•	•		•	•		•		•
scoparium (1)							•		
undulatum (41)	•	•		•	•	•	•	•	
Didymodon									
fallax (2)							•		
Distichium									
capillaceum (3)							•		
Drepanocladus									
aduncus (7)				•				•	•
var. *kneifii* (2)							•	•	
var. *polycarpus* (6)								•	•
exannulatus (6)		•		•	•				•
fluitans (6)		•		•	•		•	•	•
lapponicus (3) Fig. 4.10				•	•		•		
pseudostramineus (2)					•		•		
revolvens (33)				•	•	•	•		•
sendtneri (1)				•					
uncinatus (13)	•			•			•	•	
vernicosus (5)					•				•
Entodon									
cladorrhizans (3)								•	
Eurhynchium									
pulchellum (7)							•	•	•
Fissidens									
adianthoides (25)				•	•	•	•	•	•
cristatus (2)						•		•	
osmundoides (14)				•	•	•	•	•	

Table 4.1 (cont.). Minnesota peatland moss flora and the landform distribution of its species

Taxon	PRC	PRL	PWN	PWP	PWI	PSC	PSF	NFR	NNF
Funaria									
bygrometrica (1)							●		
Haplocladium									
microphyllum (2)				●				●	
Helodium									
blandowii (17)					●	●	●	●	●
paludosum (3)								●	
Hertzogiella									
turfacea (1)							●		
Hylocomium									
splendens (15)							●	●	
Hypnum									
fertile (6)				●	●		●		●
lindbergii (29)				●	●	●	●	●	●
pallescens (7)	●						●	●	●
pratense (26)	●			●	●	●	●	●	●
Isopterygiopsis									
muelleriana (1)							●		
Isopterygium									
pulchellum (1)							●		
Leptobryum									
pyriforme (8)				●			●	●	
Leskea									
gracilescens (2)				●					●
polycarpa (1)									●
Leucobryum									
albidum (1)					●				
glaucum (1)								●	
Lindbergia									
brachyptera (1)								●	
Meesia									
triquetra (6) Fig. 4.11				●	●	●	●		●
uliginosa (1)							●		
Mnium									
spinulosum (1)							●		
Myurella									
julacea (10)						●	●		
sibirica (1)							●		
Neckera									
pennata (1)								●	
Oncophorus									
virens (1)	●								
wahlenbergii (6)					●		●	●	
Orthotrichum									
obtusifolium (3)				●			●	●	
speciosum (3)							●	●	
var. *elegans* (3)				●			●	●	
Paludella									
squarrosa (2)						●			

Table 4.1 (cont.). Minnesota peatland moss flora and the landform distribution of its species

Taxon	PRC	PRL	PWN	PWP	PWI	PSC	PSF	NFR	NNF
Plagiomnium									
ciliare (1)								•	
cuspidatum (19)				•	•		•	•	•
drummondii (2)							•		
ellipticum (32)					•		•	•	•
medium (1)								•	
Plagiothecium									
cavifolium (1)								•	
denticulatum (24)	•			•	•		•	•	•
laetum (5)	•			•			•	•	
Platygyrium									
repens (18)				•	•		•	•	
Pleurozium									
schreberi (61)	•	•		•	•	•	•		•
Pohlia									
nutans/sphagnicola (28)	•			•	•		•	•	•
Polytrichastrum									
longisetum (4)								•	•
Polytrichum									
commune (9)								•	•
juniperinum (3)					•		•	•	
strictum (76) Fig. 4.12	•	•	•	•	•	•	•	•	•
Pseudobryum									
cinclidioides (1)								•	
Pterygynandrum									
filiforme (4)					•	•			•
Ptilium									
crista-castrensis (9)							•	•	
Pylaisiadelpha									
recurvans (9)							•	•	•
Pylaisiella									
polyantha (10)	•			•	•		•		
selwynii (13)				•	•		•	•	•
Rhizomnium									
gracile (9) Fig. 4.13						•	•		
pseudopunctatum (6)							•		
punctatum (1)								•	
Rhodobryum									
roseum (1)								•	
Rhynchostegium									
serrulatum (3)				•			•		
Rhytidiadelphus									
triquetrus (7)							•	•	
Scorpidium									
scorpioides (25)				•	•	•	•	•	•

Table 4.1 (cont.). Minnesota peatland moss flora and the landform distribution of its species

Taxon	PRC	PRL	PWN	PWP	PWI	PSC	PSF	NFR	NNF
Sphagnum									
angustifolium (89)	●	●	●	●	●	●	●	●	●
capillifolium (43)	●	●	●	●	●	●	●	●	●
centrale (38)		●	●	●	●		●	●	●
contortum (16)				●	●				●
cuspidatum (3)		●		●					
fallax (29)	●	●		●	●		●	●	●
fimbriatum (9)				●			●	●	●
flexuosum (9)	●	●			●			●	●
fuscum (70) Fig. 4.14	●	●	●	●	●	●	●	●	●
girgensohnii (4)								●	●
isoviitianum (1)		●							
jensenii (1)							●		
magellanicum (87)	●	●	●		●		●	●	●
majus (15)	●	●			●			●	●
obtusum (3)							●		●
papillosum (25) Fig. 4.15	●	●		●	●			●	●
var. *laeve* (2)								●	●
platyphyllum (7)			●	●			●		●
pulchrum (2)		●						●	
riparium (1)					●				
russowii (17)	●	●			●		●	●	●
squarrosum (16)				●	●		●	●	●
subfulvum (9)					●				●
subsecundum (27)		●	●	●	●		●	●	●
subtile (1)		●							
teres (22)				●	●		●	●	●
warnstorfii (43)		●	●	●	●	●	●	●	●
wulfianum (9)							●	●	●
Taxiphyllum									
deplanatum (2)				●				●	
Tetraphis									
pellucida (10)					●		●	●	
Thuidium									
delicatulum (12)				●			●	●	
var. *radicans* (3)							●	●	
minutulum (1)							●		
recognitum (14)				●			●	●	
Tomenthypnum									
falcifolium (14) Fig. 4.16		●			●		●	●	
nitens (29)				●	●	●	●	●	
Tortella									
fragilis (3)							●		

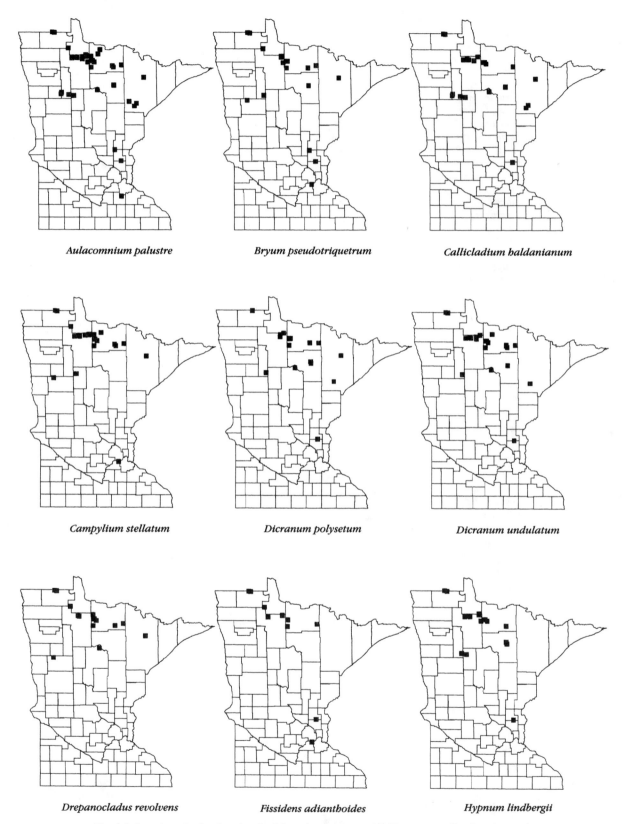

Aulacomnium palustre   *Bryum pseudotriquetrum*   *Callicladium baldanianum*

Campylium stellatum   *Dicranum polysetum*   *Dicranum undulatum*

Drepanocladus revolvens   *Fissidens adianthoides*   *Hypnum lindbergii*

Fig. 4.2. Location of collection sites for 24 peatland mosses with 25 or more collection sites each.

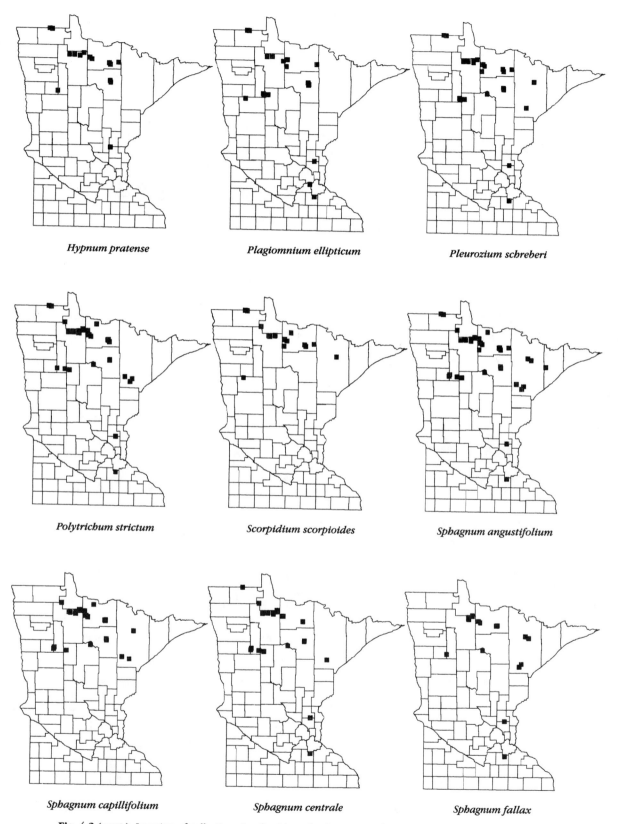

Hypnum pratense

Plagiomnium ellipticum

Pleurozium schreberi

Polytrichum strictum

Scorpidium scorpioides

Sphagnum angustifolium

Sphagnum capillifolium

Sphagnum centrale

Sphagnum fallax

Fig. 4.2 (cont.). Location of collection sites for 24 peatland mosses with 25 or more collection sites each.

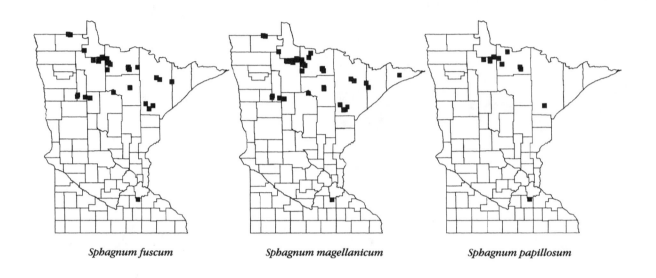

Sphagnum fuscum   *Sphagnum magellanicum*   *Sphagnum papillosum*

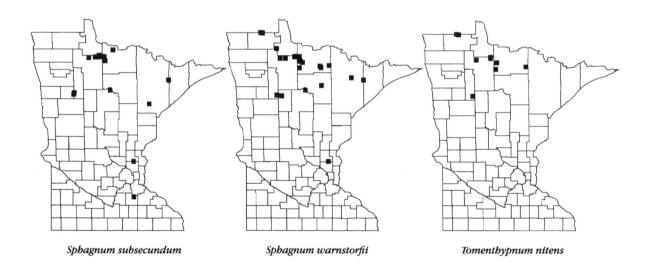

Sphagnum subsecundum   *Sphagnum warnstorfii*   *Tomenthypnum nitens*

Fig. 4.2 (cont.). Location of collection sites for 24 peatland mosses with 25 or more collection sites each.

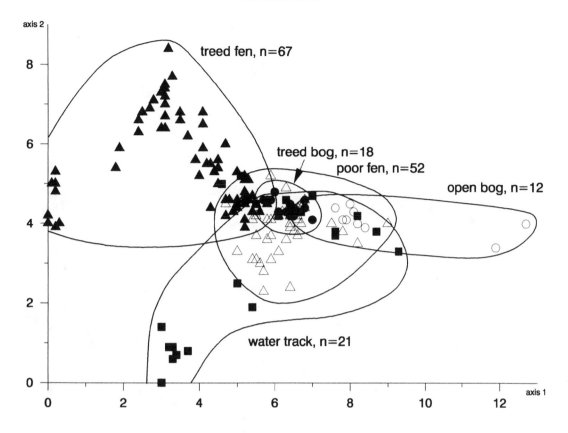

Fig. 4.3. DCA-ordination diagram of 115 bryophyte species from 170 Minnesota plots, characterized by their bryophyte abundances, with rare species down-weighted. Eigenvalues of the first axis equals 0.912, for the second axis equals 0.791.

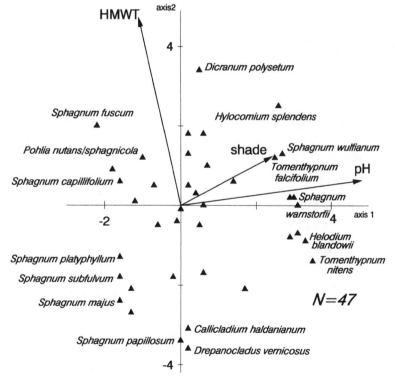

Fig. 4.4. Species-environment biplot of CCA ordination of 65 bryophyte plots and 47 species, constrained by the three environmental variables represented by *arrows*. HMWT = height above mean water table. Eigenvalues of canonical (constrained) axes are 0.575, 0.538, and 0.277.

Fig. 4.5. *Calliergon aftonianum.* An extremely rare species, closely related to *Calliergon giganteum* and *C. richardsonii* (Janssens 1990). Originally described as a Pleistocene extinct species (Steere 1942), but discovered as living populations at the Lost River spring fen (Janssens and Glaser 1986). The species is quite common in Pleistocene (Arctic Islands) and Holocene peats (Red Lake peatland). **A:** fragment (2.3x). **B:** leaves (25x). **C** and **D:** medial and alar cells (135x). (From *Canadian Journal of Botany* 64 [1986], by permission.)

Fig. 4.7. *Calliergon trifarium.* A once-common rich-fen species with an extensive fossil record, now nearly a relict in the larger patterned peatlands in Minnesota. Still commonly found in smaller, more highly minerotrophic peatlands and spring fens. **A:** fragment (8x). **B:** leaves (17x). **C** and **D:** alar and medial cells (160x). (From *Canadian Journal of Botany* 64 [1986], by permission.)

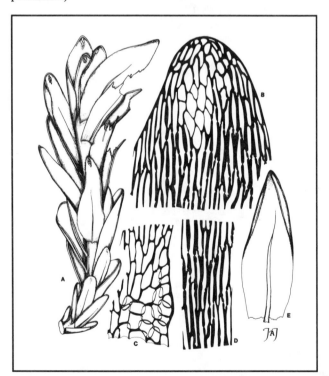

Fig. 4.6. *Calliergon stramineum.* A common species associated as a minor component with oligotrophic peat-forming mosses, in contrast to all other members of the genus. **A:** fragment (8x). **B:** leaf apex (55x). **C:** alar cells (55x). **D:** medial cells (75x). **E:** leaf (11x). (From *Canadian Journal of Botany* 64 [1986], by permission.)

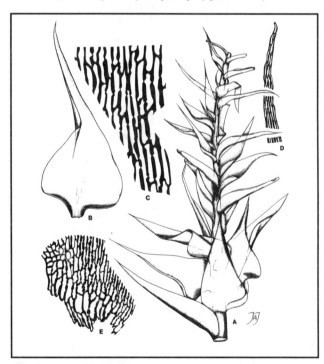

Fig. 4.8. *Campylium stellatum.* A common rich-fen indicator species, consistently present in low abundance in most minerotrophic peats and a member of all minerotrophic moss communities. **A:** fragment (19x). **B:** leaf (34x). **C:** medial cells (225x). **D:** leaf apex (88x). **E:** alar cells (88x). (From *Canadian Journal of Botany* 64 [1986], by permission.)

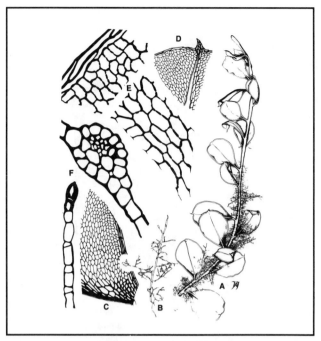

Fig. 4.9. *Cinclidium stygium*. A newly recorded species for the state of Minnesota (Janssens and Glaser 1984a; Janssens 1988a). It is an indicator of rich-fen conditions and is only known from the spring-fen drains of the Lost River peatland. **A:** fragment (3.5x). **B:** rhizoid (12x). **C:** basal cells (31x). **D:** leaf apex (21x). **E:** medial and border cells (90x). **F:** transverse section through costa and lamina (125x). (From *Michigan Botanist* 23 [1984], by permission.)

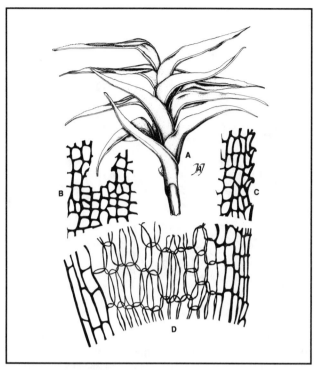

Fig. 4.11. *Meesia triquetra*. An uncommon rich-fen species, more common north of Minnesota. A very characteristic fossil in lower peat units, formed during initial stages of peatland development. **A:** fragment (15x). **B and C:** medial and marginal cells (210x). **D:** basal cells (160x).

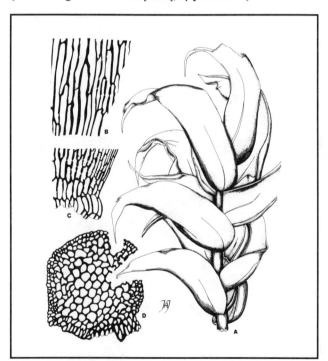

Fig. 4.10. *Drepanocladus lapponicus*. A rich-fen and intermediate-fen species only quite recently recognized for North America (Janssens 1983). It is a common fossil in some of the Red Lake peatland peats but is rare in Minnesota at present. **A:** fragment (12x). **B:** medial cells (190x). **C:** alar cells (75x). **D:** transverse section through stem (75x). (From *Canadian Journal of Botany* 64 [1986], by permission.)

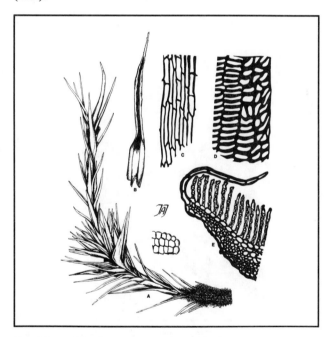

Fig. 4.12. *Polytrichum strictum*. An extremely common species in all Minnesota's peatlands, growing on top of most high *Sphagnum* hummocks. Extremely desiccation-tolerant because of the numerous xerophytic adaptations. Rarely preserved and probably strongly underrepresented in peats. **A:** fragment (3.2x). **B:** leaf (6.0x). **C, D,** and **E:** sheath cells, lamina cells, and transverse section through leaf (110x). (From *Canadian Journal of Botany* 64 [1986], by permission.)

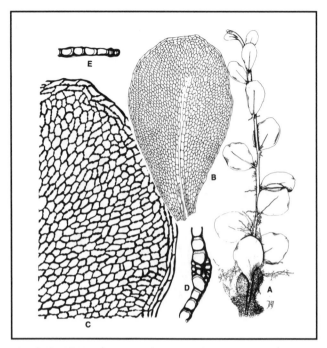

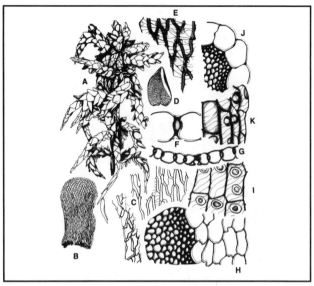

Fig. 4.13. *Rhizomnium gracile*. Extremely rare species occurring scattered in poorly to moderately minerotrophic fens, growing in highly shaded microhabitats. Known only from the Lost River spring fen in northern Minnesota (Janssens and Glaser 1984b). **A:** fragment (4.8x). **B:** leaf (15x). **C:** lamina (64x). **D:** transverse section through costa (92x). **E:** transverse section through lamina and border (97x). (From *Michigan Botanist* 23 [1984], by permission.)

Fig. 4.15. *Sphagnum papillosum*. Common carpet-forming bogmoss in poor-fen habitats. Commonly found as a fossil in Minnesota peats. Pure stands are recognizable from the air as small circular islands, expanding over sedge meadows in slowly acidifying intermediate fens. **A:** fragment (1.8x). **B:** stem leaf (11x). **C:** stem-leaf marginal and medial cells (55x). **D:** branch leaf (5.7x). **E:** branch-leaf medial cells (63x). **F** and **G:** transverse sections through branch leaf (63x and 160x). **H:** transverse section through stem and cortical cells (90x). **I:** surface view of stem-cortical cells (87x). **J:** transverse section of branch and cortical cells (66x). **K:** surface view of branch-cortical cells (63x).

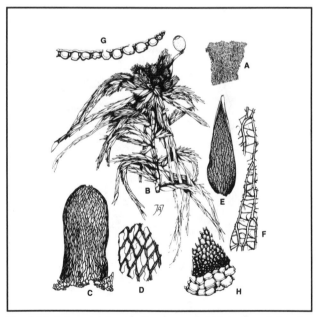

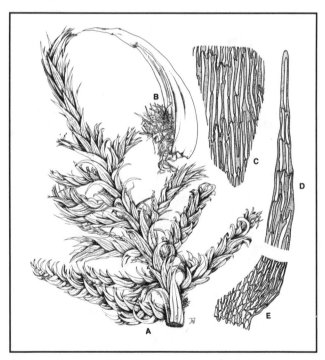

Fig. 4.14. *Sphagnum fuscum*. Very common high-hummock-forming bogmoss, occurring in all landform and peatland types. The species itself is quite stenotypic in its response to the minerotrophic and hydrological gradients but forms tall hummocks, isolated from more minerotrophic and aquatic habitats in mixed mires and rich fens. **A:** habit (0.32x). **B:** fragment (3.5x). **C:** stem leaf (25x). **D:** stem-leaf medial cells (75x). **E:** branch leaf (24x). **F** and **G:** branch-leaf medial cells and transverse section through branch leaf (100x). **H:** transverse section through stem and cortical cells (75x).

Fig. 4.16. *Tomenthypnum falcifolium*. A recently recognized species for Minnesota (Wheeler 1982) that already has a significant record. Consistently found in rare, moderately minerotrophic fens that form transitional stages to more acidified peatland types. **A:** fragment (1.1x). **B:** leaf with basal rhizoids (5.5x). **C** and **D:** medial cells and leaf apex (230x). **E:** alar cells (100x).

Acknowledgments

Funds for many aspects of this work were provided by contracts with the Peat Program of the Minnesota Department of Natural Resources and with the U.S. Department of Energy (DOE AC01-80EV10414) through H. E. Wright, and by grants from the National Science Foundation (DEB-7922142) and Andrew W. Mellon Foundation through E. Gorham.

Field assistance offered by members of the Limnological Research Center and the Department of Natural Resources is appreciated.

Literature Cited

Bowers, F. D., and S. K. Freckmann. 1979. Atlas of Wisconsin bryophytes. Museum of Natural History, Stevens Point, Wisconsin. Reports on the Fauna and Flora of Wisconsin 16.

Clymo, R. S., and P. M. Hayward. 1982. The ecology of *Sphagnum*. Pages 229-289, *in* A. J. E. Smith, editor. Bryophyte ecology. Chapman and Hall, London, U.K.

Crandall-Stotler, B. 1984. Musci, hepatics and anthocerotes — an essay on analogues. Pages 1093-1129 *in* R. M. Schuster. editor, New manual of bryology, vol. 2. Hattori Botanical Laboratory, Nichinan, Japan.

Crum, H. A. 1983. Mosses of the Great Lakes forest. 3rd edition. University of Michigan Herbarium.

———. 1984. Sphagnopsida, Sphagnaceae. *In* North American Flora, Series II, Part 11. New York Botanical Garden.

——— and L. E. Anderson. 1981. Mosses of eastern North America. Columbia University Press, New York, New York, USA.

Holzinger, J. M. 1895. A preliminary list of the mosses of Minnesota, Minnesota Botanical Studies vol. 1. Bulletin of the Geological and Natural History Survey of Minnesota 9:280-294.

Ireland, R. R., G. R. Brassard, W. B. Schofield, and D. H. Vitt. 1987. Checklist of the mosses of Canada II. Lindbergia 13:1-62.

Isoviita, P. 1966. Studies on *Sphagnum* L. I. Nomenclatural revision of the European taxa. Annales Botanici Fennici 3:199-264.

Janssens, J. A. 1983. Past and extant distribution of *Drepanocladus* in North America, with notes on the differentiation of fossil fragments. Journal of the Hattori Botanical Laboratory 54:251-298.

———. 1988a. Nonvascular plants: Mosses. Pages 219-229 *in* B. Coffin and L. Pfannmuller, editors. Minnesota's rare and endangered

flora and fauna. University of Minnesota Press, Minneapolis, Minnesota, USA.

———. 1988b. Fossil bryophytes and paleoenvironmental reconstruction of peatlands. Pages 299-306 *in* J. M. Glime, editor. Methods in bryology. Proceedings of the Bryological Methods Workshop, Mainz. Hattori Botanical Laboratory, Nichinan, Japan.

———. 1989. Ecology of peatland bryophytes and paleoenvironmental reconstruction of peatlands using fossil bryophytes. Methods manual, update December 1989. Available from the author.

———. 1990. Methods in Quaternary ecology 11 — bryophytes. Geoscience Canada, March 1990.

——— and P. H. Glaser. 1984a. *Cinclidium stygium* Sw. (Bryopsida: Mniaceae) in Minnesota. Michigan Botanist 23:19-20.

——— and P. H. Glaser. 1984b. *Rhizomnium gracile*, a moss new to Minnesota and the distribution of *Rhizomnium* in the state. Michigan Botanist 23:89-92.

——— and P. H. Glaser. 1986. The bryophyte flora and major peat-forming mosses at Red Lake peatland, Minnesota. Canadian Journal of Botany 64:427-442.

Jongman, R. H. G., C. J. F. ter Braak, and O. F. R. van Tongeren. 1987. Data analysis in community and landscape ecology. Pudoc, Wageningen, Netherlands.

Mishler, B. D., and S. P. Churchill. 1984. A cladistic approach to the phylogeny of the "bryophytes." Brittonia 36:406-424.

——— and S. P. Churchill. 1985. Cladistics and the land plants: A response to Robinson. Brittonia 37:282-285.

Scagel, R. F., R. J. Bandoni, J. R. Maze, G. E. Rouse, W. B. Schofield, and J. R. Stein. 1982. Nonvascular Plants: An evolutionary survey. Wadsworth, Belmont, California, USA.

Schofield, W. B. 1985. Introduction to bryology. Macmillan, New York, New York, USA.

Schuster, R. M. 1977 (1953). Boreal Hepaticae, a manual of the liverworts of Minnesota and adjacent regions. American Midland Naturalist 49:257-684.

Steere, W. C. 1942. Pleistocene mosses from the Aftonian Interglacial deposits of Iowa. Michigan Academy of Science, Arts and Letters, Papers, vol. XXVII:75-104.

ter Braak, C. J. F. 1987. CANOCO — A FORTRAN program for canonical community ordination by [partial] [detrended] [canonical] correspondence analysis, principal components analysis and redundancy analysis (version 2.1). ITU-TNO, Wageningen, Netherlands.

Wheeler, G. A. 1982. *Tomenthypnum falcifolium* in Minnesota. Michigan Botanist 22:66.

CHAPTER 5

Rare Vascular Plants

Paul H. Glaser

Mystique of Rare Plants

Perhaps the greatest thrill for any field botanist is to discover a new locality for a rare plant. No matter how thoroughly a region is botanized, certain species continue to be overlooked because their populations are restricted to a few isolated patches. The search for rare species then becomes a treasure hunt in which botanists pit their skills against those of previous collectors and try to discover those few species that have eluded detection. Although skill and experience are essential prerequisites for this search, the ultimate arbiters of success and failure are often luck and the vagaries of nature.

The peatlands of northern Minnesota represent a prime target for plant hunters because they are one of the least-explored areas of the state. Peatlands have long been associated with orchids and a distinctive flora, but only the most dedicated field-worker will overcome the initial dread of traversing a waterlogged quagmire. The largest peatlands, moreover, are essentially inaccessible during the growing season without the aid of a bog vehicle or helicopter. Few people venture through the largest peatlands, and they remain some of the most important remnants of pristine vegetation in the state. The recent threat to these natural areas from mining interests spurred the Department of Natural Resources to sponsor a comprehensive search for rare plants in these peatlands. This search was motivated by state and federal laws protecting rare species, but the scientific value of such species should not be underestimated.

The existence of rare plants raises some fundamental questions about plant communities and the factors that regulate the size of plant populations: Why are certain species abundant and others rare? Have recent environmental changes shifted the competitive balance among native plant species to the disadvantage of rare plants? Are rare species highly specialized for a particular habitat that is itself rare? Do rare species represent the advance wave of a migrating species or the last remnants of a declining population on the path to local extinction? Rare species are therefore important objects of study, although it is often difficult to generalize on the basis of a few rare events.

This difficulty may be overcome by a comparative approach in which the rare species within a major vegetation type are compared to each other and to associate species that have similar growth forms but are ecologically more successful. The following discussion will therefore be focused on three groups of rare species that are primarily restricted to patterned peatlands in Minnesota. A complete inventory of all rare species that occur in the peatlands of Minnesota may be found in Coffin and Pfannmuller (1988).

Rarity — What Is It?

A species may be considered rare because (1) its density or population size is very small, (2) its dominance or biomass in a community is exceptionally small, or (3) its distribution or geographic range is very small and localized. Density is often difficult to measure in plants because the exact definition of an individual is ambiguous among vegetatively reproducing species. Comparisons among species with radically different growth forms may therefore be impossible. Rare plants, however, are assumed to have very small populations relative to the more common species. The absolute size of a plant may also be a factor in determining rarity if the plants are so small and inconspicuous that they may be overlooked. Examples of such plants are the moss *Buxbaumia aphylla* and the vascular plant *Koenigia islandica*. The most common types of rare plants in Minnesota, however, are species that have very localized and restricted distributions in the state.

Most of the rare species in Minnesota peatlands are also found outside the state, where they are more abundant in terms of their density, dominance, and distribution. They reach the edge of their geographic range in Minnesota peatlands, where they have a highly localized distribution. Some of these species have extremely small populations in the state, whereas others are locally dominant where they occur. These patterns provide important clues to the factors that determine the rarity of these species.

Rarity — How Is It Determined?

Most of the rare peatland species in Minnesota were first discovered as single populations near a road. Not until a systematic survey of the patterned peatlands was made did a more extensive view of their distribution in Minnesota emerge. At first, luck played a large role in the discovery of these plants. It soon became apparent, however, that the distribution of certain species in Minnesota is intimately associated with particular peat landforms that can be spotted from a helicopter. This discovery led to the rapid mapping of rare species populations despite their inaccessible habitats.

This combination of luck and experience is illustrated by the discovery of *Xyris montana* Ries in Minnesota. *Xyris*, or yellow-eyed grass, is a small rush-like plant that can easily be mistaken for a sedge except for the inconspicuous yellow petals that protrude from scalelike bracts (Kral 1966). This species had not been collected in Minnesota until 1970, when John and Hilary Birks, visiting botanists from England, found a small population on the edge of Hornby Lake in St. Louis County. Their discovery was remarkable because Ogala Lakela had tirelessly searched the arrowhead region of Minnesota for over 30 years while compiling material for *A Flora of Northeastern Minnesota* (Lakela 1965). By 1970 this area was considered to be botanically the best-known region in Minnesota and the regional flora to be essentially complete. Seven years later Gerald Wheeler and I discovered another population of *Xyris montana* on the edge of a pond within a small fen in Itasca County. The following year we discovered three additional populations of *Xyris* in the vast Red Lake peatland in northwestern Minnesota (Wheeler and Glaser 1979). During a survey of patterned fens in northeastern Minnesota, I found two more populations of *Xyris montana*, and the most recent known record for the species was added by Norman Aaseng and Welby Smith from the Arlberg peatland in 1981. No additional stations have since been reported, despite an intensive search by botanists alerted to the existence of this rare species and its general habitat requirements.

Rare Vascular Plant Species in Minnesota Peatlands

Approximately 40% of the 174 vascular plants officially listed as rare in Minnesota are characteristic of wetland habitats (Coffin and Pfannmuller 1988). Only eight of these species, however, are associated mainly with the patterned peatlands of northern Minnesota. The distribution of these species is now relatively well known and permits some insight into their ecology within the state. Some can be compared to closely related species that are morphologically similar but have a much wider distribution in the state. Two of these rare species, *Xyris montana* Ries and *Juncus stygius* L. var. *americana* Buchenau, are very small plants that maintain exceptionally small populations in Minnesota. In terms of total plant numbers, *Juncus stygius* may be one of the rarest species in Minnesota. *Drosera anglica* and *D. linearis* are also small plants but are found in larger local populations. The other four rare peatland species reproduce by rhizomes and form clones or tussocks. *Rhynchospora fusca* (L.) Ait. f. is of intermediate size, whereas the other three — *Carex exilis*, *Cladium mariscoides*, and *Eleocharis rostellata* — are larger plants that often form clones.

Species of Low Density and Dominance

Xyris montana Ries (Yellow-eyed Grass)

Xyris montana is a small plant that can spread by a slightly forking rhizome. In Minnesota it usually forms small populations of less than 100 individuals that are highly localized (Fig. 5.1). Only in the eastern watershed at Red Lake does *Xyris* spread over a larger area. At Mud Lake it is locally a dominant in the flarks, but these landforms cover less than 0.2 ha. At other stations *Xyris* does not attain high cover values.

In Minnesota *Xyris* occurs in poor and rich fens with a pH of 4.3 to 6.5 and a Ca concentration of 0.6 to 6.4 mg l^{-1} (Table 5.1). It is usually found in patterned water tracks, although it also occurs along the margin of Hornby Lake in St. Louis County and the edge of a small pond in "Botany Bog," a small fen in Itasca County. *Xyris* seems to prefer marginal conditions where the position of the water table fluctuates periodically and where competition from the more aggressive sedges, particularly *Carex lasiocarpa*, is reduced. Thus it occurs along the edges of a bog island in the Red Lake peatland where drainage ditches have altered the water level in the adjacent water track. In the pristine eastern watershed at Red Lake, however, *Xyris* is widespread near the head of the internal water track.

In North America, *Xyris montana* is a coastal-plain species with the center of its distribution in the Maritime Provinces of Canada. It extends inland along the St. Lawrence River and the Great Lakes and reaches its western range limits in the peatlands of Minnesota. In eastern Canada, *X. montana* has a much broader range

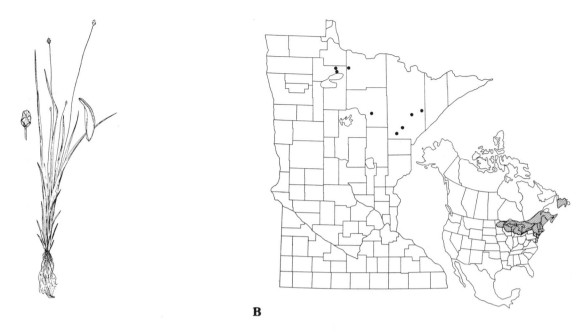

Fig. 5.1. A. *Xyris montana*, yellow-eyed grass. **B.** Distribution of *Xyris montana* in Minnesota and range in North America.

Table 5.1. Ecological characteristics of rare peatland plants in Minnesota.

Species	Water Chemistry		Landform	Number of stations	Species number
	pH	Ca (mg/l)			
Xyris montana	4.8-6.5	0.6-12.1	WT, PM	7	19
Juncus stygius	5.2-5.4	2.4-13.5	WT (m)	7	16
Drosera anglica	5.6-7.2	5.3-98.5	Flarks, SFC	16	19
Drosera linearis	5.6-6.1	5.3-31.7	Flarks, SFC	9	16
Rhynchospora fusca	4.8-6.3	5.3-12.1	WT	8	20
Carex exilis	4.9-7.2	0.9-37	WT(m), SFC	8	22
Cladium mariscoides	5.7-7.2	9.5-98.5	SFC, WT	19	19
Eleocharis rostellata	>6.8	>35	SFC	8	21

Note: The landform symbols are (WT) water track, (WTm) water track margin, (PM) pond margin, and (SFC) spring-fen channels. The species number refers to the average number of species found in a 100 m^2 plot.

of tolerance with respect to water chemistry and is most common in the mud-bottom community of acid bogs (pH < 4.2; Ca concentration < 2 mg l^{-1}) or poor fens. It also grows in brackish marshes near Fourchu, Nova Scotia.

Juncus stygius L. var. *americana* Buchenau (Bog Rush)

Juncus stygius is very similar to *Xyris* with respect to its growth habit and ecological requirements in Minnesota. It usually occurs as a few isolated individuals, although a larger number of tufts were found at Sand Lake in Lake County (Fig. 5.2). In Minnesota, *J. stygius* is restricted to water tracks, but its precise location may vary from the margins of deep pools (North Black

River) to mud-bottomed pools (Red Lake) and low clumps of *Sphagnum* (Mud Lake, Sand Lake). *J. stygius* is perhaps the most difficult rare peatland species to detect because of its very small size, inconspicuous appearance, and occurrence in very small patches.

Juncus stygius is a boreal species that is widespread throughout Canada and parts of Alaska. The Minnesota populations represent isolated colonies at the southern fringe of its geographic range. Throughout its range, *J. stygius* is a minerotrophic indicator, and in Minnesota this species has only been found in transitional rich fens (pH 5.2 to 5.4; Ca concentration 2.4 to 13 mg l^{-1}) (Table 5.1). In the boreal belt of Canada, it also occurs in poor fens, particularly in seeps draining from the edges of raised bogs.

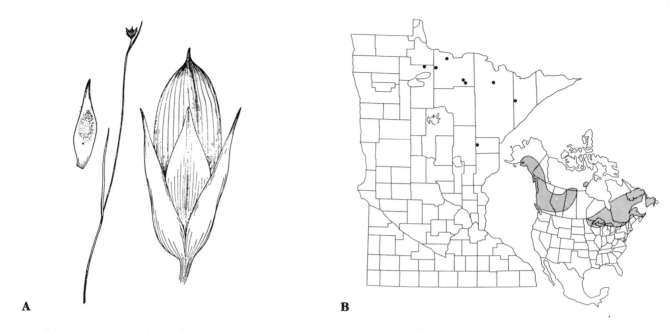

Fig. 5.2. A. *Juncus stygius*, bog rush. **B.** Distribution of *Juncus stygius* in Minnesota and range in North America.

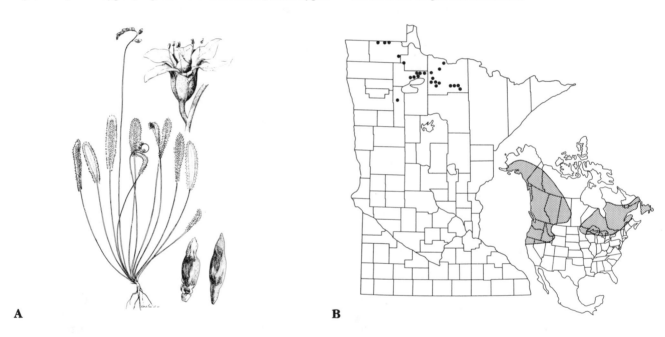

Fig. 5.3. A. *Drosera anglica*, English sundew. **B.** Distribution of *Drosera anglica* in Minnesota and range in North America. (*Drosera anglica* illustration by Vera Ming Wong, copyright Minnesota Department of Natural Resources.)

Species of Higher Density and Low Dominance

Drosera anglica Huds. (English Sundew) and *D. linearis* Goldie (Linear-Leaved Sundew)

Drosera anglica and *D. linearis* (Figs. 5.3, 5.4) have much larger populations and occur in more stations in Minnesota than *Juncus stygius* and *Xyris montana*. These two rare species of sundews are confined, however, to the deeper water of the flarks in pristine water

tracks and spring-fen channels. In the deepest water they may attain cover values of up to 20%, but they decrease in abundance as the water level drops. *Drosera linearis* is more sensitive to disturbance and never occurs in water tracks cut by drainage ditches, whereas very small populations of *D. anglica* may persist in these locations.

In North America, *D. anglica* is more widespread and has a broader ecological tolerance than *D. linearis*. *D. anglica* is a circumboreal species that grows

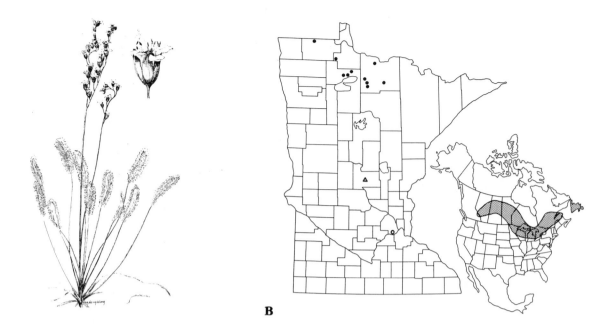

Fig. 5.4. A. *Drosera linearis,* linear-leaved sundew. **B.** Distribution of *Drosera linearis* in Minnesota and range in North America. (*Drosera linearis* illustration by Vera Ming Wong, copyright Minnesota Department of Natural Resources.)

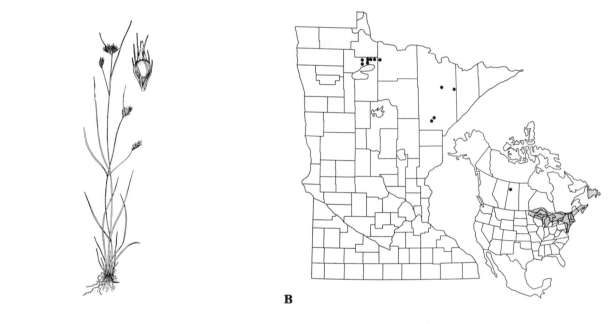

Fig. 5.5. A. *Rhynchospora fusca,* sooty-colored beak-rush. **B.** Distribution of *Rhynchospora fusca* in Minnesota and range in North America.

within the complete range of peatland types from the mud bottoms of acid raised bogs (pH < 4.2) in boreal Canada to extremely rich fens (pH > 6.8) in the continental interior. *D. linearis,* in contrast, has a more restricted range in North America and is more typical of a coastal-plain species. It extends from the Maritime Provinces of eastern Canada into the continental interior along the St. Lawrence River and the Great Lakes. *D. linearis* is restricted to minerotrophic sites across its range. In Minnesota, both of these rare species of

sundews are found in fens with a pH of 5.6 to 7.4 (Table 5.1).

Species of High Density and Dominance

Rhynchospora fusca (L.) Ait. f.
(Sooty-Colored Beak-Rush)

Rhynchospora fusca is one of the rarest species in Minnesota, although it is abundant in the few places it occurs (Fig. 5.5). This species is easy to see be-

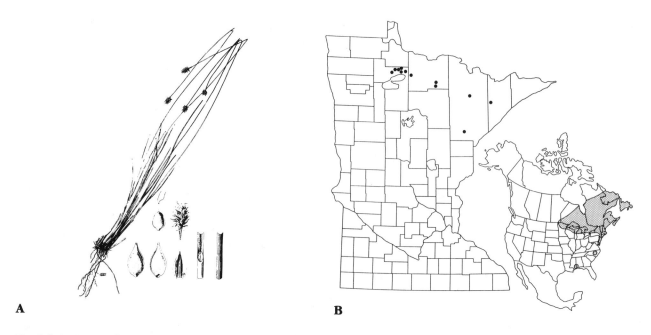

Fig. 5.6. A. *Carex exilis*, a species of sedge. **B.** Distribution of *Carex exilis* in Minnesota and range in North America.

cause of the bright green color of its spreading clones. The dark chestnut-brown spikelets of *R. fusca* quickly distinguish it in the field from the superficially similar *R. alba*, with which it is consistently associated. *R. fusca* is found in three of the four watersheds in the Red Lake peatland, where it is most common near the heads of the internal water tracks. It also occurs along the northern margin of the western water track. Outside the Red Lake peatland, *R. fusca* is a dominant in the flarks of three small patterned fens in northeastern Minnesota. The center of one of these sites has been destroyed by a powerline right-of-way, and the others have been altered by roads and drainage ditches. *R. fusca* also occurs in the Wahlenstein fen in St. Louis County, where it was first found in Minnesota by Ogala Lakela.

In Minnesota, *R. fusca* grows in poor and rich fens with a pH of 4.8 to 6.3 (Table 5.1). The water table in these areas is usually above the peat surface but is considerably lower than the deeper pools characteristic of *Drosera anglica* and *D. linearis*. In the Red Lake peatland, *R. fusca* is not associated with these two rare sundews. Instead, it is most common near the heads of the internal water tracks in which the string-flark patterns are poorly developed. *R. fusca* grows in a similar setting along the northern edge of the western water track, which contains streamlined tree islands and circular clones of *Cladium mariscoides* but no string-flark patterns. The Minnesota populations of *R. fusca* mark the western range limits for this coastal-plain species; the center of its distribution is in the Maritime Provinces of Canada.

Carex exilis Dew. (A Species of Sedge)

Carex exilis is a tussock-forming sedge that may be distinguished by its solitary spikes, which have divergent perygynia when mature. It superficially resembles *Scirpus cespitosus*, which usually grows in the same habitats but is much more widely distributed in northern Minnesota. *C. exilis* is an exceedingly rare species in Minnesota, where it is restricted to small portions of five peatlands outside the Red Lake peatland (Glaser 1983a; Fig. 5.6). Within the Red Lake peatland, it is locally a dominant along the edges of water tracks in the west-central and western watersheds. Its tussocks may coalesce to form strings in places, an effect most noticeable in the Lost Lake peatland in St. Louis County and the Sand Lake peatland in Lake County. *C. exilis* may also be important in flarks, particularly where the water table has been lowered by ditches, such as at the Alborn fen in St. Louis County. The other habitat for *C. exilis* in Minnesota is the spring-fen channels of the Lost River peatland and similar channels in the Myrtle Lake peatland. The most enigmatic feature of the distribution of *C. exilis* in Minnesota is the large number of apparently suitable sites at which it is absent.

Carex exilis is restricted to minerotrophic sites in Minnesota with a pH from 4.9 to 7.6 (Table 5.1). The Minnesota populations represent the western range limits for *C. exilis* in North America. This coastal-plain species is centered in the Maritime Provinces of Canada, where it is most common in poor fens, although it will also grow on raised bogs with a pH less than 4.2. This species is most common in patterned fens with poor-fen waters (pH 4.5 to 4.9) in eastern Canada, but its occurrence on raised bogs is generally limited to

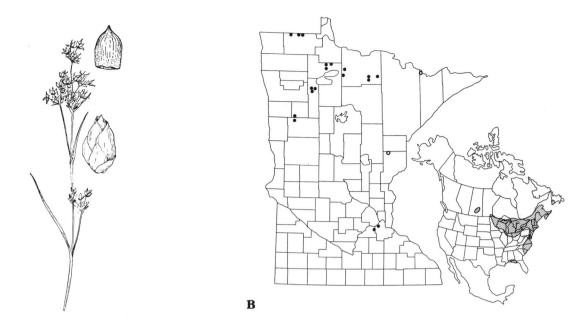

Fig. 5.7. A. *Cladium mariscoides*, twig-rush. **B.** Distribution of *Cladium mariscoides* in Minnesota and range in North America.

Nova Scotia, northern Maine, and the coastal portion of New Brunswick. It also grows in spring-fen channels in the Hudson Bay lowlands. At all these sites it is associated with *Scirpus cespitosus*, which is morphologically similar but is usually the more important dominant in the vegetation.

Cladium mariscoides (Muhl.) Torr. (Twig-Rush)

Cladium mariscoides is a stout stoloniferous sedge that is characteristic of spring-fen channels in northern Minnesota (Fig. 5.7; Table 5.1). This species is also common in the Red Lake peatland, particularly in the western water track, where it forms large circular clones. These clones are most common along the northern edge of the western water track, where the string-flark patterns are absent. The clones also occur sporadically among the streamlined tree islands and string-flark patterns in the center of the water track. The only other site for these clones in Minnesota is near the head of a system of channels in the Myrtle Lake peatland. Elsewhere, *Cladium* generally has sparse cover and does not coalesce into large patches.

C. mariscoides is generally restricted to rich and extremely rich fens in northern Minnesota that have a pH of 5.7 to 7.4 (Table 5.1). It also occurs in spring fens in western and southern Minnesota, and its range in the state may be extended as these habitats are more thoroughly explored (Fig. 5.7B). *C. mariscoides* is widely distributed along the coastal plain of eastern North America and extends inland along the St. Lawrence River and the Great Lakes. The Minnesota populations mark the western limits of its range except for disjunct populations in Saskatchewan.

Eleocharis rostellata Torr. (Beaked Spike-Rush)

Eleocharis rostellata is a tall sedge that forms stout tussocks in several spring fens in Minnesota (Glaser 1983b). It usually occurs near the heads of the spring-fen channels, which may be foci for groundwater discharge (Siegel and Glaser 1987; Glaser *et al.* 1990). In the northern peatland region, *E. rostellata* is restricted to very small localized populations; larger populations have been discovered in spring fens located south of the patterned peatland region in Minnesota (Fig. 5.8). This species shares many ecological features with the other rare species discussed here.

E. rostellata is restricted to extremely rich fens in Minnesota with a pH greater than 7 (Table 5.1). The species has a complex distribution in North America, and the Minnesota populations appear to be disjuncts from the more continuous part of its range in the eastern and western United States. *E. rostellata* superficially resembles *Scirpus acutus*, another sedge that is characteristic of the heads of the spring-fen channels in northern Minnesota. *S. acutus*, however, is readily distinguished by its more robust shoots, absence of tussocks, and crowded spikelets.

Ecological Significance of Rare Species

Rare species continue to fascinate botanists because they represent an enigmatic anomaly in nature. An examination of the eight rare species that characterize the patterned peatlands of northern Minnesota may provide some insight into the processes that produce rare populations at the edge of a species range. A comparative analysis of their distribution, habitat requirements, and associated species compensates for

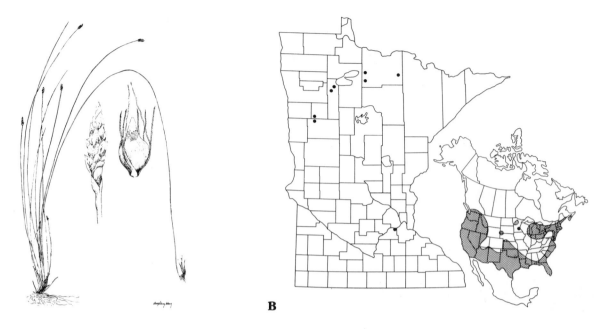

Fig. 5.8. A. *Eleocharis rostellata*, beaked spike-rush. **B.** Distribution of *Eleocharis rostellata* in Minnesota and range in North America. (*Eleocharis rostellata* illustration by Vera Ming Wong, copyright Minnesota Department of Natural Resources.)

the small data set available for each of these rare species in Minnesota.

All of these rare species share several features in common. Despite their sporadic occurrence in the patterned peatland region of northern Minnesota, most of them occur in the vast Red Lake peatland (Wheeler and Glaser 1979). Only *Eleocharis rostellata* has eluded detection at Red Lake, although it may yet be discovered when this peatland is more thoroughly explored. Several rare species tend to be found at the same locality, although only the two rare sundews consistently occur together. A partial explanation for this relationship is the fact that all of these rare species are found within a similar vegetation assemblage. The dominant species in this assemblage may change along various gradients in water chemistry, water level, and peat landform, but the composition of the assemblage is similar. This assemblage is characteristically poor in species. A 100 m² relevé plot from these stands consistently contains less than 20 species (Table 5.1), in contrast to the 40 or more species of vascular plants found in a typical swamp forest.

Another interesting feature is the consistent presence of competing species that are either taxonomically or morphologically similar to the rare plants. *Rhynchospora fusca* is consistently found with the vegetatively indistinguishable *R. alba*, whereas *Drosera anglica* and *D. linearis* are difficult to distinguish from each other and from their consistent associate *D. intermedia*. *Carex exilis* resembles *Scirpus cespitosus* in the vegetative state. In each of these cases the similar species has a much broader distribution and is consistently found in all patterned water tracks or spring-fen channels.

Factors That Influence Rarity

Five factors may have an important influence on maintaining these rare species in small, isolated populations: (1) competition, (2) availability of suitable habitat, (3) dispersal, (4) species-area effects, and (5) environmental change. Because the peat within peatlands represents an *in situ* record of past vegetation communities, the record of rare plant establishment and maintenance in these peatlands may also be reconstructed.

Competition

The rare peatland species occur in habitats that seem to be more vulnerable to invasion and establishment by new species than other sites. The low species numbers characteristic of these communities indicate that successful species must cope with a high level of physiological stress created by a waterlogged rooting medium and slightly acidic or circumneutral waters. The rarest of these species in terms of their population size seem to be *Xyris montana* and *Juncus stygius*. They are usually limited to only a few upright shoots, and their low stature places them at a competitive disadvantage with respect to the taller and more deeply rooted sedges in their habitat. These two rare species seem to have a fugitive strategy in which they are able to exploit areas where the water level has recently changed to create a break in the vegetative mat. These patches may occur along the edges of water tracks or the margins of ponds. Another example is the game trail that crosses the Alborn fen, creating a local break in the mat of sedges. *J. stygius* seems to establish populations much smaller than those of *Xyris* and would seem to have a more precarious existence in Minnesota.

Drosera anglica and *D. linearis* also avoid competition from the taller sedges by growing in the deeper water within flarks. Across the continental interior, these two species are consistently associated with this habitat within pristine patterned fens, and their rarity in Minnesota seems to be directly linked to the rarity of this specialized habitat. These two species maintain fairly large populations in the few water tracks that have deep flark pools, but they are very vulnerable to water-level fluctuations, which favor the growth of the competing sedges. The closely related species *Drosera intermedia* generally has low cover values in the flarks and probably creates less interference in the establishment of the rarer sundews than the more dominant sedges.

Carex exilis and *Rhynchospora fusca*, however, seem to compete more closely with the species that share a similar growth form. *C. exilis* and *Scirpus cespitosus* both form dense tussocks that may locally dominate a vegetation assemblage in a patterned water track or spring-fen channel. Across their geographic range, both species have broad tolerance for ranges in water chemistry, although *S. cespitosus* is more abundant and more widely distributed at both the acidic end (on raised bogs) and the alkaline end (in spring-fen channels) of the gradient than *C. exilis*. In Minnesota, *S. cespitosus* is more widely distributed and is a superior competitor, so it may prevent the establishment of *C. exilis* from all but a few sites (Glaser 1983a).

Rhynchospora fusca has a similar relationship to *R. alba*. *Rhynchospora alba* is a circumboreal species that has a broad ecological tolerance for water chemistry from acidic bogs to circumneutral spring fens. In Minnesota it attains its highest cover values within the poor and rich fens — the same range in which it competes with *R. fusca*. *Rhynchospora fusca*, however, is a coastal-plain species and is restricted to minerotrophic habitats. It is always a co-dominant with *R. alba* in Minnesota, but *R. alba* is much more widely distributed in the northern peatlands. In continental settings the coastal-plain species seem to be at a competitive disadvantage relative to their circumboreal associates. This effect has been called biological inertia by Eville Gorham.

Dispersal

The clustering of rare plant populations in the Red Lake peatland suggests that dispersal may play a role in the establishment of these species in Minnesota. The Red Lake peatland is the largest continuous peatland in the state (Glaser *et al.* 1981) and represents the largest target for a propagule — a dispersable unit — transported over long distances. The clustering of rare plant species in the spring-fen channels at Lost River supports this hypothesis because it is the closest spring fen to Red Lake. Similar spring fens that are more distant

from Red Lake, moreover, lack many of these rare plant species. According to this hypothesis, these rare plant populations first became established at Red Lake and slowly spread outward to more distant sites. The dispersal mechanisms for these rare species, however, do not seem to be significantly different from many closely related species that are more common in the Minnesota peatlands.

Species-Area Effects

The clustering of rare species in the Red Lake peatland may also be explained by the relationship between species richness and area proposed by island biogeographers (MacArthur and Wilson 1967). According to this hypothesis, the species richness of an island is proportional to its area. Larger islands or peatlands would therefore be expected to contain more habitat types and therefore more rare and common species than smaller sites. This hypothesis would also account for the large number of rare species in the Myrtle Lake peatland. However, many smaller peatlands in Minnesota contain a disproportionate number of rare species relative to their size. The Alborn fen, for example, contains three rare species despite its small size, whereas the spring fen at Lost River contains four of the rare species discussed here and several others recognized as rare by the Minnesota Department of Natural Resources. The species-area effect provides only a partial explanation for distribution of rare species in northern Minnesota peatlands.

Environmental Change

The occurrence of rare peatland species in Minnesota is directly related to the stability of the peatland environment. The two rare species of sundews, *Drosera anglica* and *D. linearis*, are restricted to one of the most stable vegetation landforms: flarks containing deep pools. These pools seem to be the end products of peatland development in the water tracks. The other rare species are found at sites that have experienced a recent change in water chemistry or water level.

The rare species of the spring-fen channels, particularly *Cladium mariscoides* and *Eleocharis rostellata*, are consistently associated with a habitat that may be of fairly recent origin. A peat core from the Lost River peatland indicates that the spring-fen channel there originated about 1,200 years ago, when groundwater began to discharge at the site (Glaser *et al.* 1990). At that time, a raised bog was converted into an extremely rich fen when calcareous groundwater was discharged at the peat surface. A core from the Nett Lake spring fen had a very similar stratigraphic profile but was not dated by ^{14}C (Glaser 1983b). Thus the occurrence of rare plants in the spring-fen channels of northern Minnesota is probably linked to a dramatic change in the peatland vegetation produced by the local discharge of calcareous groundwater at the peat surface.

The largest populations of *Carex exilis* and *Rhynchospora fusca* are found in two portions of the Red Lake peatland that have been subject to recent developmental changes. One of the most important sites for these rare species is the internal water tracks in the west-central and eastern watersheds at Red Lake. Stratigraphic cores from these sites indicate that these water tracks originated approximately 1,000 years ago (Janssens *et al.*, chapter 13), about the time that the spring-fen channels arose at Lost River. *R. fusca* is most prevalent near the heads of these water tracks, whereas *C. exilis* is more important farther downstream where the water level has been altered by drainage ditches. *C. exilis* is particularly abundant in areas where the water levels have recently been raised by the discharge of water from a drainage ditch that was dammed by beavers (Glaser, chapter 3). At these sites the margins of the ovoid islands have a scalloped pattern indicating recent flooding. The rise in the water table along these margins may have shifted the competitive balance in favor of *C. exilis*. In eastern Canada, *C. exilis* can colonize raised bogs, but it only grows in the wetter mud bottoms, where the cover of *Rhynchospora alba* and *Scirpus cespitosus* is lower. This marginal area is also a habitat for *Xyris montana*, whereas the other populations of *Xyris* and *Juncus stygius* at Red Lake occur at the heads of the internal water tracks.

The other important habitat for *Carex exilis* and *Rhynchospora fusca* is the northern edge of the western water track. In this area the characteristic string-flark patterns and tree islands appear to be actively forming (Glaser *et al.* 1981; Glaser, chapter 3). The formation of these patterns may be linked to a rise in the water table creating a temporary unstable environment. Aerial photographs demonstrate that this marginal zone of the western water track is drier than the areas immediately downslope where the string-flark patterns first appear. The alkalinity is also higher along the edge, which probably favors the establishment of *Cladium mariscoides* clones. The scattered clones of *C. mariscoides* in the center of the water track are probably remnants of more extensive populations that arose when the water chemistry was more minerotrophic.

The occurrence of *Carex exilis* elsewhere in Minnesota can be explained by its distribution in the Red Lake peatland. At Sand Lake, the water table of a small patterned fen has been artificially raised by a railroad embankment, producing conditions similar to those found in the internal water tracks at Red Lake. At Lost Lake, the oriented pools are unusual features in the peatland and may indicate a recent local rise in the water table. At Alborn, the water table was altered by road building in the early 1930s and the abandonment of this roadbed in the 1950s. The water table apparently rose after beaver dammed the drainage ditches that were no longer maintained. This effect is indicated

by the width of tree rings in the swamp forest that surrounds the fen. These abrupt changes in the hydrology may have favored the establishment of rare plants at this fen. The other stations for *C. exilis* in Minnesota are spring-fen channels in the Lost River peatland and similar features in the Myrtle Lake peatland.

The very large populations of rare plants in the Red Lake peatland may therefore be explained by dynamic developmental processes that are still operating to produce the distinctive peat landforms. Elsewhere, these processes may be simulated by localized disturbances, but the altered habitats are too limited in space and time to sustain all the potential rare species. The Red Lake peatland also is sufficiently large to contain the deep flark pools required by the rare sundews. Although each of the watersheds at Red Lake has been altered by drainage ditches, the effect of the ditches has been minimized by the great size of the peatland.

Historical Factors

The management of rare species must take into account their history and present status within a given region. A rare species may be a recent immigrant that is actively expanding its range within the area of interest. Such a species will inspire less concern among conservationists than the remnant populations of a once more-extensive species that is on the verge of local extinction. The current status of most of the world's rare species is difficult to determine because they have left little record except for the field notes of botanists. This situation is uniquely different in peatlands, which contain an *in situ* record of past vegetational changes. Many of the rare species, moreover, have diagnostic fruiting structures that should be preserved within the peat. A direct record of their past occurrence should be preserved in the peat, particularly where these species are dominants. Indirectly, the former vegetation assemblages may be reconstructed from the fossil assemblages and used to determine the past availability of suitable habitat.

The record from peat cores indicates that most rare plant habitats in Minnesota peatlands are less than 5,000 years old. Two peatlands in northeastern Minnesota, Sand Lake and Alborn fen, however, are nearly 8,000 years old. They both have rare plant populations and may have provided the initial pool for the spread of *Carex exilis, Rhynchospora fusca, Xyris montana,* and *Juncus stygius* in Minnesota. Stratigraphic work has not been conducted at Alborn to determine when these species first became established there.

The two rare sundews probably migrated into Minnesota after the mid-Holocene when water tracks with deep fen pools first developed in Minnesota. The spring-fen channels may be even younger in northern Minnesota. The only radiocarbon date obtained from these channels indicates that the habitat originated about 1,200 years ago (Glaser *et al.* 1990). Some of the spring-fen species were common in the peatlands prior to that time, however. Achenes of *Cladium mariscoides*, for example, are common at the base of cores in the Red Lake and Lost River peatlands. Apparently these peatlands initially were extremely rich fens before the peat accumulated to sufficient thickness to seal off the source of alkalinity from the underlying calcareous parent material. Stratigraphic analysis unfortunately remains an underused tool for determining the present status of rare plant populations.

Recommendations for Conservation

Preserving rare plants is a serious challenge for land managers. Rare plants can be protected from bulldozers, but natural processes may still cause their natural habitats to disappear. The exact requirements of a rare population must be known to manage it successfully, but seldom is such information available. In the absence of this knowledge, preservation of an entire ecosystem offers the hope that it contains the complete suite of required habitats. The Red Lake peatland in northwestern Minnesota seems to come closest to fulfilling these needs. The pattern-forming processes seem to be active in this peatland, producing the range of habitats required by a significant assemblage of rare species.

A few of these rare species seem to require a stable peat landform for their existence. *Drosera linearis* and *D. anglica* are stress tolerators that grow only in the deep water of the central flarks in pristine fens. These plants are vulnerable to drainage ditches that lower the water table. They are also usually absent from water tracks that have experienced any substantial disturbance from ditches, roadways, or powerline rights-of-way. Another assemblage of species including *Eleocharis rostellata* and *Cladium mariscoides* is largely restricted to spring-fen channels where the high alkalinity of the surface waters may create physiological stress. Although these habitats may have only recently originated in the northern peatlands, they appear to be stable at present.

The remaining species are characteristic of landforms that may be subject to a fluctuating water table. The species of these habitats, particularly *Xyris montana*,

Juncus stygius, Rhynchospora fusca, and *Carex exilis*, occur more sporadically and may be fugitive species that exploit local breaks in the prevailing vegetation. The exact requirements of these species is more difficult to characterize because of the large number of empty sites in northern Minnesota that otherwise seem suitable. A paleoecological study of their past record of establishment at these sites where they are now abundant may provide some revealing suggestions for preserving them successfully in Minnesota.

Literature Cited

Coffin, B., and L. Pfannmuller. 1988. Minnesota's endangered flora and fauna. University of Minnesota Press, Minneapolis, Minnesota, USA.

Glaser, P. H. 1983a. *Carex exilis* and *Scirpus cespitosus* var. *callosus* in patterned fens in northern Minnesota. Michigan Botanist 22:22-26.

————. 1983b. *Eleocharis rostellata* and its relation to spring fens in Minnesota. Michigan Botanist 22:19-21.

————, G. A. Wheeler, E. Gorham, and H. E. Wright, Jr. 1981. The patterned mires of the Red Lake peatland, northern Minnesota: vegetation, water chemistry, and landforms. Journal of Ecology 69:575-599.

————, J. A. Janssens, and D. I Siegel. 1990. The response of vegetation to chemical and hydrological gradients in the Lost River peatland, northern Minnesota. Journal of Ecology 78:1021-1048..

Janssens, J. A., B. C. S. Hansen, P. H. Glaser, and C. Whitlock. 1992. Development of a raised-bog complex in northern Minnesota. Pages 189-221 *in* H. E. Wright, Jr., B. Coffin, and N. Aaseng, editors. The patterned peatlands of Minnesota. University of Minnesota Press, Minneapolis, Minnesota, USA.

Kral, R. 1966. *Xyris* (Xyridaceae) of the continental United States and Canada. Sida 2:177-260.

Lakela, O. 1965. A flora of northeastern Minnesota. University of Minnesota Press, Minneapolis, Minnesota, USA.

MacArthur, R. H., and E. O. Wilson. 1967. The theory of island biogeography. Princeton University Press, Princeton, New Jersey, USA.

Siegel, D. I., and P. H. Glaser. 1987. Groundwater flow in a bog/fen complex, Lost River peatland, northern Minnesota. Journal of Ecology 75:743-754.

Wheeler, G. A., and P. H. Glaser. 1979. Notable vascular plants of the Red Lake peatland, northern Minnesota. Michigan Botanist 18:137-142.

PART II
Fauna

CHAPTER 6

Large Mammals

William E. Berg

Several comprehensive reviews discuss the range, biology, and ecology of mammals. Those encompassing the world (Grzimek 1972), North America (Hall 1981; Chapman and Feldhamer 1982), or geographic areas such as the north-central states (Jones and Birney 1988) are, of necessity, general. Reviews on mammals in Minnesota (Swanson *et al.* 1945; Gunderson and Beer 1953; Hazard 1982) and Wisconsin (Jackson 1961) provide more specific information, such as historical and county records, but they are nonspecific regarding habitat.

Several mammal species have been studied in Minnesota in their conventional range and habitats, but few studies have addressed peatland use. Marshall and Miguelle (1978) summarized research in Minnesota that referred to potential use of peatlands by 9 "large" and 13 "small" mammals. Of those animals, only white-tailed deer (*Odocoileus virginianus*), snowshoe hare (*Lepus americanus*), and various small mammals (Nordquist and Birney 1980) had been studied in Minnesota peatlands. The primary reasons for scant peatlands research are difficulty of access, relatively low species diversity and abundance, and lack of documentation of a need for research. The Minnesota Peat Program Final Report summarized peatland use by several animal taxa, but this summary was cursory (Minnesota Department of Natural Resources [MDNR] 1981). As a consequence, much information on mammal use of peatlands must be inferred from historical records and existing literature, and often from persons knowledgeable about wildlife and peatland relationships.

The northern forests and peatlands of Minnesota are on the edge of the Coniferous Forest Biome and its ecotone with the Deciduous Forest (Pitelka 1941). Along this edge lies the northernmost distribution for some large mammals and the southern boundary for more northern species. Although the distributions of most species are related to habitat, food, or climate, some have been modified since the immigration of Europeans; examples are the wolf (*Canis lupus*), woodland caribou (*Rangifer tarandus*), and raccoon (*Procyon lotor*). Northern Minnesota represents the southern range limits for moose (*Alces alces*), fisher (*Martes pennanti*), pine marten (*Martes americana*), wolf, snowshoe hare, Canada lynx (*Felis lynx*), and, since the early 1900s, woodland caribou and wolverine (*Gulo gulo*). Conversely, northern Minnesota represents the northern range limit for species such as gray fox (*Urocyon cinereoargenteus*), raccoon, badger (*Taxidea taxus*), cotton-tailed rabbit (*Sylvilagus* spp.), and bobcat (*Felis rufus*). The wildlife diversity of northern Minnesota is thus enriched in this transitional area.

Herbivorous mammals and birds, whether terrestrial or aquatic, have evolved to use certain habitats that provide the basic life necessities. Carnivorous mammals and birds generally occupy habitats not because they prefer those habitats, but because their prey occurs there. For example, a waterway without prey such as crayfish (*Oronectes* spp.) or minnows would be devoid of river otters (*Lutra canadensis*), and likewise a forest without suitable prey would hold no coyotes (*Canis latrans*). In some cases, a variety of habitats is used by several prey species; these habitats are in turn used by a variety of predators. In other cases, a predator may specialize on one prey species. For example, because the Canada lynx specializes on snowshoe hare, the best lynx range is in good hare habitat. One exception to this prey/predator habitat concept is that some large, elusive mammals such as wolf and cougar (*Felis concolor*) may frequent remote areas such as large peatlands for protection from human persecution, even though higher prey densities may occur elsewhere.

Fig. 6.1. Woodland caribou, a former resident of Minnesota's "big bog." The young caribou shown in this photo from the late 1930s were part of the effort to reintroduce caribou into Minnesota's Red Lake peatland. (Copyright 1990 Minnesota Department of Natural Resources.)

Many habitats such as lowland conifer, lowland grass and brush, sedge meadows, and wetlands are normally associated with peatlands. These habitats are generally lower in plant species diversity (Glaser and Duffy 1987) than upland habitats, and consequently they are also lower in wildlife diversity (Probst *et al.* 1983; U.S. Forest Service 1986) and likely population density. No single large-mammal species depends solely on peatland habitats. Small peatlands interspersed with a mosaic of other habitats, however, can support a diverse mammal fauna. As peatlands increase in size, and as distance to more favorable habitats increases, peatland use by most large-mammal species would be expected to decline (Table 6.1).

Cervids

Two of the four native members of the deer family (*Cervidae*) in Minnesota are resident: the white-tailed deer and the moose. The elk (*Cervus elaphus*) and woodland caribou are both eliminated from Minnesota (Hazard 1982). The elk subspecies introduced from Wyoming remains in limited numbers in Minnesota, however, and together with caribou is classed as "of special concern" by the Minnesota DNR (Nordquist 1988). Species accounts that follow are in approximate order of peatland dependence.

Woodland Caribou

Historical accounts indicate that woodland caribou (Fig. 6.1) were once widely distributed in northern Minnesota and ranged at least as far south as Kanabec County, but they were rare by the early 1900s (Fashingbauer 1965a). As late as 1934, occasional caribou were

sighted in remote areas of northern Minnesota, and a small band of caribou remained in the Red Lake peatlands. This band of caribou was at least somewhat migratory and used a trail network that still remains to migrate from island calving grounds in Ontario and Manitoba to winter range north of Red Lake (Bergerud and Mercer 1989; Fig. 6.2). Agricultural settlement along the Rainy River interrupted the caribou migration, thus isolating this last band. The remaining caribou suffered from poaching and continued degradation of their habitat from logging, drainage ditch construction, and wildfire (Manweiler 1939a, 1941). Increased exposure to white-tailed deer may have hastened infection of caribou with the parasitic brainworm *Parelaphostrongylus tenuis*, which is fatal to caribou and moose but not to deer (Karns 1977).

In 1932 the Red Lake Game Preserve (later the Red Lake Wildlife Management Area) was established on a large portion of the Red Lake peatlands. One of its main goals was preservation of the remaining caribou. The Federal Resettlement Administration's Beltrami Island and Pine Island resettlement projects sub-

Fig. 6.2. A mosaic of deeply trodden caribou trails persists in the patterned peatlands of northwestern Minnesota.

Table 6.1. Association indices for medium-sized and large mammals
pertaining to five peatland categories in Minnesota.

	Large unditched bogs, fens	Large ditched bogs, fens	Large forested bogs, fens, with at least some surface water	Small peatlands, usually adjacent to aquatic habitats	Small peatlands, usually adjacent to upland habitats
Cervids					
Deer	0	0	2	1	3
Caribou	2	2	2	2	3
Elk	0	0	2	0	3
Moose	0	0	2	2	3
Canids					
Coyote	1	1	2	1	3
Wolf	1	1	2	1	3
Red fox	1	1	2	1	3
Felid					
Bobcat	1	1	3	1	3
Lynx	0	0	3	0	3
Cougar	0	0	1	0	3
Mustelids					
Fisher	0	0	1	0	2
Marten	0	0	1	0	2
Weasels	1	1	2	1	3
Skunk	0	1	2	0	3
Badger	0	0	0	0	1
Wolverine	0	0	1	0	1
Mink	1	2	3	3	2
River otter	0	2	3	3	2
Bear	1	1	2	1	3
Raccoon	0	1	2	3	3
Beaver	2	3	3	3	2
Muskrat	0	2	2	3	2
Porcupine	0	0	2	0	3

Note: 0 = not found in type; 1 = rare or occasional; 2 = frequent; 3 = characteristic.

sequently moved approximately 500 families of impoverished homesteaders from the region (Manweiler
1938). With the departure of the settlers, hundreds
of dams were built to block drainage ditches, and
beaver (*Castor canadensis*) were introduced to further the work. Wolf control was instituted to save caribou. Intensive warden patrols were conducted to deter
poachers, and wallows were blasted to provide refuge
from insects for the caribou (Swanson 1936; Manweiler
1939a; Swanson *et al.* 1945).

Still the caribou decreased, and in 1937 Minnesota's
native caribou population consisted of three cows.
Funds were obtained from the U.S. Farm Security Administration to capture caribou in Canada and trans-
plant them in the "big bog." In 1938, ten caribou (two
adult bulls, four male calves, and four female calves)
were snared north of Montreal Lake, Saskatchewan,
and moved to a corral 26 km (16 miles) north of Wask-

ish, Minnesota. One adult bull was then transferred to
a 416 km² (160-square-mile) enclosure south of Hillman Lake before being released to join the three native
cows ranging nearby (Manweiler 1939b).

The remaining captives were moved to the large
enclosure in 1939 (Fig. 6.3). There the population increased to 15-20 animals by 1942, when they were
released due to high water and a scarcity of forage
(Swanson *et al.* 1945). It was difficult to count these
animals even while they were in the enclosure, and
their fate after release is unknown. None was seen
during aerial searches in 1946 (Nelson 1947) or during big-game aerial surveys in 1947-48 (Erickson 1947,
1948).

In recent years, at least one caribou, likely a lone
disperser from Ontario, has been seen in northeastern Minnesota (Peterson 1981). It is possible that an
attempt will be made to reintroduce caribou from On-

Fig. 6.3. More than 50 years old, about six miles of fence line still identifies the caribou enclosure near Hillman Lake in the "big bog." (Photo by Dan Ruda, Minnesota Department of Natural Resources.)

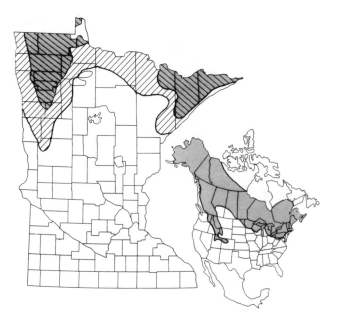

Fig. 6.4. Most moose in Minnesota occur in the disjunct northeastern and northwestern ranges (*dark shading*) that hold about 6,500 and 2,500 animals respectively. Moose occur in very low densities throughout the secondary range (*light shading*). Adapted from Fuller (1986).

tario to a remote area of northeastern Minnesota with large expanses of boreal peatlands containing few deer (and thus a low rate of *P. tenuis* larvae to infect caribou) and wolves. An area of least 6,500 km² (2,510 square miles) sufficiently removed from deer (Bergerud and Mercer 1989), and from a wolf density exceeding 10 per 1,000 km² (4 per 100 square miles), is recommended (Bergerud 1985).

Moose

Although moose occur in sparsely populated and widely scattered pockets over much of northern Minnesota (Fuller 1986), most are found in two ranges in northeastern and northwestern Minnesota that respectively hold about 6,500 and 2,500 moose (MDNR unpublished data) (Fig. 6.4). In both ranges, moose are more aquatic and make more use of peat lowlands than do deer inhabiting the same region.

Moose are indigenous to northeastern Minnesota, as evidenced by ancient Indian pictographs (Idstrom 1965). In the northeast, large peatlands are uncommon. Moose in summer utilize aquatic plants such as pondweed (*Potamogeton* spp.) and lily pad (*Nuphar* and *Numphea* spp.) that grow in flowages and backwaters in small wetlands (P. Jordan, personal communication). Use of aquatic plants is highest in early summer and early autumn (Peek *et al.* 1976). During other seasons, moose in the northeast utilize lowland edges to some extent but prefer upland deciduous-coniferous forest and cutovers (Peek *et al.* 1976).

Prior to the 1930s the primary moose range in northwestern Minnesota was the forested area near the Red Lake peatlands (Idstrom 1965), most of which is in the Red Lake Wildlife Management Area and the Beltrami Island State Forest. The remainder of the northwest range was gradually occupied by moose beginning in the 1930s, when much marginal agricultural land was abandoned by homesteaders and subsequently revegetated by deciduous brushland and forest (Berg and Phillips 1974). The habitat in this area was further enhanced by peat burnouts and by beaver dams in ditches, which created wallows for moose. Although the range of the northwest moose population has continued to expand during the 1980s as far south as Ottertail County (Fig. 6.4), the former range near the Red Lake peatlands has declined drastically in carrying capacity because expanses of shrubland dominated by willow and bog birch (*Betula pumila*) have become decadent or have succeeded to unmerchantable forest consisting of poor-quality aspen (*Populus tremuloides*) or stunted black spruce (*Picea mariana*). This habitat now supports a very low moose population of less than one per 10 km² (2.6 per 10 square miles), incapable of supporting even limited sport hunting.

White-tailed Deer

The white-tailed deer is a relatively recent resident of northern Minnesota, and prior to 1860 it was extremely rare in the northern half of the state (Swanson *et al.* 1945). The logging, settlement, and fires that began in the late 1800s created the younger forest preferred by deer. By the late 1980s deer numbered half a million in the forest zone (MDNR, unpublished data).

Despite several studies on deer in northern Minnesota, only one (Pietz and Tester 1979) has addressed peatlands use. According to this and other deer studies in north-central and northeastern Minnesota (Rongstad and Tester 1969; Wetzel *et al.* 1975; Mooty *et al.* 1987) and in northwestern Minnesota (Hunt and Magnus 1954; Carlsen and Farmes 1957), deer generally preferred mixed upland and upland/lowland edge habitats over lowlands. The one exception is that, during winter, lowlands dominated by white cedar (*Thuja occidentalis*) are preferred because they provide more favorable microclimates and easier travel (Ozoga 1968; Peek 1971). White cedar stands, especially those capable of sustaining deer in winter, are unquestionably the most important peatland habitat for deer. Large uninterrupted lowlands, whether open bogs or forested fens dominated by black spruce or tamarack (*Larix laricina*), are generally avoided by deer.

Smaller peatlands characterized by a mosaic of grass, brush dominated by willow, dogwood (*Cornus* spp.), and bog birch and forest dominated by aspen and white birch (*B. papyrifera*) provide excellent deer habitat, however. In northwestern Minnesota the mosaic of farmland and peatlands containing aspen and brushland habitat is collectively termed the aspen parkland. This area was colonized by deer during the 1920s (Hunt and Magnus 1954; Carlsen and Farmes 1957). In 1990 these parkland habitats supported deer densities of 5-9 per km² (12-23 per square mile), which are among the highest in Minnesota (MDNR, unpublished data).

Elk

It is ironic that the elk, an animal usually associated with western mountains or prairies, should make its last stand in Minnesota in the agricultural peatlands. Before settlement, elk occurred over most of Minnesota, but by 1850 it had been nearly eliminated, and in 1894 elk occurred only near Thief River Falls (Fashingbauer 1965b). Even these animals were soon gone; the last native elk killed in Minnesota was taken in 1896 (Swanson 1940).

In 1914-1915, 42 elk from Yellowstone National Park and 14 animals donated by James J. Hill were introduced to a 285 ha (700-acre) enclosure at Itasca State Park. This captive herd dwindled to 13 within a year, and after scattered stocking attempts around the state

numbered about 35 in 1935. That year Gustav Swanson, biologist for the Minnesota Department of Conservation, determined that northern Beltrami County was the best release site. Excellent habitat prevailed in the abandoned farming areas from which the Federal Resettlement Administration had moved most of the settlers (Swanson 1940). This area, west of the Red Lake peatlands, was generally abandoned upland and lowland agricultural land generously interspersed with nonagricultural peatlands. The elk flourished in their new home. "Those who live and work on the elk range" estimated the population at nearly 300 animals in 1945 (Dorer 1945). The elk gradually moved to a more heavily farmed area west of the release site, however, and began to damage crops. By 1947 poaching by irate farmers was suspected to be limiting the elk herd (Erickson 1947). Forty years later, poaching to prevent crop damage continued to reduce the herd, which in 1990 approximated 20. The conflict between elk and modern agriculture is unlikely to disappear, and Minnesota will maintain no more than a relict herd of elk in the foreseeable future.

Canids

Four wild members of the *Canidae* (dog) family occur in Minnesota: wolf, coyote, red fox (*Vulpes vulpes*), and gray fox (Hazard 1982). Only the wolf and coyote utilize peatlands to any extent. The range of the gray fox is essentially the upland parkland and hardwood habitats of central and southeastern Minnesota; thus its use of peatlands is minimal. The red fox is distributed statewide, with the highest densities in the agricultural areas of western and southern Minnesota. No studies document the habits of red fox in peatlands, and in large peatlands fox densities are probably low. Red fox hunt in any habitat harboring suitable prey, and in northern Minnesota they feed primarily on mice and voles (*Microtus pennsylvanicus, Peromyscus* spp. *Clethrionomys gapperi*, and *Synaptomys cooperi*) and deer carrion (Kuehn and Berg 1981).

Wolf

The presence of wolves in Minnesota (Fig. 6.5) is governed by an abundant ungulate food supply, large areas (at least 110 km² — 42 square miles) with minimal human disturbance (Fuller 1989), and road densities generally not exceeding 1 km of road per km² (1 mile of road per square mile) (Thiel 1985; Jensen *et al.* 1986; Mech *et al.* 1988).

The most important peatland habitat for wolves is lowland conifer, especially white cedar, because this type provides important wintering habitat for deer, the primary prey. Although large peatland complexes such as Red Lake and Pine Island generally satisfy the wolves' need for semiroadless areas, wolf use is minimal because deer are scarce. Fritts and Mech (1981) found that wolf packs frequented the area west, north, and

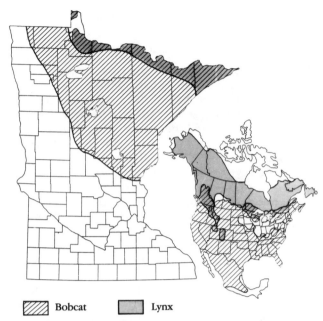

Bobcat   Lynx

Fig. 6.6. The bobcat is at its northern limit of range in northern Minnesota, while the southern portion of the Canada lynx range borders extreme northern Minnesota.

Fig. 6.5. The gray or "timber" wolf inhabits much of northern Minnesota, where prey is abundant and people and roads are few. (Wolf illustration by Nan Marie Kane, copyright Minnesota Department of Natural Resources.)

east of the Red Lake peatlands and that wolf pack territories were generally larger as they extended into the "big bog." Wolf packs generally traveled through the large muskeg-dominated peatland without hunting.

During summer, wolves frequently hunt for beaver around wetlands, many of them in small peatlands (Stenlund 1955a). During winter, wolves travel and mark scent on frozen streams and ditches bounded by peatlands, but they avoid large uninterrupted sedge meadows (Berg and Kuehn 1980).

Coyote

The coyote is Minnesota's most abundant and widely distributed large carnivore, numbering 20,000-30,000 in the northern half of the state. It occurs wherever suitable prey — such as white-tailed deer, snowshoe hare, and small mammals — is present, but it avoids areas with established wolf packs (Berg and Kuehn 1982). Coyotes were extensively studied during the 1970s (Berg and Chesness 1978) in north-central Minnesota, where small peatlands are abundant and very large peatlands are uncommon. Approximately one-third of

more than 3,000 telemetry locations were in habitats associated with wooded peatlands such as white cedar, black spruce, and tamarack. Coyotes avoided the large peatlands such as open bog and large sedge meadows, probably because of their lower prey base. Thus, despite the coyote's ability to utilize a wide range of prey species and habitats (Voigt and Berg 1987), large peatlands are probably little used.

Felids

The bobcat, Canada lynx, and cougar are all indigenous to Minnesota (Swanson *et al.* 1945). Although all occur in Minnesota, each has experienced some sort of human persecution, and only the bobcat has persisted as a viable population.

Bobcat and Lynx

The bobcat and lynx are very widely distributed in North America. The bobcat is at its northern range limit in northern Minnesota, while the lynx approaches its southern limit (Fig. 6.6). Predators with such large ranges are not likely to have rigid habitat requirements, and therefore they reside opportunistically where prey is abundant.

In almost all studies of bobcats and lynx, rabbits and hares are the most important prey types. Two studies of bobcat food habits have been conducted in Minnesota. During winter 1938-39 Rollings (1945) examined 50 bobcat stomachs, finding snowshoe hare in 52%, deer in 44%, and porcupine (*Erethizon dorsatum*) in

10%. A similar study (MDNR, unpublished data) over several winters found that in 1981, when hares were abundant, they constituted 72% by weight of the winter diet of adult female bobcats. In 1985, after the hare population crashed, this declined to 7%. Hares were less important in the diet of the larger adult male bobcats but showed a similar trend, declining from 41% to 0%. The diet of male bobcats included an average of 37% deer, compared to only 21% for females.

Population cycles of lynx and hare are closely related, as documented in Canada (Brand *et al*. 1976) and Minnesota (Henderson 1978). When hares are abundant, they may constitute almost the entire diet for lynx, but when hares are scarce, lynx pursue a wide variety of alternate prey (Saunders 1963). Most alternate prey are small, but Bergerud (1971) found that lynx killed many caribou calves in Newfoundland.

Berg (1979) and Fuller *et al*. (1985) found that bobcats used white cedar stands extensively, particularly during winter. This type is so important that the bobcat population decline in the western Great Lakes states from 1946 to 1960 is attributed to the indiscriminate logging of white cedar (Berg 1989). Black spruce and willow stands growing on peatlands were used more in summer. In addition to harboring high prey densities, dense lowland conifer stands provide an improved microclimate and shallower snow depths, which are important to the bobcat because deep, powdery snow in open habitats restricts bobcat movements (Marston 1942; McCord 1974) and reduces their ability to hunt. Petraborg and Gunvalson (1962) reported that these snow conditions resulted in bobcats starving in Minnesota in 1950, 1953, and 1954. During winter 1976-77 a similar occurrence was noted (MDNR, unpublished data). Because lynx have relatively large feet, they are much less impeded by deep snows (Nellis and Keith 1968).

Cougar

The cougar shares with the wolf the distinction of being among the most elusive large carnivores in Minnesota. Unlike the wolf, cougars were never common in Minnesota, but they may have occurred statewide. The last documented kill was in Becker County in 1897 (Swanson *et al*. 1945). One cougar was killed in Manitoba in 1973, 160 km (100 miles) north of Minnesota (Nero and Wrigley 1976). An increasing number of observations in the western Great Lakes states and interior Canada in the 1970s and 1980s suggests the presence of at least transient cougars in Minnesota (Berg 1984; Gerson 1989).

Cougars subsist mainly on deer and consequently are more likely to reside where deer occur. Deer are scarce in large open peatlands; cougars probably use these areas because they are relatively free from human disturbance. Most reported observations of cougars in Minnesota cluster in several relatively remote areas with and without peatlands but containing moderate

to high deer densities (Berg 1984; MDNR, unpublished data).

Mustelids

All northern members of the weasel family (*Mustelidae*) occur in northern Minnesota, with the exception of the wolverine (Hazard 1982). Wolverines were present in the early 1900s in northern Minnesota but were eliminated by unregulated harvest and extensive land clearing for timber and farming.

Otter and Mink

Mink (*Mustela vison*) and otter are the only large aquatic carnivores in Minnesota. Whereas otter feed mainly on fish and crayfish, mink prefer small terrestrial and aquatic prey and the eggs and young of ground-nesting birds (Sargeant *et al*. 1973). Both mink and otter are common in small peatlands, especially near beaver ponds. They are likely to be common in the very large peatlands wherever sufficient prey occurs and where ditching and dam construction in the "big bog" during the early 1900s improved the aquatic habitat (Manweiler 1939a).

Fisher and Pine Marten

Fisher and pine marten nearly shared the wolverine's fate during the early 1900s. Both were gone from most of northern Minnesota by the early 1920s. In 1928 the trapping seasons were closed for both species, and they remained so until 1977 for fisher and 1985 for marten.

In the years following the fisher and marten population depletion, the maturation of the logged-over forests and complete protection from trapping enabled both species to increase slowly. Both were still so rare in the 1950s that a verified observation was often published (Swanson *et al*. 1945; Stenlund 1955b; Balser and Longley 1966). Since the late 1960s, when fisher were generally restricted to northeastern Minnesota (Balser and Longley 1966), the entire forested area of northern Minnesota has been repopulated. Pine marten, being smaller and normally associated with more mature forests than fisher (deVos 1951), were slower to expand their range (Gunderson 1965). In the mid-1980s the pine marten range was still largely restricted to Lake, Cook, and northern St. Louis and Koochiching counties, and the large open peatlands were viewed as possible land barriers to further westward range expansion. By 1989, however, marten had skirted around the north and south sides of the Red Lake and Pine Island peatlands and inhabited the Beltrami Island State Forest south of Lake of the Woods (MDNR, unpublished data).

The main prey of the fisher is snowshoe hare, mice, and voles; mice and voles also constitute the main prey of marten. This prey is common in the smaller forested peatlands, especially in lowland conifer during winter.

Fisher and marten are uncommon in the large open peatlands.

Other Mustelids

Little is known about other mustelid species and their use of peatlands. Least, short-tailed, and long-tailed weasels (*Mustela nivalis, M. erminea, M. frenata*) prey almost exclusively on mice and voles, which are common in many small peatlands. Badgers occur statewide but prefer lighter and well-drained soils for burrowing and hunting fossorial prey. They generally avoid the heavy peat soils, but badger diggings can be found in old ditch spoils in the larger peatlands. No information on use of peatlands by striped skunks (*Mephitis mephitis*) has been gathered, but they probably occur in varied densities in all large and small peatland areas where a sufficient food base exists.

Ursid/Procyonid

Two species of large mammals in Minnesota can be classed as omnivorous — that is, their diet is a mix of vegetable and animal matter. The black bear (*Ursus americana*), sole member of the Ursidae in Minnesota, is distributed throughout the northern forest, while the raccoon is distributed statewide.

Black Bear

The historic range of the black bear in the forested regions of the 49 continental United States, Canada, and Mexico is indicative of its wide habitat tolerance (Chapman and Feldhamer 1982). Like any animal, however, the black bear is constrained by its food requirements (Rogers 1987). In Minnesota, bears forage extensively on lush green vegetation in spring and on ants from mid-June to mid-July. During early summer, bears consume white-tailed deer fawns (MDNR, unpublished data) and probably moose calves (Ballard *et al.* 1981); it is unknown how many are eaten as carrion. Fruits, berries, and hard mast such as acorns are the dominant foods as soon as they become available in late July.

One aspect of a black bear ecology study in Minnesota examined the abundance and productivity of important bear food species in 11 habitat types (Noyce and Coy 1990). The black spruce, tamarack, and white cedar types commonly associated with peatlands were among the poorest producers of bear foods. Among upland forest types only hardwoods (excluding oaks [*Quercus* spp.]) produced as little food as spruce and cedar stands, although these lowland types produced fruiting species such as swamp buckthorn (*Rhamnus alnifolia*), currant (*Ribes* spp.), and red-osier dogwood (*Cornus stolonifera*) that are not typically found in the uplands. Of the major foods, blueberries (*Vaccinium* spp.) were the most important in black spruce stands, whereas no single

species preferred by bears was available in quantity in cedar.

Low weights of bears primarily occupying lowland forest types reflect the paucity of foods in these areas. During February and March, yearling male and female "bog bears" weighed an average of 14 kg (31 pounds) and 13 kg (28 pounds) respectively, compared to 27 kg (60 pounds) and 21 kg (47 pounds) respectively for male and female yearlings in upland habitats (MDNR, unpublished data).

Although peatlands do not provide abundant food for bears, many bears use them for denning. Examination of more than 200 dens of bears radio-collared north of Grand Rapids, Minnesota, found 17% of females and 40% of males denning in cedar, tamarack, or black spruce stands. This sexual difference in denning habitat may reflect the types of dens used. Fifty-six percent of all females denned in excavated burrows, whereas 56% of males less than six years old denned under tree root-masses, and 32% of males over six years old wintered in open nests (MDNR, unpublished data). The reason for the adult male's apparent preference for nest dens in spruce or cedar bogs is unclear. These males often traveled northward 50-240 km (31-150 miles) from their summer range to den in large uninterrupted forested peatlands and then quickly returned to their original summer ranges in spring. Although these bears did not reuse dens in successive winters, each bear appeared to have chosen a specific area within the bog for denning.

Raccoon

Raccoons are more abundant in southern and central Minnesota than in the north and northeast (Timm 1975; Hazard 1982). They are often associated with lowland and marshy habitats near agricultural lands and urban areas (Dorney 1954), where their diet consists of small fish, crayfish, nuts, berries, and agricultural crops (Hazard 1982). Raccoons probably avoid the large peatlands of Beltrami, Koochiching, and Lake of the Woods counties but frequent small peatlands in north-central Minnesota, especially those with a meandering stream or ditch or adjacent to agriculture.

Large Rodents

Whereas most rodent species are "small" mammals (Nordquist 1990), three are classed as "large." The aquatic beaver and muskrat (*Ondatra zibethicus*) are statewide in distribution, while the terrestrial porcupine is limited to the nonagricultural zone of Minnesota (Hazard 1982).

Beaver

Beaver depend on an aquatic ecosystem with abundant food. They prefer stems and bark of aspen trees; brush species such as willow, alder (*Alnus rugosa*),

Fig. 6.7. A series of beaver dams in a peatland ditch. (Photograph by Dan Ruda, Minnesota Department of Natural Resources.)

and bog birch are also suitable (Longley and Moyle 1963). In nonforest areas, beaver utilize cottonwood trees (*P. deltoides*) and agricultural residue such as corn stalks.

Beaver were probably rare in large peatlands until ditches were constructed in the early 1900s in an attempt to drain the land for human settlement. During postsettlement times, beaver naturally invaded the ditch networks and were also introduced (Vesall *et al.* 1947) to raise water levels for fire prevention and wetland habitat. In these elaborately ditched systems, the absence of sufficient woody vegetation for construction of beaver dams and lodges necessitated the construction of dens in ditch spoil banks (Fig. 6.7). These bank dens often caused washouts in roads and trails. Ditch-system beaver are known to this day as "bank beaver."

Beaver are among the few wildlife species in Minnesota for which comparative population data for both peat-dominated and mixed habitats have been gathered. Aerial surveys of beaver lodges have been con-

ducted annually since 1958. Four of the twenty-four 200-800 km (125- to 500-mile) survey routes traverse large peatlands. The mean number of live beaver colonies from 1958 to 1988 in the peatland and the mixed habitats was respectively 0.19 and 0.23 per kilometer (0.30 and 0.37 per mile), indicating that the large peat expanses support fewer beaver (MDNR, unpublished data).

Muskrat

No similar population indices document muskrat population differences between peat-dominated and mixed habitats. Muskrats thrive wherever there is submergent or emergent vegetation for food and shelter (Errington 1941). Peat areas with dense stands of wild rice (*Zizania aquatica*) and cattail (*Typha* spp.) support very high muskrat densities, whereas lower densities occur in the larger ditched peat complexes. Like beaver, muskrats build bank dens in ditch systems where aquatic vegetation is lacking. Muskrats are prolific and subject to sharp seasonal fluctuations (Errington 1951); densities in Minnesota can change severalfold within a five-year span (Balser 1958). Causes include diseases, parasites, predation, freezeouts, and drought. Because water levels are relatively stable in large peatlands, population fluctuations caused by water-level changes are probably slight.

Porcupine

The porcupine's preference for upland habitats, where it feeds on the bark of coniferous and deciduous trees, precludes its presence in the large open peatlands, and there are few distribution records for Lake of the Woods, Koochiching, and the northern half of Beltrami counties (Hazard 1982). Although porcupines eat water lilies (Schoonmaker 1930), they would be expected to inhabit only the smaller peatlands adjacent to upland forests. In general, use of lowlands and especially large peatlands by porcupines is low.

Conclusions

Peatland use by large mammals and the consequences of habitat destruction in peatlands generally must be inferred from distribution records and the few studies done in Minnesota and elsewhere.

No large mammal currently residing in Minnesota is totally dependent on peatland habitats. As peatlands decrease in size, mature successionally, and meld into the northern Minnesota landscape, nearly all species of large mammals utilize peatlands to some extent (Table 6.1). Fisher, marten, badger, and porcupine are the species utilizing peatlands least.

Of the herbivores, only the woodland caribou demonstrate any reliance on the large open unditched or ditched bogs characteristic of the Red Lake and Pine Island peatland complexes. These peatlands might still be capable of sustaining caribou had not (1) the white-tailed deer range expanded, thus increasing the spread of the parasitic brainworm *P. tenuis*, and (2) the caribou migration routes to Ontario and Manitoba been interrupted. The two aquatic herbivores, beaver and muskrat, were probably uncommon until the large peatlands were ditched and occasionally dammed. The presence of these species and other aquatic organisms subsequently provided suitable habitat for aquatic predators such as otter and mink.

Mammal diversity increases in the large peatlands as woody vegetation increases, and the presence of additional prey species supports a diverse predator base: three canids, three felids, seven mustelids, bear, and raccoon. The most important of the forested peatlands is the white cedar community, which supports much of the northern Minnesota deer herd in winter and sustains snowshoe hare populations when they periodically crash in less favorable habitats.

Occasional wildfire in the open and forested peatlands historically maintained woody vegetation in the brushland stage and created peat burnouts used by many aquatic and terrestrial mammals and birds. Prevention of this natural force through wildfire detection and suppression, combined with inadequate prescribed burning, has reduced the ability of these peatlands to support some wildlife species. As a consequence, few moose occupy the once lush but now decadent brushland habitat of the Red Lake peatland complex. Peat burnouts have now become choked with cattail and support few waterfowl and furbearers. The importance of fire in perpetuating brush habitats on peatland cannot be overstated.

In many areas with large open and forested peatlands, particularly in Lake of the Woods, Roseau, Marshall, Aitkin, and southern St. Louis counties, agricultural developers have removed the surface vegetation, bulldozed the top layers of peat into windrows, and burned it in preparation for planting crops. While some of these efforts, including those for wild rice paddy development, have been successful, many have not. Continued efforts to clear and farm this marginal land have destroyed wildlife habitat and caused severe wind and water soil erosion. Much of this marginal farmland was set aside in the 1950s under the Federal Soil Bank Program and again in the late 1980s under the Federal Conservation Reserve Program. Large mammals deriving the most benefits from set-aside programs are deer throughout their range and moose in northwestern Minnesota.

The large peatlands in northwestern, northeastern, and east-central Minnesota comprise many of the relatively roadless areas of the state. Agassiz and Rice Lake National Wildlife Refuges, in addition to several large Minnesota DNR Wildlife Management Areas such as Roseau River, Red Lake, Thief Lake, Moose Willow, Kimberly, and Grayling, all originated as tax-delinquent ditched peatlands that reverted to government ownership. These areas have some of the highest red fox,

skunk, coyote, and otter densities in Minnesota — a tribute to their high wildlife-carrying capacity.

Although prey suitable for large predators does not exist in the very large uninterrupted peatlands such as the Red Lake and Pine Island complexes, these roadless areas provide a refuge for some species that require relatively inaccessible areas. These species include wolf, cougar, and male bears that migrate seasonally in search of remote winter den sites.

The large peatlands provide an important, but perhaps not vital, habitat resource for large mammals in Minnesota. The smaller peatlands, especially those interspersed with uplands, provide a portion of the habitat diversity needed by most large mammals. Whereas industrial and horticultural peat extraction, and in some cases logging, might eventually threaten the larger peatlands, the thousands of smaller peatlands seem secure. Although drainage threatens some peatlands, the primary threat to all peatlands in terms of wildlife habitat value is natural succession. In the absence of wildfire that once controlled succession, woody vegetation in peatlands gradually matures to decadent brushland or generally unmerchantable forest. The wildlife diversity and carrying capacity of these lands will subsequently decrease unless these lands are maintained by natural wildfire, prescribed burning, or mechanical treatment such as shearing, hydroaxe, or hand cutting.

Acknowledgments

I thank D. Kuehn for helping to write major portions of the manuscript for this chapter. P. Jordan provided excellent suggestions for revision, as did H. Wright, B. Coffin, N. Aaseng, and M. Lenarz. J. Summers patiently typed many drafts. D. Garshelis, K. Noyce, and P. Coy provided unpublished data from their black bear study.

Literature Cited

Ballard, W. B., T. H. Spraker, and K. P. Taylor. 1981. Causes of neonatal moose calf mortality in south central Alaska. Journal of Wildlife Management 45:335-342.

Balser, D. S. 1958. Aerial muskrat census. Minnesota Wildlife Research Quarterly 17(4):308-309.

——— and W. H. Longley. 1966. Increase of the fisher in Minnesota. Journal of Mammalogy 47:547-550.

Berg, W. E. 1984. Mountain lions in Minnesota? Minnesota Conservation Volunteer 47(274):2-7.

———. 1989. Bobcat status and management in the western Great Lakes states and central provinces. 1989 Midwest and Southeast Furbearer Workshop, Petosi, Missouri, USA.

——— and R. L. Phillips. 1974. Habitat use by moose in northwestern Minnesota with reference to other heavily willowed areas. Naturaliste Canadien 101:101-116.

———. 1979. Ecology of bobcats in northern Minnesota. Bobcat Research Conference Proceedings National Wildlife Federation Scientific and Technical Series 6:55-61.

——— and R. A. Chesness. 1978. Ecology of coyotes in northern Minnesota. Pages 229-246 *in* M. Bekoff, editor. Coyotes: Biology, behavior, and management. Academic Press, New York, New York, USA.

———— and D. W. Kuehn. 1980. A study of the timber wolf population on the Chippewa National Forest. Minnesota Wildlife Research Quarterly 40:1-16.

———— and D. W. Kuehn. 1982. Ecology of wolves in north-central Minnesota. Pages 4-11 *in* F. H. Harrington and P. Paquet, editors. Wolves: A worldwide perspective on their behavior, ecology, and conservation. Noyes Publications, Park Ridge, New Jersey, USA.

Bergerud, A. T. 1971. The population dynamics of Newfoundland caribou. Wildlife Monographs no. 25.

————. 1985. Antipredator strategies of caribou dispersion along shorelines. Canadian Journal of Zoology 63:1324-1329.

———— and W. E. Mercer. 1989. Caribou introductions in eastern North America. Wildlife Society Bulletin 17:111-120.

Brand, C. J., L. B. Keith, and C. A. Fischer. 1976. Lynx responses to changing snowshoe hare densities in central Alberta. Journal of Wildlife Management 40(3):416-428.

Carlsen, J. C., and R. E. Farmes. 1957. Movement of white-tailed deer tagged in Minnesota. Journal of Wildlife Management 21:397-401.

Chapman, J. A., and G. A. Feldhamer. 1982. Wild mammals of North America. Johns Hopkins University Press, Baltimore, Maryland, USA.

deVos, A. 1951. Recent findings in fisher and marten ecology and management. Transactions of the North American Wildlife Conference 16:498-507.

Dorer, R. J. 1945. Big game division: Elk. Minnesota Wildlife Research Quarterly 5(3):2-5.

Dorney, R. J. 1954. Ecology of marsh raccoons. Journal of Wildlife Management 18:217-225.

Erickson, A. S. 1947. Aerial survey of big game in northwestern Minnesota. Minnesota Wildlife Research Quarterly 7(1):8-13.

————. 1948. Aerial survey of big game, winter 1948. Minnesota Wildlife Research Quarterly 8(2):31-36.

Errington, P. A. 1941. Versatility in feeding and population maintenance of the muskrat. Journal of Wildlife Management 5:68-89.

————. 1951. Concerning fluctuations in populations of the prolific and widely distributed muskrat. American Naturalist 85:273-292.

Fashingbauer, B. A. 1965a. The woodland caribou in Minnesota. Pages 133-166 *in* J. B. Moyle, editor. Big game in Minnesota. Minnesota Department of Conservation Technical Bulletin no. 9.

————. 1965b. The elk in Minnesota. Pages 101-131 *in* J. B. Moyle, editor. Big game in Minnesota. Minnesota Department of Conservation Technical Bulletin no. 9.

Fritts, S. H., and L. D. Mech. 1981. Dynamics, movements, and feeding ecology of a newly protected wolf population in northwestern Minnesota. Wildlife Monographs no. 80.

Fuller, T. K. 1986. Observations of moose, *Alces alces*, in peripheral range in north-central Minnesota. Canadian Field-Naturalist 100(3):359-362.

————. 1989. Population dynamics of wolves in north-central Minnesota. Wildlife Monographs no. 105.

————, W. E. Berg, and D. W. Kuehn. 1985. Bobcat home range size and daytime cover-type use in north-central Minnesota. Journal of Mammalogy 66:568-571.

Gerson, H. B. 1989. Cougar sightings in Ontario. Canadian Field-Naturalist 102(3):419-424.

Glaser, P. H., and W. G. Duffy. 1987. The ecology of patterned boreal peatland of northern Minnesota: A community profile. U.S. Fish and Wildlife Service Biological Report no. 85 (7.14).

Grzimek, B. 1972. Grzimek's animal life encyclopedia. 13 vols. Van Nostrand Reinhold, New York, New York, USA.

Gunderson, H. L. 1965. Marten records for Minnesota. Journal of Mammalogy 46:688.

———— and J. R. Beer. 1953. The mammals of Minnesota. University of Minnesota Press, Minneapolis, Minnesota, USA.

Hall, E. R. 1981. The mammals of North America. Wiley, New York, New York, USA.

Hazard, E. D. 1982. The mammals of Minnesota. University of Minnesota Press, Minneapolis, Minnesota, USA.

Henderson, C. 1978. Minnesota Canada lynx status report, 1977. Minnesota Wildlife Research Quarterly 38(4):221-242.

Hunt, R. N., and L. M. Magnus. 1954. Deer management study: Mud Lake NWR, Holt, Minnesota. Journal of Wildlife Management 18:482-495.

Idstrom, J. M. 1965. The moose in Minnesota. Pages 57-98 *in* J. B. Moyle, editor. Big game in Minnesota. Minnesota Department of Conservation Technical Bulletin no. 9.

Jackson, H. T. 1961. Mammals of Wisconsin. The University of Wisconsin Press, Madison, Wisconsin, USA.

Jensen, W. F., T. K. Fuller, and W. L. Robinson. 1986. Wolf, *Canis lupus*, distribution on the Ontario border near Sault Ste. Marie. Canadian Field-Naturalist 100:363-366.

Jones, J. K., Jr., and E. C. Birney. 1988. Handbook of mammals of the north-central states. University of Minnesota Press, Minneapolis, Minnesota, USA.

Karns, P. D. 1977. Deer-moose relationships with emphasis on *Parelaphostrongylus tenuis*. Minnesota Wildlife Research Quarterly 37:40-61.

Kuehn, D. W., and W. E. Berg. 1981. Notes on movements, population statistics, and foods of the red fox in north-central Minnesota. Minnesota Wildlife Research Quarterly 41(1):1-10.

Longley, W. H., and J. B. Moyle. 1963. The beaver in Minnesota. Minnesota Department of Conservation Technical Bulletin no. 6.

Manweiler, J. 1938. Minnesota's "big bog." Minnesota Conservationist 63:12-13.

————. 1939a. Wildlife management in the "big bog." Minnesota Conservationist 64:14-15.

————. 1939b. Woodland caribou in the "big bog." Minnesota Conservationist 65:16-17.

————. 1941. Minnesota's woodland caribou. Minnesota Conservation Volunteer 1(4):34-40.

Marshall, W. H., and D. G. Miguelle. 1978. Terrestrial wildlife of Minnesota peatlands. Minnesota Department of Natural Resources Peat Program, St. Paul, Minnesota, USA.

Marston, M. A. 1942. Winter relations of bobcats to white-tailed deer in Maine. Journal of Wildlife Management 6:328-337.

McCord, C. M. 1974. Selection of winter habitat by bobcats (*Lynx rufus*) on the Quabbin Reservation, Massachusetts. Journal of Mammalogy 55:428-437.

Mech, L. D., S. H. Fritts, G. L. Radde, and W. J. Paul. 1988. Wolf distribution and density in Minnesota. Wildlife Society Bulletin 16:85-87.

Minnesota Department of Natural Resources. 1981. Minnesota Peat Program final report. Department of Natural Resources Division of Minerals, St. Paul, Minnesota, USA.

Mooty, J. J., P. D. Karns, and T. K. Fuller. 1987. Habitat use and seasonal range size of white-tailed deer in north-central Minnesota. Journal of Wildlife Management 57:644-648.

Nellis, C. H., and L. B. Keith. 1968. Hunting activities and success of lynxes in Alberta. Journal of Wildlife Management 32:718-722.

Nelson, U. C. 1947. Woodland caribou in Minnesota. Journal of Wildlife Management 11:283-284.

Nero, R. W., and R. E. Wrigley. 1976. Status and habits of the cougar in Montana. Canadian Field-Naturalist 91:28-40.

Nordquist, G. E. 1988. Mammals — endangered, threatened, and special concern. Pages 293-322 *in* B. A. Coffin and L. Pfannmuller, editors. Minnesota's endangered flora and fauna. University of Minnesota Press, Minneapolis, Minnesota, USA.

———. 1992. Small mammals. Pages 85-110 *in* H. E. Wright, Jr., B. A. Coffin, and N. E. Aaseng, editors. The patterned peatlands of Minnesota. University of Minnesota Press, Minneapolis, Minnesota, USA.

——— and E. C. Birney. 1980. The importance of peatland habitats to small mammals in Minnesota. Minnesota Department of Natural Resources Peat Program, St. Paul, Minnesota, USA.

Noyce, K. V., and P. L. Coy. 1990. Abundance and productivity of bear food species in different forest types of north-central Minnesota. International Conference on Bear Research and Management 8:169-181.

Ozoga, J. J. 1968. Variations in microclimate in a conifer swamp deer yard in northern Michigan. Journal of Wildlife Management 32:574-385.

Peek, J. M. 1971. Whitetails in winter. Pages 23-26 *in* M. M. Nelson, editor. The white-tailed deer in Minnesota. Minnesota Department of Natural Resources, St. Paul, Minnesota, USA.

———, D. L. Urich, and R. J. Mackie. 1976. Moose habitat selection and relationships to forest management in north central Minnesota. Wildlife Monographs no. 48.

Peterson, W. J. 1981. Coming of the caribou. Minnesota Conservation Volunteer 44(259):17-23.

Petraborg, W. H., and V. E. Gunvalson. 1962. Observations on bobcat mortality and bobcat predation on deer. Journal of Mammalogy 43:430-431.

Pietz, P. J., and J. A. Tester. 1979. Utilization of Minnesota peatland habitats by snowshoe hare, white-tailed deer, spruce grouse, and ruffed grouse. Final technical report to Minnesota Department of Natural Resources, St. Paul, Minnesota, USA.

Pitelka, F. A. 1941. Distribution of birds in relation to major biotic communities. American Midland Naturalist 25:113-137.

Probst, J. R., D. Rakstad, and K. Brosdahl. 1983. Diversity of vertebrates in wildlife water impoundments on the Chippewa National Forest. U.S. Department of Agriculture Research Paper NC 235.

Rogers, L. L. 1987. Effects of food supply and kinship on social behavior, movements, and population growth of black bears in northeastern Minnesota. Wildlife Monographs no. 97.

Rollings, C. T. 1945. Habits, foods, and parasites of the bobcat in Minnesota. Journal of Wildlife Management 9:131-145.

Rongstad, O. J., and J. A. Tester. 1969. Movements and habitat use of white-tailed deer in Minnesota. Journal of Wildlife Management 33:366-377.

Sargeant, A. B., G. A. Swanson, and H. A. Doty. 1973. Selective predation by mink, *Mustela vison*, on waterfowl. American Midland Naturalist 89:208-214.

Saunders, J. K. 1963. Food habits of the lynx in Newfoundland. Journal of Wildlife Management 27:384-390.

Schoonmaker, W. J. 1930. Porcupine eats water lily pads. Journal of Mammalogy 11:84.

Stenlund, M. H. 1955a. A field study of the timber wolf (*Canis lupus*) on the Superior National Forest, Minnesota. Minnesota Department of Conservation Technical Bulletin no. 4.

———. 1955b. A recent record of the marten in Minnesota. Journal of Mammalogy 36:133.

Swanson, G. S. 1936. The Minnesota caribou herd. Proceedings of the North American Wildlife Conference 1:416-419.

———. 1940. The American elk in Minnesota. Minnesota Conservation Volunteer 1(2):5-7.

———, T. Surber, and T. S. Roberts. 1945. The mammals of Minnesota. Minnesota Department of Conservation Technical Bulletin no. 2.

Thiel, R. P. 1985. The relationship between road densities and wolf habitat suitability in Wisconsin. American Midland Naturalist 113:404-407.

Timm, R. M. 1975. Distribution, natural history, and parasites of mammals in Cook County, Minnesota. Bell Museum of Natural History. Occasional Paper 14:1-56.

U.S. Forest Service. 1986. Wildlife habitat associations. Chippewa National Forest SFH-2609.23, Cass Lake, Minnesota, USA.

Vesall, D., R. Gersch, and R. Nyman. 1947. Beaver, timber problem in Minnesota's "big bog." Minnesota Conservation Volunteer 57(10):45-50.

Voigt, D. R., and W. E. Berg. 1987. Coyote. Pages 344-357 *in* M. Novak, J. A. Baker, M. E. Obbard, and B. Malloch, editors. Wild furbearer management and conservation in North America. Ontario Ministry of Natural Resources. Toronto, Ontario, Canada.

Wetzel, J. F., J. R. Wambaugh, and J. M. Peek. 1975. Appraisal of white-tailed deer habitats in northeastern Minnesota. Journal of Wildlife Management 36:59-66.

CHAPTER 7

Small Mammals

Gerda E. Nordquist

Introduction

Fig. 7.1. A bog from a small mammal's perspective.

A shrew peering out from a high-domed sphagnum hummock sees nothing but the surrounding bog (Fig. 7.1). Even a red squirrel, from its vantage point atop a spindly black spruce, sees only an endless sea of peat. Yet what constitutes the world in the eyes of these small mammals is only a fraction of the vast patterned peatlands of northern Minnesota. This complex of peatland communities, interspersed with northern forest habitats, encompasses a large portion of the state and adjacent Canada. Human settlers have tried to tame this exotic wilderness and failed. And many would-be animal inhabitants have been turned away at the boundaries to these bogs, fens, and swamps.

Muffled by thick, damp moss, the sounds of life in the peatlands are muted compared to the bustling scene on the upland ridges. The myriad animal trails that crisscross the dry woodlands rarely penetrate into the sodden terrain of the peatlands. Yet, life appears to thrive under the cool, saturated conditions typical of peatlands. The scolding rattle of a red squirrel declares its territory within a black spruce swamp. A ball of moss and sedge perched amidst the roots of leatherleaf quivers from the jostling of meadow vole nestlings. Beneath the low, curving boughs of white cedar rests the still, hunched form of a snowshoe hare. Across the entire range of peatland habitats a diversity of small mammals can be found. From the nutrient-poor bog to a flooded fen, small mammals have dealt successfully with the unique and often harsh conditions of the peatlands. The very conditions that have deterred many potential colonizers have been overcome if not actually exploited by many of these diminutive animals. Because they are small, secretive, and largely nocturnal, they frequently go unnoticed and consequently unappreciated. Closer inspection reveals a fascinating array of life-history strategies.

Small-Mammal Species in the Region

Among the 80-some mammals that occur in the state, more than 50 can be considered small (Table 7.1), and most of these have distributions that extend into the peatlands of northern Minnesota. Despite the potential diversity of small mammals in this area, very few species are peatlands specialists; most of them fall somewhere between the categories of common resident and occasional visitor (Table 7.2). The relative importance of peatlands to these small mammals can be inferred from the circumstances under which the animals occur. For example, the distribution of species like Richardson's ground squirrel (*Spermophilus richardsonii*) and the northern pocket gopher (*Thomomys talpoides*) barely reaches into the northern peatlands area. Unusual con-

Table 7.1. Taxonomic list of small mammals occurring in Minnesota

Order Marsupialia — Marsupials
　Family Didelphidae — New World Opossums
　　Didelphis virginiana　　　　　　　　Virginia opossum

Order Insectivora — Insectivores
　Family Soricidae — Shrews
　　Sorex arcticus　　　　　　　　　Arctic shrew
　　Sorex cinereus　　　　　　　　　Masked shrew
　　Sorex haydeni　　　　　　　　　Hayden's shrew
　　Sorex hoyi　　　　　　　　　　Pygmy shrew
　　Sorex palustris　　　　　　　　Water shrew
　　Blarina brevicauda　　　　　　　Short-tailed shrew
　　Cryptotis parva　　　　　　　　Least shrew

　Family Talpidae — Moles
　　Scalopus aquaticus　　　　　　　Eastern mole
　　Condylura cristata　　　　　　　Star-nosed mole

Order Chiroptera — Bats
　Family Vespertilionidae — Vespertilionid Bats
　　Myotis lucifugus　　　　　　　　Little brown myotis
　　Myotis septentrionalis　　　　　Northern myotis
　　Lasionycteris noctivagans　　　　Silver-haired bat
　　Pipistrellus subflavus　　　　　Eastern pipistrelle
　　Eptesicus fuscus　　　　　　　　Big brown bat
　　Lasiurus borealis　　　　　　　Red bat
　　Lasiurus cinereus　　　　　　　Hoary bat

Order Lagomorpha — Lagomorphs
　Family Leporidae — Hares and Rabbits
　　Sylvilagus floridanus　　　　　　Eastern cottontail
　　Lepus americanus　　　　　　　Snowshoe hare
　　Lepus townsendii　　　　　　　White-tailed jack rabbit

Order Rodentia — Rodents
　Family Sciuridae — Squirrels
　　Tamias minimus　　　　　　　　Least chipmunk
　　Tamias striatus　　　　　　　　Eastern chipmunk
　　Marmota monax　　　　　　　　Woodchuck
　　Spermophilus franklinii　　　　　Franklin's ground squirrel
　　Spermophilus richardsonii　　　Richardson's ground squirrel
　　Spermophilus tridecemlineatus　　Thirteen-lined ground squirrel
　　Sciurus carolinensis　　　　　　Gray squirrel
　　Sciurus niger　　　　　　　　　Fox squirrel
　　Tamiasciurus hudsonicus　　　　Red squirrel
　　Glaucomys sabrinus　　　　　　Northern flying squirrel
　　Glaucomys volans　　　　　　　Southern flying squirrel

　Family Geomyidae — Pocket Gophers
　　Thomomys talpoides　　　　　　Northern pocket gopher
　　Geomys bursarius　　　　　　　Plains pocket gopher

　Family Heteromyidae — Heteromyids
　　Perognathus flavescens　　　　　Plains pocket mouse

　Family Cricetidae — Cricetids
　　Reithrodontomys megalotis　　　Western harvest mouse
　　Peromyscus leucopus　　　　　　White-footed mouse
　　Peromyscus maniculatus　　　　Deer mouse
　　Onychomys leucogaster　　　　　Northern grasshopper mouse
　　Clethrionomys gapperi　　　　　Southern red-backed vole
　　Phenacomys intermedius　　　　Heather vole
　　Microtus chrotorrhinus　　　　Rock vole
　　Microtus ochrogaster　　　　　Prairie vole
　　Microtus pennsylvanicus　　　　Meadow vole
　　Microtus pinetorum　　　　　　Woodland vole
　　Ondatra zibethicus　　　　　　Muskrat
　　Synaptomys borealis　　　　　　Northern bog lemming
　　Synaptomys cooperi　　　　　　Southern bog lemming

　Family Muridae — Murids
　　Rattus norvegicus　　　　　　　Norway rat
　　Mus musculus　　　　　　　　　House mouse

　Family Zapodidae — Jumping Mice
　　Zapus hudsonius　　　　　　　Meadow jumping mouse
　　Napaeozapus insignis　　　　　Woodland jumping mouse

Table 7.2. Occurrence of small mammals in peatland habitats of northern Minnesota

Species	Occurrence				
	1	2	3	4	5
Arctic shrew (*Sorex arcticus*)	X				
Masked shrew (*Sorex cinereus*)	X				
Water shrew (*Sorex palustris*)	X				
Star-nosed mole (*Condylura cristata*)	X				
Southern red-backed vole (*Clethrionomys gapperi*)	X				
Heather vole (*Phenacomys intermedius*)	X				
Northern bog lemming (*Synaptomys borealis*)	X				
Southern bog lemming (*Synaptomys cooperi*)	X				
Pygmy shrew (*Sorex hoyi*)		X			
Short-tailed shrew (*Blarina brevicauda*)		X			
Snowshoe hare (*Lepus americanus*)		X			
Red squirrel (*Tamiasciurus hudsonicus*)		X			
Northern flying squirrel (*Glaucomys sabrinus*)		X			
Meadow vole (*Microtus pennsylvanicus*)		X			
Meadow jumping mouse (*Zapus hudsonius*)		X			
Franklin's ground squirrel (*Spermophilus franklinii*)			X		
White-footed mouse (*Peromyscus leucopus*)			X		
Deer mouse (*Peromyscus maniculatus*)			X		
Rock vole (*Microtus chrotorrhinus*)			X		
Eastern cottontail (*Sylvilagus floridanus*)				X	
Least chipmunk (*Tamias minimus*)				X	
Eastern chipmunk (*Tamias striatus*)				X	
Northern pocket gopher (*Thomomys talpoides*)				X	
Woodland jumping mouse (*Napaeozapus insignis*)				X	
White-tailed jack rabbit (*Lepus townsendii*)					X
Woodchuck (*Marmota monax*)					X
Richardson's ground squirrel (*Spermophilus richardsonii*)					X
Thirteen-lined ground squirrel (*Spermophilus tridecemlineatus*)					X
Gray squirrel (*Sciurus carolinensis*)					X
Fox squirrel (*Sciurus niger*)					X
Plains pocket gopher (*Geomys bursarius*)					X
Norway rat (*Rattus norvegicus*)					X
House mouse (*Mus musculus*)					X
Little brown myotis (*Myotis lucifugus*)		Unknown			
Northern myotis (*Myotis septentrionalis*)		Unknown			
Silver-haired bat (*Lasionycteris noctivagans*)		Unknown			
Eastern pipistrelle (*Pipistrellus subflavus*)		Unknown			
Big brown bat (*Eptesicus fuscus*)		Unknown			
Red bat (*Lasiurus borealis*)		Unknown			
Hoary bat (*Lasiurus cinereus*)		Unknown			

1 = characteristic; 2 = common; 3 = occasional; 4 = uncommon; 5 = rare or absent.
Relative abundance is not implied.

ditions such as a dry year or a drained and logged swamp may alter the peatland environment sufficiently to enable these species to extend their range. The Norway rat (*Rattus norvegicus*) and the house mouse (*Mus musculus*), while present in the area, almost never stray far from human dwellings or yards. Others, such as the eastern chipmunk (*Tamias striatus*) and the deer mouse (*Peromyscus maniculatus*), are common throughout the region but occur primarily in upland sites or along the periphery of adjacent peatlands. The masked shrew (*Sorex cinereus*), snowshoe hare (*Lepus americanus*), and southern red-backed vole (*Clethrionomys gapperi*) are among a number of species considered characteristic of peatlands; however, they routinely inhabit the uplands as well. Broadly distributed across the boreal and mixed forest zones of North America, these species are more generally adapted to the region than to particular habitat types. Species such as the northern bog lemming (*Synaptomys borealis*) and heather vole (*Phenacomys intermedius*) are among the relatively small group of species that occur predominantly in peatland habitats, but even

Fig. 7.2. Water shrew (*Sorex palustris*). (Copyright Don Luce, by permission.)

they occasionally are found in nonpeat sites. In short, no small mammal found in this region depends solely on the environmental conditions exclusive to peatland habitats. Yet peatlands do provide the resources important to many of these species. While not obligated to live under conditions of standing water, biting insects, and limited food resources, many small mammals clearly do so, and the factors that determine their relative preference for peatlands over uplands continues to be a source of great interest.

What Is a Small Mammal?

Bears, mountain lions, and caribou obviously are not small. It is not as clear, however, to which size category beavers, white-tailed jack rabbits, and minks belong. The distribution of size classes among Minnesota's mammals is best viewed as a continuum rather than a dichotomy of small and large. But as a matter of convenience in discussion, they are divided into two groups on the basis of size and secondarily by taxonomic affiliation (Fig. 7.2). Size corresponds reasonably well with mammalian lineages. Thus, all members of the orders Marsupialia (marsupials), Insectivora (shrews and moles), Chiroptera (bats), and Lagomorpha (rabbits and hares) and nearly all members of the order Rodentia (squirrels, gophers, mice, and voles) are included in this chapter. This encompasses species ranging in size from a pygmy shrew weighing 3 g (0.1 ounce) and measuring 7 cm (less than 3 inches) up to a woodchuck weighing 4 kg (10 pounds) and measuring 60 cm (2 feet). The remaining rodent species — beaver, muskrat, and porcupine — are addressed in the chapter on large mammals.

Separation of the state's mammals by size recognizes the qualitatively different manner in which small and large mammals respond to the conditions of their immediate surroundings. A 250-acre black spruce swamp is one of several habitats included in the range of a moose, whereas a southern red-backed vole born in this swamp may never see another habitat in its lifetime. The resources of food and shelter essential for the vole's survival must be contained within this swamp, while the moose can utilize the resources of several habitats to meet its needs. Standing water in the black spruce swamp poses no problem to the moose's travel through this habitat, but it may constitute a formidable barrier to movement by the vole. Generally speaking, the degree of dependence by mammal species on particular habitats is inversely proportional to their size. This is not meant to imply that peatland habitats are unimportant to larger mammals, but rather to emphasize the close association between small mammals and peatland environments.

Postglacial History and Zoogeography

The small mammals found on the patterned peatlands today are relatively recent arrivals to the scene. Up until about 12,000 years ago this region had been overlain by ice from repeated glacial advances of the Pleistocene epoch and later by the waters of Glacial Lake Agassiz. During this period the ancestors of Minnesota's small mammals were displaced into ice-free regions of the southern United States, where they became members of biotic communities that had little similarity to modern-day associations. The plant and animal species that made up these Pleistocene communities are today both geographically and ecologically separated (Lundelius *et al.* 1983). Among the mammal remains from Pleistocene sites in Texas, cool-adapted small mammals such as the southern bog lemming (*Synaptomys cooperi*) and water shrew (*Sorex palustris*) lived alongside such warm-adapted species as the cotton rat (*Sigmodon hispidus*) and prairie vole (*Microtus ochrogaster*) (Hoffman and Jones 1970).

The end of the Pleistocene was marked by a major climatic shift toward a more seasonal climate with greater extremes of hot and cold. The existing communities disintegrated, and the distributions of plant and animal species realigned according to physiological tolerances to new regimes of temperature, moisture, and other environmental variables (Graham and Lundelius 1984). Some species extended northward into landscapes recently exposed by retreating glaciers, while others retained their southerly distributions. Boreal, temperate, and subtropical regions became recognizable as the newly assembled communities acquired more modern characteristics. The major biomes found in the Upper Midwest, the boreal forest, deciduous forest, and prairie continue to undergo adjustment in species composition and geographic extent to the present day (Webb *et al.* 1983). Because of the close association between small mammals and habitat elements, distributional changes documented for the vegetation undoubtedly were paralleled by shifts in the ranges of small mammals.

Species ranges for most small mammals that inhabit

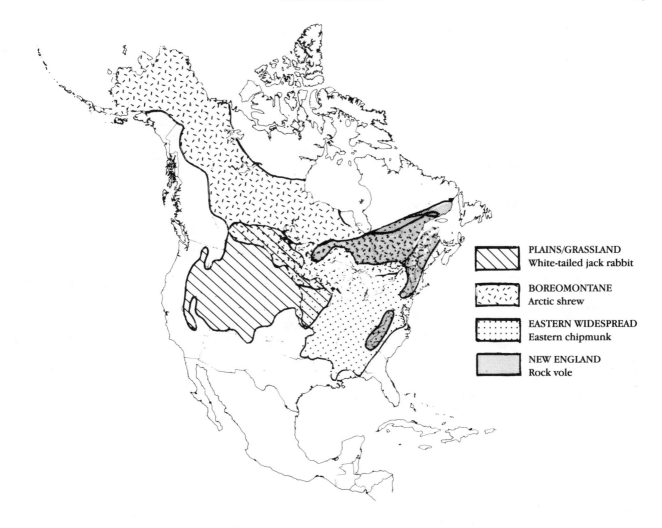

Fig. 7.3. Distributions of small mammals characteristic of North American faunal elements.

the peatlands of northern Minnesota are centered outside the boundaries of this region. For example, the woodchuck (*Marmota monax*) and eastern chipmunk (*Tamias striatus*) have distributions that encompass much of the eastern United States. The distributions of the arctic shrew (*Sorex arcticus*) and snowshoe hare (*Lepus americanus*) extend across the northern United States and Canada and reach southward into the Minnesota peatlands area. On the basis of similarities in distributions, zoogeographers recognize species with broadly overlapping ranges as comprising distinct faunal groups or elements. These elements imply similar physiological tolerances and ecological adaptations associated with the present-day geographic center of their distributions. Commonly identified by the major vegetational zone in which they occur, these faunal elements offer insight into the adaptations of the group and help to explain the patterns of distribution and abundance observed among the small-mammal species found in the peatlands.

Small mammals that occur in peatland habitats of Minnesota hail from as many as six faunal elements.

The proximity of the peatlands region to ecological tension zones between forest and prairie biomes results in distributional overlap by small mammal species with widely differing ecological affinities (Fig. 7.3). Thus, the forest-adapted masked shrew (*Sorex cinereus*) coexists with the grassland-associated Franklin's ground squirrel (*Spermophilus franklinii*) in western portions of the peatlands region, and the easterly distributed white-footed mouse (*Peromyscus leucopus*) replaces its northern relative, the deer mouse (*Peromyscus maniculatus*), along the southern edge.

Most of the small mammals found in Minnesota's northern peatlands belong to the *Boreomontane Faunal Element* (Table 7.3). Species in this group have strong affinities to northern communities, primarily coniferous forest regions, but they also extend into the mixed forest (Jones and Birney 1988). Many of these species range widely across North America, from Alaska to the eastern Canadian provinces. All the long-tailed shrews of the genus *Sorex* and most of the voles belong to this element. Both the red squirrel (*Tamiasciurus hudsonicus*) and the northern flying squirrel (*Glau-*

Table 7.3. Affinities to North American faunal elements exhibited by small mammals found in the peatland region of northern Minnesota

Boreomontane Faunal Element
  Arctic shrew *(Sorex arcticus)*[a]
  Masked shrew *(Sorex cinereus)*[a]
  Pygmy shrew *(Sorex boyi)*[a]
  Water shrew *(Sorex palustris)*[a]
  Snowshoe hare *(Lepus americanus)*[a]
  Least chipmunk *(Tamias minimus)*
  Red squirrel *(Tamiasciurus budsonicus)*[a]
  Northern flying squirrel *(Glaucomys sabrinus)*[a]
  Southern red-backed vole *(Clethrionomys gapperi)*[a]
  Heather vole *(Phenacomys intermedius)*[a]
  Meadow vole *(Microtus pennsylvanicus)*[a]
  Northern bog lemming *(Synaptomys borealis)*[a]
  Meadow jumping mouse *(Zapus budsonius)*[a]

New England Faunal Element
  Short-tailed shrew *(Blarina brevicauda)*[a]
  Star-nosed mole *(Condylura cristata)*[a]
  Northern myotis *(Myotis septentrionalis)*
  Rock vole *(Microtus cbrotorrbinus)*[a]
  Southern bog lemming *(Synaptomys cooperi)*[a]
  Woodland jumping mouse *(Napaeozapus insignis)*[a]

Eastern Widespread Faunal Element
  Eastern pipistrelle *(Pipistrellus subflavus)*
  Eastern cottontail *(Sylvilagus floridanus)*
  Eastern chipmunk *(Tamias striatus)*
  Woodchuck *(Marmota monax)*
  Gray squirrel *(Sciurus carolinensis)*
  Fox squirrel *(Sciurus niger)*
  White-footed mouse *(Peromyscus leucopus)*

Widespread Faunal Element
  Little brown myotis *(Myotis lucifugus)*
  Silver-haired bat *(Lasionycteris noctivagans)*
  Big brown bat *(Eptesicus fuscus)*
  Red bat *(Lasiurus borealis)*
  Hoary bat *(Lasiurus cinereus)*
  Deer mouse *(Peromyscus maniculatus)*[a]

Plains/Grassland Faunal Element
  White-tailed jack rabbit *(Lepus townsendii)*
  Franklin's ground squirrel *(Spermophilus franklinii)*[a]
  Richardson's ground squirrel *(Spermophilus richardsonii)*
  Thirteen-lined ground squirrel *(Spermophilus tridecemlineatus)*
  Plains pocket gopher *(Geomys bursarius)*

Great Basin Faunal Element
  Northern pocket gopher *(Thomomys talpoides)*

[a] Species found in peatland habitats.

Source: Modified from Jones et al. 1983; Jones and Birney 1988.

comys sabrinus) are closely associated with coniferous forest habitats.

The *New England Faunal Element* contains the remainder of species commonly occurring in peat habitats of Minnesota. This group also has a northerly distribution but is concentrated in the mixed forest zone of northeastern United States and southeastern Canada. Sometimes referred to as heat-sensitive, these species are generally associated with cooler environmental conditions.

The *Eastern Widespread Faunal Element* is composed of species associated with the deciduous forest region of the eastern United States. The distributions of many of these species closely follow the extent of the deciduous forest, expanding into other zones by way of riparian forest routes or vegetational changes caused by logging or other human activities. None of these species is found in peatland habitats to any great degree.

Species of the *Widespread Faunal Element* occur over much of North America, and thus their area of origin is obscure. These species tend to be highly mobile, as in the case of bats, or extreme ecological generalists, like the deer mouse *(Peromyscus maniculatus)*.

The *Plains/Grassland Faunal Element* consists of species with clear affinities to grassland habitats, some extending into savannahs or open disturbed sites (Jones and Birney 1988). Their presence on peat sites is largely along the western border, where agricultural practices have greatly altered the local peatland habitats.

The northern pocket gopher *(Thomomys talpoides)* is the only representative of the *Great Basin Faunal Element* and barely reaches the northwestern boundary of the patterned peatlands region. This group is centered in the sagebrush and montane woodlands of the Great Basin area (Jones *et al.* 1983).

Small Mammals Found in Other Holarctic Peatlands

Plant communities resembling the swamps, bogs, and fens of Minnesota can be found among the boreal forest and tundra zones of both North America and Eurasia. Hummocky topography covered with leatherleaf, other ericaceous shrubs, and sparse conifers occurs in Alaska, Minnesota, Fennoscandia (which includes Sweden, Norway, and Finland), and the Soviet Union, as do wet shrub thickets and saturated expanses of sedges and other graminoids. The Nearctic and Palearctic peatlands often share the same or closely related plant species and are comparable in structure. The circumboreal plant communities established after the Pleistocene have developed under similarly cool climatic conditions, and the geophysical processes at work in the peatlands of Minnesota act upon other northern regions in an equivalent manner.

Small mammals are closely associated with vegetational and structural components of the habitat. Therefore, it is not surprising to find that small mammals inhabiting peatlands in the Holarctic region closely resemble one another in lifeform and community composition (Table 7.4). The similarities in small-mammal communities can be explained in terms of common ancestral lineages, as well as parallel adaptive histories among more distantly related species.

Table 7.4. Comparisons of the number of species of shrews and small rodents among similar peatland habitats throughout the Holarctic region.

HABITAT Group Genus	Holarctic Region			
	AK	MN/CC	ME	FS
CONIFER SWAMP				
Shrews				
Blarina		1	1	
Neomys				1
Sorex	3	4	4	1
Small rodents				
Apodemus				1
Clethrionomys	2	1	1	1
Lemmus				1
Microtus	2	2	1	
Napaeozapus			1	
Peromyscus	1	1	1	
Synaptomys		1	1	
Zapus			1	
Total species	8	10	11	5
SHRUB SWAMP				
Shrews				
Blarina		1	1	
Sorex	3	3	2	2
Small rodents				
Apodemus				1
Clethrionomys	1	1	1	2
Lemmus	1			1
Microtus	2	1	1	1
Napaeozapus			1	
Sicista				1
Synaptomys			1	
Zapus		1	1	
Total species	7	7	8	8
FEN				
Shrews				
Blarina		1	1	
Sorex	2	4	3	
Neomys				1
Small rodents				
Arvicola				1
Clethrionomys	1	1		2
Lemmus	1			1
Microtus	3	1	1	1
Napaeozapus			1	
Sicista				1
Synaptomys	1	2	1	
Zapus	1	1	1	
Total species	9	10	8	7
BOG				
Shrews				
Sorex	2	2	3	2
Small rodents				
Clethrionomys	1	1	1	2
Lemmus	1			1
Microtus	2	2	1	1
Napaeozapus			1	
Peromyscus			1	
Synaptomys	1	2	1	
Zapus			1	
Total species	7	7	9	6

Note: AK = Alaska; MN/CC = Minnesota and Central Canada; ME = Maine; FS = Fennoscandia.

Sources: Bee and Hall 1956; Buckner 1957; Manville and Young 1965; van den Brink 1968; Haukioja and Koponen 1975; Jonsson 1975; Sölhoy *et al.* 1975; Nordquist and Birney 1980; Stockwell and Hunter 1985; and Angelstam *et al.* 1987.

Glacial events of the Pleistocene had a significant impact on the distribution of small mammals. Whereas glacial advances displaced many northern forms southward, the Beringian land connection between northeastern Siberia and Alaska served as an ice-free refugium for a number of small-mammal species. As recently as 12,000 years ago, this land bridge provided habitat for cold-tolerant species and facilitated faunal interchange between the North American and Eurasian continents (Macpherson 1965). The repeated transfer of small-mammal species between the continents and subsequent adjustments of their ranges under similar climatic and floristic regimes resulted in broadly adapted species with circumboreal distributions. Two important components of small-mammal communities in the boreal peatlands are the long-tailed shrews (*Sorex* spp.) and the meadow voles (*Microtus* spp.). These two highly successful groups are distributed throughout the Holarctic (Fig. 7.4). Two shrews that are characteristic of the peatlands in Minnesota, the masked shrew (*Sorex cinereus*) and the arctic shrew (*Sorex arcticus*), occur throughout northern Europe and the Soviet Union (Hoffman and Peterson 1967). The meadow vole (*Microtus pennsylvanicus*), found in the fens of Minnesota, is replaced by the closely related field vole (*Microtus agrestis*) in equivalent habitats in Fennoscandia.

The small-mammal communities in the bogs of Alaska, the pine barrens of New Jersey, or the mires of Scotland are composed of species unique to each region in addition to those species that they share in common. Typically, the structure of the small-mammal communities in equivalent habitats is much alike among widely separated areas. Table 7.4 shows the number of shrews and small rodents that may occur on broadly defined peatland habitats in Alaska, northern Minnesota and central Canada, Maine, and Fennoscandia. Among habitats and trophic levels the number of species is roughly equivalent. Present in all peatland habitats are shrews, insectivorous small mammals that feed largely on insect larvae and other ground invertebrates. In open peatlands, dominated by sedges and grasses, the small-grazers guild, represented by microtine rodents (*Microtus, Lemmus, Synaptomys*), predominates. Omnivorous small mammals (*Apodemus, Clethrionmys, Peromyscus*) are typically associated with forested peatlands, where the diversity of fruits, seeds, and fungi is generally greater.

The similarities between small-mammal communities in Holarctic peatlands are more apparent than their differences. Without close scrutiny, the small mammals inhabiting a bog in Minnesota would be virtually indistinguishable from those found in Fennoscandia. Even ecological equivalents with separate Old World and New World distributions, such as the woodland mice (*Apodemus v. Peromyscus*) and water shrews (*Neomys v. Sorex*) are strikingly similar in morphology and adap-

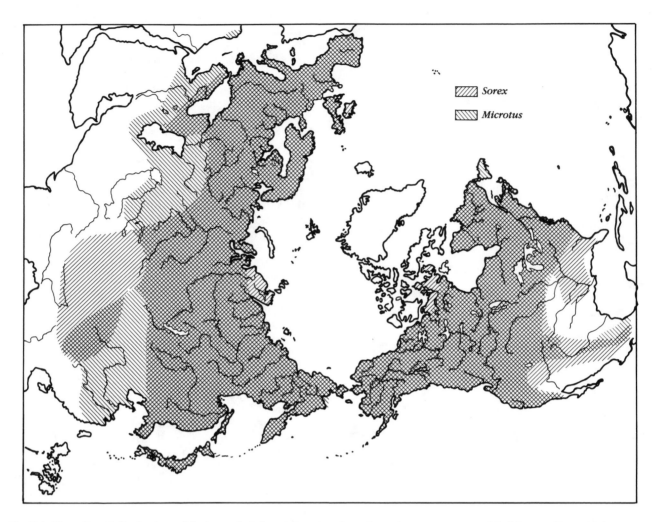

Fig. 7.4. Circumboreal distributions of the long-tailed shrews (*Sorex*) and meadow voles (*Microtus*). (Modified from Hoffman 1984.)

tations. Clearly, the conditions of peatlands restrict the kinds of small mammals capable of successfully inhabiting them, yet the small-mammal communities characteristic of these habitats are recognizable throughout the Holarctic region.

The Peatland Environment from a Small-Mammal Perspective

Distinguished by poorly decomposed organic substrate that frequently is flooded by acidic and nutrient-poor water, peatland environments differ considerably from adjacent uplands, where decomposing plant materials enrich the well-drained mineral soil. These underlying conditions are further manifested in vegetational differences between peat and upland habitats. Compared to uplands, peatlands typically have lower plant-species richness, slower rates of growth, and higher concentrations of secondary plant compounds. These factors exert a profound influence on habitat selection by small mammals in the patterned peatlands region.

Critical Environmental Conditions

Water and temperature have a direct impact on small mammals. The consistently cool temperatures and wet conditions characteristic of most peatlands play an important role in small-mammal temperature regulation. Shrews, because of their relatively small size and high metabolic rate, must conserve heat, reduce evaporative water loss, and find sufficient food to maintain their high levels of activity. Whereas cool temperatures may mean that small mammals have to eat more, the saturated environment of peatlands helps them conserve water. Additionally, bog waters act as breeding and egg-laying sites for amphibians (Karns and Regal 1979) and support large and diverse insect larval populations (Judd 1961). Both amphibian eggs and larval insects are important food resources for insectivores. Thus it is not surprising that moist habitats, including peatlands, have been strongly associated with an abundance of shrews (Getz 1961b; Buckner 1966). Some vole species, on the other hand, are susceptible to heat stress (Wunder 1985). Cool, damp peatlands allow them to avoid high summer temperatures and provide a means for dumping excess body heat.

Very few small mammals are aquatic, and flooded conditions present a formidable obstacle to movement in wet peatland habitats. Even in moderately wet sites, the saturated peat substrate and the typically high water table thwart attempts to establish recessed runways, burrows, and underground nest sites. Thus the subterranean element of small-mammal habitats is largely absent in most peatlands. High sphagnum hummocks and, to a lesser extent, grass and sedge tussocks provide an alternative medium for protected travel and nest sites. A masked shrew was observed "swimming" through the fronds of sphagnum moss in an open bog, and vole nests were found amidst sphagnum hummocks in black spruce swamps and grass tussocks in fens (personal observation). Clearly, small mammals are capable of selecting suitable microhabitat within otherwise inhospitable sites. For example, in a severely flooded fen, above-water vegetation was restricted to isolated hummocks of moss and sedge and root masses of bog birch, yet a diverse assemblage of small-mammal species was present in high numbers (Nordquist and Birney 1980).

Perhaps the greatest impact on small mammals is through the indirect effect of the peatland environment on the composition, diversity, and structure of the vegetation. Peatlands, especially ombrotrophic habitats, exhibit a flora of low species richness, as well as low abundance. The response of plant species under nutrient stress is to reduce leaf growth and increase inpalatability to herbivores through the production of secondary compounds (Grime 1978). Many of the plants growing in the more ombrotrophic peatlands have smaller leaf area and fewer leaves per plant than plants of the same species growing in more minerotrophic peatlands or in upland sites (unpublished data). The fewer numbers of small mammals observed in open and forested bogs suggests that the carrying capacity for small-mammal herbivores is greatly reduced in these habitats (Fig. 7.5).

The conifers and ericaceous plants characteristic of peatland habitats are typically high in phenolic compounds, and willow shrubs are rich in tannins. Graminoids, particularly the abundant sedges of fens, contain high levels of silica. These and other compounds have been implicated in reducing the digestibility of plant proteins in herbivorous rodents (Batzli 1985), and Labrador tea was shown to be toxic to microtines (Jung and Batzli 1981). In contrast, plants in upland habitats have lower levels of secondary compounds and greater diversity and abundance of highly palatable fruit- and seed-bearing plants.

Upland habitats are utilized by most small-mammal species in the region and are clearly preferred by some. Not surprisingly, these habitats are among the highest in species number and overall small-mammal abundance (Fig. 7.5). Peatland habitats, however, are vitally important to a number of small-mammal species whose

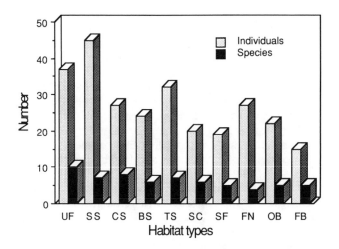

Fig. 7.5. Average number of individuals and species of small mammals in habitats of the patterned peatlands region. UF, upland forest; SS, shrub swamp; CS, cedar swamp; BS, black spruce swamp; TS, tamarack swamp; SC, shrub carr; SF, shrub fen; FN, fen; OB, open bog; FB, forested bog.

ecological requirements are best met by their unique conditions.

Habitats

Small mammals are present in every habitat type represented in the patterned peatlands region. Bog lemmings reside in the open expanse of sedge fen and black spruce islands north of Upper Red Lake. Shrews forage in the concentric bands of sphagnum, tamarack, black spruce, and balsam fir that surround a basin lake in the Boundary Waters. Deer mice scurry along deadfall beneath the dark shade of a cedar swamp in the Agassiz Peatlands Natural Area. Across the patterned peatlands region, from the flat lake bed of Glacial Lake Agassiz to the more rugged topography of the Arrowhead region, a diverse array of peat and upland habitats is found. Although peatland plant communities such as tamarack swamp, shrub fen, and open bog are recognizable throughout, local differences in landscape and hydrological conditions produce regional variation that is reflected in the distributional patterns of small mammals in the area.

The names given to habitat categories in this chapter may not correspond with those used to describe the hydrogeological and floristic characteristics of peatlands, but they serve to identify broad habitat types that are meaningful in the study of small mammals. The habitat categories used in this chapter are forested bog, open bog, fen, shrub fen, tamarack swamp, black spruce swamp, cedar swamp, shrub carr, shrub swamp, and upland forest. The following descriptions focus on attributes that are important to small mammals.

Bogs. Both forested bogs and open bogs are characterized by a continuous sphagnum mat that often forms high hummocks with standing water in the intervening troughs. The diversity and density of forbs, low erica- ceous plants, and graminoids (sedges and grasses) are generally low. In forested bogs, stunted black spruce and tamarack trees provide an additional layer of cover that may vary from an open canopy of small, widely spaced trees to a nearly closed canopy of trees that are tall but small-crowned. Small mammals residing in the largely treeless habitat of the open bog rely on the thick sphagnum layer as the primary protective cover. Both of these habitats are typically oligotrophic and can be components of raised bogs and bog drains (see Fox *et al.* 1977).

Fens. These habitats are among the wettest peatland sites. Sedges and grasses form a dense ground layer that is important to small mammals for both cover and food. Trees and shrubs are largely absent, and herbaceous plants tend to be diverse but not abundant. Included under this category are poor fens, characterized by relatively more sphagnum moss, and rich fens, where sedge mats form the predominant ground cover. Shrub fens have patches of low woody plants, and shrub carrs have a significant layer of low- to mid-height shrubs, particularly willow. These minerotrophic habitats often intergrade into one another in large peatland tracts.

Swamps. Conifer swamps are a major cover type in the northeast and constitute significant portions in the central area of the patterned peatlands region. These forested peatlands occur as large tracts or as iso- lated forest islands; canopy coverage ranges from 50 to 100%. Conifer swamps are further divided accord- ing to the predominant canopy species — tamarack, black spruce, or white cedar. Tamarack swamps exhibit a rich understory of woody and herbaceous plants be- neath both open and closed canopies. On sites with larger trees, a tall shrub layer is often present, and moss-covered roots lying close to the surface form a semi-subterranean structure to the habitat.

Black spruce swamps are forests dominated by black spruce but frequently contain tamarack, balsam fir, and cedar in the canopy. The habitats included under this category may be sufficiently distinct as to be considered separate habitat types. Stands of large, widely spaced trees typically support a diverse shrub and ground layer. Where this habitat adjoins the upland, the bound- ary between conifer swamp and coniferous upland for- est is not well defined in terms of the vegetation or the small mammals inhabiting the area. Also considered as black spruce swamp are stands of tall, top-crowned trees that form a nearly closed canopy. Here, the for- est floor is characteristically open, with feathermoss predominating and the ground vegetation restricted to isolated patches.

The canopy of cedar swamps generally provides dense cover, resulting in very open ground with patches devoid of vegetation. Forbs are diverse but nu- merous only beneath breaks in the canopy. The low, curving growth form typical of cedar provides ample horizontal surfaces along which scansorial and semi- arboreal small mammals can travel.

In shrub swamps, tall alders or willows typically form dense high cover. Ground vegetation varies widely among sites. Examples include an unbroken layer of moss with patches of ericaceous plants; a dense multiple layer of tall herbaceous plants, grasses, and ground forbs; and a sparsely vegetated understory with bare mucky soils and pools of standing water.

Forested Upland. This habitat generally consists of mixed stands of deciduous and coniferous trees. Stands made up entirely of second-growth deciduous species can be found throughout the patterned peatlands re- gion, but all conifer stands also contain deciduous trees to varying degrees. For the purposes of this chapter, these diverse habitats have been lumped into a sin- gle category, upland forest. Most of the upland habitat is composed of deciduous forest, typically dominated by trembling aspen; birch, elm, and ash are present in lesser amounts. Included in this category are damp sites that contain species associated with hardwood swamps. The understory of the deciduous forest is characterized by a dense layer of high shrubs and an open but diverse forest floor, including many fruit- and seed-bearing plant species. Moss coverage is min- imal, and the relatively thick litter layer and abundant ground debris provide cover for small mammals unlike that found in peatland habitats. Also included in the upland forest category are the coniferous forests, in- cluding balsam fir, white cedar, and jack, red, and white pines. High shrub understory is often present over di- verse but patchy ground vegetation. Ground debris is less than in deciduous forests, and ground vegetation is sparse and localized, particularly on sites where pine needles accumulate.

Distinct as peatland habitats are, they share a number of characteristics with uplands. The ranges in soil pH, moisture, habitat structure and coverage, and plant- species composition overlap between minerotrophic peatlands and uplands. It is not surprising, therefore, to see a corresponding overlap among the resident small-mammal species. While peatland conditions may restrict some small mammals, most species in north- ern Minnesota have habitat requirements sufficiently broad to utilize both peat and nonpeat environments to varying degrees.

Distribution and Abundance of Small Mammals in the Minnesota Peatlands

Varying patterns of occurrence and abundance are ob- served among the small-mammal species inhabiting the peatlands region of northern Minnesota. The south- ern red-backed vole (*Clethrionomys gapperi*) and the masked shrew (*Sorex cinereus*), for instance, can be found in nearly all habitat types, and they usually are the most abundant species in those habitats. At the opposite end of the spectrum are species such as the water shrew (*Sorex palustris*) and the northern bog lemming (*Synaptomys borealis*) that are recorded spo-

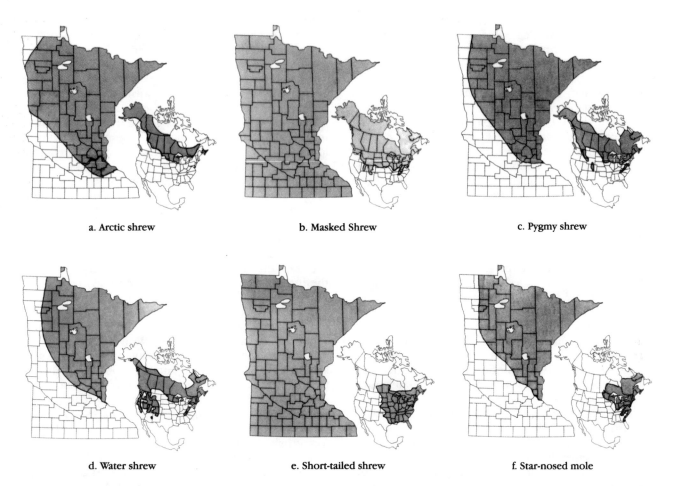

Fig. 7.6. Distributions of insectivores occurring in the peatland region of northern Minnesota (state distributions represent currently available information on local occurrences and may differ from the North American distributions for the species).

radically, are found in relatively few habitats, and always occur in low numbers. In general, the majority of small-mammal species in this region are numerous in only a few habitats and occur in low numbers or are absent from the rest. Examination of the range of habitat types where individual species are found and their relative abundance in these habitats affords a better understanding of the ecological requirements for each small-mammal species. It is extremely difficult, if not impossible, to clearly identify the factor critical to a species survival: dietary requirements, nesting and protective shelter, and physiological limitations. In practice, these criteria are inferred by identifying common characteristics among the habitats in which each species is most numerous.

Species Descriptions

Arctic shrew (*Sorex arcticus*). Of all the shrews occurring in this region, the arctic shrew exhibits the strongest preference for particular peatland habitats (Fig. 7.6a). It is characteristically associated with habi-

tats that are minerotrophic, wet, and open, with dense high shrubs and a ground layer dominated by graminoids. It is one of the most common species in shrub swamps, shrub carrs, and fens and typically occurs in tamarack swamps. It is less well represented in forested peatland sites and is rare in ombrotrophic peatlands and forested upland habitats. Kalin (1976) and Timm (1975) found that the arctic shrew preferred rich, open habitats dominated by graminoids and tall shrubs. It was reported to be most abundant in tamarack swamps (Bailey 1929; Gunderson and Beer 1953), grass-sedge meadows (Heaney and Birney 1975), and heathlike wetlands (Batten 1980).

Masked shrew (*Sorex cinereus*). The masked shrew can be found in nearly all peat and nonpeat habitats in this region of northern Minnesota (Fig. 7.6b). It is frequently the most numerous small mammal occurring on forested peatland sites and is relatively less abundant in shrub swamps, fens, and nonpeat habitats. The masked shrew shows no strong preference for specific habitats and appears to be successful over a wide range of environmental conditions. The widespread and abundant nature of this species has been described by others (Bailey 1929; Getz 1961b; Timm 1975). Although the degree of preference for wet or dry sites varies among published accounts, the masked shrew is generally most abundant in peatlands or in damp nonpeat habitats.

Pygmy shrew (*Sorex hoyi*). The pygmy shrew has a widespread distribution similar to that of the masked shrew (Fig. 7.6c), but it is rarely abundant. No strong preference is exhibited by this species for any peat or nonpeat habitat type in northern Minnesota. Published reports of preferred habitat for this species vary greatly and are often contradictory. Pygmy shrews have been associated with wet marshes (Long 1972), spruce-fir bogs (Brown 1967), hardwood forests (Schmidt 1931), and dry, open woods (Cahn 1937).

Water shrew (*Sorex palustris*). Documentation of water shrews in Minnesota is rare and sporadic (Fig. 7.6d). The few records of this shrew from peatlands are not consistently associated with particular habitats or conditions other than standing water. They have been found in tamarack swamps, open bogs, and shrub fens, sites that are wet, minerotrophic, and characterized by a dense low shrub layer and graminoid ground cover. Never abundant, water shrews are recorded from a variety of habitats, including wooded and nonwooded sites (Whitaker and Schmeltz 1973; Kalin 1976), sphagnum swamp, maple thicket, and yellow birch forest (Hamilton 1943). It is considered highly aquatic in its habits (Swanson *et al.* 1945).

Short-tailed shrew (*Blarina brevicauda*). The short-tailed shrew is widespread among nearly all habitats in the region but is abundant in only a few (Fig. 7.6e). It reaches its greatest relative abundance in upland habitats, particularly forested sites with deciduous trees. In the peatland habitats, the short-tailed shrew occurs regularly and in higher numbers on more minerotrophic sites such as shrub swamps, tamarack swamps, and cedar swamps. The short-tailed shrew reportedly displays a strong preference for heavily shaded or forested habitats (Pruitt 1959), and the amount and type of cover is important to the distribution of this species (Leraas 1942). The species is considered widespread and common throughout its range. It was the most common small mammal found in the habitats examined by Kalin (1976), Cahn (1937), and Rand (1933).

Star-nosed mole (*Condylura cristata*). The star-nosed mole is detected irregularly on peatlands in Minnesota, but it may be more common than records imply (Fig. 7.6f). It has been found in cedar swamp and shrub swamp habitats with mucky peat soils, standing water, and a relatively open ground layer, but also in muskeg and upland forest. Suitable burrowing substrate and abundant soil arthropods for food may be limiting factors in the distribution of this species. Swanson *et al.* (1945) felt that this mole may be present in all forested habitats in northern Minnesota, while Kalin (1976) found it numerous in

open bogs and considered it characteristic of peatlands.

Bats. During the summer, the seven species of bats in Minnesota have nearly statewide distributions that include the patterned peatlands region (Fig. 7.7), but little is known about what habitats are selected by bats to forage, raise young, or roost during this time. All bats in Minnesota feed on flying insects, and anecdotal information suggests that bats concentrate where these insects are most numerous and where the habitat allows relatively unobstructed flight. Most observations of bats in this region have been at the perimeter of forest clearings or along the banks of lakes and streams. Large trees in upland deciduous forests have been reported as roost sites for the hoary bat (*Lasiurus cinereus*) (Fig. 7.7g). Three species — the hoary bat, red bat (*Lasiurus borealis*) (Fig. 7.7f), and silver-haired bat (*Lasionycteris noctivagans*) (Fig. 7.7c) — leave the state during winter, but the remaining species — the little brown myotis (*Myotis lucifugus*) (Fig. 7.7a), northern myotis (*Myotis septentrionalis*) (Fig. 7.7b), big brown bat (*Eptesicus fuscus*) (Fig. 7.7e), and eastern pipistrelle (*Pipistrellus subflavus*) (Fig. 7.7d) — hibernate in caves and mines. With the exception of a few mines in the northeast, such hibernating sites are largely absent in the patterned peatlands region, forcing bats to migrate seasonally into this area.

Eastern cottontail (*Sylvilagus floridanus*). The eastern cottontail is largely absent from the patterned peatlands region and is found in brushy and open forest habitats in the southern two-thirds of the state (Fig. 7.8a). Although wooded swamps are listed among the preferred habitats for this species (Hazard 1982), most records for the eastern cottontail are from nonpeat habitats (Chapman *et al.* 1980).

Snowshoe hare (*Lepus americanus*). Snowshoe hares at times of population highs are common and widespread throughout most of the peatland and upland habitats of northern Minnesota, with highest numbers occurring on forested sites (Fig. 7.8b). A dense shrub layer that provides both food and cover is positively correlated with high numbers of this species. Minnesota records of snowshoe hare indicate that it occurs most frequently in forested swamps (Burt 1946; Bailey 1929) and is typically found in habitats with thick underbrush (Johnson 1922). Pietz and Tester (1983) found that snowshoe hare preferred lowland habitats, particularly conifer swamps, alder fens, and jack pine-alder edges; tall shrub cover, however, appeared to be the crucial factor determining habitat use.

White-tailed jack rabbit (*Lepus townsendii*). White-tailed jack rabbits may be found along the western boundary of the patterned peatlands region but are absent from most of the area (Fig. 7.8c). Associated with open habitats, this species has followed agricultural expansion into this portion of the state (de Vos 1964).

Least chipmunk (*Tamias minimus*). The least chipmunk occurs primarily in the forested uplands

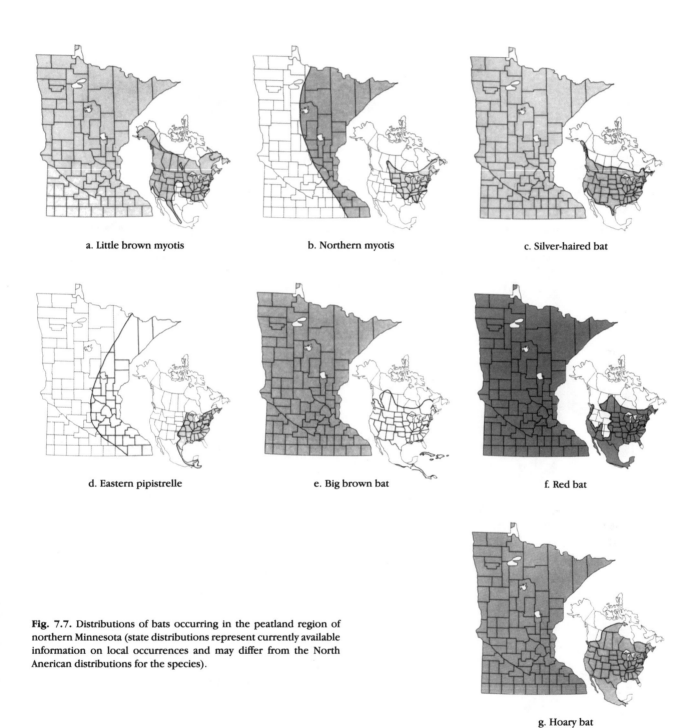

a. Little brown myotis

b. Northern myotis

c. Silver-haired bat

d. Eastern pipistrelle

e. Big brown bat

f. Red bat

Fig. 7.7. Distributions of bats occurring in the peatland region of northern Minnesota (state distributions represent currently available information on local occurrences and may differ from the North Anerican distributions for the species).

g. Hoary bat

of northern Minnesota, where it is seldom numerous (Fig. 7.9a). Records of this species in peatland habitats appear to be wanderers from adjacent uplands. The highest relative abundance for this species occurred in dry, forested habitats with dense shrub cover and substantial amounts of ground debris (Nordquist and

Birney 1980). Records indicate that the least chipmunk is common in disturbed areas, open and broken country, and at wooded edges (Burt 1946; Forbes 1966; Kalin 1976).

Eastern chipmunk (*Tamias striatus*). The eastern chipmunk is common in all forested upland sites in the

a. Eastern cottontail   b. Snowshoe hare   c. White-tailed jackrabbit

Fig. 7.8. Distributions of lagomorphs occurring in the peatland region of northern Minnesota (state distributions represent currently available information on local occurrences and may differ from the North American distributions for the species).

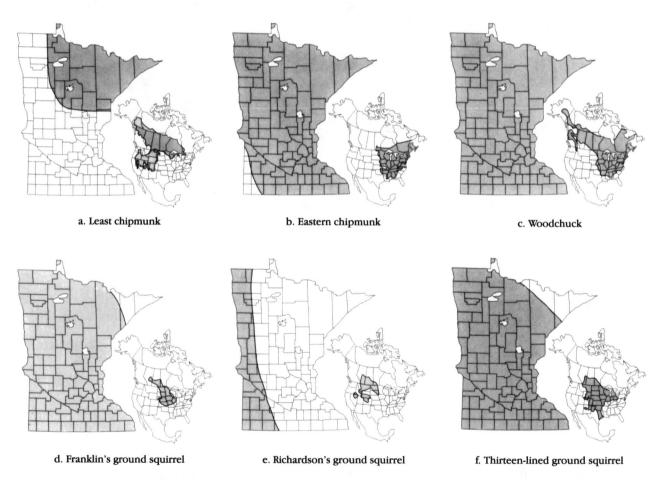

a. Least chipmunk   b. Eastern chipmunk   c. Woodchuck

d. Franklin's ground squirrel   e. Richardson's ground squirrel   f. Thirteen-lined ground squirrel

Fig. 7.9. Distributions of chipmunks (a and b) and ground squirrels (c-f) occurring in the peatland region of northern Minnesota (state distributions represent currently available information on local occurrences and may differ from the North American distributions for the species).

region (Fig. 7.9b). It seems to prefer dry, deciduous forests with a dense, tall shrub layer and open forest floor. Occasionally it ventures onto peatland habitats bordering forested uplands, but there is little mention in the literature of the eastern chipmunk in association with peatland habitats. Forbes (1966) commonly found it in woods bordering bogs, and Brown and Lanning (1954) took it from tamarack swamps. Nevertheless, the primary habitats for this species appear to be dry, open, upland forests (Hamilton 1943; Kalin 1976).

Woodchuck (*Marmota monax*). Woodchucks occur throughout the region but are largely restricted to nonpeat soils and conditions that allow them to excavate burrows (Fig. 7.9c). The species is primarily associated with openings in upland forests, both deciduous and coniferous (Hazard 1982).

Franklin's ground squirrel (*Spermophilus franklinii*). Franklin's ground squirrel has been found in drier, grassy peat sites and along elevated, grassy roadways adjacent to shrub carrs and sedge fens, but apparently it is not a regular inhabitant of peatlands (Fig. 7.9d). Others record this species as typical of parkland habitat (Soper 1961), wetland edges, and open brushy areas (Johnson 1930; Hamilton 1943).

Richardson's ground squirrel (*Spermophilus richardsonii*). Richardson's ground squirrel barely extends into the patterned peatlands region, where the species reaches its eastern limit (Fig. 7.9e). It is primarily restricted to grazed grasslands and croplands and has not been recorded in peatland habitats.

Thirteen-lined ground squirrel (*Spermophilus tridecemlineatus*). Although this species occurs in the region and was found occasionally in deciduous uplands, it has never been recorded in a peatland site (Fig. 7.9f). The thirteen-lined ground squirrel is usually associated with grassland habitats, including pasture, and cultivated fields (Burt 1946; Soper 1948).

Gray squirrel (*Sciurus carolinensis*) and Fox squirrel (*Sciurus niger*). These tree squirrels are largely associated with upland deciduous forests (Fig. 7.10a, b). Fox squirrels typically are found in open parklands, while gray squirrels are more characteristic of mature hardwood and mixed forests (Hazard 1982). Both species are largely absent from the patterned peatlands region and have not been recorded in forested swamps.

Red squirrel (*Tamiasciurus hudsonicus*). The red squirrel is widespread and common among northern forested habitats, both peat and nonpeat (Fig. 7.10c). It is a frequent and conspicuous inhabitant of conifer swamps, occurring less abundantly in deciduous forest habitats. This species was recorded as common in coniferous forests (Kalin 1976), tamarack swamps (Bailey 1929), cedar swamps (Conner 1959), and hardwood forests (Johnson 1930; Burt 1946).

Northern flying squirrel (*Glaucomys sabrinus*). The northern flying squirrel occurs in forested habitats of both peat and nonpeat (Fig. 7.10d). It has been recorded from black spruce and cedar conifer swamps as well as mixed upland forests in the patterned peatland region. Records of occurrence for this species vary according to habitat type, but most authors agree that it is strongly associated with forested areas, particularly larger trees (Bailey 1929; Burt 1946). Relative preference for peatland versus upland sites is not clear. Kalin (1976) and Dice (1921) found the northern flying squirrel in mixed upland forests, whereas Hamilton (1943), Bailey (1929), and Johnson (1922) indicated that it preferred conifer swamps. It is replaced by its cogener, the southern flying squirrel (*Glaucomys volans*), over the southern half of the state (Fig. 7.10e). The southern flying squirrel is strongly associated with upland forests and has not been documented in peatlands in northern Minnesota.

Northern pocket gopher (*Thomomys talpoides*) and plains pocket gopher (*Geomys bursarius*). The distribution of pocket gophers is limited largely to areas suitable for burrowing: well-drained soils free of large roots and rocks. The northern pocket gopher barely enters the northwest corner of the state, where mounds probably created by this species were found in drained and cultivated peat soils and heavy mineral soils (Fig. 7.11a). The plains pocket gopher occurs throughout most of the state but does not extend far into the patterned peatlands region, where it appears to be restricted to areas of lighter, sandy soils (Fig. 7.11b). Both species are primarily found in open, dry sites and are largely absent from peatland habitats.

White-footed mouse (*Peromyscus leucopus*). The white-footed mouse replaces the deer mouse (*P. maniculatus*) in the southern and western edge of the patterned peatlands region, where it occurs over a variety of habitats (7.12a). The species is most numerous in coniferous upland forest, but it is also found in tamarack conifer swamp. Most records of this species are from forested sites. The white-footed mouse has been found in moist hardwood swamps (Getz 1961c), pine barrens and conifer swamps (Green 1925), and dry mixed forests (Conner 1953).

Deer mouse (*Peromyscus maniculatus*). The woodland subspecies of the deer mouse (*P. m. gracilis*) occurs in peat habitats, but it is significantly more abundant on nonpeat sites, especially upland coniferous and mixed forests. The prairie form (*P. m. bairdii*) replaces it in open grassland habitats to the west and south of the patterned peatlands region (Fig. 7.12b shows the combined distribution of both forms). Cedar swamps are the only peatland site where this species occurs in any number. The deer mouse characteristically is found in forested habitats with open understory and abundant ground debris. This species is considered abundant and widespread over most habitats in its range (Hamilton 1943; Timm 1975). Brown and Lanning (1954) report the deer mouse common in low woody areas and tamarack swamps, whereas Batten (1980) found this species to avoid wet sites.

Norway rat (*Rattus norvegicus*) and house mouse (*Mus musculus*). These commensal species oc-

a. Gray squirrel   b. Fox squirrel   c. Red squirrel

d. Northern flying squirrel   e. Southern flying squirrel

Fig. 7.10. Distributions of tree squirrels occurring in the peatland region of northern Minnesota (state distributions represent currently available information on local occurrences and may differ from the North American distributions for the species).

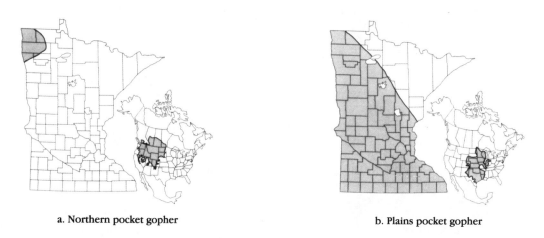

a. Northern pocket gopher   b. Plains pocket gopher

Fig. 7.11. Distributions of pocket gophers occurring in the peatland region of northern Minnesota (state distributions represent currently available information on local occurrences and may differ from the North American distributions for the species).

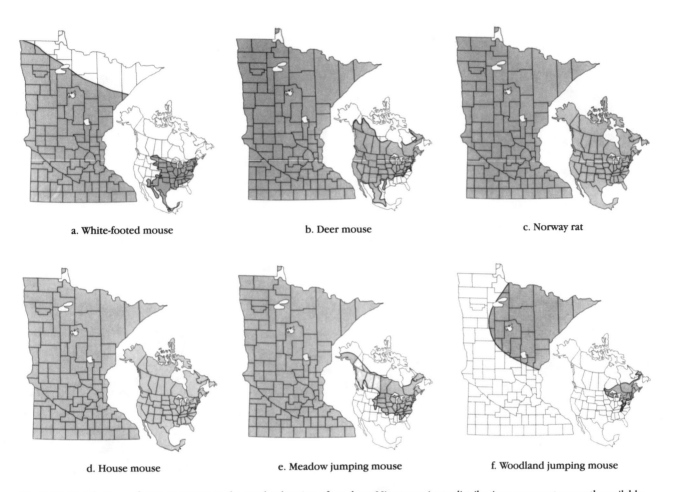

a. White-footed mouse b. Deer mouse c. Norway rat

d. House mouse e. Meadow jumping mouse f. Woodland jumping mouse

Fig. 7.12. Distributions of mice occurring in the peatland region of northern Minnesota (state distributions represent currently available information on local occurrences and may differ from the North American distributions for the species).

cur throughout the patterned peatlands region but rarely far from human habitation (Fig. 7.12c, d). Neither species has been taken from peatland habitats.

Meadow jumping mouse (*Zapus hudsonius*). The meadow jumping mouse is found in a wide variety of habitats, both peat and nonpeat (Fig. 7.12e). In the patterned peatlands region, high relative densities have been attained in mixed deciduous upland forest, black spruce swamp, and shrub swamp. Stockwell and Hunter (1985) found this species to be among the most abundant small mammals in shrub thicket, streamside meadow, and pools. Timm (1975), Kalin (1976), and Burt (1946) suggested that this species exhibits little habitat preference and may be found in open and closed habitats with dry or wet conditions. Other authors, however, generally associate it with moist, open habitats having dense ground vegetation (Whitaker 1972). It has also been recorded in tamarack and cedar swamps (Brown and Lanning 1954; Getz 1961c) and mixed upland forest (Kalin 1976).

Woodland jumping mouse (*Napaeozapus insignis*). The woodland jumping mouse is detected infrequently in this region (Fig. 7.12f). It occurs in forested upland sites, both coniferous and deciduous, with open forest floor. Miller and Getz (1977) reported that this species has very narrow habitat breadth compared to other small mammals. It has been recorded primarily from moist, forested habitats (Schmidt 1931; Hamilton 1943) but also from open, grassy areas (Green 1925; Manville 1949).

Southern red-backed vole (*Clethrionomys gapperi*). The southern red-backed vole is one of the most abundant and widespread small mammals found in northern Minnesota, second only to the masked shrew (Fig. 7.13a). Highest relative abundances are in peatland sites, particularly black spruce and mixed conifer swamps, but this vole is also numerous in upland forested sites. Diverse vertical structure and ground debris are habitat components frequently associated with this species (Batten 1980; Nordquist and Birney 1980). The southern red-backed vole has been recorded in a wide variety of habitats and environmental conditions. Most published records imply a strong association with moist conifer forests.

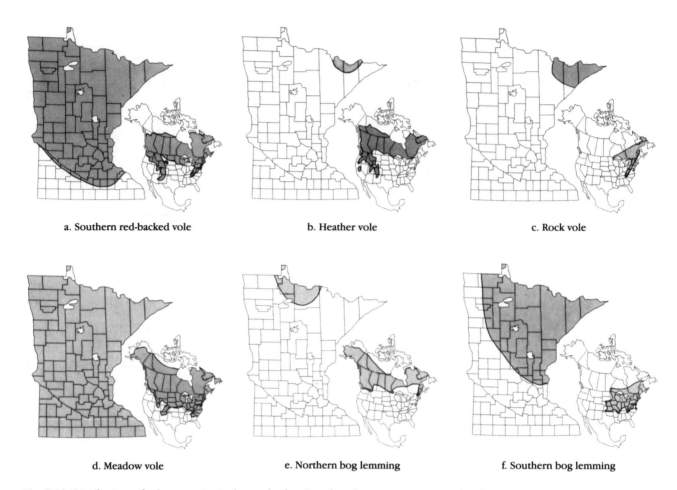

a. Southern red-backed vole   b. Heather vole   c. Rock vole

d. Meadow vole   e. Northern bog lemming   f. Southern bog lemming

Fig. 7.13. Distributions of voles occurring in the peatland region of northern Minnesota (state distributions represent currently available information on local occurrences and may differ from the North American distributions for the species).

Heather vole (*Phenacomys intermedius*). The heather vole barely extends into the state in the northeast corner of the patterned peatlands region (Fig. 7.13b). Recent records for this species have found it in open bog and muskeg (Etnier 1989). The rarity of this species in Minnesota makes it difficult to describe its habitat preference for the region. Elsewhere, it can be found in a variety of habitats and moisture conditions and occurs in open and forested habitats with heatherlike ground vegetation (McAllister and Hoffman 1988).

Rock vole (*Microtus chrotorrhinus*). The rock vole is found only in the northeast corner of the patterned peatlands region, where it reaches its western distributional limit (Fig. 7.13c). The species is found as small, isolated populations that may experience periodic local extinctions (Christian and Daniels 1985). Most accounts of this species associate it with moist conditions and moss-covered boulder fields in or adjacent to forested habitats. It has been reported in both deciduous and coniferous upland forests and adjacent to a black spruce lowland forest (Timm *et al.*

1977). The species does not show a strong preference for any particular habitat type (Kirkland and Jannett 1982).

Meadow vole (*Microtus pennsylvanicus*). The meadow vole may be found in a number of habitats in this region, but it is common in relatively few (Fig. 7.13d). It is most abundant in open, grassy sites and poorly represented in forested habitats. Peatlands with a dense graminoid component such as fens, shrub fens, and shrub swamps typically contain greater meadow vole densities. Standing water apparently is not a significant deterrent to this species as long as moss hummocks or grass tussocks provide above-water sites. While this species shows no apparent selection for particular species of grass or sedge, Getz (1961a) suggested that abundance and clumping characteristics were important to nesting and feeding sites, and Birney *et al.* (1976) found that the amount of graminoid cover is positively correlated with population levels of this species.

Northern bog lemming (*Synaptomys borealis*). The northern bog lemming is rare in the state (Fig.

Table 7.5. Relative abundance of small-mammal species among different areas of the patterned peatlands region. (Note the variation in abundances among areas)

Species	Region			
	West	Central	East	South
Arctic shrew	++	+	+	+++
Masked shrew	++	++	++	++
Pygmy shrew	+++	+	+	++
Water shrew		+++++		
Short-tailed shrew	++	+	++	++
Star-nosed mole	+++			+++
Snowshoe hare	++	+	++	+++
Least chipmunk	+++++	+		
Eastern chipmunk	+++	+	++	+
Franklin's ground squirrel	++	++		++
13-lined ground squirrel	+++++			
Red squirrel	++	++	+	++
Northern flying squirrel	+++++	+		
Deer mouse	++	++	+	+++
Southern red-backed vole	++	+	+++	+
Meadow vole	+++	+	++	++
Northern bog lemming	++	++++		
Southern bog lemming		++	+++	++
Meadow jumping mouse	+++	+	+	+++
Woodland jumping mouse				+++++

Note:
+	less than 20%
++	20-40%
+++	40-60%
++++	60-80%
+++++	more than 80%

7.13e). The few state records of this species are from the patterned peatlands region and constitute the southern limit of its range in central North America. The species appears to have a highly restricted distribution in this area and has been found in open bog and shrub carr — wet, open conditions with a dense low shrub layer of ericaceous plants. Elsewhere, this species has been found in spruce woods (Soper 1948), tamarack-black spruce forest (Smith and Foster 1957), and sedge-grass meadow (Gunderson and Beer 1953).

Southern bog lemming (*Synaptomys cooperi*). The southern bog lemming occurs sporadically throughout the patterned peatland region, and its distribution appears to be patchy and variable from year to year (Fig. 7.13f). It has achieved high relative abundance in black spruce swamps and forested bogs, and a few individuals have been found in other conifer swamps, open bogs, and fens. Nonpeat habitats in the region are largely avoided by this species, but Kalin (1976) and Iverson *et al.* (1967) found it in upland habitats. Most published reports indicate that the southern bog lemming occurs in greatest numbers in moist, open habitats with abundant grasses and sedges (Conner 1953; Getz 1961a; Timm 1975; Linzey 1983) and in conifer swamps (Brown and Lanning 1954; Buckner 1957; Pruitt 1959).

Patterns of Distribution

Patterns of distribution among small-mammal species in the patterned peatlands region may be examined on a number of levels. On a regional scale, Nordquist and Birney (1980) found that species occurred in different frequencies in western, central, eastern, and southern portions of the patterned peatlands (Table 7.5). Both species richness and individual abundance of small mammals exhibited a marked increase in peatland areas adjacent to other vegetational regions such as the grasslands to the west and the deciduous forest to the south. The major factor underlying the observed patterns apparently is the local enhancement of plant-species richness. This may enable the invasion of small-mammal species not regularly occurring in peatland habitats, and result in higher densities of species than are usually associated with peatlands.

Some species like the masked shrew (*Sorex cinereus*), red squirrel (*Tamiasciurus hudsonicus*), and snowshoe hare (*Lepus americanus*) are found in equivalent numbers across the patterned peatlands region, while species such as the woodland jumping mouse (*Napaeozapus insignis*), northern bog lemming (*Synaptomys borealis*), and water shrew (*Sorex palustris*) occur predominantly in one or two areas. Explanations for such patterns are problematic. For a species at its distributional limits, such as the thirteen-lined ground squirrel (*Spermophilus tridecemlineatus*), lack

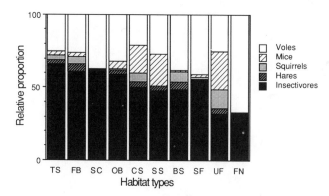

Fig. 7.14. Relative composition of small-mammal groups in peat and upland habitats of northern Minnesota. TS, tamarack swamp; FB, forested bog; SC, shrub carr; OB, open bog; CS, cedar swamp; SS, shrub swamp; BS, black spruce swamp; SF, shrub fen; UF, upland forest; FN, fen.

of representation in the central and eastern portions of the region may reflect relative scarcity at the edge of its range. But this does not explain the distributional pattern observed for the northern flying squirrel (*Glaucomys sabrinus*) at its range limits. For the meadow vole (*Microtus pennsylvanicus*) and southern bog lem-

ming (*Synaptomys cooperi*), representative habitats were sampled in all regions, and the relative abundance patterns must be indicative of regional differences in the quality and distribution of preferred habitat.

The relative composition of small-mammal groups in the various habitats provides another view of these communities (Fig. 7.14). Shrews are a major component of small-mammal communities in peatlands and a relatively smaller component in upland habitats. They are the dominant small mammal in forested bog, tamarack swamp, and shrub fen and have the least representation in upland forest. In general, habitats with high proportions of shrews show no common pattern of characteristics such as canopy cover, nutrient level, or predominant plant species. The relative abundance of shrews, however, must be considered in the context of the rodent species also occurring in these habitats. Forested bog typically has few voles or mice, whereas the fen can at times be overrun by voles. Fig. 7.15 illustrates that no shrew species occurs to any great extent in forested bog. Compared to the relative abundance of all other small mammals in this habitat, however, they are the most numerous.

Voles and mice occur in greatest proportions in fens and upland habitats. The high relative abundance of the meadow vole (*Microtus pennsylvanicus*) accounts for

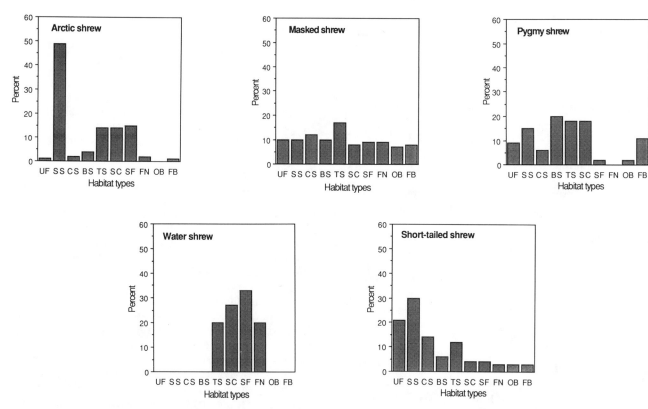

Fig. 7.15. Percent occurrence of shrew species among habitats in the peatlands region of northern Minnesota. UF, upland forest; SS, shrub swamp; CS, cedar swamp; BS, black spruce swamp; TS, tamarack swamp; SC, shrub carr; SF, shrub fen; FN, fen; OB, open bog; FB, forested bog.

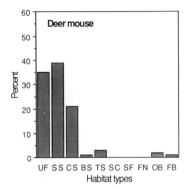

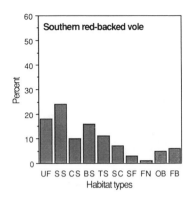

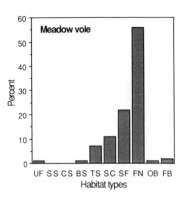

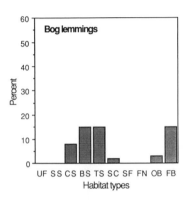

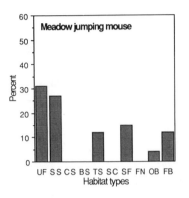

Fig. 7.16. Percent occurrence of mice and vole species among habitats in the peatlands region of northern Minnesota. UF, upland forest; SS, shrub swamp; CS, cedar swamp; BS, black spruce swamp; TS, tamarack swamp; SC, shrub carr; SF, shrub fen; FN, fen; OB, open bog; FB, forested bog.

the dominance of the fen by rodents (Fig. 7.16). Alternatively, the large rodent component observed in the upland habitats is made up primarily of the deer mouse (*Peromyscus maniculatus*) and the two jumping mice (*Zapus hudsonius and Napaeozapus insignis*). Squirrels, both the red squirrel (*Tamiasciurus hudsonicus*) and northern flying squirrel (*Glaucomys sabrinus*), are strongly associated with forested habitats of all types (Fig. 7.14). Never as abundant as other small-mammal groups, they attain their greatest relative abundance in nonpeat forests. The snowshoe hare (*Lepus americanus*) is broadly distributed across most habitats and absent only from the fen. Its higher occurrences may be associated with abundant shrub forage, as found in the shrub swamp or deciduous upland forest, and with habitat affording protection and concealment, such as the black spruce conifer swamp.

Any conclusions concerning the habitat preferences of small-mammal groups are based upon widely differing individual species responses (Figs. 7.15, 7.16). Species such as the arctic shrew (*Sorex arcticus*), meadow vole (*Microtus pennsylvanicus*), and meadow jumping mouse (*Zapus hudsonius*), for example, exhibit narrow habitat preferences, whereas the masked shrew (*Sorex cinereus*), pygmy shrew (*Sorex hoyi*),

and southern red-backed vole (*Clethrionomys gapperi*) are habitat generalists. The short-tailed shrew (*Blarina brevicauda*) and deer mouse (*Peromyscus maniculatus*) show strong preference for upland habitats, while the water shrew (*Sorex palustris*) and bog lemmings (*Synaptomys borealis* and *S. cooperi*) select only peatland habitats.

Each small mammal, with its peculiar adaptations, physiological limitations, and behavioral traits, is uniquely distributed among the various peatland habitats (Fig. 7.17). In habitats that best fulfill its requirements, a species is expected to occur regularly and in greatest relative abundance. Some species, such as the masked shrew (*Sorex cinereus*), appear to meet their needs under most peatland conditions, whereas the southern bog lemming (*Synaptomys cooperi*) is found among relatively few of the available habitats. In habitats that are secondarily suitable, small mammals occur sporadically or in smaller relative numbers. Such is the case of the southern red-backed vole (*Clethriono-mys gapperi*) that is recorded in all peatland habitats in northern Minnesota but shows noticeable variation in reliability of occurrence and relative abundance among the different peatland types. Small mammals cannot be viewed entirely as independent entities, however; their

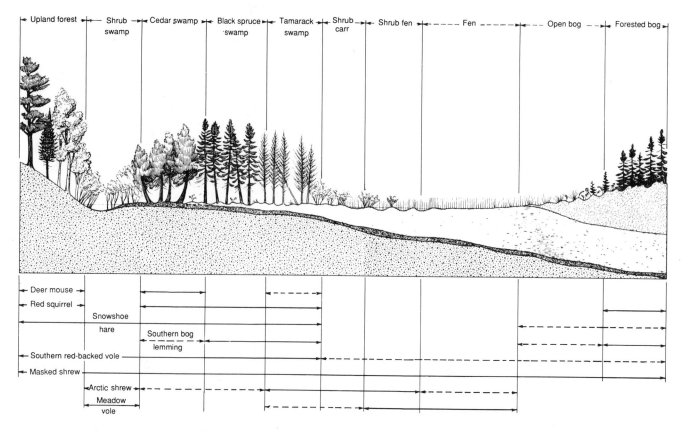

Fig. 7.17. Habitat associations among representative small-mammal species occurring in the patterned peatlands of northern Minnesota. (*Solid lines* represent primary habitat; *broken lines* represent secondary habitats.)

patterns of distribution and abundance are modified by interactions with other small mammal inhabitants. The small-mammal communities characteristic of particular peatland habitats are the sum of these individual and communal responses to the environment (Table 7.6).

The Importance of Peatlands to Small Mammals in Northern Minnesota

The small-mammal species residing in the peatlands of northern Minnesota are less diverse than those found in the adjoining uplands, and their densities generally are lower in the peatland habitats. Characteristics of the peatland environment such as the high water table, the absence of suitable burrowing substrate, and the limited vegetational diversity and productivity lead to the conclusion that peatlands are largely inhospitable environments for small mammals that require dry nest sites, protective shelter, and sufficient food resources to sustain viable populations. In fact, when these needs are amply met in the nearby uplands, it seems surprising to find small mammals in the peatlands at all. Yet they are there, at times in great numbers. What determines their presence in peatlands? Do the seemingly harsh peatlands offer some advantages — perhaps fewer predators or fewer competitors — over the up-

lands? Or do the peatlands constitute ecological "sinks" (*sensu* Stenseth 1983) into which the naive erroneously stray or the infirm are driven? Simple answers to these questions are not to be found in existing information about small mammals living in peatland habitats. This underscores the complexity of interactions between small mammals and the peat environment, as well as how much remains to be learned about the importance of peatlands to small mammals in northern Minnesota and elsewhere.

No obligatory relationship has been demonstrated between small mammals and peatlands in Minnesota; all can and do survive in nonpeat environments. On a species-by-species basis, however, the relative costs and benefits of living in peatlands vary widely. Clearly, some small-mammal species thrive in peatland conditions and appear to prefer peatlands over regional upland choices. The long-tailed shrews (*Sorex* spp.) occur consistently, in relatively high numbers, and in the case of the water shrew (*S. palustris*) may occur preferentially in peatlands in this region of Minnesota. Alternatively, some species are more successful in uplands. The woodland deer mouse (*Peromyscus maniculatus gracilis*) appears to be quite at home in upland habitats and cedar conifer swamps but rarely occurs in forested bogs. The appearance

Table 7.6. Small-mammal associations for habitats in the patterned peatlands of northern Minnesota (modified from Nordquist and Birney 1980).

Habitat	Small-mammal associations		
	Characteristic	Common	Occasional
Forested bog	Masked shrew Red squirrel Southern red-backed vole	Pygmy shrew Snowshoe hare Southern bog lemming	
Open bog	Masked shrew Southern red-backed vole Meadow vole		Pygmy shrew Water shrew Heather vole Southern bog lemming Northern bog lemming
Fen	Arctic shrew Masked shrew Pygmy shrew Meadow vole Southern red-backed vole		Water shrew Short-tailed shrew Meadow jumping mouse
Shrub fen	Masked shrew Arctic shrew Meadow vole	Short-tailed shrew Southern red-backed vole Meadow jumping mouse	Pygmy shrew Water shrew
Shrub carr	Masked shrew Arctic shrew Southern red-backed vole Meadow vole	Pygmy shrew Short-tailed shrew	Water shrew Northern bog lemming
Tamarack swamp	Arctic shrew Masked shrew Short-tailed shrew Snowshoe hare Red squirrel Southern red-backed vole	Pygmy shrew Deer mouse Meadow vole Meadow jumping mouse	
Black spruce swamp	Masked shrew Snowshoe hare Red squirrel Southern red-backed vole	Pygmy shrew Short-tailed shrew Southern bog lemming	Rock vole Heather vole
Cedar swamp	Masked shrew Short-tailed shrew Snowshoe hare Red squirrel Deer mouse Southern red-backed vole	Arctic shrew	Pygmy shrew Star-nosed mole Northern flying squirrel Least chipmunk
Shrub swamp	Arctic shrew Masked shrew Short-tailed shrew Snowshoe hare Southern red-backed vole Meadow vole	Pygmy shrew Meadow jumping mouse	Star-nosed mole
Upland forest	Masked shrew Short-tailed shrew Snowshoe hare Eastern chipmunk Red squirrel Deer mouse Southern red-backed vole	Pygmy shrew Least chipmunk Northern flying squirrel Rock vole Meadow jumping mouse Woodland jumping mouse	Franklin's ground squirrel

Fig. 7.18. Northern bog lemming (*Synaptomys borealis*). (Copyright Don Luce, by permission.)

of subadults in forested bogs during late summer suggests that this habitat, as well as other peatland types, may constitute dead-end habitats into which individuals wander but rarely if ever establish breeding populations. This very same habitat, on the other hand, supports sizable populations of the southern red-backed vole (*Clethrionomys gapperi*) and masked shrew (*S. cinereus*).

While species such as the woodland deer mouse wander widely in their exploration of potential habitats, other species appear tightly associated with particular peatland conditions. Bog lemmings (*Synaptomys* spp.) rarely occur outside the peatlands in northern Minnesota and typically are found in graminoid-dominated habitats, where they coexist with, and often are numerically dominated by, the meadow vole (*Microtus pennsylvanicus*). Southern bog lemmings (*S. cooperi*) are displaced from preferred habitat in Virginia by meadow voles during periods of high vole densities (Linzey 1984). Similar observations in Minnesota suggest that bog lemmings retreat to suboptimal peatland habitats rather than adjacent upland sites when meadow voles are numerous (unpublished data). The relatively rare occurrences of southern bog lemmings in some habitats, such as shrub fens and cedar swamps, may be the result of displacement by competing species rather than an indication of preference.

Whether they are wetland oases or soggy graves, peatlands affect the life histories of small mammals in ways as diverse as the habitats and the small mammals themselves. The impact of peatland environments on these small, secretive inhabitants of northern Minnesota is clearly profound, yet a thorough understanding of the underlying factors remains elusive.

The sedge "haystacks" piled in the sun by a bog lemming (Fig. 7.18), the heaps of spruce cone bracts dismantled by a red squirrel, and the alder stems diag-

onally pruned by a snowshoe hare are subtle reminders of the presence of these small residents of the Minnesota peatlands.

Literature Cited

Angelstam, P., L. Hansson, and S. Pehrsson. 1987. Distribution borders of field mice *Apodemus:* The importance of seed abundance and landscape composition. Oikos 50:123-130.

Bailey, B. 1929. Mammals of Sherburne County, Minnesota. Journal of Mammalogy 10:153-164.

Batten, E. M. 1980. Habitat ecology of small mammals in northeastern Minnesota and a critique of its application to assessment of metal mining impacts. Thesis. University of Minnesota, Minneapolis, Minnesota, USA.

Batzli, G. O. 1985. Nutrition. Pages 779-811 *in* R. H. Tamarin, editor. Biology of New World *Microtus*. American Society of Mammalogists, Special Publication no. 8.

Bee, J. W., and E. R. Hall. 1956. Mammals of northern Alaska on the Arctic Slope. Miscellaneous Publication, Museum of Natural History, University of Kansas 8:1-309.

Birney, E. C., W. E. Grant, and D. D. Baird. 1976. Importance of vegetative cover to cycles of *Microtus* populations. Ecology 57:1043-1053.

Brown, L. N. 1967. Ecological distribution of six species of shrews and comparison of sampling methods in the central Rocky Mountains. Journal of Mammalogy 48:617-623.

Brown, N. R., and R. G. Lanning. 1954. The mammals of Renfrew County, Ontario. Canadian Field-Naturalist 68:171-180.

Buckner, C. H. 1957. Population studies on small mammals of southeastern Manitoba. Journal of Mammalogy 38:87-97.

———. 1966. Populations and ecological relationships of shrews in tamarack bogs of southeastern Manitoba. Journal of Mammalogy 45:181-194.

Burt, W. H. 1946. The mammals of Michigan. University of Michigan Press, Ann Arbor, Michigan, USA.

Cahn, A. R. 1937. The mammals of the Quetico Provincial Park of Ontario. Journal of Mammalogy 18:19-30.

Chapman, J. A., J. G. Hockman, and M. M. Ojeda C. 1980. *Sylvilagus floridanus*. American Society of Mammalogists, Mammalian Species 136:1-8.

Christian, D. P., and J. M. Daniels. 1985. Distributional records of rock voles, *Microtus chrotorrhinus*, in northeastern Minnesota. Canadian Field-Naturalist 91:413-414.

Conner, P. F. 1953. Notes on the mammals of a New Jersey pine barrens area. Journal of Mammalogy 34:227-235.

———. 1959. The bog lemming *Synaptomys cooperi* in southern New Jersey. Publications of the Museum, Michigan State University, Biological Series 1(5):161-248.

de Vos, A. 1964. Range changes of mammals in the Great Lakes region. American Midland Naturalist 71:210-231.

Dice, L. R. 1921. Notes on the mammals of interior Alaska. Journal of Mammalogy 2:20-28.

Etnier, D. A. 1989. Small mammals of the Boundary Waters Canoe Area, with a second Minnesota record for the heather vole, *Phenacomys intermedius*. Canadian Field-Naturalist 103:353-357.

Forbes, R. B. 1966. Studies of the biology of Minnesota chipmunks. American Midland Naturalist 76:290-308.

Fox, R., T. Malterer, and R. Zarth. 1977. Inventory of peat resources in Minnesota. Minnesota Department of Natural Resources, Division of Minerals, Progress Report January 1977.

Getz, L. L. 1961a. Factors influencing the local distribution of *Microtus* and *Synaptomys* in southern Michigan. Ecology 42:110-119.

———. 1961b. Factors influencing the local distribution of shrews. American Midland Naturalist 65:67-88.

———. 1961c. Notes on the local distribution of *Peromyscus leucopus* and *Zapus hudsonius*. American Midland Naturalist 65:486-500.

Graham R. W., and E. L. Lundelius, Jr. 1984. Coevolutionary disequilibrium and Pleistocene extinctions. Pages 223-249 *in* P. S. Martin and R. G. Klein, editors. Quaternary extinctions, a prehistoric revolution. University of Arizona Press, Tucson, Arizona, USA.

Green, M. M. 1925. Notes on some mammals of Montmorency County, Michigan. Journal of Mammalogy 6:173-178.

Grime, J. P. 1978. Plant strategies and vegetation processes. Wiley, Chichester, U.K.

Gunderson, H. L., and J. R. Beer. 1953. The mammals of Minnesota. Occasional Papers of the Minnesota Museum of Natural History, University of Minnesota 6:1-190.

Hamilton, W. J., Jr. 1943. The mammals of eastern United States. Comstock, Ithaca, New York, USA.

Haukioja, E., and S. Koponen. 1975. Faunal structure of investigated areas at Kevo, Finland. Pages 19-28 *in* F. E. Wielgolaski, editor. Fennoscandian tundra ecosystems. Part 2: Animals and systems analysis. Springer-Verlag, New York, New York, USA.

Hazard, E. B. 1982. The mammals of Minnesota. University of Minnesota Press, Minneapolis, Minnesota, USA.

Heaney, L. R., and E. C. Birney. 1975. Comments on the distribution and natural history of some mammals in Minnesota. Canadian Field-Naturalist 89:29-34.

Hoffman, R. S. 1984. Small mammals in winter: The effects of altitude, latitude and zoogeographic region. Special Publication of the Carnegie Museum of Natural History 10:9-23.

——— and J. K. Jones, Jr. 1970. Influence of late-glacial and postglacial events on the distribution of recent mammals on the northern Great Plains. Pages 355-394 *in* W. Dort, Jr., and J. K. Jones, Jr., editors. Pleistocene and recent environments of the central Great Plains. University of Kansas Department of Geology, Special Publication 3.

——— and R. S. Peterson. 1967. Systematics and zoogeography of *Sorex* in the Bering Strait area. Systematic Zoology 16:127-136.

Iverson, S. L., R. W. Seabloom, and J. M. Hnatink. 1967. Small mammal distributions across the prairie-forest transition of Minnesota and North Dakota. American Midland Naturalist 78:188-197.

Johnson, C. E. 1922. Notes on the mammals of northern Lake County, Minnesota. Journal of Mammalogy 3:33-39.

———. 1930. Recollections of the mammals of northwestern Minnesota. Journal of Mammalogy 11:435-452.

Jones, J. K. Jr., D. M. Armstrong, R. S. Hoffman, and C. Jones. 1983. Mammals of the northern Great Plains. University of Nebraska Press, Lincoln, Nebraska, USA.

——— and E. C. Birney. 1988. Handbook of mammals of the north-central states. University of Minnesota Press, Minneapolis, Minnesota, USA.

Jonsson, S. 1975. Faunal structure of IBP tundra site and its surroundings, Abisko, Sweden. Pages 46-54 *in* F. E. Wielgolaski, editor. Fennoscandian tundra ecosystems. Part 2: Animals and systems analysis. Springer-Verlag, New York, New York, USA.

Judd, W. W. 1961. Studies of the Byron Bog in southwestern Ontario. XII. A study of the population of insects emerging as adults from Redmond's Pond in 1951. American Midland Naturalist 65:89-100.

Jung, H. G., and G. O. Batzli. 1981. Nutritional ecology of microtine rodents: Effects of plant extracts on the growth of arctic microtines. Journal of Mammalogy 62:286-292.

Kalin, O. T. 1976. Distribution, relative abundance, and species richness of small mammals in Minnesota, with an analysis of some structural characteristics of habitats as factors influencing species richness. Dissertation. University of Minnesota, Minneapolis, Minnesota, USA.

Karns, D. R., and P. J. Regal. 1979. The relationship of amphibians and reptiles to peatland habitats in Minnesota. Minnesota Department of Natural Resources, Division of Minerals, St. Paul. Peat Program Final Report December 1, 1979.

Kirkland, G. L. Jr., and F. J. Jannett, Jr. 1982. Microtus chrotorrhinus. American Society of Mammalogists, Mammalian Species 180:1-5.

Leraas, H. J. 1942. Notes on mammals from west central Minnesota. Journal of Mammalogy 23:343-345.

Linzey, A. V. 1983. *Synaptomys cooperi*. American Society of Mammalogists, Mammalian Species 210:1-5.

———. 1984. Patterns of coexistence in *Synaptomys cooperi* and *Microtus pennsylvanicus*. Journal of Mammalogy 65:382-393.

Long, C. A. 1972. Notes on habitat preference and reproduction in pigmy shrews, *Microsorex*. Canadian Field-Naturalist 86:155-160.

Lundelius, E. L. Jr., R. W. Graham, E. Anderson, J. Guilday, J. A. Holman, D. W. Steadman, and S. D. Webb. 1983. Terrestrial vertebrate faunas. Pages 311-353 *in* S. C. Porter, editor. Late-Quaternary environments of the United States, vol. 1, The Late Pleistocene. University of Minnesota Press, Minneapolis, Minnesota, USA.

Macpherson, A. H. 1965. The origin of diversity in mammals of the Canadian arctic tundra. Systematic Zoology 14:153-173.

Manville, R. H. 1949. A study of small mammal populations in northern Michigan. Miscellaneous Publications, Museum of Zoology, University of Michigan 73:1-83.

——— and S. P. Young. 1965. Distribution of Alaskan mammals. Bureau of Sport Fisheries and Wildlife, U.S. Fish and Wildlife Service Circular 211:1-74.

McAllister, J. A., and R. S. Hoffman. 1988. *Phenacomys intermedius*. American Society of Mammalogists, Mammalian Species 305:1-8.

Miller, D. H., and L. L. Getz. 1977. Factors influencing local distribution and species diversity of forest floor small mammals in New England. Canadian Journal of Zoology 55:806-814.

Nordquist, G. E., and E. C. Birney. 1980. The importance of peatland habitats to small mammals in Minnesota. Minnesota Department of Natural Resources, Division of Minerals, St. Paul. Peat Program Final Report.

Pietz, P. J., and J. R. Tester. 1983. Habitat selection by snowshoe hares in north central Minnesota. Journal of Wildlife Management 47:689-696.

Pruitt, W. O. Jr. 1959. Microclimates and local distribution of small mammals on the George Reserve, Michigan. Miscellaneous Publication of the Museum of Zoology, University of Michigan 109:1-27.

Rand, A. L. 1933. Notes on the mammals of the interior of western Nova Scotia. Canadian Field-Naturalist 47:41-50.

Schmidt, F. J. W. 1931. Mammals of western Clark County, Wisconsin. Journal of Mammalogy 12:99-117.

Smith, D. A., and J. B. Foster. 1957. Notes on the small mammals of Churchill, Manitoba. American Midland Naturalist 90:228-231.

Sölhoy, T., E. Östbye, H. Kauri, A. Hagen, L. Lien, and H.-J. Skar. 1975. Faunal structure of Hardangervidda, Norway. Pages 29-45 *in* F. E. Wielgolaski, editor. Fennoscandian tundra ecosystems. Part 2: Animals and systems analysis. Springer-Verlag, New York, New York, USA.

Soper, J. D. 1948. Mammal notes from the Grande Prairie-Peace River Region, Alberta. Journal of Mammalogy 29:49.

———. 1961. Field data on the mammals of southern Saskatchewan. Canadian Field-Naturalist 75:23-41.

Stenseth, N. C. 1983. Causes and consequences of dispersal in small mammals. Pages 63-101 *in* I. R. Swingland and P. J. Greenwood, editors. The ecology of animal movement. Oxford University Press, Oxford, U.K.

Stockwell, S. S., and M. L. Hunter. 1985. Distribution and abundance of birds, amphibians and reptiles, and small mammals in peatlands of central Maine. Maine Department of Inland Fisheries and Wildlife Report.

Swanson, G., T. Surber, and T. S. Roberts. 1945. The mammals of Minnesota. Technical Bulletin of the Minnesota Department of Conservation 2:1-108.

Timm, R. M. 1975. Distribution, natural history, and parasites of mammals of Cook County, Minnesota. Occasional Papers of the Bell Museum of Natural History, University of Minnesota 14:1- 56.

———, L. R. Heaney, and D. D. Baird. 1977. Natural history of rock voles (*Microtus chrotorrhinus*) in Minnesota. Canadian Field-Naturalist 91:177-181.

van den Brink, F. H. 1968. A field guide to the mammals of Britain and Europe. Houghton Mifflin, Boston, Massachusetts, USA.

Webb, T. III, E. J. Cushing, and H. E. Wright, Jr. 1983. Holocene changes in the vegetation of the Midwest. Pages 142-165 *in* H. E. Wright, Jr., editor. Late-Quaternary environments of the United States, vol. 2, The Holocene. University of Minnesota Press, Minneapolis, Minnesota, USA.

Whitaker, J. O. Jr. 1972. *Zapus hudsonius*. American Society of Mammalogists, Mammalian Species 11:1-7.

——— and L. L. Schmeltz. 1973. Food and external parasites of *Sorex palustris* and food of *Sorex cinereus* from St. Louis County, Minnesota. Journal of Mammalogy 54:283-285.

Wunder, B. A. 1985. Energetics and thermoregulation. Pages 812-844 *in* R. H. Tamarin, editor. Biology of New World Microtus. American Society of Mammalogists, Special Publication no. 8.

CHAPTER 8

Bird Populations

*Gerald J. Niemi and
JoAnn M. Hanowski*

There has long been a mystique among birders and naturalists about the rare, exotic, or unusual birds that might be found in the vast and remote peatlands of northern Minnesota. The source of this speculation is unclear, but perhaps the soggy feel and musty smell of *Sphagnum* moss are contributing factors. Because peatlands are reputed to be inhospitable places where one might sink through the peat at any step or be devoured by mosquitos and flies, the expectation is that a new "species" of bird might even be discovered, because surely no one has spent much time exploring these areas.

Minnesota peatlands can certainly be inhospitable at times, and they are definitely a different kind of place. They feel soggy and smell unusual because you may be walking on 10 m (30 feet) or more of decaying plant and animal remains. Yet in the spring the color of the Labrador tea, bog rosemary, and cottongrass in full bloom and the sights and sounds of the sandhill crane (Fig. 8.1), sharp-tailed grouse, and palm warbler (Fig. 8.2) are unforgettable experiences. But what about those nasty mosquitos and flies? Certainly, you should not visit the peatland poorly clothed, especially in early morning or late afternoon in forested portions of the peatland where they congregate. But at other times they are not bad, especially balanced against the pleasure of seeing the elusive Connecticut warbler (Fig. 8.3) or Lincoln's sparrow.

Research in Peatlands

Discovery of a new bird species in Minnesota peatlands is unlikely. Today this is possible only in the remote forests of South America, Africa, or New Guinea. The amount of ecological research on Minnesota peatlands is limited, however, in comparison with that on habitats like deciduous forests. We have a good idea of the

bird species found within peatlands as well as the habitats where they occur, the times of the year they are found, and their abundance. Our knowledge of how important Minnesota peatlands are to the survival of several rare bird species like the sandhill crane,[1] yellow rail, and great gray owl (Fig. 8.4) is more limited. Questions about the role of birds in processing nutrients and in controlling insect abundances are still relatively unexplored. As pressure to use peatlands for agriculture, forestry, and energy increases, the need for greater knowledge also will increase.

Information on Minnesota's peatland birds cannot be found in a single document. Radisson, Hennepin, Carver, and other 18th-century explorers undoubtedly observed birds in Minnesota peatlands, but their records are too vague to be of much use. In the 19th century, scientists like Keating, Say, Blakiston, Trippe, and Hatch traversed Minnesota observing and collecting birds. Fortunately, much of the information on Minnesota ornithology up to the 1930s was summarized in the classic treatise *The Birds of Minnesota* by T. S. Roberts (1932). This two-volume book on Minnesota birds contains a wealth of information on the life history and distribution of birds, including those that occur in peatlands. Since the 1930s, a number of observational accounts of birds that occur in peatland habitats have been published (Swanson 1940; Beer and Frenzel 1956; Hofslund 1956; Kendeigh 1956; Cottrille 1964; Hickey *et al.* 1965; Green 1971; Green and Niemi 1978; Niemi and Pfannmuller 1979; Eckert 1983), but most have not focused particularly on peatland birds. Among the more detailed accounts, Green and Janssen (1975) summarized all available data re-

---

[1] All scientific names of birds regularly found in Minnesota peatlands are listed in Appendix 1.

Fig. 8.2: The palm warbler is a common breeding species of muskeg and black spruce forests. Males can be seen perched near the top of spruce trees singing their buzzy, slow trill. A distinctive feature of this species is its constantly bobbing tail. (Copyright Don Luce, by permission.)

series of studies on how disturbances affect peatland bird species.

Because many of the birds that occur in Minnesota's peatland also are found in peatlands from Maine to Alaska, we can use information gathered by researchers like Erskine (1977), who has studied

Fig. 8.1: The sandhill crane, a species of special concern in Minnesota, breeds in open fen areas of the Red Lake peatland. Adults stand over 1m (3 feet) tall; they are distinguished by their gray plumage and red forehead. They can often be seen feeding in agricultural fields adjacent to peatlands. Wetland drainage and fragmentation have contributed to the decline of this species throughout its breeding range in northwestern Minnesota. (Copyright Don Luce, by permission.)

garding the phenology and distribution of birds in the state, including all birds that occur in peatlands.

Although Minnesota has more peatland area than any other U.S. state except Alaska, less than ten years ago there were no comprehensive studies of peatland birds. Inspired by the potential economic development of peatlands, the Minnesota Department of Natural Resources (DNR) initiated a series of biological studies on peatlands. Warner and Wells (1984) established a network of line transects and mist-nets to observe birds within the major habitats of the Red Lake peatland. Marshall and Miquelle (1978) initiated a comprehensive review of information available on many of the birds found in peatland habitats. Pietz and Tester (1979) gathered more detailed data on how spruce grouse and ruffed grouse used peatland habitats. More recently, Niemi and Hanowski (1984, 1986a, 1987) initiated a

Fig. 8.3: The Connecticut warbler, a common breeding species of forested peatlands in northern Minnesota is often heard but seldom seen. Look for the male who usually sings his loud and distinctive song on a perch about one meter below the top of spruce trees. The bold white eye ring and gray head and nape distinguish this species. (Copyright Don Luce, by permission.)

Fig. 8.4: The great gray owl is a permanent resident of northern Minnesota peatlands. It nests in forested peatlands but is probably more commonly seen in the winter, especially during "invasion" years. (Copyright Don Luce, by permission.)

peatlands throughout Canada; Stockwell and Hunter (1985) in Maine; Anderson (1982), Brewer (1967), and Ewert (1982) in Michigan; Weise (1973), Idzikowski (1982), and Niemi and Hanowski (1986b) in Wisconsin; Carbyn (1971) in the Northwest Territories; and Spindler and Kessel (1980) in Alaska.

An examination of these studies indicates that bird communities found in peatlands of northern Minnesota are transitional between those of the boreal zone to the north, the temperate deciduous forest to the southeast, and the prairies to the west. For instance, boreal species found in peatlands of southern Wisconsin include the winter wren, white-throated sparrow, and swamp sparrow (Weise 1973). Other boreal species such as the boreal chickadee, hermit thrush, ruby- and golden-crowned kinglets, palm warbler, and Lincoln's sparrow, which are well represented in Minnesota peat-

lands, are rare in peatlands of southern Wisconsin. Minnesota peatlands, especially the Red Lake peatland area, have several species that are predominantly found in prairie regions. These include LeConte's sparrow, bobolink, and sharp-tailed grouse, which are rare in peatlands in the boreal zone of Canada (Erskine 1977). Minnesota peatlands also contain many species that are common in the eastern deciduous forests, but most, such as the veery, rose-breasted grosbeak and song sparrow, are associated with shrubs or with edge habitats near roads or drainage ditches within the peatland.

More information is available on peatland birds in Finland and Sweden than in any other countries. Research by Sammalisto (1955, 1957), Merikallio (1958), Hildén (1967), Häyrinen (1970), Järvinen and Sammalisto (1976), Väisänen and Järvinen (1977), Nilsson (1977, 1980, 1982), Kouki and Järvinen (1980), and many others is respected by ornithologists throughout the world. In other countries with peatlands, research has also been relatively intense, including the work of Dyrcz *et al.* (1972, 1973) in Poland and Kumari (1972) and Renno (1958) in Estonia.

Among the most striking contrasts between peatland birds of North America and those in the Old World is that the predominant species in Minnesota peatlands are small birds like warblers, sparrows, and wrens, whereas the predominant species in the Old World are wagtails, pipits, and waders, including the yellow wagtail, meadow pipit, lapwing, ruff, wood sandpiper, whimbrel, godwit, and snipe.[2] Even gulls like the herring gull and the black-headed gull are now relatively common in many peatlands in the Old World. In Minnesota peatlands, waders are rare except for the common snipe and Wilson's phalarope; the latter is primarily found in open areas of the Red Lake peatland. Birders would find it unusual to see a gull in a Minnesota peatland area, but in the Old World gulls have been recent colonists of peatlands. The black-headed gull, for instance, was not found in the Biebrza marshes of Poland in the late 1800s but is now a common breeder there (Dyrcz *et al.* 1972). The species has colonized the virgin peatlands of Finland since World War II (Järvinen and Ulfstrand 1980). Gulls are expanding both in population and in range throughout the upper midwestern United States, and it is possible that their range expansion will be similar to that already observed in the Old World.

Among the major reasons for the differences in the kinds of birds found in Minnesota and Old World peatlands are differences in the structure of the vegetation: peatlands in the Old World have less vegetation and more open water than those in Minnesota (Niemi *et al.* 1984). Although there are "open" sedge fen and muskeg areas, Minnesota tends to have denser vegetation and few water pools.

---

[2] See Appendix 2 for scientific names of European species and nonpeatland birds of North America mentioned in the text.

Evolution of Peatland Birds

The evolution of peatland birds can be examined at two broad levels: (1) the process by which a species came to reside in peatland habitats, which involves the study of biogeography, and (2) possible adaptations for use of peatland habitats. Because no information on biogeographic evolution of peatland bird communities in the New World exists, ideas on biogeographic patterns must be inferred from those proposed for the North American avifauna as a whole. Discussions on the adaptation of individual species to peatlands are limited to morphological patterns.

Several scientists have theorized about why birds are distributed across the North American landscape as we see them today. The estimated age of bird species that occur in peatland habitats is greater than the age of peatlands themselves. The Minnesota peatlands, for instance, are less than 5,000 years old, and most evolutionary biologists believe that more than 5,000 years are necessary for a new species to evolve. The species found in peatlands today probably have evolved in areas that were precursors of today's peatlands or in habitats similar to those found in today's peatlands.

Scientists have speculated on how some bird species began using peatlands, but there is no consensus on this issue. One fact about speciation that is usually agreed upon, however, is the importance of geographic isolation.

Simply stated, the evolution of a new species depends on the separation of a sufficient number of individuals of that species from the parent stock. Separation is usually achieved by the development of a boundary such as a mountain range or body of water that is large enough to inhibit genetic exchange between the separated populations. The boundary separating two populations must not only be large enough to maintain isolation, it must exist long enough for the populations to evolve sufficient differences to achieve the status of two distinct species (either two new species or one old and one new).

If species did not evolve in peatlands, then the biogeographic patterns we observe today must have been a consequence of isolating events that occurred prior to the formation of our present-day peatlands. These isolating events are likely to have been associated with the formation of glacial refugia (e.g., Mengel 1964). As glacial ice from the north expanded and contracted several times during the past million years, populations of species became isolated. If conditions were suitable, these populations could become two species. Certain refugia could have peatland or peatland-like habitats where the isolated populations evolved specific characteristics for survival. Among the most common examples of this form of speciation are eastern and western species counterparts like the mourning warbler and MacGillivray's warbler (*Oporonis tolmiei*) (Mengel 1964) or the chickadee species complex (Lack 1969). Other examples include species with distinct subspecies in the east and west such as the flicker complex (Miller 1956), the northern oriole (Corbin 1979), and the song sparrow (Aldrich 1984). We know of no eastern and western examples for peatland species, but the yellow-rumped warbler, a common species in forested peatlands, was once considered a separate species in the east and west (Hubbard 1969); the two forms were found to interbreed in areas where their ranges overlapped. This is now used as an example of a situation where differences evolved (e.g., in plumage), but the populations were not isolated long enough to achieve sufficient differences to prevent reproductive isolation. Because birds are so mobile, studies of speciation on a large continental land mass are difficult.

The second level of evolution is concerned with the mechanisms that a species uses to survive in peatland habitats. We can consider the peatland environment a test to which the body form of a species has been subjected. Survival (presence) or extinction (absence) of the species in the peatland is a measure of whether the species passed or failed the test.

The morphological forms of birds found in peatland habitats are diverse, ranging from ducks' webbed feet and flattened bills, which aid them in aquatic areas, to kinglets' small, needlelike bills and delicate feet, which allow them to glean small insects hiding in the needles of spruce trees. Of course these structural features of peatland birds are also relevant to non-peatland habitats with vegetational or landform features in common with peatlands. We presume these features are adaptations, but a convincing argument that a feature is an adaptation with a genetic and evolutionary base requires more evidence. Evidence that an adaptation has a biological base is more convincing if we observe it in many different organisms, especially those that are genetically unrelated (e.g., different families). This is called convergent evolution. An extreme example is the similarity between fish and mammals (dolphins, whales) that inhabit marine environments. A similar analogy, but on a finer scale, can be made in peatland habitats. Niemi (1985) compared peatland birds and habitats in Finland with those in Minnesota. He wanted to know if species of the same genus (congeners) that occurred in both shrub and coniferous forest peatlands had similar patterns of morphological differences that could be associated with habitat. The assumptions were that congeners have evolved from a common ancestor and therefore that morphological differences among congeners were attributable to their use of different habitats. Species found in peatland shrub habitats had longer and thinner bills, longer legs, and shorter wings (e.g., yellow warbler and swamp sparrow) than their congeners found in coniferous forests (e.g., yellow-rumped warbler and Lincoln's sparrow) regardless of continent. This indicated that morphology of birds is closely related to the habitats they use. Most of these differences would be imperceptible to the human eye, but they probably represent important adaptations for survival in peatland habitats.

Ecology of Peatland Birds

Seasonal and Annual Patterns

Appendix 1 is a list compiled from all systematic inventories of Minnesota peatlands of the bird species regularly found in Minnesota's peatlands during some portion of the year and the approximate times they can be observed. Species that are seen only occasionally in the peatlands or that have been seen only in the riparian zones or near rice paddies close to Minnesota peatlands are not included. Although many of the latter species are commonly observed near peatlands (e.g., common grackle and marsh wren), our experience indicates that many of them are associated only with habitats that are not found in natural peatlands. A total of 110 species — 36% (110 of 304) of species regularly found in Minnesota — are listed as major users of peatland habitats.

Birds are easier to see and count in some seasons than in others. Most birds (especially males) are conspicuous from late May to early July, when they are actively singing to defend territories and to attract mates. During this period it is easiest to see and hear birds, and we get an impression that more species and individuals are present than during other periods. Most species and individuals should be present during migration season, however, especially during fall, when a habitat would include resident breeding species, migratory species, and young-of-the-year birds of both resident and migratory species. Because of these differences in census efficiency among seasons, we have refrained from making comparisons of bird activity between migration periods and breeding seasons. In northern latitudes, of course, the number of bird species and number of individuals we observe in peatlands is higher in spring, summer, and fall than in winter.

Winter's hostile weather and fewer daylight hours have a drastic effect on the availability of food, especially insects, and many birds leave the state until spring returns. In a census of 40 km (27 miles) of transects during four days in the winter of 1979 in the Red Lake peatland we observed only 4 species and 12 individuals. One Christmas bird count in the Sax and Zim area of northern Minnesota (Eckert 1978) recorded 25 species in a circle with a 12.5-km (7.5-mile) radius, but the total count area included only 30% peatland habitats. Although the number of species and number of individuals may be very low during winter, there are some unusual birds. During a field trip to the Red Lake peatland in late December 1980, Niemi and Finnish ornithologist Olli Järvinen on one morning observed a gyrfalcon, rough-legged hawks, sharp-tailed grouse, a snowy owl, gray jays, boreal chickadees, and pine grosbeaks. Similarly, among the 25 species observed in the Christmas bird count of 1977 (Eckert 1978), the goshawks, great gray owls, gray jays, and white-winged crossbills probably were seen in peatland areas.

Patterns in bird communities change with the sea-

son. During summer, for example, more species and individuals can be observed in shrub areas and along ditches and rivers (riparian zones), while during the winter more are observed in forested areas and fewer in shrub areas or open fens. Shifts in availability and accessibility of food and cover are likely reasons for these changes. During colder months, forested areas offer as food seeds and some insects and provide shelter from hostile weather. In contrast, open habitats such as fen and shrub areas probably provide little accessible food and only meager cover during the months of snow cover.

Because of the difficulty in comparing the number of species and individuals observed between migration and breeding seasons, we focus most of our remaining discussions on annual variation of peatland bird communities during the breeding season. Birds usually occur in similarly structured habitats during migration (Parnell 1969; Bairlein 1981) and on wintering grounds (Lack 1976; Fitzpatrick 1980; Terborgh 1980). However, Warner and Wells (1984) in their studies in the Red Lake peatland observed that many species used habitats that were different from their breeding habitats. Whether birds in peatlands occupy different habitats during migration than they do during the breeding season must still be regarded as an unresolved issue that deserves additional study.

The density of birds breeding in peatlands varies considerably from year to year, but the species found in peatland areas are relatively consistent (Fig. 8.5). For example, four species commonly are found every year in open fen areas: LeConte's sparrow, bobolink, sedge wren, and savannah sparrow (Table 8.1). Densities of these common species, however, can double or be halved from one year to the next. A census of 40 ha (100 acres) of sedge-fen habitat in the Red Lake peatland in three consecutive years found that density of LeConte's sparrow ranged from 60 to 108 breeding pairs in this area. Similar fluctuations in density occurred in all peatland habitats we examined (low shrub, high shrub, muskeg, and lowland conifer forests).

Most fluctuations in the number of species breeding in peatlands are due to the disappearance or reappearance of uncommon or rare species in a community. These species probably cannot maintain constant or stable populations because the habitat may be marginal or because their populations are limited by other factors such as high winter mortality in areas away from breeding sites. As habitats change in the natural process of succession, bird community composition will change. Successional changes are relatively slow in peatlands, however, and few changes occur over the course of a human lifetime.

Stability in the presence of common species in breeding bird communities is probably due to the ability of these species to maintain population levels adequate to allow them to recolonize these areas successfully year after year. Factors affecting density often are classified into two types: density-dependent factors like the

A. Number of breeding pairs

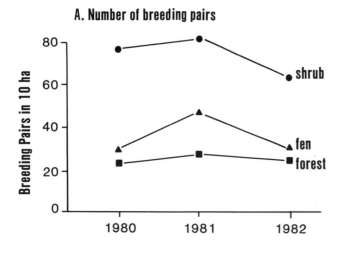

B. Number of species

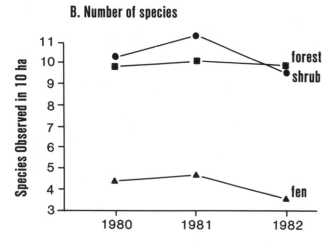

Fig. 8.5: Annual variation of three peatland bird communities (shrub, fen, and forest) from 1980 to 1982, based on breeding-bird censuses for (A) breeding density and (B) species richness.

Table 8.1. Range of abundances in number of breeding pairs observed in a 10-ha (25-acre) area for primary species found in five habitats of the Red Lake peatland during the breeding seasons from 1980 to 1982

Species	Peatland Habitats				
	Sedge fen	Low shrub	High shrub	Muskeg	Spruce
Alder Flycatcher			1-5		1-2
Yellow-bellied Flycatcher					
Sedge Wren	5-14	15-24	19-30		
Hermit Thrush					1-2
Nashville Warbler				0-2	4-6
Yellow Warbler		2-3			
Yellow-rumped Warbler				0-2	1-3
Palm Warbler				0-2	5-6
Common Yellowthroat		6-14	13-16		
Connecticut Warbler					3-6
Bobolink	5-6	4-5	3-4	0-1	
Savannah Sparrow	1-2			8-14	
LeConte's Sparrow	15-27	7-14	7-11	0-2	
Dark-eyed Junco					1-2
Chipping Sparrow					0-1
Clay-colored Sparrow		2-9	0-3	0-2	
Lincoln's Sparrow				2-8	
Swamp Sparrow		3-10	10-18		

Source: Niemi and Hanowski 1984.

availability of nest sites and density-independent factors like weather and catastrophic events. Bird species have relatively complex life histories, so it is likely that several factors affect population levels. One factor may limit populations one year and another factor the next. Bobolinks, for instance, breed in fen and shrub peatlands but spend their winters on the pampas in South America. Populations of bobolinks could be reduced because of food shortages on the wintering grounds one year and reduced another year because of poor reproductive success on breeding grounds. Progress in identifying and predicting population fluctuations is hampered by our inability to measure all these variables in the field.

Bird Habitats of Peatlands

Bird species are distributed in peatlands according to the structure and species composition of the vege-

tation. The Connecticut warbler is usually found in semi-open to closed coniferous forests dominated by black spruce or tamarack, for example, while LeConte's sparrow is found in fens and sedge-shrub areas. Because of this range of habitat preferences, two areas rarely have the same combination of species. The presence of a tree with a nest hole may determine whether a tree swallow will occur in a shrub area, while the density of black spruce may determine the presence of palm warblers. On the basis of the vegetational composition, it is possible to make some reasonable predictions about the species likely to be present in an area.

Scientists have debated whether bird species form discrete communities or whether they are distributed along gradients of the environment independently of each other. We view species as independent entities that occupy a selected portion of an environmental gradient. The concept of a bird community (a collection of species that occupies the same segment of the gradient) is simply a convenient way of communicating our ideas on bird distributions. In Figs. 8.6 and 8.7 we have drawn a gradient that ranges from sedge-shrub peatland on the far left to dense coniferous forest on the far right. Both areas emanate from the central area of open-water pools and short, sparse sedge vegetation. The left side of the figure represents minerotrophic, or nutrient-rich, peatlands, which receive nutrients from

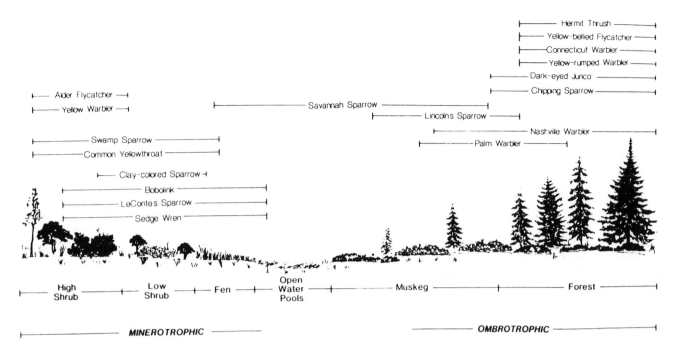

Fig. 8.6: Visual description of a peatland habitat gradient representing the predominant peatland habitat types. The gradient represents peatlands from highly vegetated minerotrophic (productive) habitats on the left to forested ombrotrophic (less productive) habitats on the right. The 18 most abundant species (Table 8.1) of Minnesota peatlands are shown in relation to their distribution in these peatland habitats.

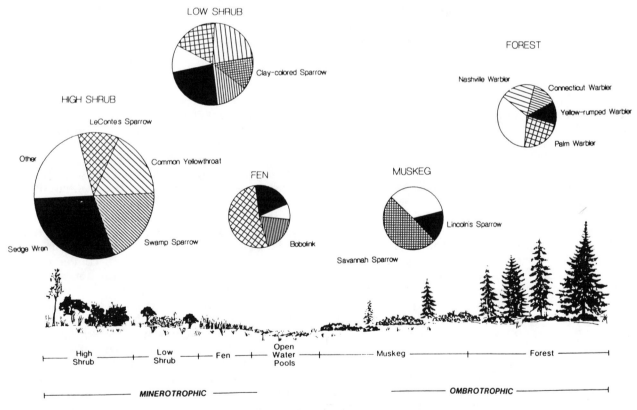

Fig. 8.7: Community composition of breeding birds for five segments of the peatland environment. The size of each circle is proportionate to the density of the bird community, and the predominant species of that community are shown in proportion to their abundance.

rainfall and the surrounding land mass; the right side represents ombrotrophic, or nutrient-poor, peatlands, which receive nutrients primarily from rainfall.

In Fig. 8.6 we focus on the distribution of the 18 most abundant breeding species found in Minnesota's peatlands within the peatland gradient. In our discussion we cite other species that characterize selected portions of this gradient. Information on habitat associations for all species is listed in Appendix 1. In Fig. 8.7 we summarize the predominant community of breeding birds found along this gradient in conjunction with their relative abundance within the community.

Beginning from the central portion of Fig. 8.6 or 8.7 and moving to the left, the open portions of the peatland are dominated by the savannah sparrow. Other characteristic species are the northern harrier and sandhill crane; these areas need to be quite large (at least several hundred acres) for these two species to occur. As vegetation becomes denser and dominated by sedges, as in fens, the LeConte's sparrow, sedge wren, and bobolink become more abundant, and densities of the savannah sparrow decrease. In these areas one also can find the short-eared owl, Wilson's phalarope, and yellow rail. In areas where shrubs such as bog birch appear with sedges, the clay-colored sparrow, common yellowthroat, and swamp sparrow appear. Finally, as sedge areas become dominated by willow and other shrubs, species such as the alder flycatcher and yellow warbler appear, but predominant species are still the sedge wren, swamp sparrow, and common yellowthroat. Other species less commonly found in these areas are the rose-breasted grosbeak, veery, American bittern, and Nashville warbler. The Nashville warbler is one of only two species we have commonly observed in both ombrotrophic and minerotrophic peatlands.

Beginning again from the central portion of Fig. 8.6 or 8.7 and moving to the right, the matted sedge and heath vegetation of the ombrotrophic areas is dominated by the savannah sparrow (the other species that occurs in both peatland types). Where a few scattered black spruce or tamarack trees are present, the eastern kingbird can be found. As these areas become dominated by taller heath shrubs such as Labrador tea, bog rosemary, and leatherleaf and by higher densities of black spruce or tamarack, the Lincoln's sparrow, palm warbler, and a few gray jays appear. These areas are often referred to as muskeg, and they are relatively low in breeding-bird species. Along with fens, they are used as dancing grounds by the sharp-tailed grouse (Hanson 1953). Finally, as these semi-open areas become forests, one finds a variety of species, including the yellow-bellied flycatcher, hermit thrush, Nashville warbler, yellow-rumped warbler, palm warbler, chipping sparrow, dark-eyed junco, and white-throated sparrow. Other species less commonly found in these forested peatlands include the gray jay, blue jay, boreal and black-capped chickadees, red-breasted nuthatch, Swainson's thrush, and ruby- and golden-crowned kinglets. These forested areas are also the best places to

find some of the rarest species of peatlands birds such as the sharp-shinned hawk, northern hawk-owl, great gray owl, spruce grouse, black-backed woodpecker, and red and white-winged crossbills.

The species and habitats described here primarily are indicative of the natural patterned peatlands of Minnesota. To gain access to these peatlands one must travel on roads that cross peatlands or canoe along ditches such as those in the Red Lake peatland, which were created in attempts to drain the peatlands in the early 1900s (chapter 17; Soper 1919). Roads and ditches have created riparian zones that consist of thick shrubs such as alder and willow, trees such as aspen, and bridges that cross ditches or streams. In these areas one commonly finds mallard, blue-winged teal, northern flicker, eastern kingbird, eastern phoebe, cliff swallow, barn swallow (the phoebe and swallows near bridges or dwellings), American crow, American robin, warbling vireo, red-winged blackbird, Brewer's blackbird, brown-headed cowbird, and song sparrow. Finally, within large peatlands like the Red Lake peatland, one can often find patches of deciduous forest with an avifauna typical of deciduous forests, including the broad-winged hawk, least flycatcher, red-eyed vireo, black-throated green warbler, chestnut-sided warbler, and American redstart.

Bird Communities of Peatlands

Characteristics of bird communities are described by several variables, including species richness and total density, and by a variety of indices (e.g., Shannon-Wiener diversity index, Simpson's Index) that combine density and species richness. We focus here on species richness and density because they are easiest to interpret and because particular indices usually are highly correlated with species richness or density (Hurlbert 1971; Peet 1974; James and Wamer 1982).

We consider peatland habitats and communities as segments of a gradient across the peatland landscape. Communities are snapshots of this continuous gradient. Here we consider five such snapshots: sedge fen, low shrub, high shrub, muskeg or open conifer forest, and closed conifer forest (Fig. 8.6). The number of species that can be observed in an area during summer is highest in high shrub and conifer forests of the peatland. In a typical year about 15 species could be commonly observed in a 10-ha (25-acre) area in the high-shrub habitat and about 11 species in the closed conifer forest, but only about 4 species in the same-sized sedge-fen habitat (Table 8.1, Fig. 8.7). The number of individuals varies considerably among habitats. More individuals would be observed in an area of minerotrophic peatland than in a similarly sized area of an ombrotrophic peatland. Although the number of individuals observed varies from year to year, more than 120 individuals typically could be found in a 10-ha area in the high-shrub habitats during the breeding season. In contrast, probably fewer than 80 individuals would be found in 10 ha of the muskeg or closed conifer forests.

There are two primary reasons for these differences. Species richness has been correlated with the complexity of the vegetation. *Complexity* is a difficult term to define, but we interpret it in terms of different vegetational structures available (e.g., nest sites, foraging areas, and cover) in a habitat (i.e., more niches). Areas with a larger number of habitat structures ("greater complexity") can support a greater number of bird species. Species-rich high-shrub habitats and closed conifer forests are the most complex areas of peatlands, while species-poor muskeg or fen habitats are least complex. Shrub areas usually consist of some trees, shrubs, sedges, and some forbs, but fens consist primarily of sedges with few shrubs and forbs.

Density, on the other hand, seems to be related to productivity of the habitat (i.e., increased food availability). Minerotrophic peatland habitats are more productive than ombrotrophic peatlands, and bird densities are higher.

Comparisons with Other Habitats

Peatlands have often been referred to as wastelands. An old-timer we met near Waskish, a town in the heart of the Red Lake peatland, when we first began counting birds in that area, was very surprised when he learned that we were going to count birds in the bog. "There aren't any birds out there!" he told us. He could not have been more wrong: there are many birds out there. In Fig. 8.8 we compare the densities and number of species commonly observed in five peatland habitat types with four non-peatland habitats: grasslands, mixed deciduous/coniferous forests, deciduous forests, and upland coniferous forests. It is apparent that the peatland habitats are comparable in densities with other habitats. Muskeg and closed coniferous forests tend to have relatively low densities, but high-shrub habitats have some of the highest densities found in Minnesota habitats. Densities in sedge-fen habitats are similar or slightly higher on average than those found in grassland habitats.

In general, the number of species found in peatlands tends to be slightly lower than in upland forests, although species richness in grasslands is about the same as in sedge fen. The largest number of species found in peatlands is in high-shrub and closed-conifer habitats, and species richness in these habitats is similar to that in many of the earlier successional stages of upland forests.

Future of Peatlands Birds

There is much speculation on the potential economic development of Minnesota peatlands for energy, horticultural peat, agriculture, and forestry. What effect will this development have on the peatland birds? Are there "unique" peatland birds? What kinds of peatland

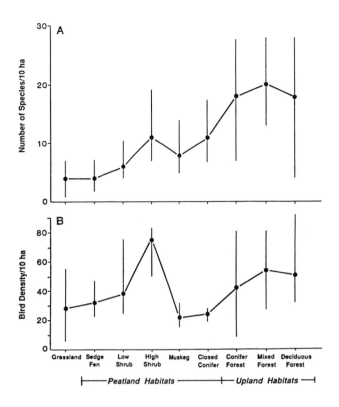

Fig. 8.8: Comparisons of bird-community density and number of breeding species in five peatland habitats with four common upland habitats of Minnesota. Mean values are shown by the dots, and the ranges by vertical lines. The data for the non-peatland habitats was obtained from Back (1979), Capen (1979), Crawford and Titterington (1979), Temple *et al.* (1979), and Wiens and Dyer (1975).

habitats must be preserved to maintain populations of the species that use peatlands? Unfortunately, there are no easy answers to these questions, nor do we think adequate information is available to address them.

No bird species currently recognized by the state of Minnesota or by the federal government as endangered or threatened is exclusively found in peatlands. A few federally recognized endangered species such as the Bald Eagle (*Haliaeetus leucocephalus*) and Peregrine Falcon (*Falco peregrinus*) (USFWS 1986) occasionally hunt in peatlands, but it is unlikely that peatlands are or were important areas for them. Several bird species found in Minnesota peatlands have "special concern" status either in the state (Minnesota DNR 1983) or on a national level (e.g., the Blue List of the National Audubon Society) or are listed among the most-sought-after North American species by the nation's birders (ABA 1979) (Table 8.2). Even though there are many species with special-concern status in Minnesota peatlands, most of them also occur in non-peatland habitats. The spruce grouse, three-toed woodpecker, and black-backed woodpecker also occur in upland and lowland coniferous forests, and the sand-

Table 8.2. Rare bird species in Minnesota peatlands.

Species	Source
American Bittern	1, 2
Gyrfalcon	3
Spruce Grouse	3
Sharp-tailed Grouse	1
Yellow Rail	2, 3
Sandhill Crane	2
Wilson's Phalarope	2
Snowy Owl	3
Northern Hawk-Owl	3
Great Gray Owl	3
Short-eared Owl	1, 2
Three-toed Woodpecker	3
Black-backed Woodpecker	3
Connecticut Warbler	3

Note: Rarity was defined on the basis of (*1*) consistent inclusion on the *American Birds* Blue List for species with potentially declining populations or ranges, (*2*) the Minnesota Department of Natural Resources state list of endangered, threatened, or special-concern species, or (*3*) the list of bird species recognized as desirable for observation by bird watchers in the United States.

Source: American Birders Association 1979.

hill crane, yellow rail, and short-eared owl in a variety of open habitats throughout the prairie regions of the United States and Canada. To evaluate the vulnerability of these species, we need estimates of the proportion of the species' population that is in peatlands and estimates of their overall abundance throughout their range. Many of these species are candidates for threatened or endangered status, so a realistic plan to develop peatlands will need to consider the potential effect on these species.

Peatlands throughout the world — particularly in Finland, Sweden, and Ireland — have been altered for economic purposes. Peatlands in northern Minnesota have been relatively unaffected by development, even though the largest and most pristine, the Red Lake peatland, has been disturbed by ditching. That attempt was unsuccessful, but more shrub habitats, which support the most species and the highest densities of birds in peatlands, have developed adjacent to the drainage ditches (Glaser *et al.* 1981). These shrublands are used regularly by three of the rare species listed in Table 8.2, including the American bittern, sharp-tailed grouse, and yellow rail. Most of these shrublands, however, developed from open fen habitats (Glaser *et al.* 1981). Open fens support fewer species and lower populations than shrublands, yet five of the species listed in Table 8.2 (the sharp-tailed grouse, yellow rail, sandhill crane, Wilson's phalarope, and short-eared owl) use open fens. A reduction in open fens in the Red Lake peatland would diminish habitat available for the sandhill crane, Wilson's phalarope, and short-eared owl.

Our observations indicate that the sharp-tailed grouse and yellow rail prefer fens to shrublands and, of the rare species listed in Table 8.2, only the American bittern prefers shrubland to fen. Therefore, the net loss of fens to shrublands may have had a detrimental effect on as many as five bird species of concern in Minnesota and a beneficial effect on just one.

Drainage ditches may have created an ornithological dilemma. Collectively, there are more bird species and more individuals in the peatlands of Minnesota today because of the increase in shrub habitats and riparian zones. This increase, however, has reduced the acreage of fen habitat and the populations of several species of special concern that are associated with those habitats. This is why managing for maximum diversity of species is not necessarily a wise strategy. Management of peatlands to increase the acreage of shrub habitat would increase species diversity but would further reduce populations of several recognized rare and unique species.

Suppression of fire is another human influence on peatland habitats that has resulted in a shift from fen to shrublands. Without fire, many of the open peatlands succeed to shrub areas or are invaded by tree species. Any of the species that prefer fen peatlands are affected by fire suppression. Populations of the sharp-tailed grouse have been particularly affected. Estimates by W. Berg (1989) indicate that hunter harvest of the sharp-tailed grouse decreased from more than 150,000 in 1949 to fewer than 10,000 birds annually in the 1980s. A more thorough examination of the effects of shifts within peatlands from fen to shrublands is urgently needed, especially in light of proposed economic development.

Summary

Minnesota's peatlands house an interesting and diverse avifauna. Most of the species inhabiting peatlands have been known for more than a hundred years, yet our understanding of how birds interact with the ecological functions of a peatland is limited. Indeed, the only systematic studies of Minnesota's peatland birds were initiated less than twenty years ago. In the United States we lag behind our Canadian neighbors and especially far behind the Nordic countries of Finland and Sweden in our understanding of peatland birds.

Little research has focused on the evolutionary history of peatland birds. Because most of the bird species are older than the present-day peatlands, it is unlikely that any species has evolved there. The species presently found in Minnesota peatlands probably evolved in habitats that were similar to those found in today's peatlands or were precursors of peatland habitats. Morphological analyses support the concept that species are adapted to structural configurations of the habitat.

We list 110 of the 304 species (36%) regularly found in Minnesota as major users of peatland habitats. Most species and most individuals are likely to be seen during the breeding season (late May to early July); fewest species are observed during winter.

The number of bird species breeding within a peatland area is higher in shrub and forested areas than in open fens and muskeg, probably owing to the complexity of the vegetation. The number of individual birds is higher in the more nutrient-rich minerotrophic habitats, especially shrub peatlands, than in ombrotrophic habitats like muskeg and spruce-forested areas. These patterns are similar during migration, but as snow cover makes food less available and accessible and as the weather becomes hostile, more species and individuals can be observed in the forested habitats than in the open shrub and fen areas, although populations are drastically reduced from those observed during the breeding season.

No officially designated federal or state endangered or threatened bird species are found regularly in Minnesota's peatlands. Fourteen bird species were identified as species of special concern based on (1) status as "special-concern" species in the state of Minnesota (DNR 1983), (2) inclusion on the *American Birds* Blue List, or (3) designation as "most-sought-after" species for observation by North American birders. The most critical peatland habitats for these species are open fen, muskeg, and coniferous forests. The most urgent concern is the conversion of open fen to shrubland as a result of ditching and suppression of fire. This conversion may have resulted in reductions in habitats available for up to 5 of the 14 species of special concern.

Minnesota has more peatlands than any other of the lower 48 United States. Efforts to study and protect these areas have not been proportionate to their size within the state. Peatlands have been called wastelands, a term with negative connotations: a wasteland is uncultivated, barren, without vegetation. The role of Minnesota's peatlands as natural filters of water pollutants and as recharge areas for groundwater aquifers is only beginning to be appreciated. We need to get on with the business of understanding the peatlands and their fragile nutrient economy. Ignorance is now our guide.

Acknowledgments

We greatly appreciate the critical readings on earlier versions of this manuscript by John Blake, Kim Eckert, Dick Green, Janet Green, Pershing Hofslund, Adeline Nunez, and Lee Pfannmuller. We also thank Northern States Power, the Minerals Division and Nongame Program of the Minnesota Department of Natural Resources, and Robert Wallace and Associates for the financial support that has given us the opportunity to work in Minnesota peatlands.

APPENDICES: Chapter 8

Appendix 1. Bird species that regularly use peatland habitats in Minnesota. Species that use only the riparian zones have been excluded because they do not represent native peatland species.

Species	Date[a]	General[b] abundance	High shrub	Low shrub	Sedge fen	Muskeg	Closed conifer forest	Open pools	Riparian zones
Herons and Bitterns									
American Bittern (*Botaurus lentiginosus*)	B	U	X		X				X
Waterfowl									
Mallard (*Anas platyrhynchos*)	B	C	X	X	X	X	X	X	X
Blue-winged Teal (*Anas discors*)	B	U			X			X	X
Ring-necked Duck (*Aythya collaris*)	B	U						X	
Hawks									
Northern Harrier (*Circus cyaneus*)	B	C	X	X	X	X			
Sharp-shinned Hawk (*Accipiter striatus*)	B	R					X		
Northern Goshawk (*Accipiter gentilis*)	P	R				X	X		
Rough-legged Hawk (*Buteo lagopus*)	W	C	X	X	X	X			
Gyrfalcon (*Falco rusticolus*)	W	R	X	X	X	X			
Grouse									
Spruce Grouse (*Dendragopus canadensis*)	P	R				X	X		
Ruffed Grouse (*Bonaso umbellus*)	P	U					X		X
Sharp-tailed Grouse (*Tympanuchus phasianellus*)	P	U	X	X	X	X			
Rails and Grouse									
Yellow Rail (*Coturnicops noveboracensis*)	B	C	X	X	X				
Sora (*Porzana carolina*)	B	U		X					
Sandhill Crane (*Grus canadensis*)	B	U			X	X		X	
Shorebirds									
Greater Yellowlegs (*Tringa melanoleuca*)	M	U			X			X	
Lesser Yellowlegs (*Tringa flavipes*)	M	U			X			X	

Appendix 1 (cont.). Bird species that regularly use peatland habitats in Minnesota.

Species	Date[a]	General[b] abundance	Habitat[c]						
			High shrub	Low shrub	Sedge fen	Muskeg	Closed conifer forest	Open pools	Riparian zones
Shorebirds (cont.)									
Solitary Sandpiper (*Tringa solitaria*)	B	R			X				
Common Snipe (*Gallinago gallinago*)	B	C	X	X	X	X		X	
Wilson's Phalarope (*Phalaropus tricolor*)	B	U			X			X	
Cuckoos									
Black-billed Cuckoo (*Coccyzus erythropthalmus*)	B	U	X						X
Owls									
Snowy Owl (*Nyctea scandiaca*)	W	R				X			
Northern Hawk-Owl (*Surnia ulula*)	B	R				X	X		X
Great Gray Owl (*Strix nebulosa*)	P	R					X		X
Short-eared Owl (*Asio otus*)	B	U	X	X	X	X			
Hummingbird									
Ruby-throated Hummingbird (*Archilochus colubris*)	B	U	X						
Woodpeckers									
Downy Woodpecker (*Picoides pubescens*)	P	U					X		X
Hairy Woodpecker (*Picoides villosus*)	P	U					X		X
Three-toed Woodpecker (*Picoides tridactylus*)	W	R					X		
Black-backed Woodpecker (*Picoides arcticus*)	P	U					X		
Northern Flicker (*Colaptes auratus*)	B	U	X						X
Flycatchers									
Olive-sided Flycatcher (*Contopus borealis*)	B	U				X	X		
Yellow-bellied Flycatcher (*Empinonax flaviventris*)	B	C				X	X		
Alder Flycatcher (*Empidonax alnorum*)	B	C	X						X
Eastern Kingbird (*Tyrannus tyrannus*)	B	C	X						X
Swallows									
Tree Swallow (*Tachycineta bicolor*)	B	C	X			X			X

Appendix 1 (cont.). Bird species that regularly use peatland habitats in Minnesota.

Species	Date[a]	General[b] abundance	High shrub	Low shrub	Sedge fen	Muskeg	Closed conifer forest	Open pools	Riparian zones
Jays and Crows									
Gray Jay (*Perisoreus canadensis*)	P	C				X	X		
Blue Jay (*Cyanocitta cristata*)	P	C				X	X		X
Black-billed Magpie (*Pica pica*)	P	U	X			X			
American Crow (*Corvus brachyrhynchos*)	B	U					X		X
Common Raven (*Corvus corax*)	P	U					X		X
Chickadees and Nuthatches									
Black-capped Chickadee (*Parus atricapillus*)	P	C	X				X		X
Boreal Chickadee (*Parus hudsonicus*)	P	U					X		
White-breasted Nuthatch (*Sitta carolinensis*)	P	U					X		X
Red-breasted Nuthatch (*Sitta canadensis*)	P	C					X		
Creepers									
Brown Creeper (*Certhia americana*)	B	U					X		X
Wrens									
House Wren (*Troglodytes aedon*)	B	U	X				X		X
Winter Wren (*Troglodytes troglodytes*)	B	C					X		X
Sedge Wren (*Cistothorus platensis*)	B	A	X	X	X				
Kinglets									
Golden-crowned Kinglet (*Regulus satrapa*)	B	C					X		
Ruby-crowned Kinglet (*Regulus calendula*)	B	C					X		
Thrushes									
Veery (*Catharus fuscescens*)	B	C	X						X
Gray-cheeked Thrush (*Catharus minimus*)	M	U					X		X
Swainson's Thrush (*Catharus ustulatus*)	B	U					X		

Appendix 1 (cont.). Bird species that regularly use peatland habitats in Minnesota.

Species	Date[a]	General[b] abundance	Habitat[c]						
			High shrub	Low shrub	Sedge fen	Muskeg	Closed conifer forest	Open pools	Riparian zones
Thrushes (cont.)									
Hermit Thrush (*Catharus guttatus*)	B	C				X	X		X
American Robin (*Turdus migratorius*)	B	C					X		X
Gray Catbird (*Dumetella carolinensis*)	B	U	X						X
Brown Thrasher (*Toxostoma rufum*)	B	U	X						X
Waxwings and Shrikes									
Bohemian Waxwing (*Bombycilla garrulus*)	W	U					X		X
Cedar Waxwing (*Bombycilla cedrorum*)	B	U					X		X
Northern Shrike (*Lanius excubitor*)	W	U	X	X					X
Vireos									
Solitary Vireo (*Vireo solitarius*)	B	U					X		
Philadelphia Vireo (*Vireo philadelphicus*)	B	R	X						
Warblers									
Golden-winged Warbler (*Vermivora chrysoptera*)	B	U	X						
Tennessee Warbler (*Vermivora peregrina*)	B	U	X						X
Orange-crowned Warbler (*Vermivora celata*)	M	U	X						X
Nashville Warbler (*Vermivora ruficapilla*)	B	A	X			X	X		X
Northern Parula (*Parula americana*)	B	U					X		
Yellow Warbler (*Dendroica petechia*)	B	C	X						X
Magnolia Warbler (*Dendroica magnolia*)	B	U					X		
Cape May Warbler (*Dendroica tigrina*)	B	U					X		
Yellow-rumped Warbler (*Dendroica coronata*)	B	A				X	X		
Black-throated Green Warbler (*Dendroica virens*)	B	U					X		
Blackburnian Warbler (*Dendroica fusca*)	B	C					X		

Appendix 1 (cont.). Bird species that regularly use peatland habitats in Minnesota.

Species	Date[a]	General[b] abundance	Habitat[c]						
			High shrub	Low shrub	Sedge fen	Muskeg	Closed conifer forest	Open pools	Riparian zones
Warblers (cont.)									
Pine Warbler (*Dendroica pinus*)	B	R					X		
Palm Warbler (*Dendroica palmarum*)	B	A				X	X		
Bay-breasted Warbler (*Dendroica castenea*)	B	U					X		
Black-and-white Warbler (*Mniotilta varia*)	B	C	X						X
American Redstart (*Setophaga ruticilla*)	B	U	X						X
Northern Waterthrush (*Seiurus noveboracensis*)	B	U	X						
Connecticut Warbler (*Oporornis agilis*)	B	C				X	X		
Common Yellowthroat (*Geothlypis trichas*)	B	A	X	X					X
Wilson's Warbler (*Wilsonia pusilla*)	B	R	X						X
Sparrows and Blackbirds									
Rose-breasted Grosbeak (*Pheucticus ludovicianus*)	B	C	X						X
American Tree Sparrow (*Spizella arborea*)	M	U	X	X					X
Chipping Sparrow (*Spizella passerina*)	B	C				X	X		
Clay-colored Sparrow (*Spizella pallida*)	B	A	X	X		X			
Savannah Sparrow (*Passerculus sandwichensis*)	B	A	X	X	X	X			
LeConte's Sparrow (*Ammodramus leconteii*)	B	A	X	X	X	X			
Sharp-tailed Sparrow (*Ammodramus caudacutus*)	B	U			X				
Lincoln's Sparrow (*Melospiza lincolnii*)	B	C		X		X			
Swamp Sparrow (*Melospiza georgiana*)	B	A	X	X					
White-crowned Sparrow (*Zonotrichia leucophrys*)	M	U	X	X					X
White-throated Sparrow (*Zonotrichia albicollis*)	B	C	X				X		X
Dark-eyed Junco (*Junco hyemalis*)	B	C				X			
Snow Bunting (*Plectrophenax nivalis*)	W	U	X	X					X

Appendix 1 (cont.). Bird species that regularly use peatland habitats in Minnesota.

Species	Date[a]	General[b] abundance	Habitat[c]						
			High shrub	Low shrub	Sedge fen	Muskeg	Closed conifer forest	Open pools	Riparian zones
Sparrows and Blackbirds (cont.)									
Bobolink (*Dolichonyx oryzivorus*)	B	A	X	X	X	X			
Red-winged Blackbird (*Agelaius phoeniceus*)	B	C	X						X
Rusty Blackbird (*Euphagus carolinus*)	B	R	X						
Brewer's Blackbird *Euphagus cyanocephalus*)	B	U				X			
Brown-headed Cowbird (*Molothrus ater*)	B	C	X			X	X		X
Pine Grosbeak (*Pinicola enucleator*)	V	U					X		X
Purple Finch (*Carpodacus purpureus*)	B	U					X		X
Red Crossbill (*Loxia curvirostra*)	P	U					X		
White-winged Crossbill (*Loxia leucoptera*)	P	U					X		
Common Redpoll (*Carduelis flammea*)	W	C	X	X					X
Hoary Redpoll (*Carduelis hornemannil*)	W	R	X	X					X
Pine Siskin (*Carduelis pinus*)	P	U					X		
American Goldfinch (*Carduelis tristis*)	B	C	X						X
Evening Grosbeak (*Coccothraustes vespertinus*)	P	U					X		X

[a]B = Breeding species
P = Permanent resident
W = Winter visitor
M = Spring and fall migrant

[b]A = abundant
C = common
U = uncommon
R = rare

[c]Habitats are as shown in Fig. 8.2 of text, except riparian zones, which represent vegetation zones along ditches and roadsides.

Appendix 2. Bird species and scientific names not listed in Appendix 1 in the order they appear in the text.

SPECIES		SPECIES	
Yellow Wagtail	(*Motacilla flava*)	Common Grackle	(*Quiscalus quiscalus*)
Meadow Pipit	(*Anthus pratensis*)	Marsh Wren	(*Cistothorus palustris*)
Lapwing	(*Vanellus vanellus*)	Eastern Phoebe	(*Sayornis phoebe*)
Ruff	(*Philomachus pugnax*)	Cliff Swallow	(*Hirundo pyrrhonota*)
Wood Sandpiper	(*Tringa glareola*)	Barn Swallow	(*Hirundo rustica*)
Whimbrel	(*Numenius phaeopus*)	Broad-winged Hawk	(*Buteo platypterus*)
Godwit	(*Limosa limosa*)	Least Flycatcher	(*Empidonax minimus*)
Snipe	(*Gallinago gallinago*)	Chestnut-sided Warbler	(*Dendroica pensylvanica*)
Herring Gull	(*Larus argentatus*)	Bald Eagle	(*Haliaeetus leucocephalus*)
Black-headed Gull	(*Larus ridibundus*)	Peregrine Falcon	(*Falco peregrinus*)

Literature Cited

Aldrich, J. W. 1984. Ecogeographical variation in size and proportions of song sparrows (*Melospiza melodia*). Ornithological Monographs no. 35, American Ornithologists' Union, Allen Press, Lawrence, Kansas, USA.

American Birders Association. 1979. ABA poll of most-wanted birds. Birding 11:54-58.

Anderson, S. H. 1982. Effects of the 1976 Seney National Wildlife Refuge wildfire on wildlife and wildlife habitat. United States Department of the Interior, Fish and Wildlife Service, Resource Publication 146, Washington, D.C., USA.

Back, G. N. 1979. Avian communities and management guidelines of the aspen-birch forest. Pages 67-79 *in* R. M. DeGraaf and K. E. Evans, editors. Workshop proceedings management of north central and northeastern forests for nongame birds. USDA Forest Service, General Technical Report NC-51, St. Paul, Minnesota, USA.

Bairlein, F. 1981. Ökosystemanalyse der Rastplätze von Zugvögeln: Beschreibung und Deutung der Verteilungsmuster von ziehenden Kleinvögeln in verschiedenen Biotopen der Stationen des, "Mettnau-Reit-Illmitz-Programmes." Ecology of Birds 3:7-137 (English summary).

Beer, J. R., and L. D. Frenzel. 1956. Additional bird records for the Quetico-Superior Wilderness Area. Flicker (Loon) 28:40.

Berg, W. E. 1989. Sharp-tailed grouse management problems in the lake states: Does the sharptail have a future? Paper presented at the Prairie Grouse Technical Meeting, September 12-15, 1989, Escanaba, Michigan, USA.

Brewer, R. 1967. Bird populations of bogs. Wilson Bulletin 79:371-396.

Capen, D. E. 1979. Management of northeastern pine forests for nongame birds. Pages 90-109 *in* R. M. DeGraaf and K. E. Evans, editors. Workshop proceedings management of north central and northeastern forests for nongame birds. USDA Forest Service, General Technical Report NC-51, St. Paul, Minnesota, USA.

Carbyn, L. N. 1971. Densities and biomass relationships of birds nesting in boreal forest habitats. Arctic 24:51-61.

Corbin, K. W., C. G. Sibley, and A. Ferguson. 1979. Genic changes associated with the establishment of sympatry in orioles of the genus *Icterus*. Evolution 33:624-633.

Cottrille, B. 1964. Lake County nesting records, 1962; Lake County nesting records, 1963. Loon 36:22-23.

Crawford, H. S., and R. W. Titterington. 1979. Effects of silvicultural practices on bird communities in upland spruce-fir stands. Pages 110-119 *in* R. M. DeGraaf and K. E. Evans, editors. Workshop proceedings management of north central and northeastern forests

for nongame birds. USDA Forest Service, General Technical Report NC-51, St. Paul, Minnesota, USA.

Dyrcz, A., J. Okulewicz, L. Tomialojc, and J. Witkowski. 1972. Breeding avifauna of the Biebrza marshes and adjacent territories. Acta Ornithologica 13:185-252.

———, J. Okulewicz, and B. Wiatr. 1973. Birds breeding in the Łeczna-Włodawa Lake District (including a quantitative study on low peats). Acta Zoologica Cracov 18:399-474.

Eckert, K. 1978. Sax-Zim, Minnesota. American Birds 32:714.

———. 1983. A birders guide to Minnesota. Kim Eckert. Duluth, Minnesota, USA.

Erskine, A. J. 1977. Birds in boreal Canada: Communities, densities, and adaptations. Canadian Wildlife Service, Report Series no. 41.

Ewert, D. 1982. Birds in isolated bogs in central Michigan. American Midland Naturalist 108:41-50.

Fitzpatrick, J. W. 1980. Wintering of North American tyrant flycatchers in the Neotropics. Pages 67-78 *in* A. Keast and E. S. Morton, editors. Migrant birds in the Neotropics: Ecology, behavior, distribution, and conservation. Smithsonian Institution Press, Washington, D.C., USA.

Glaser, P. H., G. A. Wheeler, E. Gorham, and H. E. Wright, Jr. 1981. The patterned mires of the Red Lake peatland, northern Minnesota: Vegetation, water chemistry, and landforms. Journal of Ecology 69:575-599.

Green, J. C. 1971. Summer birds of the Superior National Forest, Minnesota. Loon 43:103-107.

——— and R. B. Janssen. 1975. Minnesota birds: Where, when, and how many. University of Minnesota Press, Minneapolis, Minnesota, USA.

——— and G. J. Niemi. 1978. Birds of the Superior National Forest. Superior National Forest, USDA Forest Service, Duluth, Minnesota, USA.

Hanowski, J. M., and G. J. Niemi. 1987. Breeding bird populations in a proposed wetland treatment area of northern Minnesota. Journal of the Minnesota Academy of Science 53:7-10.

Hanson, H. C. 1953. Muskeg as sharp-tailed grouse habitat. Wilson Bulletin 65:235-241.

Häyrinen, U. 1970. Suomen suolinnuston regionaalisuudesta ja soiden suojelusta. Pages 84-110 *in* E. Kumari, editor. Linde Kahel Pool Soome Lahte. Vulgus, Tallinn, Estonia.

Hickey, J. J., J. T. Emlen, and S. C. Kendeigh. 1965. Early-summer bird-life of Itasca State Park. Loon 37:27-39.

Hildén, O. 1967. Lapin pesimälinnusto tutkimuskohteena. Luonnon Tutkija 71:152-162.

Hofslund, P. B. 1956. The birds of Gooseberry Falls State Park. Flicker (Loon) 28:62-70.

Hubbard, J. P. 1969. The relationships and evolution of the *Dendroica coronata* complex. Auk 86:393-432.

Hurlbert, S. H. 1971. The nonconcept of species diversity: A critique and alternative parameters. Ecology 52:577-586.

Idzikowski, J. H. 1982. Summer birds reaching the margins of their range at the Cedarburg Bog and the UWM Field Station. University of Wisconsin, Milwaukee, Field Stations Bulletin 15:1-15.

James, F. C., and N. O. Wamer. 1982. Relationship between temperate forest bird communities and vegetation structure. Ecology 63:159-171.

Järvinen, O., and L. Sammalisto. 1976. Regional trends in the avifauna of Finnish peatland bogs. Annales Zoologica Fennici 13:31-43.

—— and S. Ulfstrand. 1980. Species turnover of a continental bird fauna: Northern Europe, 1850-1970. Oecologia 46:186-195.

Kendeigh, S. C. 1956. A trail census of birds at Itasca State Park, Minnesota. Flicker (Loon) 28:90-104.

Kouki, J., and O. Järvinen. 1980. Single-visit censuses of peatland birds. Ornis Fennica 57:134-136.

Kumari, E. 1972. Changes in the bird fauna of Estonian peat bogs during the last decades. Aquilo Series Zoology 13:45-47.

Lack, D. 1969. Tit niches in two worlds; or, homage to Evelyn Hutchinson. American Naturalist 103:43-50.

——. 1976. Island biology, illustrated by the land birds of Jamaica. University of California Press, Berkeley, California, USA.

Marshall, W. H., and D. G. Miquelle. 1978. Terrestrial wildlife of Minnesota peatlands. Minnesota Department of Natural Resources, St. Paul, Minnesota, USA.

Mengel, R. M. 1964. The probable history of species formation in some northern wood warblers (Parulidae). Living Bird 3:9-43.

Merikallio, E. 1958. Finnish birds. Their distribution and numbers. Fauna Fennica 5:1-181.

Miller, A. H. 1956. Ecologic factors that accelerate formation of races and species of terrestrial vertebrates. Evolution 10:262-277.

Minnesota Department of Natural Resources. 1983. Review draft: Plan for the management of nongame wildlife in Minnesota. Minnesota Department of Natural Resources Nongame Division. St. Paul, Minnesota, USA.

Niemi, G. J. 1985. Patterns of morphological evolution in bird genera of new world and old world peatlands. Ecology 66:1215-1228.

—— and J. M. Hanowski. 1984. Effects of a transmission line on breeding bird populations in the Red Lake peatland, northern Minnesota. Auk 101:487-498.

—— and J. M. Hanowski. 1986a. Bird species and communities in altered and natural peatlands of northern Minnesota. Pages 25-59 *in* Habitat characteristics for bird species of special concern and their importance in habitat reclamation. Technical Report to the Minnesota Department of Natural Resources by the University of Minnesota, Duluth, Minnesota, USA.

—— and J. M. Hanowski. 1986b. ELF communications system ecological monitoring program. *In* compilation of 1985 annual reports of the Navy ELF communication system ecological monitoring program, vol. 3, Section J. National Technical Information Service. Technical Report E065499-26.

—— and L. Pfannmuller. 1979. Avian communities: Approaches to describing their habitat associations. Pages 154-178 *in* R. M. DeGraaf and K. E. Evans, editors. Workshop proceedings management of north central and northeastern forests for nongame birds. USDA Forest Service, General Technical Report NC-51, St. Paul, Minnesota, USA.

——, J. M. Hanowski, J. Kouki, and A. Rajasarkka. 1984. Intercontinental comparisons of habitat structure as related to bird distribution in peatlands of eastern Finland and northern Minnesota, USA. Pages 59-73 *in* C. W. Fuchsman and S. A. Spigarelli, editors. Proceedings of the International Symposium on Peat Utilization. Bemidji State University, Bemidji, Minnesota, USA.

Nilsson, S. G. 1977. Composition and density of the bird community on raised peat-bogs in southwestern Sweden. Fauna och Flora 72:227-233.

——. 1980. Quantification of the effects of drainage on the birds breeding on raised peat bogs. Fauna och Flora 75:256-260.

——. 1982. Seasonal changes in census efficiency of birds at marshes and fen mires in southern Sweden. Holarctic Ecology 5:55-60.

Parnell, J. F. 1969. Habitat relations to the Parulidae during spring migration. Auk 86:505-521.

Peet, R. K. 1974. The measurement of species diversity. Annual Review of Ecology and Systematics 5:285-307.

Pietz, P., and J. Tester. 1979. Utilization of Minnesota peatland habitats by snowshoe hare, white-tailed deer, spruce grouse, and ruffed grouse. Minnesota Department of Natural Resources, St. Paul, Minnesota, USA.

Renno, O. 1958. Environmental influences on the numbers of the avifauna inhabiting the marshes of western Estonia. Ornitholoogiline Kogumik 1, Tarta (English summary).

Roberts, T. S. 1932. The birds of Minnesota, vols. 1 and 2. University of Minnesota Press, Minneapolis, Minnesota, USA.

Sammalisto, L. 1955. Suomenselän vedenjakajaseudun nevalinnustosta. Ornis Fennica 32:1-8.

——. 1957. The effect of the woodland-open peatland edge on some peatland birds in south Finland. Ornis Fennica 34:81-89.

Soper, E. K. 1919. The peat deposits of Minnesota. Minnesota Geological Survey Bulletin 16:1-261.

Spindler, M., and B. Kessel. 1980. Avian populations and habitat use in interior Alaska taiga. Syesis 13:61-104.

Stockwell, S. S., and M. L. Hunter. 1985. Distribution and abundance of birds, amphibians and reptiles, and small mammals in peatlands of central Maine. A report to the Maine Department of Inland Fisheries and Wildlife, Wildlife Department, University of Maine, Orono, Maine, USA.

Swanson, G. 1940. Wildlife in the canoe country (Lake County). Flicker (Loon) 15:25-28.

Temple, S. A., M. J. Mossman, and B. Ambuel. 1979. The ecology and management of avian communities in mixed hardwood-coniferous forests. Pages 132-153 *in* R. M. DeGraaf and K. E. Evans, editors. Workshop proceedings management of north central and northeastern forests for nongame birds. USDA Forest Service, General Technical Report NC-51, St. Paul, Minnesota, USA.

Terborgh, J. 1980. The conservation status of Neotropical migrants: Present and future. Pages 21-30 *in* A. Keast and E. S. Morton, editors. Migrant birds in the Neotropics: Ecology, behavior, distribution, and conservation. Smithsonian Institution Press, Washington, D.C., USA.

United States Fish and Wildlife Service. 1986. Endangered and threatened wildlife and plants. Department of the Interior, U.S. Fish and Wildlife Service 50 CFR 17.11 and 17.12.

Väisänen, R., and O. Järvinen. 1977. Structure and fluctuation of the breeding bird fauna of a north Finnish peatland area. Ornis Fennica 54:143-153.

Warner, D., and D. Wells. 1984. Bird population structure and seasonal habitat use as indicators of environmental quality of peatlands. Minnesota Department of Natural Resources, St. Paul, Minnesota, USA.

Weise, C. M. 1973. Breeding birds of the forested portions of Cedarburg Bog. University of Wisconsin, Milwaukee Field Stations Bulletin 6:1-9.

Wiens, J. A., and M. I. Dyer. 1975. Rangeland avifaunas: Their composition, energetics, and role in the ecosystem. Pages 146-182 *in* D. R. Smith, editor. Proceedings of the symposium on management of forest and range habitats for nongame birds. USDA Forest Service, General Technical Report WO-1, Washington, D.C., USA.

CHAPTER 9

Amphibians and Reptiles

Daryl R. Karns

Boreal peatlands fall in a zone between 43° and 68° north latitude (Heinselman 1975). Large areas in northern North America, Europe, and Asia are dominated by this unique and variable landscape. The amphibian and reptile community of the peatlands of northern Minnesota is the subject of this chapter. The Minnesota peatlands are a mosaic of habitats that can differ greatly in terms of nutrient availability, water quality, and vegetation (Heinselman 1963, 1970). These characteristics of peatland habitats affect the amphibians and reptiles living there in diverse ways.

The scientific literature dealing with peatlands is largely botanical; terrestrial vertebrates have not been well studied. The existing literature on the amphibians and reptiles of peatland habitats deals mainly with small lake-basin sites and with aspects of the ecology of one or a few species (Marshall and Buell 1955; Bellis 1959, 1965; Heatwole and Getz 1960; Heatwole 1961; Gosner and Black 1957; Blackith and Speight 1974; Strijbosch 1979; Beebee and Griffin 1977; Freda and Dunson 1986).

There appear to be only two localities where amphibian and reptile communities of extensive North American boreal peatlands have been studied: northern Minnesota (this study) and central Maine (Stockwell 1985; Stockwell and Hunter 1985, 1989). These studies were commissioned by various agencies in advance of proposed peatland development. Neither European nor Soviet studies on the herpetofaunal communities of boreal peatlands have been located. The situation is somewhat better for the study of mammal and bird communities in boreal peatlands (chapters 6, 7, and 8 in this volume). The limited scientific literature dealing with the herpetofauna is not really surprising, because amphibians and reptiles are often the least well studied of the terrestrial vertebrates in an area.

This chapter summarizes the results of three years of research in the peatlands of northern Minnesota (Karns and Regal 1978, 1979; Karns 1984). After a brief introduction to the biology and study of amphibians and reptiles, data on the composition of the herpetofauna, patterns of habitat utilization, and seasonal changes in activity in the Minnesota peatlands are reported. These data show that the peatlands are relatively low in the number of species present compared to surrounding areas; this is especially true in nutrient-poor bogs where the herpetofauna is low in both the number of species present and in population numbers. The herpetofauna of the peatlands is overwhelmingly dominated by amphibians; the species present are typical of eastern woodlands, and none of the species is restricted to peatland habitats. Although low in species richness, amphibians are abundant and constitute an important component of the peatlands fauna. The last section of this chapter is a discussion of the physical, biotic, and historical factors that may have influenced the characteristics of the herpetofauna of the peatlands. Available information on the peatlands elsewhere in the United States and on other continents is examined in an attempt to achieve a global perspective on the amphibians and reptiles of peatlands.

A caveat concerning this analysis: the peatlands of northern Minnesota cover a large geographic area; this one study obviously could not survey the entire region in detail. The general areas of investigation were the peatlands of central and southern Koochiching County and the area of western Beltrami County north of Upper Red Lake (Fig. 9.1). This region represents the eastern arm of Glacial Lake Agassiz and is the area where the most extensive peatlands in Minnesota are found. Research was concentrated in the general area of Big Falls in central Koochiching County, with

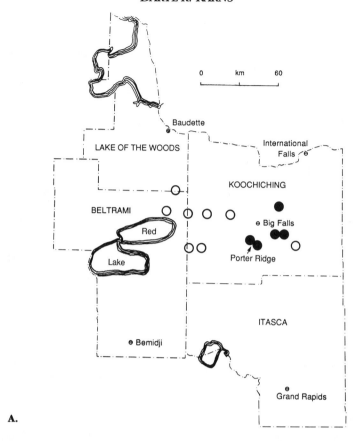

A.

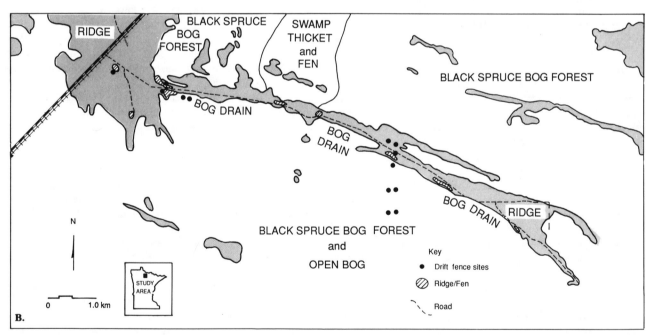

Fig. 9.1. A. A county map of north-central Minnesota showing the location of amphibian and reptile survey study sites. The focal point of the survey was Big Falls in central Koochiching County. Porter Ridge was the primary study site. *Closed circles* indicate drift-fence trapping sites. *Open circles* indicate other collecting areas visited regularly. **B.** A schematic map of the Porter Ridge site near Big Falls showing the major landscape features and location of drift-fence sites. Upland areas are shaded. Thirteen drift fences were located in this study area: two in upland forest on the ridge, four near fens associated with the ridge, three in the bog drain that separates the ridge from the bog forest to the south, and four in the interior bog forest. See Fig. 9.5 for a view of the southern edge of the ridge and Fig. 9.2 for a picture of a drift-fence site. Drift-fence data from Porter Ridge are summarized in Fig. 9.6 and Table 9.3.

regular collecting and observation trips to other localities. Porter Ridge Bog in central Koochiching County, 11.7 km (7 miles) south-southwest of Big Falls, was the primary study site.

Thus the portrait of the herpetofauna of the northern Minnesota peatlands presented here is based on intensive study of one small geographic area supplemented by observations from a larger area. Available information indicates that it is reasonable to use the Big Falls study area as a general model for amphibian and reptile community structure and dynamics in the Minnesota peatlands. Further, this analysis of the physical and biotic factors that contribute to the particular nature of the peatlands herpetofauna studied should be applicable not only to the Minnesota peatlands but to boreal peatlands in general. Peatlands are a complex and variable environment, however, and caution must be exercised in such extrapolation.

Amphibians and Reptiles

Although amphibians and reptiles are lumped together in the scientific discipline of herpetology (literally the study of "crawling things"), they represent very different grades of biological organization. They differ from each other as much as birds and mammals differ, although the study of birds and mammals is recognized as the separate disciplines of ornithology and mammalogy. The association of amphibians and reptiles under the herpetology label came about for historical and practical reasons and should not be taken as an indication of close similarity between these two groups (Porter 1972).

Amphibians have moist, glandular skin, which is permeable to both water and gases. They have no claws and may or may not have limbs. There are three major groups of amphibians: (1) salamanders, (2) anurans (frogs and toads), and (3) caecilians (limbless burrowing forms found in the tropics). Reptiles may or may not have limbs; they have claws and dry, scaly skin. There are six major groups of reptiles: (1) turtles, (2) crocodiles and alligators, (3) the tuatara (a lizardlike reptile found on a few islands off the coast of New Zealand), (4) amphisbaenians (worm-lizards, snakelike burrowing reptiles of the tropics and subtropics), (5) lizards, and (6) snakes.

Aside from external physical characteristics, there are important differences between amphibians and reptiles in terms of life cycles and physiology. As is usual in biology, there are exceptions and variations to any generalizations, but in general the amphibians and reptiles of the peatlands conform to the following descriptions.

All peatland amphibians exhibit the textbook amphibian life cycle: the eggs are laid in water, embryos develop and hatch in water and then go through an aquatic larval feeding stage (tadpoles or salamander larvae), followed by metamorphosis into the mainly terrestrial adult form. Amphibians are the only vertebrates with this type of metamorphic life cycle. Thus ecological study of amphibians in peatlands must take into account both the aquatic and the terrestrial phases of their lives.

Reptile eggs have protective extraembryonic membranes not found in amphibians; they are laid on land or retained by the female. Fertilization is internal. Reptile embryos develop and either hatch from the egg or emerge from the female as miniature versions of the adults. There is no larval stage or metamorphosis: the young develop directly into adults.

Amphibians, because of their permeable skin, lose water and are in danger of desiccation if conditions become too dry. This physiological characteristic restricts most amphibians to habitats with sufficient moisture (although behavioral and physiological "tricks" allow some amphibians to live in deserts). It may seem that the amphibian skin is a handicap, but it is used in gas exchange, and its rich assortment of glands can produce toxic secretions, protective mucus, and sexual attractants. The dry scaly skin of reptiles, on the other hand, is an effective barrier to water loss. Because the skin is largely impermeable to water, reptiles must rely on lungs for gas exchange. The skin and reproductive physiology of reptiles make them much less dependent on water than amphibians.

Both amphibians and reptiles are ectotherms (cold-blooded); they are not capable of maintaining a high constant body temperature by metabolic means. Unlike most endothermic (warm-blooded) birds and mammals, they must rely on the external environment for their body temperature. These differences in thermoregulatory capabilities have important ecological implications (Pough 1983). Amphibians and reptiles do not have to put energy into heat production and can put more of the energy obtained from their food directly into growth and reproduction. A given habitat can support more amphibians and reptiles compared to energy-"wasteful" birds and mammals.

The ectothermic life-style also allows amphibians and reptiles to cope effectively with shortages of food and water. A drought, low levels of food, or the enforced inactivity of the long northern winter are not great hardships for them as long as suitable shelters can be found. In general, north-temperate-zone amphibians are more cold-tolerant and less selective about their body temperature than reptiles and are active over a wider range of temperatures.

Studying Amphibians and Reptiles

Because of their secretive habits, sensitivity to environmental fluctuations, and highly seasonal patterns of activity, amphibians and reptiles pose a variety of problems for the researcher (Campbell and Christman 1982; Vogt and Hine 1982; Karns 1986). Several techniques were employed to obtain information on community composition and structure in the northern Minnesota peatlands.

Fig. 9.2. Drift-fence site in a black-spruce forest. Drift fences were the primary sampling tool in this study. Each fence consisted of a 15-m (50-foot) length of aluminum flashing partially buried in the substrate. Amphibians and reptiles are trapped by pitfall and funnel traps placed along the fence. The majority of animals caught at bog sites were wood frogs and American toads. Captured animals were removed from traps, weighed, measured, sexed, toe-clipped, and released.

Drift fences have become a standard tool for sampling herpetological communities. They are artificial barriers against which an array of pitfall and funnel traps is placed (Fig. 9.2). Drift fences are effective at sampling any organism that moves along the ground — small mammals and invertebrates in addition to amphibians and reptiles.

A total of 26 drift fences were employed over the three-year study in a variety of peatland habitats. Each fence was a strip of .019 gauge rolled aluminum flashing 15 m long and 50 cm wide (50 feet long and 20 inches wide). Either one or two fences were set up at each site; fences were open periodically from April to mid-October in 1978 and 1979 and from April to May in 1980. The seven-month trapping period covered the entire period of herpetofaunal activity. Karns (1986) discusses drift-fence construction.

Traps were checked at least twice a week, depending on the weather. For each amphibian and reptile trapped, the following information was recorded: species, length, weight, sex (if possible), side of fence, and type of trap. Animals were toe-clipped (one digit) to allow recognition of previously trapped animals, and released on the opposite side of the fence. The drift-fence network was the primary source of information for this study.

Drift-fence sampling has inherent biases: treefrogs and larger snakes can escape from pitfall traps, extremely sedentary species probably will not encounter the fence, and predators (e.g., raccoons or humans) may remove animals from the fence. Thus some species are underrepresented by drift-fence sampling. In this study, for example, breeding-call surveys indicated

that gray treefrogs were common at some sites, yet they were not caught at drift fences. For these reasons, a combination of sampling techniques employed over an extended period of time provides the best assessment of the herpetofauna in a previously unsurveyed area (for a discussion of herpetological sampling techniques, see Campbell and Christman 1982; Vogt and Hine 1982; Gibbons and Semlitsch 1981; and Karns 1986).

The drift-fence trapping program was supplemented with traditional "search-and-seize" hand collecting and nighttime road cruising to search for animals crossing roads (particularly effective on secondary roads at night after rains) and for road kills. Breeding activity was monitored by breeding-call surveys in which frog and toad breeding sites were visited throughout the season, the calling species identified, and the intensity of breeding activity noted. The suitability of breeding sites for amphibian growth and development was checked by periodic sampling of larval populations.

With these techniques, three fundamental parameters of community structure were assessed:

(1) Species richness: the number of species present.

(2) Relative abundance and diversity: diversity measurements take into account both the species richness and the relative abundance of each species present in a community. Out of the entire number of animals in a community, are there relatively equal numbers of individuals of each species, or do most of the animals present belong to one or two dominant species?

(3) Habitat utilization: the habitats used by individuals of a given species and when they are used.

A representative sample (voucher collection) of the peatland amphibians and reptiles collected were preserved and are deposited at the Bell Museum of Natural History at the University of Minnesota.

The Herpetofauna of the Minnesota Peatlands

Community Composition

Approximately 3,000 specimens obtained by drift-fence sampling, hand collecting, and road cruising were examined in 1978 and 1979. Table 9.1 presents a list of the 11 species collected. The majority of specimens (n = 2,468) were collected in 1,699 days of drift-fence trapping.

The herpetofauna of the peatlands is dominated by amphibians. Six frog species, one salamander, two snakes, and two turtles were collected. Of the total number of animals trapped in drift fences, 99.5% were amphibians; two species (wood frog, *Rana sylvatica*, and Eastern American toad, *Bufo a. americanus*) constituted 77.6% of the total (47.0% and 30.6% respectively).

The gray treefrog (*Hyla versicolor*) and Cope's treefrog (*Hyla chrysocelis*) are morphologically difficult to distinguish but can be differentiated on the ba-

Table 9.1. Checklist of the amphibians and reptiles of the patterned peatlands of northern Minnesota: central and southern Koochiching and western Beltrami counties

Amphibians

  Anurans (frogs and toads)

Rana sylvatica	Wood frog
Rana pipiens	Northern leopard frog
Bufo a. americanus	Eastern American toad
Hyla versicolor	Gray treefrog
Pseudacris c. crucifer	Northern spring peeper[a]
Pseudacris triseriata maculata	Boreal chorus frog

  Salamanders

Ambystoma laterale	Blue-spotted salamander

Reptiles

  Turtles

Chelydra s. serpentina	Snapping turtle[b]
Chrysemys picta bellii	Western painted turtle

  Snakes

Thamnophis s. sirtalis	Eastern garter snake
Storeria o. occipitomaculata	Northern redbelly snake

[a]The spring peeper (formerly *Hyla crucifer*) has recently been placed in the genus *Pseudacris* by Hedges (1986).

[b]Snapping turtles, which were not collected during this survey, are included based on the accounts of reliable observers.

sis of breeding vocalizations (Jaslow and Vogt 1977). Breeding-call surveys indicated that only the gray treefrog was found in the study area.

The snapping turtle (*Chelydra s. serpentina*) cannot be considered a resident peatland species. Snappers are associated with permanent rivers, streams, and ponds in the area. Snappers were not collected during this study, but snapper shells and accounts of reliable observers demonstrate their presence. The other species listed were found in peatland habitats and are typical woodland residents of the eastern United States. Chorus frogs (*Pseudacris triseriata*) and northern leopard frogs (*Rana pipens*) are also commonly associated with grasslands and marshes.

Peatlands in general and *Sphagnum* bogs in particular are notable for their unique flora and fauna; many species of plants and animals are restricted to bog habitats. Among North American amphibians and reptiles, the Pine Barrens treefrog (*Hyla andersonii*), the carpenter frog (*Rana virgatipes,* also known as the sphagnum frog), the four-toed salamander (*Hemidactylium scutatum*), and the bog turtle (*Clemmys muhlenbergii*) are notable for their strong association with bog habitats (Conant 1975; Noble and Noble 1923; Wright and Wright 1949). These species tend to have spotty distributions reflecting their presence in disjunct bog habitats. No species found in the boreal peatlands of northern Minnesota could be classified as a bog specialist. The herpetofauna of the peatlands is composed of wide-ranging generalist species of the eastern woodlands that are compatible with the special habitat features of the boreal peatlands.

Habitat Utilization

The boreal peatlands are a complex mosaic of habitat types. They are notable for their wide spectrum of nutrient regimes (ranging from ombrotrophic, nutrient-poor bogs to minerotrophic, nutrient-rich fens), and the large variation in water quality associated with these different sites (acidic bog water compared to circumneutral fen water). Important vegetational differences are associated with these changes in nutrient status and water quality (*Sphagnum* bogs compared to sedge-dominated fens). Are there differences in the patterns of species richness, relative abundance, species diversity, and activity of amphibians and reptiles correlated with these distinctive variations in the quality of peatland habitats?

The survey of the herpetofauna of peatland habitat types in 1978 (Fig. 9.3 and Table 9.2) was a limited sampling program based on one drift fence per site and collateral hand collecting and road cruising. The goal was to make a preliminary assessment of the herpetofauna by placing drift fences in six representative peatland habitats: open bog, raised-bog spruce forest, open fen/swamp thicket, black-spruce forest, tamarack swamp, and cedar swamp. One fence was set at a sandy upland site adjacent to a small fen.

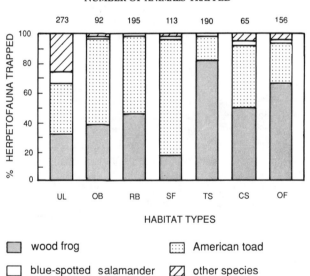

Fig. 9.3. Composition of the herpetofauna of different peatland habitats (described in Table 9.2). Data are based on drift-fence trapping only. The relative abundance of each species is shown as a percentage. The number above each histogram bar is the total number of animals trapped at that site. Note the dominance of wood frogs and American toads at all sites. Less common species are pooled under "other species" and discussed in the text. See Table 9.2 for additional information (UL = upland fen, OB = open bog, RB = raised bog, SF = black spruce forest, TS = tamarack swamp, CS = cedar swamp, OF = open fen/swamp thicket).

Table 9.2. Species richness, relative abundance, and diversity of amphibians and reptiles in different peatland habitats.

Site	Trophic status[a]	Number species	Total no. trapped	Trap rate[b]	Diversity (H')[c]	No. trap days
Upland/fen (UL)	M	10	273	2.1	1.59	131
Open bog (OB)	O	4	92	0.7	0.82	129
Raised bog forest (RB)	O	4	195	1.6	0.77	120
Black spruce forest (SF)	WM	5	113	1.3	0.51	86
Tamarack swamp (TS)	WM	4	190	1.7	0.63	112
Cedar swamp (CS)	M	4	65	0.6	1.01	103
Open fen/swamp thicket (OF)	WM	5	156	1.6	0.83	100

Note: Species richness and diversity are based on both drift-fence trapping and hand collecting; total number trapped and the trap rate are based on drift-fence trapping only. One drift fence was employed at each site. The data were collected in 1978.

[a]M = minerotrophic (fen: nutrient-rich, high productivity); O = ombrotrophic (bog: nutrient-poor, low productivity); WM = weakly minerotrophic (intermediate between ombrotrophic and minerotrophic sites in nutrients and productivity).

[b]Trap rate = number of animals trapped divided by number of trap days. Trap day = one drift fence (15 m long) open for 24 hours.

[c]H' = Shannon diversity index = $-\Sigma p_i \log_e p_i$. This index takes into account both species richness and the number of individuals present. A low value indicates that one or two species dominate the community (Magurran 1988).

The upland site had the highest species richness (10), trap rate, and diversity index values of the sites sampled. The fen associated with this site was an active amphibian breeding area. The other sites, except the swamp thicket/fen, were all *Sphagnum* sites of varying trophic status. All of the peatland sites sampled were similar in species richness (four or five species). Two sites (open bog and cedar swamp) had relatively low trap rates. The other peatland sites exhibited roughly similar trap rates. The diversity index value was highest at the fen/upland site and varied among the peatland sites without a clear pattern.

These data do not reveal a strong correlation between habitat and herpetofaunal community structure in the peatland sites sampled. The cedar swamp was the most minerotrophic site surveyed, for example, and the open bog was ombrotrophic, yet these sites were similar in species richness and trapping success. The ombrotrophic raised bog was similar to the nutrient-rich tamarack swamp and swamp thicket/fen in species richness, species diversity, and trapping success. There was no obvious relationship between species richness and abundance and distance from the nearest upland. The limited nature of the 1978 trapping program makes it difficult to draw conclusions regarding habitat utilization from these data.

Wood frogs (n = 532) and American toads (n = 437) were clearly the dominant species, comprising 49.1% and 40.3% respectively of the 1,084 animals trapped (Figs. 9.3, 9.4). Although drift-fence trapping underrepresents some species (e.g., treefrogs), there is no doubt that these two species are dominant peatland species. The remaining species noted in Fig. 9.3 include the following (in order of trapping success): chorus frog (n = 68), blue-spotted salamander (n = 28), northern leopard frog (n = 6), northern redbelly snake (*storeria o. occipitomaculata*) (n = 6), Northern spring peeper (*Pseudacris c. crucifer*) (n = 5), and eastern garter snake (*Thamnophis s. sirtalis*) (n = 2).

A difference in habitat utilization between wood frogs and American toads was observed between extremely wet and dry sites. At the wet tamarack and fen/swamp thicket sites, characterized by large pools of water, wood frogs were clearly the dominant species. At the much drier spruce bog-forest, where the terrain was mainly *Sphagnum* hummocks without pools, American toads were dominant. At sites intermediate in wetness the two species were trapped in roughly equivalent numbers.

In spite of the lack of pattern observed among the peatland sites sampled in 1978, the large difference ob-

Fig. 9.4. The wood frog, *Rana sylvatica,* deserves the title of "bog frog." It is the most abundant amphibian in the Minnesota peatlands and is found in all habitats. Wood frogs' characteristics make them well suited to survive and reproduce in the ecologically rigorous peatland environment. Wood frog embryos and tadpoles exhibit the greatest tolerance to acidic bog water among peatland amphibians, and wood frogs can tolerate partial freezing of body fluids during winter dormancy. (Wood frog illustration by Don Luce, Bell Museum of Natural History.)

Fig. 9.5. Porter Ridge Bog. This east-facing view shows the sedge- and *Sphagnum*-dominated channel (bog drain) that separates the raised-bog forest to the south (right) from the mixed conifer-hardwood forest of the sandy Porter Ridge to the north (left). The ombrotrophic bog dominates the environmental chemistry of the many pools of water found in the bog drain. The water in the drain is acidic (pH < 4.5), darkly colored, and toxic to amphibian embryos and larvae. Some amphibians do attempt (unsuccessfully) to breed in the bog drain, but adjacent "safe" fen areas are more heavily utilized.

served between the minerotrophic upland/fen site and the peatland sites suggested that the trophic status of the peatland environment might have an influence on the community structure of the herpetofauna. A more detailed investigation of the relationship between the trophic status of peatland sites and the herpetofauna was done in 1979 at Porter Ridge, where ombrotrophic bog and minerotrophic fen and upland habitats are in close proximity.

Porter Ridge is a former beachline of ancient Lake Agassiz (Fig. 9.1.B.). The sandy upland ridge vegetation consists of mixed conifer-hardwood forest. In the poorly drained regions adjacent to Porter Ridge, typical water-saturated peatlands have evolved over the past 3,000 years. On the south side of the ridge the bog forest is separated from the upland forest on the ridge by a moatlike channel (bog drain) dominated by sedges and *Sphagnum* mounds. The bog dominates the environmental chemistry of the bog drain, and bog-water pools (pH < 4.5, low calcium content, high organic content) are abundant. Abrupt transitions between bog-water areas and fens (pH > 5.0, high calcium content) occur along the ridge (Figs. 9.1.B, 9.5).

The placement of the Porter Ridge drift fences was designed to test the hypothesis that there was a difference in herpetofaunal community structure correlated with ombrotrophic and minerotrophic habitats. The prediction was that there would be fewer species and lower population sizes at the ombrotrophic sites. Fences were placed in three nutrient "grades" of habitat: the minerotrophic upland ridge and associated fens, the intermediate bog drain, and the ombrotrophic raised-bog forest. Bog fence sites were located up to

1.3 km (0.8 miles) from the ridge. Fig. 9.1.B shows the location of the 13 drift fences; Fig. 9.6 and Table 9.3 and summarize drift-fence data from Porter Ridge.

Ridge sites, particularly the drift fences placed near fen areas, exhibited the highest species richness, species diversity, and trap rate. Six species of amphibians and two species of snakes were trapped. Wood frogs (n = 532), blue-spotted salamanders (*Ambystomia laterale*) (n = 207), and American toads (n = 169) were the dominant species and accounted for 48.1%, 18.7%, and 15.3% respectively of the 1,105 animals trapped at ridge sites. The "other species" indicated on Fig. 9.6.C were (in order of trapping success): chorus frog (n = 135), spring peeper (n = 59), northern redbelly snake (n = 2), and eastern garter snake (n = 1). The gray treefrog was present but not trapped in drift fences. The exceptionally high fen/ridge trap rate reflects the fact that the fens were centers of activity in the spring (breeding) and in the late summer (emergence of metamorphosed young).

The weakly minerotrophic bog drain separating the ridge from the raised bog was intermediate in species richness, species diversity, and trap rate compared to the ridge sites and the black-spruce forest of the raised bog. American toads (n = 99), wood frogs (n = 40), and blue-spotted salamanders (n = 22) were the dominant species and accounted for 60.0%, 24.2%, and 13.3% respectively of the 165 animals trapped. Only four other animals were trapped in the bog drain (two chorus frogs, one spring peeper, and one northern redbelly snake).

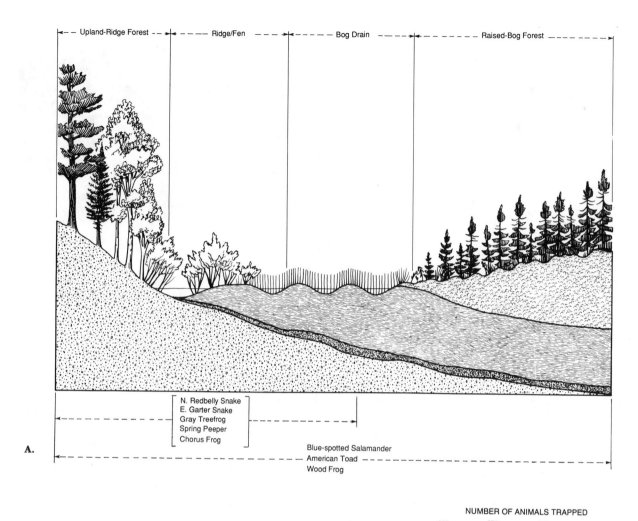

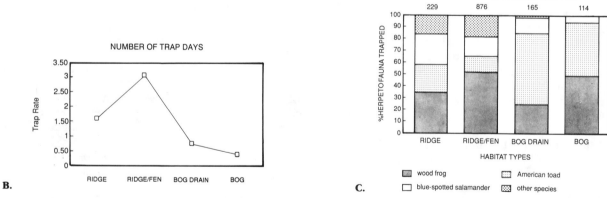

A.

B.

C.

Fig. 9.6. Changes in species composition and abundance of amphibians and reptiles correlated with habitat at Porter Ridge Bog. The data are from a series of 13 drift fences located in the Porter Ridge study area. Table 9.3 summarizes the drift-fence data. Fig. 9.1.B shows the location of the drift fences. The Porter Ridge site is described in the text and in Fig. 9.1 and 9.5. **A.** Change in habitat types and associated changes in the herpetofauna along a transect traveling south from the upland-ridge forest into the adjacent raised-bog forest. **B.** Trapping success in the different habitats along this transect (trap rate = number of animals trapped divided by number of trap days). **C.** Composition of the herpetofauna in the different habitats along the transect. The relative abundance of each species is shown as a percentage. The number above each histogram bar is the total number of animals trapped in that habitat. Less common species (pooled under "other species") are discussed in the text. In B and C note the reduction in trap rate and species richness moving from the nutrient-rich ridge and fen environment to the nutrient-poor bog. Wood frogs and American toads dominated the raised-bog sites. These graphs are based on drift-fence data collected in 1979.

Table 9.3. Species richness, relative abundance, and diversity of amphibians and reptiles in different habitats associated with Porter Ridge Bog.

Site (no. drift fences)[a]	Trophic status	Number species[b]	Total no. trapped	Trap rate	Diversity (H′)[c]	No. trap days
Ridge (2)	M	8	229	1.6	1.51	140
Fen/ridge (4)	M	8	876	3.1	1.33	286
Bog drain (3)	WM → O	7	165	0.8	1.03	214
Raised bog (4)	O	3	114	0.4	0.88	278

Note: Species richness is based on both drift-fence trapping and hand collecting; total number trapped, trap rate, and diversity are based on drift-fence trapping only. One drift fence was employed at each site. The data were collected in 1979.

[a]M = minerotrophic (fen: nutrient-rich, high productivity); O = ombrotrophic (bog: nutrient-poor, low productivity); WM = weakly minerotrophic (intermediate between ombrotrophic and minerotrophic sites in nutrients and productivity).

[b]Trap rate = number of animals trapped divided by number of trap days. Trap day = one drift fence (15 m long) open for 24 hours.

[c]H′ = Shannon diversity index = $-\Sigma p_i \log_e p_i$. This index takes into account both species richness and the number of individuals present. A low value indicates that one or two species dominate the community (Magurran 1988).

The raised bog exhibited the lowest species richness, species diversity, and trap rate. Only three species were encountered: wood frogs (n = 56), American toads (n = 51), and blue- spotted salamanders (n = 7) accounted for 49.1%, 44.7%, and 6.1% respectively of the 114 animals trapped. Wood frogs and American toads were clearly the dominant species. Minerotrophic ridge sites thus supported five more species than raised bog sites, and the trap rate was almost seven times as great. By all measures employed, the raised-bog herpetofauna at Porter Ridge was depauperate compared to the herpetofauna of the ridge and associated fens. Fig. 9.6 clearly illustrates these differences.

Activity Patterns and Habitat Utilization of Juveniles and Adults

Amphibians exhibit a characteristic seasonal activity cycle: spring movement by adults to breeding sites, postbreeding dispersal to summer feeding ranges, emergence of recently metamorphosed amphibians later in the summer, and overwintering movements in the late summer and fall. What is the relationship between herpetological activity and peatland habitat types for adults and juveniles? Figure 9.7 presents seasonal activity data for the Porter Ridge site in 1979. Differences in trapping success between bog and ridge sites noted in the preceding section are clearly shown.

At the bog-drain and raised-bog trapping sites there was a conspicuous delay in the initiation of spring activity compared to ridge sites. Early-breeding amphibians (chorus frogs, wood frogs, spring peepers) were trapped at ridge drift-fence sites as soon as most of the snow cover disappeared (about April 20 in 1979). Chorusing was heard almost immediately thereafter in

the fens and near-ridge pools of the bog drain where thawing had occurred. Wood-frog eggs were first noted on April 27 at sites along the ridge. By comparison, no amphibians were trapped at the bog-drain or raised-bog drift-fence sites (located at the edge or in the interior of the black-spruce raised-bog forest) until about May 20, a full month after the initiation of activity on the ridge.

The reason for this delay is that bog habitats thaw later than ridge sites because of their water-saturated

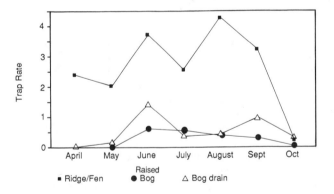

Fig. 9.7. Seasonal patterns of herpetofaunal activity at Porter Ridge Bog. Data from drift fences located in different habitats along Porter Ridge (Table 9.3) were used to construct this chart. Ridge/Fen includes both upland-forest sites and fen sites associated with the ridge (six fences, 426 trap days, 1,105 animals trapped). The raised-bog forest sample consisted of four fences (278 trap days, 114 animals trapped) and the bog drain of three fences (214 trap days, 165 animals trapped). All species are included in the tabulation. Trap rate is the number of animals trapped divided by the number of trap days. Note the marked difference between bog and fen sites.

soils. This delay in activity has obvious ecological implications for early-spring-breeding amphibians. Adults may miss or be late for the initiation of breeding if they overwinter in bogs. Also, the delay means that the relatively short northern Minnesota herpetological activity season is further reduced in bog habitats.

Different age-size classes of animals often utilize the environment in different ways. At Porter Ridge, wood frog adults dominated the ridge/fen sites and juveniles dominated the bog sites. Adults were active and dominant throughout the season (April-October) at ridge/fen sites (74.2% adults; 25.8% juveniles; n = 566). The trap rate for juveniles was low until August, when the recently metamorphosed froglets emerged from the breeding ponds. The bog drain became active in June, and a low level of activity was exhibited by both juveniles and adults (50% adults; 50% juveniles; n = 40). Activity also began in June in the interior bog sites, but here there was a clear dominance of juveniles (90.9% juveniles; 9.1% adults; n = 40).

At Wisner Bog (similar in general topography to Porter Ridge and located 25 km north), juvenile wood frogs dominated the site (one adult trapped, n = 23) early in the season (May and June). There was a conspicuous increase in adult frogs beginning in June (51.4% adults; 48.6% juveniles; n = 111 for July through October). This summer "invasion" of adult wood frogs may represent a post-breeding-season dispersal from the adjacent upland.

American toads at Porter Ridge became active about one month later (mid-May) than wood frogs. Adults and juveniles were well represented at all sites: fen/ridge (56.5% adults; 43.5% juveniles; n = 138), bog drain (60.9% adults; 39.1% juveniles; n = 110), raised bog (44.2% adults; 55.8% juveniles; n = 52). In contrast to the Porter Ridge Bog site, juvenile toads dominated the Wisner Bog throughout the season (87.7% juveniles; 12.3% adults; n = 154). Adult toads exhibited an increase in activity in the latter part of the season somewhat similar to, but not as pronounced as, that of wood frogs at the same site.

Thus the two dominant amphibian species in the peatlands exhibited differential patterns of habitat utilization by juveniles and adults and variation among sites. Overall, both wood frog and American toad juveniles dominated the bog sites (Wisner Bog and Porter Ridge bog drain and raised bog: 65.3% juveniles; n = 545) and adults dominated the upland ridge/fen sites at Porter Ridge (70.7% adults; n = 704).

A possible explanation for juvenile dominance at peatland sites and reasons for the species differences noted are discussed later. These activity data combined with the habitat utilization data show that the peatlands, and bog habitats in particular, affect the distribution, abundance, and activity patterns of resident amphibians and reptiles.

Regional Comparison with Other Herpetofaunas

Species Richness and Diversity

Species richness refers to the number of species present; species diversity takes into account both the number of species present and the relative abundance of each species. The Shannon index employed here is a measure of diversity (Magurran 1988). A community with 10 species in which one species accounted for 90% of the individuals would have a low index value. In contrast, a community with 10 species in which there were equal numbers of individuals of each species would have a high index value.

Table 9.4 reports species richness and diversity values from localities in the midwest and northeastern United States and Canada. In this comparison, all 11 species believed to be found in the general study area in Minnesota were used (Fig. 9.1 and Table 9.1). Comparisons of species richness and diversity must be considered in the context of more general regional and continental trends. Both amphibians and reptiles show a regular decline in species richness as one travels northward from the tropics. Amphibians also exhibit an east-west gradient in the United States correlated with decreasing annual rainfall toward the west (Kiester 1971). Temperate-zone amphibians are generally more cold-tolerant than reptiles, and several species range north of the arctic circle (Hodge 1976). Reptiles are better suited for drier environments than amphibians and dominate the grasslands and deserts. These trends are also observed on a smaller scale within the upper midwest (Fig. 9.8). The actual number and identity of species occurring in a given area is a complicated consequence of climate, vegetation, and historical factors. When all these factors and trends are taken into account, the peatlands of northern Minnesota stand out as a herpetologically depressed area in the region in terms of both species richness and diversity.

Minnesota and Maine Peatlands

There is only one study of an extensive boreal peatland herpetofauna comparable to the Minnesota study. Stockwell and Hunter (1985, 1989) and Stockwell (1985) employed a drift-fence network in a survey of the herpetofauna of the peatlands of central Maine (Table 9.4). They trapped in three regions of the state and at eight peatland vegetation types within each region: *lagg* (bog drain), forested bog, wooded heath, shrub heath, moss-*Chamaedaphne* lawn, bog pool, streamside meadow, and shrub-thicket. The trapping season (April-September) was similar to the Minnesota study. The Maine study put more effort into sampling representative peatland types over a wide region and did not sample upland sites. The Minnesota study reported here consisted of a limited one-season sampling program of diverse peatland habitat types followed by

Table 9.4. Regional comparison of the species richness and diversity of amphibians and reptiles at various localities in the upper midwest and northeastern United States and in south-central Canada

Site	General habitat type	Species richness			Shannon diversity index range and (mean)	Source
		total	amphibians	reptiles		
Minnesota	Peatlands	11	7	4	0.51-1.03 (0.81)	Karns, this study
Maine	Peatlands	13	12	1	0.38-1.58 (0.82)	Stockwell and Hunter (1989)
SE Minnesota	Forest, marsh, prairie along Mississippi River	41	28	13	–	Lang (1982), MHS (1985)
SW Minnesota	Former prairie, now largely farmland	25	10	15	–	Lang (1982), MHS (1985)
NC Minnesota (Lake Itasca)	Forest/prairie transition, lakes	22	14	8	–	Lang (1982), MHS (1985)
SE Ontario	Coniferous forest, lakes	17	13	4	–	Cook (1984)
SE Manitoba	Mixed forest, grassland	20	13	7	–	Preston (1982)
NC Manitoba	Coniferous forest	14	11	3	–	Preston (1982)
New Hampshire	Forest	–	–	–	0.77-1.39 (1.15)	Rudis (1984)
SE Indiana	Deciduous forest	49	25	24	1.32-2.26 (1.8)	Karns (1992)

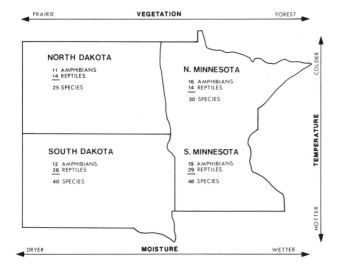

Fig. 9.8. Geographic variation in the species richness of amphibians and reptiles in Minnesota and the Dakotas. Note the general reduction in the number of species as one travels north. The number of amphibian species declines from east to west but stays fairly constant from south to north. The number of reptile species declines from south to north but remains constant from east to west. These patterns reflect basic differences in amphibian and reptile biology and are correlated with the changes in temperature, moisture, and vegetation shown on the map. See text for further discussion. The drawing is modified from Karns (1986), by permission.

a focused comparison of ombrotrophic and minerotrophic sites in one area. Given these differences in objectives, how does the Maine peatland herpetofauna compare to that of Minnesota?

Species richness was low at both Minnesota and Maine sites (13 species in Maine, 11 in Minnesota). Not unexpectedly, the composition of the herpetofauna differed. The snapping turtle, chorus frog, western painted turtle (*Chrysemys picta belli*), and redbelly snake were not found in Maine peatlands but were present in Minnesota. The spotted salamander (*Ambystoma maculatum*), central newt (*Notophthalmus viridescens*), dusky salamander (*Desmognathus fuscus*), two-lined salamander (*Eurycea bislineata*), bullfrog (*Rana catesbeiana*), green frog (*Rana clamitans*), and pickerel frog (*Rana palustris*) were recorded in Maine but not in Minnesota. These species differences in Maine and Minnesota peatlands reflect differences in the herpetofauna of the northeastern and upper midwestern United States. The herpetofauna of both peatlands was composed of wide-ranging generalist species. None of the species encountered was restricted to peatland habitats.

Species richness was comparable in the Maine and Minnesota peatland habitat types sampled (five to seven species per habitat type in Maine compared to three to seven species in Minnesota). Bog drain, forested bog, wooded heath, and shrub heath were the most productive trapping sites in the Maine study. A streamside meadow site in Maine had the highest species richness, and the inclusion of several rare species there elevated the species count to 10. The highest species counts in Minnesota were upland sites adjacent to peatland (8-10 species). The results of the 1978 Minnesota survey in which six representative peatland habitat types were sampled (Table 9.2) and the Maine survey were similar in that no consistent patterns of species composition and relative abundance in relationship to peatland habitat type were observed. Species diversity as indicated by the Shannon diversity index was similar for the two peatlands (Maine: $\overline{X} = 0.82$; range $= 0.38$-1.58; Minnesota: $\overline{X} = 0.81$;

range = 0.51-1.03). This comparison excludes Minnesota upland sites (H' = 1.51, 1.59, 1.33; Tables 9.2 and 9.3).

Drift-fence data show that both faunas are overwhelmingly dominated by amphibians (Maine: 94.0% anurans, 4.5% salamanders, 0.14% snakes; n = 2,179. Minnesota: 88.8% anurans, 10.7% salamanders, 0.5% snakes; n = 2,468). At both sites wood frogs were the dominant species (59% of all captures in Maine, 47% in Minnesota) followed by the green frog (30%), northern leopard frog (5%), and blue-spotted salamander (4%) in Maine and the American toad (30.6%), blue-spotted salamander (10.7%), and chorus frog (8.3%) in Minnesota.

The importance of green frogs in the Maine peatlands is a major difference between the two areas. Green frogs were conspicuously absent from Minnesota peatlands according to hand collecting, breeding-call surveys, and road cruising as well as drift-fence trapping. Conversely, American toads were an important component of the herpetofauna in the Minnesota peatlands but not in Maine. Green frogs and American toads are common amphibians in both the upper midwestern and northeastern United States. Green frogs are a relatively aquatic species compared to the more terrestrial American toad (Conant 1975). This ecological difference suggests that the freshwater pools and ponds that green frogs frequent are a more common feature of the Maine peatlands.

Trapping success was higher in Minnesota than in Maine peatlands. During the 1978-79 trapping seasons in Minnesota, 1,090 animals were trapped in 1,142 trap days (trap rate = 0.95 animals/day) at peatland sites (upland sites excluded). In Maine (1983-84), 2,179 animals were trapped in 8,733 trap days (trap rate = 0.25 animals/day) at peatland sites. It is difficult to say if this fourfold difference in herpetofaunal abundance is due to differences in the properties of the two peatlands, to differences in the trapping programs, or to natural variability in population numbers over time.

Juvenile amphibians were the dominant animals trapped in both Minnesota and Maine peatlands. Stockwell and Hunter (1989) reported that 82.0% of all captures were juveniles. In Minnesota peatlands, 65.3% of the two dominant peatland amphibians (wood frog and American toad, 77.6% of the total animals trapped) were juveniles.

In general, the Minnesota and Maine peatland studies present a similar portrait of their respective herpetofaunas: communities with relatively few species that are low in diversity and dominated by two species of anurans (wood frogs and American toads in Minnesota and wood frogs and green frogs in Maine). Juveniles constituted the majority of animals captured.

The most significant difference between the studies is found in the comparison of minerotrophic and ombrotrophic sites and the relationship of uplands to peatland sites. Stockwell and Hunter found no correlation between the number of species trapped and the distance of the trap site from the nearest upland. They found that amphibians were no more abundant near uplands than in the center of peatlands and concluded that distance from upland does not influence the abundance of amphibians. In marked contrast, the Porter Ridge data presented here show a marked decline in species richness (from eight to three species), a reduction in species diversity (from 1.51 to 0.88), and a reduction in abundance (from 2.6 to 0.4 animals/trap day) with distance (up to 1.3 km) from the minerotrophic upland into the adjacent ombrotrophic bog.

Differences in objectives and methods between the 1978 and 1979 collecting seasons in Minnesota and the Maine study may account for the differences and similarities between the two studies. As noted above, the 1978 Minnesota sample of a group of six representative peatland sites and one upland site produced similar results to the Maine study in that no strong correlation between the composition and abundance of the herpetofauna and peatland habitats was observed. Based on the observation that the Minnesota upland site was the richest site sampled in 1978, the Porter Ridge investigation was specifically designed as a comparison of community structure between adjacent minerotrophic upland sites and ombrotrophic bog sites, and significant differences were observed. The Maine study cannot be directly compared to the Porter Ridge study because upland forests adjacent to peatland sites were not sampled in Maine. Also, the distance of sampling sites from the upland were different: most of the Maine sites (20 of 24) were within 300 m (1000 ft.) of the upland, and four sites were between 400 (~1200 ft.) and 600 m (~1800 ft.) from the upland (Stockwell 1985). The Porter Ridge sites were located up to 1,300 (~4000 ft.) from adjacent upland.

Clearly, more comparisons are needed to substantiate the pattern observed at Porter Ridge. My qualitative observations from a variety of localities throughout the Minnesota peatlands, however, are consistent with the quantitative assessment from Porter Ridge: there is a marked difference in the composition and abundance of the herpetofauna in ombrotrophic bog sites compared to upland sites.

Another factor to consider in assessing these differences is geographic variation in peatland habitats. Glaser and Janssens (1986) document important differences in physiography between the midcontinental forested bogs of Minnesota and the nonforested maritime and semiforested continental bogs found in Maine that could influence the distribution and abundance of the herpetofauna.

Nonpeatland Comparisons

How does the herpetofauna of the peatlands compare to those found elsewhere in and near Minnesota (Table 9.4)? The herpetological "hotspot" of Minnesota is the Mississippi River valley in the southeastern corner of the state. This area offers a rich assortment of habitats supporting 41 species. Twenty-five species (15 reptiles) are found in the drier southwestern part of the state in the Renville County area. The Lake Itasca area is only 75 miles south of the Big Falls study area, but 22 species are found there, double the number of species recorded in the peatlands. The high number of species at Itasca is a result of the transition between woodland and prairie found there (Lang 1982; MHS 1985). Similarly, more species (20) can be found in the mixture of woodland and grassland habitats of the "southern transition" of southeastern Manitoba (Preston 1982; Cook 1984). There are 17 species in the lake and conifer forest zone of southeastern Ontario (Cook 1984); here the prairie transition cannot be invoked as an explanation for increased species richness. In Manitoba one must travel to the boreal coniferous forest zone, which runs through the center of the province, to find a low species richness (14 species) comparable to the Minnesota and Maine peatlands (Preston 1982).

Some species (green frog, mink frog — *Rana septentrionalis*, northern prairie skink — *Eumeces septentrionalis*) were conspicuously absent from the Minnesota peatlands but are found at nonpeatland localities both north and south of the peatland study areas. These biogeographic patterns suggest that the peatland habitat in some way restricts some amphibian and reptile species and that northern latitude and climate alone cannot satisfactorily explain the impoverished nature of the Minnesota peatlands. Possible reasons for the restrictive nature of the peatlands as a habitat for amphibians and reptiles are discussed below. It is important to stress that the relatively low species richness and diversity of the peatlands does not mean that amphibians and reptiles are uncommon; they are very abundant. There are simply many individuals of a few species, as indicated by the relatively low diversity values for both the Minnesota and Maine peatlands.

In New Hampshire forest habitats, Rudis (1984) typically found 9 to 11 species. Of all New Hampshire drift-fence captures, 57% were wood frogs, 17% were American toads, and 15% were redback salamanders (*Plethodon cinerus*). In a drift-fence survey of forests and old fields in southeastern Indiana, Karns (1992) typically found 10 to 17 species at forest sites. Zigzag salamanders (*Plethodon dorsalis*) (20.1%), wood frogs (17.9%), redback salamanders (13.0%), and green frogs (7.3%) dominated the herpetofauna (n = 667). Salamanders were a much more conspicuous component of the herpetofauna in these northern hardwood forests than in the peatlands studied, where anurans are dominant. Diversity index scores were generally higher in Indiana and New Hampshire forests than in Maine and Minnesota peatlands (Table 9.4). Herpetofaunal abundance as measured by drift-fence trapping success varied considerably: Minnesota peatlands = 0.95 animals/day; Maine peatlands = 0.25; southeastern Indiana forest = 0.37.

This analysis of the herpetofauna of the peatlands of Minnesota and Maine indicates that these communities are herpetologically rather depauperate in terms of species richness and diversity compared to other north-temperate communities of the region. The peatlands do support large numbers of a few species of amphibians, especially anurans, however.

Bog Lakes

The flora and fauna of *Sphagnum*-dominated bogs associated with successional lake changes have been better studied than those of major peatlands. Amphibians and reptiles, however, have received little attention (Marshall and Buell 1955; Bellis 1959, 1965; Heatwole and Getz 1960; Saber and Dunson 1978). Bog lakes are usually a mixture of habitat types in a relatively small area undergoing succession; several kinds of peatland habitat are often present. The bog lakes in the studies cited above are small isolated islands of peatland surrounded by minerotrophic upland. This is the opposite of the situation found in extensive peatlands.

All these bog-lake studies found that amphibians were the primary component of the herpetofauna. Wood frogs, northern leopard frogs, green frogs, and American toads were dominant species; at different sites, different members of this quartet dominated. Heatwole and Getz (1960) and Saber and Dunson (1978) noted that their bog sites in Michigan and Pennsylvania were depauperate compared to other habitat sites studied and suggested that the low pH of these sites might be a cause. Biebighauser (personal communication) did not collect green frogs in drift fences set near bog lakes in Minnesota, although they bred in a nearby lake. He also reported that redback salamanders were collected in forested areas but not at bog sites. These observations suggest that peatland affects herpetofaunal habitat utilization even on the relatively small scale of bog lakes.

Limiting Factors Controlling Peatland Habitat Utilization

This analysis of the distribution and abundance of amphibians and reptiles in northern Minnesota peatland habitats shows that amphibians, particularly anurans, dominate the herpetofauna, and that peatlands in general and *Sphagnum* bogs in particular are not suitable habitats for a number of species. As suggested in other community studies (Campbell and Christman 1982; Karns 1992), the particular plant associations found in an area do not seem to be a primary factor in determining the community structure of amphibians and reptiles. It is the physical characteristics that are of primary importance. In the peatlands such factors as moisture, pH, temperature, and trophic status seem to be of particular significance.

An ecological concept of classical importance in understanding biogeographic patterns is that of limiting factors. The ability of an organism to utilize a particular habitat depends on a complex array of physical and biotic conditions. The presence or absence of an organism can be controlled by the deficiency or excess of any of a number of factors that approach the tolerance limits of that organism. This concept was formalized in 1913 by V. E. Shelford as the "law" of tolerance and has inspired a great deal of research (Krebs 1985). What are the key limiting factors that have been involved in the creation and the maintenance of the patterns of herpetofaunal distribution and abundance observed in the Minnesota peatlands?

Distribution of Water Resources in Peatlands

The peatlands of northern Minnesota are a collection of poorly drained landforms and blocked drainages. The area is similar to the Florida Everglades in this respect. There are few deep lakes, and surface water is found in networks of shallow pools. Casual observation suggests that these extensive wetlands should be an amphibian paradise, and amphibians do dominate the herpetofauna and are found in high numbers. This merger between terrestrial and aquatic environments presents ecological restrictions for some species, however.

The water-saturated soils discourage the presence of terrestrial burrowing species. Both the northern prairie skink and the tiger salamander (*Ambystoma tigrinum*), for example, are found to the south, west, and northwest of the peatlands. They apparently find the water-saturated soils of the peatlands unsuitable for burrowing, despite the occurrence of suitable patches of sandy upland habitat within the peatlands.

The scarcity of permanent, deeper bodies of water, on the other hand, may prohibit the colonization of peatlands by more aquatic species. The mink frog is a true northern frog and ranges far into Canada. This species is usually associated with lakes and lily pads, but it was not found in the lake-poor peatlands. Tadpoles and salamander larvae are also affected by the scarcity of deeper lakes and ponds; these bodies of water do not freeze to the bottom and provide a year-round habitat. Tiger salamanders, for example, produce slow-growing carnivorous larvae, which often overwinter in their natal ponds; this is not possible in the shallow pools typical of the peatlands. As noted earlier, the quality and distribution of water resources may also help explain the rather mysterious absence of the aquatic green frog, which is common in Maine peatlands and found to the south, north, and east of the peatlands in Minnesota.

Microclimate

Temperature and moisture are major factors controlling habitat utilization by amphibians and reptiles (Duellman and Trueb 1986). Because amphibians and reptiles are ectotherms, they depend on the environment for their body temperature. As opposed to reptiles,

most amphibians can be active over a fairly wide range of temperatures; activity cannot be initiated until temperatures fall within the thermal activity range for a given species, however.

A northern Minnesota logger will tell you that the bogs are the first to freeze and the last to thaw. This observation has profound ecological consequences for ectothermic amphibians and reptiles. As described earlier, bogs were a month behind adjacent upland/fen areas in the initiation of spring activity because the water-saturated peat soil of the bogs took much longer to thaw. This is especially important for early-breeding amphibians.

Substrate temperature stations at the Porter Ridge site indicated that the bogs were more thermally variable than the adjacent upland, with greater extremes throughout the summer and fall (Karns and Regal 1979). Very cool microclimates are always available in the bog in the summer, yet exposed sites can be quite warm. *Sphagnum* sites also showed the signs of oncoming winter earlier than upland forests. *Sphagnum* consists largely of water, and the individual *Sphagnum* strands at the surface easily freeze. Thus the already short herpetological activity season of northern Minnesota is further compressed in the peatlands and particularly in bog habitats.

Precipitation, humidity, and the general availability of water are important ecological factors for amphibians. The moist, glandular skin of amphibians is water-permeable; unless they have access to water or moist microhabitats, they will desiccate. Peatlands and other wetlands, except perhaps in times of extreme drought, are hydric refuges in which favorable moisture conditions can always be found. Because the peatlands are water-saturated habitats, precipitation and humidity may be less important ecological factors than in drier environments. Even in the water-saturated peatlands, however, drift-fence data indicate that rainfall and humidity were important in initiating and controlling local movements of amphibians and reptiles. Numerous studies (Vogt and Hine 1982; Gibbons and Semlitsch 1981) indicate that amphibians and reptiles move in greater abundance during periods of rain and high humidity. These cues appear to be important in the peatlands even though ambient conditions are normally wet and humid.

The wet, humid conditions of the peatlands would be especially valuable for juvenile amphibians, which are at a water-balance disadvantage because of their high ratio of surface area to volume (Spotila 1972). Bellis (1965) noted that juvenile wood frogs in northern Minnesota dispersed from an upland breeding pond into a small bog during July and August and apparently stayed there for the remainder of the summer. Stockwell and Hunter (1985) also noted juvenile dispersal into peatlands in Maine. In Minnesota, juveniles were generally the dominant component of the bog herpetofauna. At Wisner Bog in 1978 a conspicuous movement of adult wood frogs and American toads into bog sites occurred in the later summer. Dry upland conditions may have initiated this movement.

Overwintering

In northern Minnesota, peatland amphibians and reptiles spend over half the year in overwintering sites in a dormant state. Lack of proper sites and unusually severe weather conditions can cause high mortality. Snow, for example, serves as an insulating layer, and in years of low snowfall mortality would probably increase. Suitable overwintering sites are clearly a key limiting factor (Hodge 1976).

Water-saturated habitats impose constraints on overwintering amphibians. Experimental studies (Schmid 1965) demonstrate that amphibians differ in their ability to tolerate sustained immersion in water (hydration stress). Field studies (Hodge 1976; Tester and Breckenridge 1964) indicate that overwintering mortality is correlated with the wetness of the site for some species (wood frog, Canadian toad — *Bufo hemiophrys*). Mortality is higher in overwintering sites that are too moist. Overwintering simulation experiments indicated that American toads and to a lesser extent wood frogs cannot tolerate water-saturated peat as an overwintering medium (Karns and Regal 1979).

It is not known with certainty where peatland amphibians overwinter. Indirect evidence from drift-fence trapping and observation (Karns and Regal 1979; Stockwell and Hunter 1985; Bellis 1965) indicates that some species (wood frogs, green frogs, American toads) do overwinter in wet peatland habitats, in contradiction to the studies noted above. Part of the explanation of this discrepancy may lie in the existence of suitably dry microhabitats that do not freeze inside of *Sphagnum* mounds or within the root systems of trees. Alternatively, some species may overwinter in frozen *Sphagnum.* Until recently it was thought that amphibians could not tolerate freezing temperatures, but recent studies (Schmid 1982; Storey 1990) demonstrate that through a combination of biochemical "antifreeze" and frost tolerance some species of frogs (spring peeper, wood frog, gray treefrog) can tolerate prolonged freezing temperatures.

Several *Sphagnum* mounds were dissected in the early spring in a vain search for overwintering amphibians. The upper portions of the mound, corresponding to the living and recently dead layers of moss, were frozen, and the ice structure was porous and granular. Below the upper layers, as the peat became more compacted, so did the structure of the ice. The experimental work cited above suggests that this upper layer of frozen peat is probably a perfectly suitable overwintering environment for some species. Thus it may not be necessary to postulate the existence of special overwintering refuges that do not freeze in peatlands. Some species may remain partially frozen and protected by biochemical antifreeze in frozen *Sphagnum* until the spring thaw. Current technology (microradio-transmitters, radioactive tagging) offers the possibility of tagging individuals and monitoring their overwintering behavior and physiology in peatlands.

Productivity and Resource Availability

The peatlands exhibit a wide range of nutrient regimes. Nutrient-rich minerotrophic fen sites are more productive and exhibit greater plant-species richness than nutrient-poor ombrotrophic bog sites (Boelter and Verry 1977; Moore and Bellamy 1974). Levels of nutrients and corresponding productivity may affect the quantity, the temporal and spatial availability, and the quality of food resources, which in turn influence the number of species and number of individuals in a given area (Krebs 1985). At Porter Ridge, major differences in species richness and abundance of amphibians and reptiles were found between nutrient-rich fen/ridge sites and nutrient-poor bog sites (Table 9.3). Do differences in resources help account for these changes in the herpetofauna?

The main diet of all adult peatland amphibians and reptiles except turtles is arthropods (insects, spiders, etc.). A short-term (one-week) arthropod-trapping experiment using sticky boards in bog and upland habitats (four traps at each site) was performed at Porter Ridge (Karns and Regal 1979). Approximately four times as many arthropods were trapped at ridge sites than at bog sites. Obviously, more research is needed to verify this difference, but these preliminary results suggest that bog habitats may offer fewer food resources than nutrient-rich sites, and this may affect the composition and size of amphibian and reptile populations.

Nutrient availability may also affect the quality of food resources. Janzen (1974) suggested that in nutrient-poor environments (like ombrotrophic bogs) plants will devote more resources to chemical defenses against predation than in nutrient-rich environments, because each gram of tissue in the nutrient-poor environment is more difficult to replace. This could affect not only the number of insects present but also the quality of the insects as food. Insects feeding on plants rich in chemical defenses may retain some of these chemicals, making the insects less palatable. This is speculation but is certainly a testable hypothesis.

The physiology of amphibians and reptiles is an important consideration in a discussion of peatlands resources. Amphibians and reptiles are low-energy ectothermic systems (Pough 1983) that do not have the high bioenergetic demands of endotherms (birds and mammals). In nutrient-poor bog environments, where resources may be a limiting factor, ectothermy would be a definite advantage.

Ectothermy also has important implications for amphibian and reptile population size; more ectotherms than endotherms can be supported on a given input of energy. Burton and Likens (1975) estimated that salamander biomass in a New Hampshire forest was

Table 9.5. Developmental and hatching success of amphibian embryos in bog and control water

Species	Bog water (pH 4.2)				Control water (pH 7.5)			
	Hatch	Curl	n	N	Hatch	Curl	n	N
Wood frog	$11.0 + 1.15$	77.2 ± 2.12	1758	53	96.3 ± 0.61	0.4 ± 0.21	2121	61
Blue-spotted salamander	0	59.1 ± 13.51	164	4	97.3 ± 0.29	0	152	4
American toad	0	14.6 ± 2.76	3038	58	93.0 ± 2.21	0.2 ± 0.16	1921	33
N. leopard frog	0	0.9 ± 0.66	435	12	98.5 ± 0.66	0	987	8
Chorus frog	0	0	285	8	64.5 ± 10.38	0	232	7
Spring peeper	0	0	325	8	71.5 ± 7.13	0.3 ± 0.30	419	8

Note: Mean percentage hatch or curl (± 1 SE), total number of eggs tested (n), and total number of egg masses sampled (N) are shown. Curl refers to embryos that develop but hatching is inhibited and the embryo dies within the egg (see Fig. 9.9).

Source: Karns (1984).

2.6 times higher than that of birds during the peak of the breeding season and about equal to that of the small-mammal community, yet the salamanders consumed only about 20% of the resources consumed by the endotherms. The herpetofauna, mainly amphibians, may well have the greatest biomass of any group of vertebrates in the peatlands; more individuals can survive and reproduce on fewer resources because of their low energy requirements. The peatlands herpetofauna may be relatively low in the number of species present, but the number of individuals is very high. Because of their abundance, amphibians and reptiles are an important source of high-quality food for a wide range of resident and transient peatland dwellers in both the terrestrial and aquatic domains of the peatlands. The study of animal ecology in the peatlands cannot be complete without serious consideration of the herpetofauna, trophic levels, and resource availability. Much more information is needed.

Bog-Water Toxicity

The darkly colored acidic water that is associated with bogs and other blackwater habitats has long been suspected of having toxic properties. Janzen (1974) reviewed numerous reports from both tropical and temperate areas indicating that blackwater habitats are deleterious to a wide range of aquatic and semi-aquatic taxa. These reports show a reduction in species diversity, abundance, or general productivity in blackwater areas. The experimental and field studies of Gosner and Black (1957), Dunson and Connell (1982), Saber and Dunson (1978), and Freda and Dunson (1986) in bog habitats in New Jersey and Pennsylvania, Strijbosch (1979) in Dutch wetlands, Beebee and Griffin (1977) in British heathlands, and Karns (1984) in Minnesota indicate that bog water is a deleterious breeding medium for most species of amphibians. Thus bog water may help explain the patterns of species distribution and abundance observed in the peatlands.

Six species of amphibians taken from peatland study sites in Minnesota (American toad, chorus frog, northern leopard frog, wood frog, spring peeper, and blue-spotted salamander) were tested in bog water (Karns 1984). Eggs of these species could be fertilized in bog water, but development and hatching was affected in all species tested. Only the wood frog exhibited a moderate level of hatching success in these experiments (Table 9.5).

Most wood frog embryos that did not hatch exhibited a peculiar condition called the "curling effect" by Freda and Dunson (1986): the embryo develops normally but the perivitelline chamber surrounding the embryo fails to expand, and the embryo is curled within the egg and cannot hatch (Fig. 9.9). Most embryos of other species and some wood frog embryos simply fail to develop normally and exhibit a variety of developmental abnormalities associated with low pH. Field experiments in bog habitats in northern Minnesota, New Jersey, and central Pennsylvania produced similar results (Karns 1984; Freda and Dunson 1986).

Laboratory experiments, field larval censuses, and monitoring of breeding ponds for the emergence of metamorphosed amphibians show that the wood frog was the most tolerant of bog water of all Minnesota peatland species tested (Karns 1984). Even wood frogs, however, did not survive through metamorphosis at ombrotrophic bog-water sites (pH < 4.5). Advanced wood frog larvae were found in weakly minerotrophic poor-fen sites (*Sphagnum*-dominated, pH 4.5-5.0), but never larvae of other species. Larvae of all species were regularly found at minerotrophic fen sites (pH > 5.0). Bog water associated with *Sphagnum* bogs in northern Minnesota peatlands is clearly a deleterious developmental medium for resident amphibians.

Why is bog water toxic to amphibian embryos and larvae? The low pH of bog water is important but does not completely explain the phenomenon. Studies by Freda and Dunson (1984, 1985, 1986), Dunson and Connell (1982), and Karns (1983, 1984) suggest that

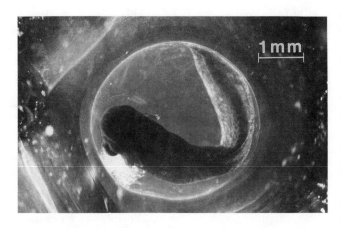

Fig. 9.9. Wood frog embryo reared in bog water. This embryo is in developmental stage 23 (approximately seven days old). Normally, in a suitable developmental medium, it already would have hatched. The bog water inhibited hatching, and the embryo died within the egg. The well-defined membrane surrounding the embryo is the vitelline membrane, which becomes tough and inelastic in bog water. The bright spot near the embryo's head is a tuft of material that may be hatching enzyme coagulated by chemical reaction with bog water. In bog water, the gelatinous outer layers of the egg become milky white in color and abnormally firm. The embryo has been bent into a U shape (the curl condition) by the failure of the membranes surrounding it to soften and expand normally.

organic compounds in bog water, various ions, temperature, and pH produce the toxic effects of bog water in a complex and synergistic fashion.

Peatland amphibians are often confronted with a mosaic of possible breeding sites of varying water quality, some of which are toxic. How do amphibians respond to this situation? At the Porter Ridge site amphibians exhibited a differential utilization of sites correlated with water quality (Karns 1984). Bog-water pools dominated the environmental chemistry along the southern edge of Porter Ridge (approximately 70% of the available water was toxic bog water). Censuses of breeding frogs and egg masses, however, showed that breeding activity was approximately five times greater at the relatively uncommon nontoxic-fen sites. Amphibians breeding at toxic bog-water sites failed to reproduce successfully.

The differential utilization of sites could be due to active habitat selection by amphibians in either of two ways: (1) active behavioral avoidance of bog-water sites or (2) high fidelity to natal fen sites where hatching and development occurred (Karns 1984; Freda and Dunson 1986). If amphibians cannot successfully breed at bog-water sites, bog-breeding populations must be maintained by dispersal from fen areas. Experiments show that American toads and wood frogs can discriminate between bog and fen sites based on odor alone (Karns 1984); however, these experiments did not distinguish between bog avoidance and fen preference. If the natal-site-fidelity hypothesis is correct, there is no need to invoke a special adaptation for bog-water

avoidance. The relatively small number of bog breeders may be dispersers from natal fens that simply do not discriminate against bog sites. Natal-site fidelity is a well-documented phenomenon among amphibians (Duellman and Trueb 1986).

Thus water quality in peatlands can affect local distribution and abundance of amphibians. Because bog water directly affects the reproductive success of peatland amphibians, it is a powerful source of selection. The toxicity of bog water would completely exclude amphibians from an area if only bog-water breeding sites were available, but the natural heterogeneity of peatland habitats will usually result in some "safe" breeding sites. The importance of bog water as a physiological, ecological, and evolutionary factor for amphibians is further discussed by Karns (1984) and Freda and Dunson (1986).

Historical Factors

The Minnesota peatlands are a geologically "new" environment, which may have contributed to the impoverished nature of the herpetofauna. Griffin (1977) has documented a series of vegetational changes following the retreat of the Wisconsin ice sheet and the gradual disappearance of Glacial Lake Agassiz from what are now the peatlands of northern Minnesota. Around 9,000 years ago, boreal forest invaded the former lake bed, followed by a mixed conifer-hardwood forest grading into oak savannah, with sedge marshes in low areas. Around 8,000 years ago, the climate became drier and warmer, and prairie vegetation moved into the former lake plain, with marshes occupying the poorly drained areas. Not until 4,000 years ago did climatic conditions begin to become cooler and wetter, permitting the development of peatlands. *Sphagnum* moss accumulation began in the Porter Ridge area about 3,000 years ago, according to radiocarbon dates from the Lake Agassiz peatlands 25 km (15 miles) east of Porter Ridge (Heinselman 1970; Janssen 1968). *Sphagnum* moss accumulation began in the Red Lake peatland 65 km (40 miles) northeast of Porter Ridge somewhat later (around 2,000 years ago; Finney 1966; Griffin 1975, 1977).

Minnesota was almost entirely glaciated at the height of the Wisconsin glaciation and obviously inhospitable to amphibians and reptiles. The herpetofauna of Minnesota today represents an invasion of amphibians and reptiles from refuges to the south of the glaciation (Porter 1972). Presumably, the herpetofauna in northern Minnesota has shifted with the climatic and vegetational changes. During the drier, warmer period from 8,000 to 4,000 years ago, for example, community composition may have been similar to the prairie herpetofauna found today in the grasslands of northwestern Minnesota, North Dakota, and southern Manitoba. With the formation of the peatlands, prairie species (e.g., northern prairie skink) would have been excluded and forest species typical of the eastern woodlands colonized the area. Some of these lake and woodland species were apparently excluded from the peatlands as a result of their special requirements (mink frog, green frog). Thus the amphibians and reptiles that we find in the peatlands today are the product of very recent (on a geological scale) ecological events, and the peatlands herpetofauna may still be adjusting to these changes.

Conclusions

The herpetofauna of the Minnesota peatlands has been influenced by a variety of factors. Some restrictive factors help explain the relatively low species richness of the herpetofauna, especially in nutrient-poor bogs, compared to other environments. In spite of limiting factors, the herpetofauna is an abundant and conspicuous component of the peatlands. In terms of biomass and numbers of individuals present, there are probably more amphibians in the peatlands than all other vertebrates combined. They are extremely important in the ecological dynamics of the peatlands in both aquatic and terrestrial environments. Peatland habitats appear to be very productive environments for those species able to tolerate the special conditions found there. The distribution and abundance of the peatlands herpetofauna is due to a complex combination of physical, biotic, and historical factors; single-factor explanations for the observed patterns are not satisfactory.

A Global Perspective

In addition to the vast boreal peatlands of North America and Eurasia, there are southern peatlands in the temperate latitudes of Tasmania, Australia, and New Zealand and subtropical and tropical peatlands in North America, South America, and Asia (Moore and Bellamy 1974; Gore 1983).

The southeastern coastal plain contains some of the largest peatlands in the United States, including the New Jersey Pine Barrens, the Great Dismal Swamp of Virginia and North Carolina, the Okefenokee Swamp in Georgia, and the Pocosins of North Carolina. Farther south are the extensive peatlands of the Everglades (Hofstetter, 1983). Some information is available on the amphibians and reptiles of these peatlands.

The Pine Barrens is the best known of these localities. Conant (1979) presents a zoogeographical review of the herpetofauna of southern New Jersey with particular attention to the Pine Barrens. He records 58 species from the area (25 amphibians, 33 reptiles), which includes a variety of habitats other than peatlands. Ten of these species are restricted to the Pine Barrens.

Conant suggests that the acid waters of the Pine Barrens may restrict colonization of the area by amphibians. He describes a colony of gray treefrogs that failed to establish a permanent population because they could not breed in the area. Gosner and Black (1957) and Freda and Dunson (1986) have experimen-

tally investigated the phenomenon of the toxicity of bog water to amphibian embryos and larvae in the Pine Barrens. Conant notes that the acid-tolerant Pine Barrens treefrog and the carpenter frog do well in bog habitats and are confined to the Pine Barrens in New Jersey.

Little more than species lists is available for the Great Dismal Swamp (58 species: 28 amphibians, 30 reptiles; Delzell 1978), the Okefenokee Swamp (101 species: 37 amphibians, 64 reptiles; Laerm *et al.* 1984), and the Florida Everglades (80 species: 17 amphibians, 63 reptiles; George 1972). Not even this minimal information is available for the Pocosin wetlands of North Carolina; Wilbur (1981) describes a fruitless search of the scientific literature for studies on the community or population ecology of the fauna of the Pocosins. He notes that the area needs investigation of nongame animals on the most descriptive of levels.

One finds the same poverty of information on the herpetofauna of peatlands outside the United States. A recent multiauthor volume on the ecology of mire (swamp, bog, fen, and moor) ecosystems of the world (Gore 1983) contains reviews of the animal communities of the peatlands of Sweden, the Soviet Union, Australia, the tropical peat swamps of western Malaysia, and the Magellanic tundra complex of southern South America. Few references to amphibians and reptiles are found, and no community-level analyses of the herpetofauna are cited. For example, Botch and Masing (1983) comment that in Soviet mire ecosystems "... of the vertebrates, only birds are numerous." Tantalizing bits of information on individual species are noted: there is the corroboree frog (*Pseudophryne corroboree*), which lives above 1,370 m (4,500 feet) in the *Sphagnum* bogs of highland regions of Australia. These frogs are buried under snow for several months of the year and lay eggs in burrows up to 25 cm (10 inches) deep in the decaying vegetation of *Sphagnum* mounds (Campbell 1983).

Why is there so little information on the herpetofauna of the peatlands, and on peatlands animal ecology in general? The obvious answer is logistics: wetlands are not easy environments in which to work. The dearth of information on the herpetofauna of the peatlands of the coastal plain of the eastern United States, which are all located near populated areas and major universities, attests to this. It is not surprising that more isolated peatlands have not been investigated.

A common theme that runs throughout the scientific literature on peatlands is the importance of peatlands as habitat for species restricted to peatlands and as refuges for formerly wide-ranging species that are now confined to peatland areas as a result of the destruction of suitable habitat elsewhere. This in itself is a strong argument for the preservation of peatlands. As is true of wetlands in general, these areas have traditionally been considered wastelands that demand to be drained and developed. Alderman (1965) commented on the extensive peatland of central Canada: "It just lies there, smeared across Canada like leprosy." We hope that this attitude is changing and that the peatlands will come to be seen as valuable beyond their potential for industrial exploitation.

Recent studies of extensive peatlands in Maine and Minnesota were prompted by plans for large-scale mining of peat, which would have a devastating impact on peatland communities on a local if not regional basis. The economic situation then changed, and the impetus for peatland development was reduced (until the next oil crisis). The legacy of these peatland development schemes may be the knowledge we have gained from the research programs that were generated. In Minnesota, at least, the peatlands have been thrust into the public awareness and are now more fully recognized as an important and beautiful part of our natural heritage.

Literature Cited

Alderman, T. 1965. It's a nuisance. Imperial Oil Review 49(3):6-10.

Beebee, T. J. C., and J. R. Griffin. 1977. A preliminary investigation into Natterjack toad (*Bufo calamita*) breeding site characteristics in Britain. Journal of Zoology, London 181:341-350.

Bellis, E. D. 1959. A study of movement of American toads in a Minnesota bog. Copeia 1959(2):173-174.

——. 1965. Home range and movements of the wood frog in a northern bog. Ecology 46:92-98.

Blackith, R. M., and M. C. D. Speight. 1974. Food and feeding habits of the frog *Rana temporaria* in bogland habitats in the West of Ireland. Journal of Zoology, London 172:67-72.

Boelter, D. H., and E. S. Verry. 1977. Peatland and water in the northern lake states. USDA Forest Service General Technical Report NC-31. North Central Forest Experiment Station, St. Paul, Minnesota, USA.

Botch, M. S., and V. V. Masing. 1983. Mire ecosystems in the USSR. Pages 95-152 *in* A. J. P. Gore, editor. Ecosystems of the world 4B, Mires: Swamp, bog, fen and moor. Elsevier, New York, New York, USA.

Burton, T. M., and G. E. Likens. 1975. Energy flow and nutrient cycling in salamander populations in the Hubbard Brook Experimental Forest, New Hampshire. Ecology 58:1068-1080.

Campbell, E. O. 1983. Mires of Australasia. Pages 153-180 *in* A. J. P. Gore, editor. Ecosystems of the world 4B, Mires: Swamp, bog, fen and moor. Elsevier, New York, New York, USA.

Campbell, H. W., and S. P. Christman. 1982. Field techniques for herpetofaunal community analysis. Pages 193-200 *in* N. J. Scott, Jr., editor. Herpetological communities. Wildlife Research Report no. 13, U.S. Department of the Interior, Fish and Wildlife Service.

Conant, R. 1975. A field guide to reptiles and amphibians of eastern and central North America. Houghton Mifflin, Boston, Massachusetts, USA.

——. 1979. A zoogeographical review of the amphibians and reptiles of southern New Jersey, with emphasis on the Pine Barrens. Pages 467-488 *in* R. T. T. Forman, editor. Pine Barrens: Ecosystem and landscape. Academic Press, New York, New York, USA.

Cook, F. R. 1984. Introduction to Canadian amphibians and reptiles. National Museum of Natural Sciences, National Museums of Canada, Ottawa, Canada.

Delzell, D. E. 1978. A provisional checklist of amphibians and reptiles in the Dismal Swamp area, with comments on their range of distribution. Pages 244-260 *in* P. W. Kirk, Jr., editor. The Great Dismal Swamp. University Press of Virginia, Charlottesville, Virginia, USA.

Duellman, W. E., and L. Trueb. 1986. Biology of amphibians. McGraw-Hill, New York, New York, USA.

Dunson, W. A., and J. Connell. 1982. Specific inhibition of hatching in amphibian embryos by low pH. Journal of Herpetology 16(3):314-316.

Finney, H. R. 1966. Some characteristics of four raised bogs in northern Minnesota: Stratigraphic relationships and mineralogical properties of the inorganic fraction. Ph.D. thesis, University of Minnesota, Minneapolis, Minnesota, USA.

Freda, J., and W. A. Dunson. 1984. Sodium balance of amphibian larvae exposed to low environmental pH. Physiological Zoology 57(4):435-443.

—— and W. A. Dunson. 1985. The influence of external cation concentration on hatching of amphibian embryos in low pH water. Canadian Journal of Zoology 63:2649-2656.

—— and W. A. Dunson. 1986. Effects of low pH and other chemical variables on the local distribution of amphibians. Copeia 1986(2):454-466.

George, J. C. 1972. Everglades wildguide. Natural History Series, Office of Publications, U.S. National Park Service.

Gibbons, J. W., and R. D. Semlitsch. 1981. Terrestrial drift fences with pitfall traps: An effective technique for quantitative sampling of animal populations. Brimleyana no. 7:1-16.

Glaser, P. H., and J. A. Janssens. 1986. Raised bogs in eastern North America: Transitions in landforms and gross stratigraphy. Canadian Journal of Botany 64:395-415.

Gore, A. J. P., editor. 1983. Ecosystems of the world 4B, Mires: Swamp, bog, fen and moor. Elsevier, New York, New York, USA.

Gorman, E. 1967. Some chemical aspects of wetland ecology. National Research Council of Canada, Association Committee on Geotechnical Research, Technical Memoirs, no. 90:20-83.

Gosner, K. L., and I. H. Black. 1957. The effects of acidity on the development and hatching of New Jersey frogs. Ecology 38(2):256-262.

Griffin, K. O. 1975. Vegetation studies and modern pollen spectra from the Red Lake Peatland, northern Minnesota. Ecology 56(3):531-546.

——. 1977. Paleoecological aspects of the Red Lake Peatland, northern Minnesota. Canadian Journal of Botany 55(2):172-192.

Heatwole, H. 1961. Habitat selection and activity of the wood frog, *Rana sylvatica* Le Conte. American Midland Naturalist 66(2):301-313.

—— and L. L. Getz. 1960. Studies on the amphibians and reptiles of Mud Lake Bog in southern Michigan. Jack Pine Warbler 38:107-112.

Hedges, S. B. 1986 An electrophoretic analysis of holarctic hylid frog evolution. Systematic Zoology 35:1-21.

Heinselman, M. L. 1963. Forest sites, bog processes, and peatland types in the Glacial Lake Agassiz region, Minnesota. Ecological Monographs 33:327-374.

——. 1970. Landscape evolution, peatland types, and the environment in the Lake Agassiz Peatlands Natural Area, Minnesota. Ecological Monographs 40:235-261.

——. 1975. Boreal peatlands in relation to environment. Pages 93-103 *in* A. D. Hasler, editor. Coupling of land and water systems. Springer-Verlag, New York, New York, USA.

Hodge, R. P. 1976. Amphibians and reptiles in Alaska, the Yukon, and Northwest Territories. Alaska Northwest, Anchorage, Alaska, USA.

Hofstetter, R. H. 1983. Wetlands in the United States. Pages 201-244 *in* A. J. P. Gore, editor. Ecosystems of the world 4B, Mires: Swamp, bog, fen and moor. Elsevier, New York, New York, USA.

Janssen, C. R. 1968. Myrtle Lake: A late- and post-glacial pollen diagram from northern Minnesota. Canadian Journal of Botany 46:1397-1408.

Janzen, D. H. 1974. Tropical blackwater rivers, animals, and mast fruiting by the Dipterocarpaceae. Biotropica 6(2):69-103.

Jaslow, A. P., and R. C. Vogt. 1977. Identification and distribution of *Hyla versicolor* and *Hyla chrysoscelis* in Wisconsin. Herpetologica 33(2):201-205.

Karns, D. R. 1983. The role of low pH in the toxicity of bog water to amphibian embryos. Bulletin of the Ecological Society of America 64(2):188.

————. 1984. Toxic bog water in northern Minnesota peatlands: Ecological and evolutionary consequences for breeding amphibians. Unpublished Ph.D. thesis. University of Minnesota, Minneapolis, Minnesota, USA.

————. 1986. Field herpetology: Methods for the study of amphibians and reptiles in Minnesota. Occasional Paper no. 18, Bell Museum of Natural History. University of Minnesota, Minneapolis, Minnesota, USA.

————. 1992. The herpetofauna of Jefferson County: Analysis of an amphibian and reptile community in southeastern Indiana. Proceedings of the Indiana Academy of Sciences for 1988. Vol. 98.

———— and P. J. Regal. 1978. Relationship of amphibians and reptiles to peatland habitats in Minnesota. Progress Report V. Minnesota Department of Natural Resources, Division of Minerals, St. Paul, Minnesota, USA.

———— and P. J. Regal. 1979. The relationship of amphibians and reptiles to peatland habitats in Minnesota. Final Report. Minnesota Department of Natural Resources, Division of Minerals, St. Paul, Minnesota, USA.

Kiester, A. R. 1971. Species density of North American amphibians and reptiles. Systematic Zoology 20:127-137.

Krebs, C. J. 1985. Ecology. Harper and Row, New York, New York, USA.

Laerm, J., B. J. Freeman, L. J. Vitt, and L. E. Logan. 1984. Checklist of vertebrates of the Okefenokee Swamp. Pages 682-701 *in* A. D. Cohen, D. J. Casagrande, M. J. Andrejko, and G. R. Best, editors. The Okefenokee Swamp: Its natural history, geology, and geochemistry. Wetland Surveys, Los Alamos, New Mexico, USA.

Lang, J. W. 1982. The reptiles and amphibians of Minnesota: Distribution maps, habitat preferences, and selected references. Final Report. Nongame Program, Minnesota Department of Natural Resources, St. Paul, Minnesota, USA.

Magurran, A. E. 1988. Ecological diversity and its measurement. Princeton University Press, Princeton, New Jersey, USA.

Marshall, W. H., and M. F. Buell. 1955. A study of the occurrence of amphibians in relation to a bog succession, Itasca State Park, Minnesota. Ecology 36(3):381-387.

MHS (Minnesota Herpetological Society). 1985. Distribution maps for reptiles and amphibians of Minnesota. MHS publication, Minneapolis, Minnesota, USA.

Moore, P. D., and D. J. Bellamy. 1974. Peatlands. Springer-Verlag, New York, New York, USA.

Noble, G. K., and R. C. Noble. 1923. The Anderson tree frog, observations on its habits and life history. Zoologica 2:417-455.

Porter, K. R. 1972. Herpetology. W. B. Saunders, Philadelphia, Pennsylvania, USA.

Pough, F. H. 1983. Amphibians and reptiles as low-energy systems. Pages 141-188 *in* W. P. Aspey and S. I. Lustick, editors. Behavioral energetics. Ohio State University Press, Columbus, Ohio, USA.

Preston, W. B. 1982. The amphibians and reptiles of Manitoba. Manitoba Museum of Man and Nature, Winnipeg, Manitoba, Canada.

Rudis, D. D. 1984. Amphibian and reptile habitat associations in three New England forest cover types. Unpublished M.S. thesis. University of Massachusetts, Amherst, Massachusetts, USA.

Saber, P. A., and W. A. Dunson. 1978. Toxicity of bog water to embryonic and larval anuran amphibians. Journal of Experimental Zoology 204:33-42.

Schmid, W. D. 1965. Some aspects of the water economies of nine species of amphibians. Ecology 46(3):261-269.

————. 1982. Survival of frogs in low temperature. Science 215:697-698.

Spotila, J. R. 1972. Role of temperature and water in the ecology of lungless salamanders. Ecological Monographs 42:95-125.

Stockwell, S. S. 1985. Distribution and abundance of amphibians, reptiles, and small mammals in eight types of Maine peatland vegetation. Unpublished M.S. thesis. University of Maine, Orono, Maine, USA.

———— and M. L. Hunter, Jr. 1985. Distribution and abundance of birds, amphibians and reptiles, and small mammals in peatlands of central Maine. Report to the Maine Department of Inland Fisheries and Wildlife.

———— and M. L. Hunter, Jr. 1989. Relative abundance of herpetofauna among eight types of Maine peatland vegetation. Journal of Herpetology 23(4):409-414.

Storey, K. B. 1990. Life in a frozen state: Adaptive strategies for natural freeze tolerance in amphibians and reptiles. American Journal of Physiology. 258(3):559-568.

Strijbosch, H. 1979. Habitat selection of amphibians during their aquatic phase. Oikos 33:363-372.

Tester, J. R., and W. J. Breckenridge. 1964. Winter behavior patterns of the Manitoba toad, *Bufo hemiophrys*, in northwestern Minnesota. Annales Academae Scientiarum Fennicae, Series A, IV. Biologica:423-431.

Vogt, R. C., and R. L. Hine. 1982. Evaluation of techniques for assessment of amphibian and reptile populations in Wisconsin. Pages 201-217 *in* N. J. Scott, editor. Herpetological Communities. Wildlife Research Report no. 13. U.S. Department of the Interior, Fish and Wildlife Service.

Wilbur, H. M. 1981. Pocosin fauna. Pages 62-68 *in* C. J. Richardson, editor. Pocosin wetlands: An integrated analysis of coastal plain freshwater bogs in North Carolina. Hutchinson Ross, Stroudsburg, Pennsylvania, USA.

Wright, A. H., and A. A. Wright. 1949. Handbook of frogs and toads of the United States and Canada. 3rd edition. Comstock, Ithaca, New York, USA.

PART III
Hydrology

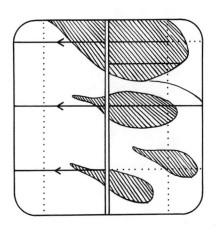

CHAPTER 10

Surface Hydrology

Kenneth N. Brooks

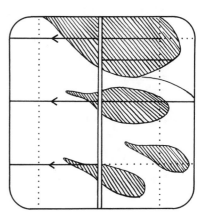

The patterned peatlands of northern Minnesota illustrate the close association between peat and water. Peatlands develop where there is an excess of water, particularly shallow groundwater (Granlund 1932; Heinselman 1961; Romanov 1961; Paivanen 1973; Ahti 1977; Boelter and Verry 1977; Ivanov 1981; Ingram 1983). The intricate patterns of peatland vegetation in northern Minnesota result from the distinctive interactions of precipitation and groundwater flow. Overlooked, and sometimes misunderstood, is the relationship between these peatlands and the streamflow that originates from them.

The perceived hydrologic role of peatlands, as with wetlands in general, has been somewhat controversial and not always correct. Are peatlands high-water-yielding areas? Do they sustain high levels of streamflow throughout the year? In contrast to other ecosystems, do peatlands reduce flooding? How is streamflow changed when peatlands are ditched, mined, or altered in vegetative cover? Conflicting answers to these questions are based on a lack of understanding and an inability to quantify hydrologic processes of peatlands.

Minnesota peatlands are of particular interest to hydrologists and water-resource managers because they occur extensively in the headwater areas of many streams and rivers. Even outside the main area of patterned peatlands, a substantial component of most watersheds in northern Minnesota is peatland.

The ability to quantify peatland-streamflow relationships is helpful in determining the hydrologic value of peatlands and estimating the hydrologic impact of peatland development. Such a capability requires a model that can simulate the important hydrologic functions of peatlands. Any hydrologic model, to be useful, must be conceptually correct and represent hydrologic processes with adequate mathematical formula-

tion. In this chapter on the hydrology of peatlands, I describe a hydrologic model developed to advance our understanding of peatland hydrology and to predict the streamflow response from peatlands.

The Hydrologic Function of Peatlands

The streamflow response of a peatland to rainfall or snowmelt depends largely on the size of the watershed and the depth of the water table. Streamflow volumes throughout the year depend on the extent of linkage between peatland and regional groundwater. Ombrotrophic peatlands (bogs) are so named because they can be considered hydrologically isolated from regional groundwater; the upper layers of peat are nourished primarily by dissolved ions in precipitation (Gorham 1957). Minerotrophic peatlands (fens) are linked to regional groundwater and receive nutrients and minerals from groundwater as well as precipitation (Gorham 1957; Moore and Bellamy 1974; Boelter and Verry 1977). As a result, vegetation and streamflow patterns differ in the two peatland types (Boelter and Verry 1977; Clausen and Brooks 1983a, 1983b). The nutrient-poor bogs exhibit stunted black spruce, a low heath shrub understory, and generally a floor of *Sphagnum* moss; streamflow from bogs is erratic and mirrors patterns of rainfall and snowmelt. Fens, in contrast, offer a more diverse vegetative cover with a variety of tall shrub and overstory species; streamflow is less erratic and is sustained at higher levels during dry seasons because of the connection with regional groundwater. Complex watersheds that contain bogs, fens, and locally mineral-soil uplands have streamflow characteristics that fall somewhere between these extremes.

Reviews of peatland hydrology by Boelter and Verry

(1977), Clausen and Brooks (1980), Ingram (1983), Carter (1986), and Gafni (1986) reveal that:

(1) Annual evapotranspiration far exceeds annual streamflow from peatlands.

(2) The depth to the water table governs the magnitude of both streamflow response to moisture input and evapotranspiration losses.

(3) The majority of horizontal groundwater flow in peatlands occurs through the upper layers of peat soil, commonly called the acrotelm; flow in the deeper, more decomposed anaerobic peat layers, collectively called the catotelm, is much less and can be considered a negligible contribution to streamflow from a peatland. The exception to this is an artesian spring near the outflow channel of a peatland.

(4) The greatest percentage of annual streamflow from natural peatlands in northern Minnesota occurs in the spring as a result of snowmelt.

(5) Peatlands, like most wetlands, temporarily store much of the precipitation, so that stormflow peak discharge is lowered.

(6) Streamflow from peatlands is generally reduced during summer because of high rates of evapotranspiration; the reduction in streamflow is more pronounced in bogs than in fens, in which flow is sustained because of the connection to groundwater aquifers.

A description of the water budget of peatlands and a discussion of the respective storages and flow processes follow as a context for understanding these points.

Water Budget

The water budget of any watershed system is defined as:

$$Q = P - ET - \Delta S + GW_i - GW_o$$

where

Q	=	streamflow
P	=	precipitation
ET	=	evapotranspiration
S	=	change in storage ($S_2 - S_1$)
GW_i	=	groundwater inflow
GW_o	=	groundwater outflow

The streamflow from minerotrophic fens can be described with this relationship. Although regional groundwater inflow and outflow represent important components of the water budget of a fen, the flow pathways and boundary conditions affecting groundwater exchange are still poorly understood. Because regional groundwater does not contribute to the streamflow response of ombrotrophic bogs, the water budget of bogs can be simplified to:

$$Q = P - ET - \Delta S$$

Streamflow from a bog therefore depends only on the balance between precipitation and evapotranspiration from the bog itself.

Evapotranspiration

Like most wetland systems, the major pathway by which water is lost from a peatland is evapotranspiration (Carter 1986). Of the 510 to 760 mm (20 to 30 inches) of precipitation that falls annually on peatlands in northern Minnesota, 455 to 610 mm (18 to 24 inches) are evaporated or transpired (Verry and Boelter 1978). The rate of evapotranspiration (AET), expressed as a ratio of potential evapotranspiration (PET), is inversely related to the depth of the water table below the hollow bottoms or the mined surface, as illustrated in Fig. 10.1 (Virta 1966; Juusela et al. 1970; Boelter 1972). When the water table is near the soil surface, evapotranspiration rates approach potential rates. The more commonly seen AET/PET relationship for mineral-soil uplands is shown in Fig. 10.1 for comparison.

Daily and weekly ET rates in peatlands can be approximated with pan-evaporation measurements and climatological formula used to estimate potential evapotranspiration (Bay 1966; Virta 1966). The relationships between actual and potential evapotranspiration are quite variable, however. Bonde et al. (1961) indicated that ET losses from different wetland plant communities ranged from 93 to 288 percent when expressed as a percent of open-water evaporation. Heikurainen (1963) reported that transpiration rates of conifer and birch stands were similar in Finland. Dooge (1975) found that evapotranspiration from fens was 5 to 15 percent higher than that from bogs. This difference may be explained more by the greater stability of the water table in a fen (in contrast to that of a bog) than by differences in vegetation type. The depth of water table appears to be more important in determining ET losses than the type of vegetation.

Evapotranspiration losses from open *Sphagnum* bogs with sedges, grass, and Ericaceae cover are greatest when the water table is about 10 cm (4 inches) below the bottom of the hollows (Boelter and Verry 1977). Evaporation from *Sphagnum* moss is highest at this level, and roots of other plants have adequate aeration to transpire actively. Once the water table drops to about 33 cm (13 inches) below the hollow bottom, a sharp reduction in evapotranspiration is observed.

Although the type of vegetation seems to have little effect on the total annual ET loss, changes in peatland vegetation can alter the pattern of ET losses. Verry (1980) reported that clear-cutting spruce from a small bog in northern Minnesota did not change annual water yields but did cause a change in water-table

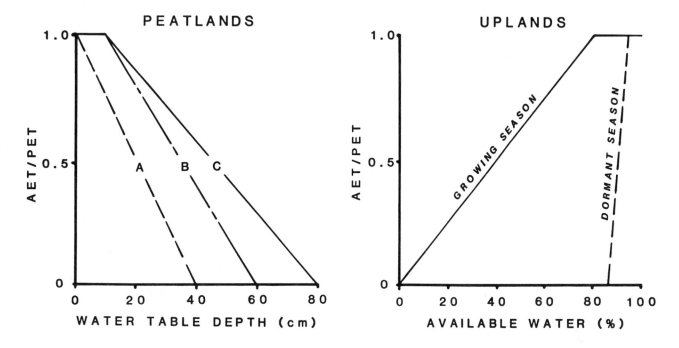

Fig. 10.1. Ratios of actual evapotranspiration (AET) to potential evapotranspiration (PET) for peatlands and upland mineral-soil areas. The functions for peatlands are: A for > 25 percent *Sphagnum* moss (Romanov 1962), B for mined areas, and C for sedges and < 25 percent *Sphagnum* (Juusela *et al.* 1970).

behavior. Following clear-cutting, the water table in the clear-cut bog was as much as 10 cm (4 inches) higher than predicted during wet periods and 19 cm (7.5 inches) lower than predicted during dry periods. With removal of the spruce overstory, as much as 30 percent of the annual precipitation that would otherwise be intercepted and evaporated was then available to the groundwater in the clear-cut bog. During wet periods this difference in interception apparently causes higher water tables. During dry periods, the increased wind, temperature, and solar radiation at the peat surface in the clear-cut areas produce a greater evaporative demand at the surface. In addition, an increase in grass and sedge cover following clear-cutting apparently results in a greater transpiration loss than that from the spruce overstory. Over the year, the net effect was no change in water yield, implying that annual evapotranspiration was unchanged.

When the water table is at least 30 to 40 cm (12 to 16 inches) below the bottom of the hollows, the total ET loss is similar to that of mineral-soil systems. The moisture content of the unsaturated soil profile then governs evapotranspiration rather than the depth of the water table.

Shallow Groundwater Flow

Once excess water reaches the water table of a peatland, the relationship between shallow groundwater flow and the peat soil determines the streamflow re-

sponse of peatlands. The basic law that governs shallow groundwater flow in peatlands is Darcy's Law:

$$Q = -K A \, dh/dl$$

where

Q = rate of flow (volume/unit time)
K = hydraulic conductivity (volume/unit area/unit time)
A = cross-sectional area of flow
dh/dl = hydraulic gradient (change in hydraulic head over horizontal distance; approximated by the slope of the water table)

The hydraulic gradient is the driving force for flowing water, and it can vary with time and space. Heinselman (1961) reported gradients ranging from 0.06 to 0.46 percent in Lindford Bog in Minnesota. Siegel (1983) reported a regional hydraulic gradient of 0.05 percent in northern Minnesota. Gafni (1986) found gradients to average less than 0.10 percent in bogs and fens, although gradients become steeper downslope from the center of a raised bog toward its lagg. Eggelsmann and Schuch (1976) reported similar gradient changes in Germany; the gradient changed from 0.03 percent in the center of a bog to nearly 7 percent next to its lagg. Drainage ditches steepen the hydraulic gradient of peatlands to over 20 percent near the ditch (Gafni and Brooks 1986). In all but extensively ditched peatlands, the low hydraulic gradients constrain groundwater velocity.

Both hydraulic conductivity (K) and groundwater velocity decrease with increasing humification and usually with depth. Gafni and Brooks (1990) estimated an average K of 1764 cm/hr (694.5 inches/hr) in the upper 10-cm (4-inch) layer of a fibrous peat soil, but the highest groundwater velocity was 0.74 cm/hr (0.3 inch/hr). At a depth of 40 to 50 cm (16 to 20 inches), groundwater velocities dropped to an average of 0.019 cm/hr (0.007 inch/hr). The von Post degree of humification was found to be a good predictor ($r^2 = 0.81$) of K for von Post values from 1 to 7, as illustrated in Fig. 10.2. Because peat is more decomposed at von Post humification values of 8 to 10 (von Post and Granlund 1926), one would expect the hydraulic conductivity to be extremely low at these values.

Low hydraulic gradients result in low flow velocities even in the upper fibrous layers of peat. Neglecting pipe flow, groundwater velocities generally become very small as depth and humification increase. Verry (1984) reported that lateral flow through the peat soil of a *Sphagnum* bog ceased when the water table was one-third the distance from the mean annual minimum level to the mean annual maximum level. This occurred at the interface between von Post values 2 and 3, a zone with adequate hydraulic conductivity. Verry further refined the definition of acrotelm as the upper layer of peat through which the water table oscillates in all but severe drought years. Therefore, the upper acrotelm and free water in the hollows contribute significantly to streamflow. The bottom of the acrotelm parallels the peat surface as traced by connecting hollow bottoms, and it may be 20 to 130 cm (8 to 51 inches) below the water table surface where lateral flow stops, the point where streamflow ceases (Verry 1984). As the water table drops into the more decomposed peat soil, or the layer called the catotelm, the volume of groundwater flow becomes negligible relative to the total annual

flow. This catotelm maintains a constant water content in all but severe drought periods.

As a peatland is ditched and mined, the water table drops to the underlying layers. The acrotelm and catotelm relationships of the peat soil are then altered. Even though the hydraulic gradient is steepened by ditching, the low hydraulic conductivities of deeper peat soil tend to restrict flow rates (Gafni and Brooks 1986).

Streamflow

Like evapotranspiration rates, streamflow from peatlands is governed primarily by the depth of the water table at the time of rainfall or snowmelt. Normally the water table is at or near the soil surface in early spring and recedes as ET losses become greater during the summer months. The high levels of flow during the spring and low levels during summer from all peatlands mirror such water-table changes.

Most streamflow from Minnesota peatlands occurs in the spring as a result of snowmelt (Verry and Boelter 1978). Bay (1969) reported that two-thirds of the annual streamflow from a small bog was derived from snowmelt, but that snow represented only 20 to 25 percent of the annual precipitation. The dominance of snowmelt runoff is apparent in the seasonal streamflow pattern of a Minnesota bog (Fig. 10.3) compared to a bog in Scotland (Fig. 10.4). In Minnesota, streamflow from small bogs may cease in the winter as a result of freezing. Following snowmelt, streamflow diminishes rapidly in response to higher evapotranspiration losses during the summer months. This pattern of reduced summer streamflow is evident in most peatlands.

Because of their active linkage with regional groundwater systems, fens exhibit more uniform streamflow throughout the year than do bogs (Fig. 10.5). A comparison of flow-duration curves for a fen and a bog show that streamflow is nearly constant 70 percent of the time for the fen (Fig. 10.6). Streamflow from the bog is more variable and even ceased 20 percent of the time.

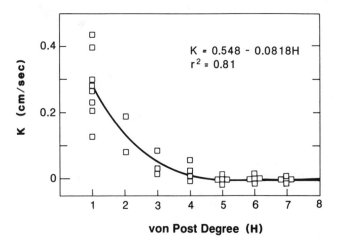

Fig. 10.2. Relationship between hydraulic conductivity and von Post degree of decomposition (H) for the S-2 bog in northern Minnesota from Gafni and Brooks (1990).

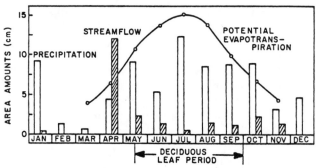

Fig. 10.3. Monthly precipitation, streamflow, and potential evapotranspiration for a perched bog watershed, 1969 (modified from Verry and Boelter 1978).

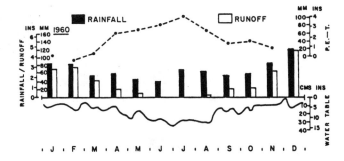

Fig. 10.4. Rainfall, runoff, evapotranspiration, and water-table levels for a raised bog in Scotland (from Robertson *et al.* 1968).

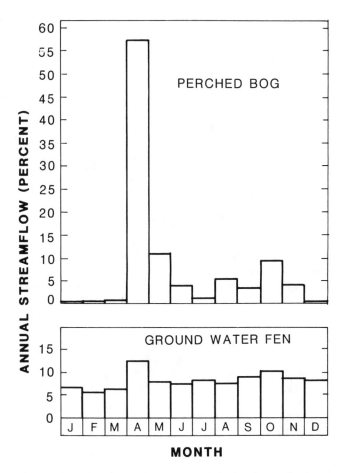

Fig. 10.5. Monthly streamflow from a perched ombrotrophic bog and a minerotrophic fen in northern Minnesota for 1969 (from Boelter and Verry 1977).

The stormflow response of peatlands to large rainstorms, like that of any wetland system, is quite different from that of mineral-soil watersheds. The stormflow response of peatlands is largely the result of the physical characteristics of peatland watersheds. The lack of topographic relief, the absence of well-defined drainage channels (and the extremely low drainage density in contrast to mineral-soil watersheds), and the

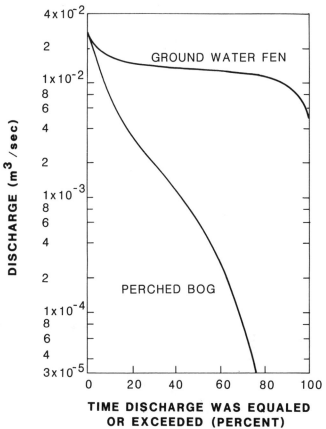

Fig. 10.6. Flow-duration curves for a groundwater fen and a perched bog watershed (from Boelter and Verry 1977).

shallow water table all combine in such a way that peatlands behave hydrologically like unregulated shallow reservoirs (Fig. 10.7). When the water table is shallow, the corresponding storage in the peatland and discharge from the peatland are both high (Boelter and Verry 1977). As the water table drops, both storage and discharge diminish. Such relationships coincide with reservoir routing methods used in surface hydrology (Linsley *et al.* 1982). Unlike reservoirs, however, the relationship of water-table depth to discharge is determined by the rate of flow through the acrotelm near the outflow point.

Modeling Approach

Models help to identify and define what is known and what needs to be known about a particular system. The process of developing a model helps to determine the type of research needed and how to organize a research program. The hydrologic relationships of the model must be theoretically sound and mathematically formulated to represent the integrated response or hydrologic function of peatlands. A conceptual model of

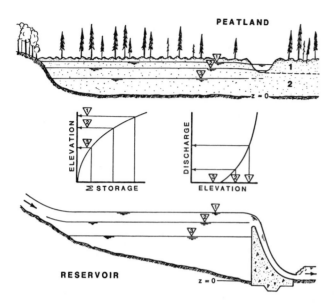

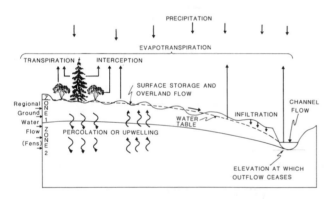

Fig. 10.7. Comparison of peatland-streamflow relationship to that of a reservoir storage-elevation-outflow relationship (from Guertin *et al.* 1987).

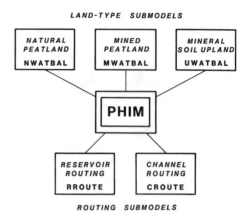

Fig. 10.9. Submodels of the peatland hydrologic impact model (PHIM).

quent field and laboratory research (Brooks *et al.* 1982), which in turn led to the development of a functional model, the peatland hydrologic impact model, referred to as PHIM (Guertin 1984). By comparing model responses to actual streamflow responses of peatland watersheds, the model was tested and verified (Guertin and Brooks 1985; Barten 1985; Guertin *et al.* 1987).

Peatland Hydrologic Impact Model

The concepts and relationships discussed in this chapter are the basis for PHIM and have largely been verified with field research (Guertin 1984; Barten 1985). The model can mathematically represent the physical relationships of a watershed that contains combinations of natural peatland, mined peatland, and mineral-soil upland areas (Fig. 10.9). It is capable of routing streamflow through either channels or reservoirs. The three independent land-type submodels represent the different watershed conditions that can be found in northern Minnesota. In applications of the model, the submodels can be used to apportion and link together different land types to represent the physical condition of the watershed.

Natural Peatland Submodel (NWATBAL)

The hydrologic processes modeled in NWATBAL are represented in Fig. 10.8. Precipitation and air-temperature data are entered in the model to determine rainfall, snow accumulation, or snowmelt over the watershed. Rainfall interception by forest canopy and understory shrubs is calculated from existing interception relationships (Verry 1976). Net precipitation is then added to the water table. If the peatland is a fen, regional groundwater (from a groundwater model or estimated from dry-period flow in a peatland stream) also must be added to the peatland groundwater system.

Evapotranspiration is determined from the relation-

Fig. 10.8. Hydrologic processes of peatlands related to streamflow prediction within the peatland hydrology model; regional ground-water interactions such as upwelling must be provided as input (from Predmore and Brooks, 1980).

peatland hydrology was initially developed by Predmore and Brooks (1980).

The development of a conceptual model required that the hydrologic cycle be defined for a peatland system and that all relevant processes be identified and quantified to reflect the responses discussed here (Fig. 10.8). For Minnesota conditions, such a model must be able to simulate the streamflow response of a peatland that may be a mosaic of bogs, fens, and mineral-soil uplands. The difference between a bog and a fen (Fig. 10.8) would be denoted by the absence or presence of regional groundwater flow into the peatland.

The conceptual model was used to design subse-

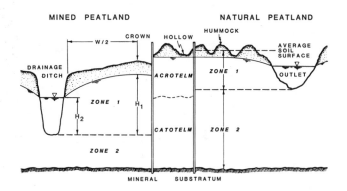

Fig. 10.10. Zones of peat soils used to model groundwater-streamflow response from natural and mined peatlands (from Guertin *et al.* 1987).

ship in Fig. 10.1; potential evapotranspiration (PET) is calculated from a method by Hamon (1961). Curve A, adapted from Romanov (1962), is applied for peatlands in which *Sphagnum* moss covers more than 25 percent of the area. Curve C represents *Carex*-dominated peatlands in which *Sphagnum* is less than 25 percent of the area (Juusela *et al.* 1970). Curve B represents mined peatlands and is used in the mined peatland submodel.

The residual of net precipitation and evapotranspiration is then added to or subtracted from storage in the peat soil (Fig. 10.10). When zone 2 is saturated, percolation from zone 1 is approximated by the hydraulic conductivity at the zone boundary, which can be estimated from the von Post Classification of peat at that depth (Gafni 1986). If the water table recedes into zone 2, percolation is calculated as a function of soil-water storage (Huggins and Monke 1968). The exchange of water between the peat and the underlying mineral soil is determined from the hydraulic gradient and the hydraulic conductivity of the most decomposed peat layer or that of the mineral soil, whichever is smaller. The corresponding change in water-table depth is calculated, and discharge is determined from the elevation-discharge relationship in the reservoir-routing method (Fig. 10.7).

Mined Peatland Submodel (MWATBAL)

Peat mining transforms a poorly drained, rough-surfaced, vegetated, natural peatland into a system of ditched, crowned fields that are devoid of vegetation. This type of peatland is represented mathematically in the submodel MWATBAL. The most significant hydrologic changes that occur as a result of mining (Brooks 1986; Gafni and Brooks 1986; Leibfried and Berglund 1986) are as follows:

(1) Interception by vegetation is eliminated.

(2) Infiltration rates are reduced; long dry periods in the summer promote hydrophobic surface-soil

conditions that result in surface runoff from rainfall; the formation of an impervious frost layer (concrete frost) is more prevalent in mined peat soils, resulting in greater surface runoff from snowmelt.

(3) Evapotranspiration is reduced.

(4) As mining proceeds, the hydraulic conductivities of the surface soil diminish; field studies indicate K (conductivity) values averaging 2.4 cm/hr (1 inch/hr) for a mined soil surface layer compared to 697 cm/hr (274 inches/hr) for the acrotelm of an undisturbed soil. As a result, average groundwater velocities can be reduced a hundredfold.

(5) Ditching steepens the hydraulic gradient and lowers the water table within mined fields and in adjacent unmined peatlands. Removing water previously stored in the peat of mined and adjacent areas increases the volume of streamflow while dewatering occurs. A more efficient conveyance system causes water to leave the peatland more quickly, resulting in higher stormflow peaks.

These changes are simulated with the MWATBAL submodel by altering interception storage coefficients, infiltration relationships, and evapotranspiration relationships (curve B in Fig. 10.1) and by physically representing the channel (ditch) and field configuration to properly route water from the mined fields. The effects of mining on downstream locations can be estimated by specifying the location of the mined area within the boundaries of a larger watershed system. Streamflow from the different land types can be combined and routed at the appropriate locations to simulate the total system response.

Mineral-Soil Upland Submodel (UWATBAL)

Mineral-soil upland components in a peatland watershed are modeled with a separate submodel, UWATBAL. This submodel represents unsaturated systems and simulates the processes of interception, infiltration, subsurface flow, soil-moisture storage, and evapotranspiration (Fig. 10.1). In essence, a water budget is calculated, and excess water is routed to a peatland, a channel, or a reservoir, as determined by the user of the model. The details of this model are not described here.

Model Test and Verification

Testing and verification of the peatland hydrologic impact model have been accomplished for three peatland watersheds in northern Minnesota: (1) the Toivola fen, a 3,758-ha (9,286-acre) transitional peatland that is largely a fen with some raised bogs and less than 2 percent uplands; (2) a 155-ha (383-acre) mined bog near Cromwell; and (3) the 9.72-ha (24-acre) S-2 upland-peatland watershed near Marcell (Guertin 1984; Barten 1985; Guertin *et al.* 1987).

Table 10.1. Summary of stormflow simulations
for an undisturbed peatland (Toivola)
and a mined peatland (Corona)

	Ratio of predicted/observed[a]		
Site	Stormflow volume	Peak discharge	Combined[b]
Toivola (undisturbed peatland)			
Calibration (n = 6)	0.91 (0.10)	0.80 (0.22)	0.86 (0.15)
Test (n = 6)	0.86 (0.16)	0.84 (0.22)	0.85 (0.18)
Corona (mined peatland)			
Calibration (n = 5)	0.96 (0.10)	0.86 (0.35)	0.91 (0.22)
Test (n = 4)	0.91 (0.08)	0.65 (0.09)	0.78 (0.07)

[a]Ratio (1 standard deviation).
[b]Combined Ratio = (Volume Ratio/2) + (Peak Flow Ratio/2)
  Pooled standard deviation.

Source: Guertin *et al.* 1987.

Table 10.2. Summary of annual water yield simulation for
the S-2 upland-peatland watershed near Marcell, Minnesota

Year	Observed water yield	Predicted water yield	Predicted / Observed
	(mm)	(mm)	
Calibration			
1970	163	144	0.89
1973	154	160	1.06
1977	167	161	0.96
1978	229	222	0.97
1979	285	215	0.76
Test			
1971	174	191	1.09
1972	181	181	0.99
1974	181	176	0.98
1975	215	188	0.87
1976	57	63	1.10
1980	87	89	1.02

Source: Guertin *et al.* 1987.

The model was tested on the basis of hourly stormflow events for the Toivola and Cromwell peatlands (Table 10.1) and daily streamflow for the S-2 watershed (Table 10.2), which allowed comparison of annual water yield. Examples of simulated and observed hydrographs are presented for these three watersheds in Fig. 10.11. The simulated stormflow volumes and peak discharges for the Toivola fen tracked the observed values quite well but were consistently low. The differences between simulated and observed streamflow were partly due to unrepresentative precipitation data for the watershed. Because of limited accessibility, only one precipitation measuring station was available during the period of study; this station apparently underestimated the mean depth of precipitation over the entire watershed.

Simulated stormflow events at the mined bog site were likewise consistently lower than observed events. In this case precipitation data were adequate, but the mined fields and ditches were not maintained in a condition that promoted effective drainage. As a result, there was a backwater effect caused by obstructed ditches and retention storage in concave portions of some of the fields. The model simulated fields and ditches that were designed to promote efficient drainage. Even with such sources of error, simulated stormflow volumes averaged only 10 percent less than observed.

Daily streamflow simulations over one-year periods for the S-2 upland-peatland watershed averaged 93 percent and 101 percent of observed values for the calibration and test years, respectively. Because excellent climatological and discharge records were available for this site, these results more accurately reflect the modeling errors than data errors that confounded the results of the other two sites.

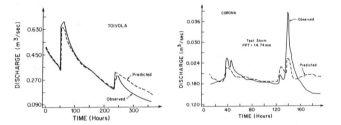

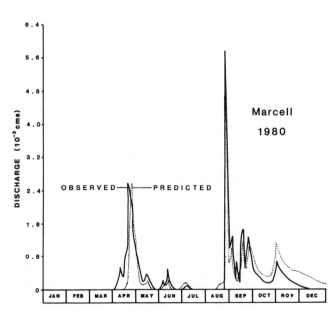

Fig. 10.11. Streamflow simulated (predicted) by the peatland hydrologic impact model compared with observed streamflow for three peatland watersheds in northern Minnesota (from Guertin 1984).

Conclusions

The process of developing a conceptual model of peatlands and the subsequent field research, model refinement, and testing has advanced our knowledge of the hydrologic function of peatlands. The peatland hydrologic impact model has been tested with field data from an undisturbed fen, a mined bog, and an upland-peatland watershed. Annual water yield and stormflow events were simulated with PHIM, and the model results are generally within 10 percent of observed events. Further testing and development is warranted, although these results are comparable with most operational hydrologic models. The results of this work allow us to predict streamflow response from either natural or mined peatlands in northern Minnesota if we have sufficient data to characterize the watershed and the climatological inputs.

PHIM can be used to examine the effects of potential peatland development schemes on streamflow. For example, peatland sites being considered for horticultural development (mining) or conversion to agricultural croplands can be evaluated with respect to the downstream changes in stormflow, snowmelt runoff, dry-period flow, and annual water yield. Ecological and economic factors can then be brought into the analysis to help decide which development schemes are most suitable and which may not be acceptable.

As our knowledge of peatland hydrology and the interaction between peatlands and regional groundwater flow improves, this model can be improved. The value of peatlands as functional hydrologic systems can then be better understood and quantified, enabling suitable planning for peatland conservation or peatland development.

Literature Cited

Ahti, E. 1977. Runoff from open peatlands as influenced by ditching. I. Theoretical analysis. Commun. Inst. For. Fenn. 92(3):1-16.

Barten, P. K. 1985. Testing and refinement of the Peatland Hydrologic Impact Model. Unpublished M.S. paper, College of Forestry, University of Minnesota, St. Paul, Minnesota, USA.

Bay, R. R. 1966. Evaluation of an evapotranspirometer for peat bogs. Water Resources Research 2(3):437-442.

———. 1969. Runoff from small peatland watersheds. Journal of Hydrology 9:90-102.

Boelter, D. H. 1972. Water table drawdown around an open ditch in organic soils. Journal of Hydrology 15:329-340.

——— and E. S. Verry. 1977. Peatland and water in the Northern Lake States. General Technical Report. U.S. Department of Agriculture Forest Service NC-31.

Bonde, A. N., J. D. Ives, and D. B. Lawrence. 1961. Ecosystem studies at Cedar Creek Natural History Area III: Water use studies. Minnesota Academy of Science Proceedings 29:190-198.

Brooks, K. N. 1986. Hydrologic impacts of peat mining. Pages 160-169 in D. D. Hook et al., editors. The ecology and management of wetlands. Timber Press, Portland, Oregon, USA.

———, J. C. Clausen, D. P. Guertin, and T. C. Stiles. 1982. The water resources of peatlands — final report. Minnesota Department of Natural Resources.

Carter, V. 1986. An overview of the hydrologic concerns related to wetlands in the United States. Canadian Journal of Botany 64:364-374.

Clausen, J. C., and K. N. Brooks. 1980. The water resources of peatlands; a literature review. Department of Forest Resources, College of Forestry, University of Minnesota, St. Paul, Minnesota, USA.

——— and K. N. Brooks. 1983a. Quality of runoff from Minnesota peatlands: I. A characterization. Water Resources Bulletin 19(5): 763-767.

——— and K. N. Brooks. 1983b. Quality of runoff from Minnesota peatlands: II. A method for assessing mining impacts. Water Resources Bulletin 19(5):769-772.

Dooge, J. 1975. The water balance of bogs and fens. Pages 69-76 in Hydrology of marsh-ridden areas. Proceedings of the Minsk Symposium 1972. UNESCO Press, Paris, France.

Eggelsmann, R., and M. Schuch. 1976. Moorhydrologie. Pages 153-162 in K. Gottlich, editor. Moor- und Turfkunde. Schweizerbartsche, Stuttgart, Germany.

Gafni, A. 1986. Field tracing approach to determine flow velocity and hydraulic conductivity in saturated peat soils. Ph.D. thesis, University of Minnesota, St. Paul, Minnesota, USA.

——— and K. N. Brooks. 1986. Hydrologic properties of natural versus mined peatlands. Proc. Symp. on Advances in Peatlands Engineering, August 25-26. Ottawa, Canada, National Research Council of Canada, in press.

——— and K. N. Brooks. 1990. Hydraulic characteristics of four peatlands in Minnesota. Canadian Journal of Soil Science 70:239-253.

Gorham, E. 1957. The development of peatlands. Quarterly Review of Biology 32:145-166.

Granlund, E. 1932. De svenska hegmossarnas geologi. Sveriges Geologiska Undersökningen Serie C. no. 373:1-193.

Guertin, D. P. 1984. Modeling streamflow response from Minnesota peatlands. Ph.D. thesis, University of Minnesota, Department of Forest Resources, St. Paul, Minnesota, USA.

———, P. K. Barten, and K. N. Brooks. 1987. The peatland hydrologic impact model: Development and testing. Nordic Hydrology 18:79-100.

——— and K. N. Brooks. 1985. Modeling streamflow response of Minnesota peatlands. Pages 123-131 in E. B. Jones and T. J. Ward, editors. Watershed management in the eighties. American Society of Civil Engineers Symposium, Denver, Colorado, USA.

Hamon, W. R. 1961. Estimating potential evapotranspiration. Journal of the Hydrology Division, Proceedings of the American Society of Civil Engineers. 87(HY3):107-120.

Heikurainen, L. 1963. On using groundwater table fluctuations for measuring evapotranspiration. Acta Forestalia Fennica 76(5):5-16.

Heinselman, M. L. 1961. Black spruce on the peatlands of former Glacial Lake Agassiz and adjacent areas in Minnesota: A study of forest sites, bog processes, and bog types. Ph.D. thesis, University of Minnesota, St. Paul, Minnesota, USA.

Huggins, L. F., and E. J. Monke. 1968. A mathematical model for simulating the hydrologic response of a watershed. Water Resources Research 4(3):529-539.

Ingram, H. A. P. 1983. Hydrology. Pages 67-158 in A. J. P. Gore, editor. Ecosystems of the world 4A, Mires: Swamp, bog, fen and moor. Elsevier, Amsterdam, Netherlands.

Ivanov, K. E. 1981. Water movement in mirelands. Academic Press, New York (trans. from Russian by A. Thomson and H. A. P. Ingram), London, U.K.

Juusela, T., S. Kaunisto, and S. Mustonen. 1970. Turpeesta Tapahtuvann Haidhduntaan Vaikuttavista Tekijoista (On factors affecting evapotranspiration from peat). Communicationes Instituti Forestalis Fenniae 67(1):45.

Leibfried, R. T., and E. R. Berglund. 1986. Groundwater hydrology of a fuel peat mining operation. *In* Proceedings of Symposium on Advances in Peatlands Engineering, August 25-26, 1986, Ottawa, Canada, National Research Council Canada, *in press*.

Linsley, R. K., M. A. Kohler, and J. L. H. Paulhus. 1982. Hydrology for Engineers. 3rd edition. McGraw-Hill, New York, New York, USA.

Moore, P. D., and D. J. Bellamy. 1974. Peatlands. Springer-Verlag, New York, New York, USA.

Paivanen, J. 1973. Hydraulic conductivity and water retention in peat soils. Acta Forestalia Fennica 129:1-70.

Predmore, S. R., and K. N. Brooks. 1980. Predicting peat mining impacts on water resources — a modeling approach. Pages 655-661 *in* Proceedings of the 6th International Peat Congress, Duluth, Minnesota, USA.

Robertson, R. A., I. A. Nicholson, and R. Hughes. 1968. Runoff studies on a peat catchment. Proceedings of the Second International Peat Congress, Leningrad, USSR 1:161-166.

Romanov, V. V. 1961. Hydrophysics of bogs. Israel Program for Scientific Translations, 1968, U.S. Department of Agriculture and the National Science Foundation.

———. 1962. Evaporation from bogs in the European territory of the USSR. Israel Program for Scientific Translations, 1968, U.S. Department of Agriculture and National Science Foundation.

Siegel, D. I. 1983. Groundwater and evolution of patterned mires, Glacial Lake Agassiz peatlands, northern Minnesota. Journal of Ecology 71:913-922.

Verry, E. S. 1976. Estimating water yield differences between hardwood and pine forests. Research paper NC-128, U.S. Forest Service, North Central Forest Experiment Station, St. Paul, Minnesota, USA.

———and D. H. Boelter. 1978. Peatland hydrology. Pages 389-402 *in* Wetland functions and values. American Water Resources Association.

———. 1980. Water table and streamflow changes after stripcutting and clearcutting an undrained black spruce bog. Pages 493-498 *in* Proceedings of the 6th International Peat Congress, Duluth, Minnesota, USA.

———. 1984. Microtopography and water table fluctuation in a sphagnum mire. Pages 11-31 *in* Proceedings of the 7th International Peat Congress, vol. 2, Dublin, Ireland.

Virta, J. 1966. Measurement of evapotranspiration and computation of water budget in treeless peatlands in the natural state. Communications Physico-Mathematicae 32(1):70.

von Post, L., and E. Granlund. 1926. Sodra Sveriges torvtillgangar. I. Sveriges Geologiska Undersökningen, Serie C no. 335.

CHAPTER 11

Groundwater Hydrology

Donald I. Siegel

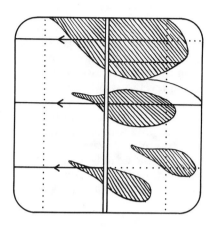

The interactions between groundwater and surface water in the patterned peatlands of Minnesota are poorly understood, although groundwater seepage has been qualitatively suggested as a factor that controls differences in surface-water chemistry and the two major types of vegetation communities: raised bogs and fens (Heinselman 1963, 1970; Boelter and Verry 1977; Gorham 1957; Glaser *et al.* 1981). Raised bogs are topographically higher than fens and consist mostly of Sphagnum mosses and spruce trees (*Picea mariana*). Fen vegetation consists largely of sedges (*Carex*) and marsh herbs. Some of the fens, called water tracks, are major avenues for surface-water runoff.

Most nutrients and mineral salts found in the surface water of raised bogs come from precipitation. The bog surface is convex and therefore cannot receive runoff from the side (Ingram 1967; Gorham 1957). The concentration of total dissolved inorganic solids in bog surface water is typically less than 5 mgL^{-1}. Dilute bog waters also have pH less than 4.2, because mineral acids are added by precipitation, and organic acids are released by the biodegradation and oxidation of peat (Thurman 1985).

In contrast, fen surface water typically has concentrations of dissolved inorganic solids greater than 50 mgL^{-1} and pH greater than 5.7 (Glaser *et al.* 1981; Boelter and Verry 1977). Such fen water is termed minerotrophic, in contrast to mineral-poor (ombrotrophic) bog water. Although groundwater seepage may contribute dissolved solids to fen water (Heinselman 1963, 1970; Boelter and Verry 1977; Sjörs 1963; Ingram 1983; Ivanov 1981), it is uncertain why water chemistry so abruptly changes at the interface between bog and fen and why pH and concentrations of dissolved solutes gradually increase down surface-water flow paths on bogs.

One hypothesis to explain differences in surface-water pH is that acidity in the water is neutralized by the oxidation of organic material in the upper part of the peat column ("oxidative corrosion" of Sjörs 1963). Gorham *et al.* (1985) experimentally found, however, that most of the organic matter in unpolluted bog water would have to be destroyed by vigorous photo-oxidation before the pH increases from 4.0 to over 5.7. Such vigorous oxidation does not occur in natural bogs and fen waters, which have concentrations of dissolved organic carbon (largely organic acids) in the tens of parts per million (Thurman 1985). Dissolved organic carbon constitutes the largest part of total dissolved solids in these dilute waters. Furthermore, the very accumulation of peat in mires indicates that the rate of net biomass production, an acid-releasing process, exceeds the rate of biomass oxidation that might consume acid.

An alternative hypothesis to explain neutralization of bog acids is mineral alkalinity added by groundwater seepage (Siegel 1983). Calculations show that if 5% groundwater with a composition typical of that found under the patterned peatlands is added to bog water, it may be sufficient to cause all the major chemical differences between bog and fen waters (Siegel 1983).

The purpose of this chapter is to review the current state of knowledge on groundwater flow in the patterned peatlands of northern Minnesota. Groundwater flow is discussed within the context of flow systems (Domenico 1972). This is a powerful conceptual approach to evaluate how groundwater movement may control peatland vegetation patterns and surface-water chemistry (Siegel 1983; Siegel and Glaser 1987; Glaser *et al.* 1986). In the past, peatland hydrologists viewed water in peat (peat pore water) as physically separated from water in the underlying "mineral" soils (groundwater). This distinction is inappropriate because pores in peat and underlying geologic materials are connected. Pore water in peat is best studied as part of the entire hydrogeologic system (Siegel 1981).

Groundwater Hydrology

A brief discussion of groundwater flow is useful as a context of the groundwater hydrology of the patterned peatlands. Groundwater moves in three dimensions from high to low hydraulic head, which operationally is the level to which water will naturally rise in a pipe open below the water table. Hydraulic head is usually measured as the elevation of water in a piezometer, a tube or pipe with an opening near the bottom to allow the entrance of water.

Hydraulic head is related to the water's potential energy in the flow field and consists of both fluid pressure and elevation (Hubbert 1940):

$$h = P + z \qquad (1)$$

where

  h = hydraulic head at the measuring point
  P = fluid pressure of the water
  z = elevation above a reference datum, usually sea level.

Precipitation infiltrating downward moves deep into the groundwater flow system, wherever hydraulic head decreases with depth below the water table. These places are called recharge zones. Conversely, groundwater discharge (upwelling) occurs wherever hydraulic head increases with depth below the water table. In discharge areas, groundwater moves up to the land surface, emerges at seeps, and is evaporated or is used by plants (Fig. 11.1).

The amount of groundwater moving through the porous material is calculated from Darcy's Law:

$$Q = -KIA \qquad (2)$$

where

  Q = volume of groundwater moving per unit time
  K = hydraulic conductivity (permeability of the porous material to water)
  I = hydraulic gradient (change in hydraulic head over the distance the water moves)
  A = the cross-sectional area through which the groundwater moves.

Finally, the spatial distribution of hydraulic head can be theoretically calculated at "steady-state" by solving appropriate mathematical (differential) equations (Freeze and Cherry 1979). Steady-state implies that long-term recharge by precipitation equals long-term groundwater discharge and that the elevation of the water table at any point is constant. A steady-state assumption is probably valid in the peatlands as a first approximation, because the water table is at or within

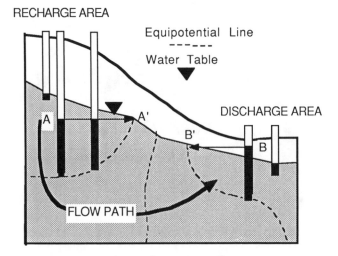

Fig. 11.1. The major aspects of a groundwater flow system. *Dashed lines* are equipotential lines. *Arrows* indicate the general direction of groundwater flow. Water levels rise to the same level in all piezometers (monitoring wells) open to a given equipotential line. The elevation of the water line is the elevation where equipotential lines intersect the water table (A′ and B′). Groundwater recharge occurs when water levels in deep piezometers are below the water table; discharge occurs when water levels at depths are above the water table.

a meter of the land surface during much of the year (Siegel 1981).

The equations governing hydraulic head can be solved directly (by integration) or approximately (but very close to the direct solution) by computer. Both direct solutions and computer approximations show that groundwater occurs in flow systems at different scales (Toth 1963; Freeze and Witherspoon 1966, 1967; Fig. 11.2). A fundamental principle of modern hydrogeology is that groundwater at different depths in the same location can occur in completely different flow systems. The size and depth of these flow systems can be predicted by computer model experiments that are compared (calibrated) to field data on hydraulic head (Wang and Anderson 1982). Computer simulations of groundwater flow are now a fundamental research tool for groundwater studies in areas where extensive fieldwork is difficult and costly.

Regional Geologic Setting of the Patterned Peatlands

The patterned peatlands occupy the former lake plain of Glacial Lake Agassiz, which extended from northeastern Minnesota well into Manitoba (Fenton *et al.* 1983). Glacial Lake Agassiz was formed about 11,700 years ago as the Laurentide ice sheet retreated north into Canada (Wright 1972). The lake withdrew about 11,000 years ago from the area now occupied by the Red Lake peatland. The lake formed beach ridges of

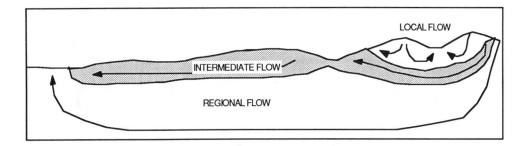

Fig. 11.2. Groundwater flow systems at different scales. Local-scale flow systems are recharged at topographic highs on the water table and discharged at immediately adjacent lows. Intermediate-scale flow systems discharge beyond immediately adjacent lows on the water table. Regional-scale flow systems are recharged at the major regional topographic highs on the water table and discharge at the major regional topographic lows.

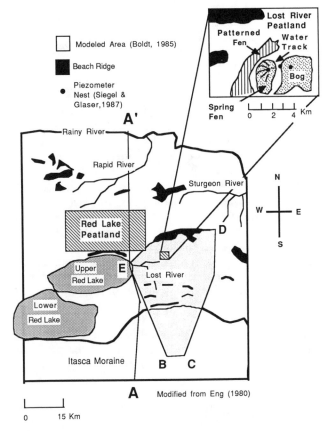

Fig. 11.3. Location of the patterned peatlands of northern Minnesota and areas for which computer simulations of groundwater flow have been made. Transect A-A' is the line for the two-dimensional groundwater models (Siegel 1981, 1983; Boldt 1986). Area B-C-D-E is the area for the three-dimensional groundwater model (Boldt 1986). The insert map of the Lost River peatland shows the location of clusters of piezometers installed on the spring fen (*left*), water track (*center*), and raised bog (*right*).

sand and gravel that cross the peatlands approximately from east to west (Eng 1980a, b; Fig. 11.3).

The Lake Agassiz peatlands can roughly be divided into two major parts, the almost completely peat-covered Red Lake Peatlands and the Lost River and Black River peatlands, which contain local outcrops of mineral soils. The peatlands are extremely flat, with a regional topographic gradient of only about 0.1%, equivalent to the angle formed if a thin coin is placed under the end of a meter-long measuring stick. Peat is more than 3 m (10 feet) thick in some places (Fox *et al*. 1977).

Glacial drift and lake sediments under the peat are about 30 m (100 feet) thick (Siegel 1981). Precambrian igneous and metamorphic rocks or Cretaceous shale constitute the basement (Sims 1970; Siegel 1981).

Groundwater Flow in the Peatlands

Field Studies

Groundwater in the patterned peatlands was first indirectly considered by Bidwell *et al*. (1970), Hegelson *et al*. (1975), and Lindholm *et al*. (1976), who determined that the Red Lakes, Rainy River, and Sturgeon River are major discharge areas for regional groundwater flow. The Itasca moraine south of the area is the principal recharge area for regional groundwater flow to the peatlands. Regional groundwater discharge is shown by artesian domestic wells along the eastern shores of Upper Red Lake and at the northern edge of the Itasca moraine. Groundwater recharge on the Itasca moraine is shown by progressively deeper water levels in piezometers on the moraine (Siegel 1981).

Water-level measurements in the peatlands proper are scant. Water levels in piezometers on a beach ridge in the Red Lake peatland indicate that the beach ridge is a groundwater discharge area (Siegel 1981). There the water level in glacial till above the bedrock surface was 156 cm higher than the water table, showing that groundwater moves upward.

Detailed studies of the groundwater flow system of the Lost River peatland (Fig. 11.3) involved a raised peat mound characterized by surface-water channels draining from its crest The mound is separated from a large raised bog by an intervening fen (Siegel and Glaser 1987). The chemical composition of water in

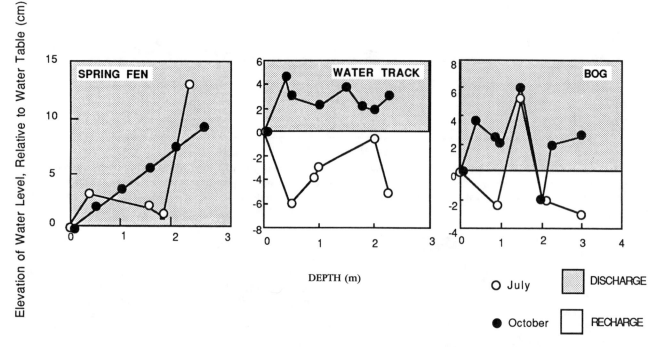

Fig. 11.4. Water-level measurements from piezometers at the Lost River peatland, 1983. All measurements at the spring fen are above the water table, indicating continuous groundwater discharge. Water levels on the raised bog and water track show that the direction of groundwater flow reversed during the year. In October, the sites were discharge areas; in July, they were recharge areas.

the peat-mound channels is very similar to the chemical composition of groundwater in mineral soils found beneath the peat (Siegel and Glaser 1987). This similarity suggests that the peat mound is a focus for groundwater discharge, and the mound is termed a spring fen. Water levels in piezometers inserted to depths between 2 and 3 m at the crest of the raised bog, in the water track, and at the mouth of a channel on the spring fen were at least several centimeters higher than the water table in the fall of 1983 (Siegel and Glaser 1987), indicating that groundwater moves upward (Fig. 11.4).

Water levels in the peat-mound channel, in particular, directly increase with depth — a "certain" indication of groundwater discharge (Freeze and Cherry 1979). On the bog and water track, however, the water levels were lower than the water table in the spring of 1983. This shows that a component of groundwater flow was downward as recharge. Such a reversal from discharge to recharge has never been previously reported in mires, although it is common along stream banks.

Water levels at the bog and water track did not vary linearly with depth in a simple fashion as they did at the spring fen. Rather, water-level changes varied irregularly with depth and document local heterogeneities in the peat. Head reversals with depth commonly occur when local zones of higher permeability are found within porous media with lower permeability (Freeze and Cherry 1979).

Pore water chemistry in the peat column also shows that groundwater discharge occurs in all peat land-

forms. For example, pH is less than 4.0 in the surface water on the bog and water track and over 7.0 at a depth of only 1 m (Fig. 11.5). In the spring-fen channels, however, groundwater discharge is sufficiently large to produce surface water with pH the same as that in groundwater from the underlying mineral soils. Specific conductance, a measure of total dissolved inorganic solids, also increases with depth at all sites, reaching values typically found for groundwater in the underlying mineral soils (Fig. 11.6). The increases in pH and specific conductance would not occur if the bog and water track were predominantly recharge areas. In recharge areas, peat pore water is dilute and acidic throughout the peat column except immediately above the base of the peat, where dissolved minerals can move upward against the hydraulic gradient because of chemical diffusion.

Computer Simulations of Peatland Groundwater Flow

Computer simulations ("models") of groundwater flow were made to determine the predominance of local or regional groundwater flow (Siegel 1981, 1983; Boldt and Siegel 1985; Boldt 1986). Models were made by the finite-difference method, which approximates the distribution of hydraulic head in the groundwater flow field at discrete points called nodes (Wang and Anderson 1982). The accuracy of the models in simulating the real system depends upon the hydrologic bound-

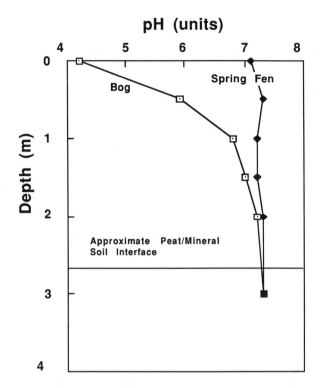

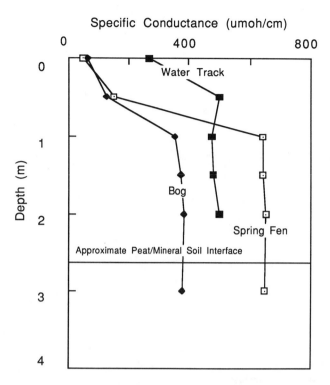

Fig. 11.5. Variations of pH with depth at the Lost River peatland, July 1983.

Fig. 11.6. Variations of specific conductance with depth at the Lost River peatlands, July 1983.

aries, the variations in hydraulic conductivity chosen as model parameters, and the number of nodes.

Two types of computer models have been designed: two-dimensional models (Siegel 1981, 1983; Boldt and Siegel 1985; Boldt 1986) and three-dimensional models (Boldt and Siegel 1985; Boldt 1986). Cross-sectional models better determine details on vertical flow directions of groundwater, whereas three-dimensional models better predict areas of recharge and discharge. Computer models of complicated groundwater flow problems often are "nonunique." Different assumed hydrogeologic boundaries and aquifer properties often result in identical solutions. Therefore, models show only what *cannot* be present; they can only suggest what *may* be present. Additional field verification should be done to test model validity.

Fig. 11.7 illustrates the hydraulic boundaries chosen for the simulations. The bedrock underlying the mineral soil was assumed to be impermeable, and thus the no-flow lower boundary of groundwater systems. The Rainy River, Itasca moraine, and selected local groundwater divides were also considered as no-flow boundaries. Regionally, the surficial aquifers end where the Rainy River cuts through surficial materials to bedrock. The Itasca moraine is a regional groundwater divide: north of the moraine, groundwater and surface water move to Hudson Bay; south of the divide, water moves to the Mississippi River drainage system.

Cross-sectional models and north-south boundaries of the three-dimensional models in the Lost River peat-

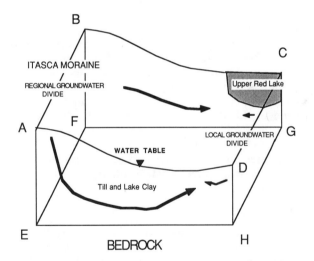

Fig. 11.7. Hydrogeologic boundaries used in computer simulations of groundwater flow. Groundwater divides on the Itasca moraine, at the Rainy River, and on a large beach ridge north of the Lost River peatland were considered no-flow hydraulic boundaries. Selected flow lines east of Upper Red Lake were also considered no-flow hydraulic boundaries in three-dimensional simulations. The metamorphic rocks and the shales under the glacial deposits were considered a geologic no-flow boundary that defined the bottom of the hydrogeologic system. Portions of models that define the water table and lakes were considered constant-head boundaries, where water levels did not change with time. This is an appropriate assumption for long-term steady-state simulations.

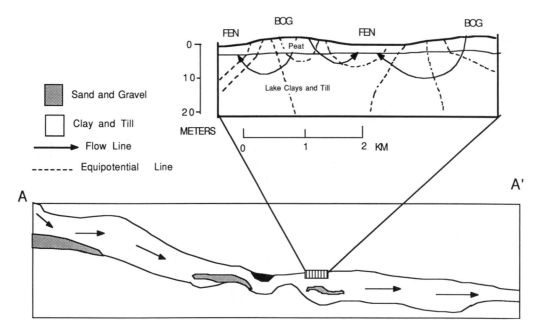

Fig. 11.8. Generalized results of two-dimensional computer simulations of groundwater flow along the north-south transect A-A′ (Fig. 11.3) from the Itasca moraine across the Red Lake peatlands to the Rainy River. *Dashed lines* are equipotential lines. *Arrows* are flow paths. The ratio of horizontal to vertical hydraulic conductivity used in the models was 1000:1. The detailed inset shows that pronounced local flow systems develop when the water-table mounds under raised bogs are applied to the models. These flow systems penetrate through the peat and into the underlying glacial sediments. Local recharge areas are the raised bogs, and local discharge areas are the fens (modified from Siegel 1983 and Boldt 1986).

land were constructed along regional groundwater flow paths on the water table, which, by definition, are no-flow boundaries (Bear 1972). The water-table topography and elevation of Upper Red Lake were held constant during the simulations to model long-term steady-state conditions. The hydraulic conductivity was varied in the models depending on the type of sediment at each node: till, peat, lake clay, or sand. Variations in types of surficial materials with depth were determined from test holes and by extrapolation from published data (Siegel 1981; Boldt 1986).

Red Lake Peatlands

Two-dimensional cross-sectional models were made for (1) a south-north regional transect from the Itasca moraine across Upper Red Lake and the Red Lake peatland to the Rainy River (Siegel 1981, 1983) and (2) across two raised bogs in the peatland (Boldt 1986; Boldt and Siegel 1985) (Fig. 11.6). The results of the regional model showed that horizontal (or lateral) flow predominates when a simple linear regional gradient on the water table is used as the top of the modeled area. Under horizontal flow conditions, groundwater does not move toward the land surface, and the results of these simulations are at odds with the observed upward hydraulic gradients described by Siegel (1981, 1983).

The reason why the computer-calculated heads and the observed heads do not match is related to the

distribution of the vegetation landforms. The water table under the raised bogs is also raised in a "mound" (Siegel 1981; Ingram 1982) up to several meters higher than the water table in the adjacent fens. When these water-table mounds are incorporated in the models, local groundwater flow systems about 1 to 2 km wide developed under and adjacent to the raised bogs (Fig. 11.8). Groundwater theoretically moves downward under the raised bogs and upward toward the land surface at adjacent fens (Siegel 1981, 1983). As a result, the water levels in piezometers should decrease with depth under raised bogs and increase with depth at adjacent fens. These head relationships have been documented in bogs and fens in northeastern Minnesota (Department of Natural Resources, unpublished data) and in Maine (Nicols 1983). Water-budget determinations in a raised-bog watershed near the Lake Agassiz peatlands in Minnesota also suggest that groundwater flow under a raised bog is downward (Verry and Timmons 1982).

The potentially profound influence of the water-table mounds under the bogs acting on groundwater flow is identical to that determined for groundwater mounds in mineral soils where regional water-table slopes are gentle (Toth 1963). Depending on the ratio of vertical to horizontal hydraulic conductivity, downward flow from the bogs can theoretically penetrate more than 10 m below the peat into the underlying mineral soils (Boldt and Siegel 1985; Boldt 1986; Fig. 11.8). If the raised bogs were not present, groundwater would move only sluggishly from the peatland centers toward Upper Red Lake and major rivers. In a sense, the hydrogeologic system of the Lake Agassiz peatlands might more accurately be called a hydrobiologic system: the growth of vegetation and accumulation of peat in raised bogs establish the local groundwater flow systems.

Lost River Peatland

Three-dimensional computer models were made to determine why continuous groundwater discharge occurs at the spring fen and only seasonal discharge occurs at the bog and water track (Boldt and Siegel 1985; Boldt 1986). Neither discharge at the spring fen nor the reversed hydraulic gradients at the bog and water track are explained by groundwater flow driven by water-table mounds under raised bogs. The water table under the spring fen and the adjacent raised bog are both about 1 m higher than that in the adjacent water track (Almendinger *et al.* 1986). Therefore, the spring fen and raised bog should both theoretically be recharge areas, not discharge areas.

Discharge mounds have previously been identified in wetlands where geologic materials with low hydraulic conductivity are breached to allow groundwater upwelling from underlying confined aquifers. One well-documented example of such a discharge mound is a wetland in the Indiana Dunes National Park, where lake clays are locally absent above a confined sand and gravel aquifer (Wilcox *et al.* 1986). The discharge zone is covered by a mound of fen peat and is sustained by nutrients found in the groundwater.

The computer simulations suggest that a conceptually similar hydrogeologic setting may control the groundwater discharge at the spring fen and adjacent landforms. The only computer simulation that calculated discharge at the field site was one that included an east-west-trending zone of low permeability located immediately south of the spring fen, combined with a buried sand and gravel lens extending south of a major beach ridge north of the Sturgeon River (Fig. 11.9). A buried bedrock topographic high is located west of the Lost River peatland (Siegel 1981) and may extend as a ridge to the east. Most geologic structures in north-central Minnesota trend roughly east to west (Sims 1970). The predominant form of glacial deposition in Glacial Lake Agassiz was underflow fans (Fenton *et al.* 1983). This environment of deposition is consistent with a subsurface extension of a sand and gravel lens from the beach ridge north of Sturgeon River south to the peat mound at the Lost River peatland.

The recharge zone for groundwater upwelling at the Lost River peatland is probably the beach ridge located about 8 km to the north. Groundwater flow theoretically can be driven by the difference between the elevation of the water table on the beach ridge and that on the water-table mounds under the Lost River raised bog and spring fen. The hydraulic gradients cal-

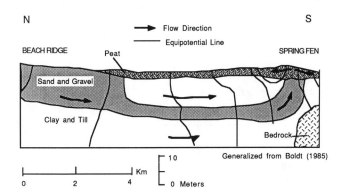

Fig. 11.9. Generalized north-south hydrogeologic cross-section derived from a three-dimensional model of groundwater flow in the Lost River peatland (area B-C-D-E of Fig. 11.3). In this simulation, a bedrock no-flow boundary and confined sand lens were used to generate upward groundwater flow at the spring fen and raised bog. Although these geologic features have yet to be identified in the field, this was the only geologic scenario in the computer modeling experiments that calculated the observed seasonal discharge at the peatland (Boldt 1986).

culated by the model are very sensitive to the elevation of the water table at the beach ridge. A 1 m decrease in water-table elevation at the beach ridge reverses the hydraulic gradient at the raised bog from up to down (Boldt 1986).

Groundwater Flow Rates

Knowing the vertical direction of groundwater movement, of course, does not indicate the amount of groundwater seepage at the land surface in the fens or the amount of deep infiltration under the bogs. The amount of groundwater movement is calculated from values of hydraulic conductivity, hydraulic gradient, and cross-sectional area in Darcy's Law.

The hydraulic conductivity (K) of peat is poorly known at the scale of a typical local groundwater flow system in the peatlands. Laboratory and field experiments on K of peat in northern Minnesota and elsewhere around the world have generally neglected decomposed peat deeper than 1 m (see Chason and Siegel 1986 for a review). On the basis of these studies, the upper 50 cm of the peat column, called the acrotelm (Ingram 1967), has a K as high as 30 m d^{-1}, whereas decomposed peat (the catotelm) at depths greater than 1 m depth is assumed to be almost impermeable, with K orders of magnitude smaller than that of the acrotelm. However, recent field and laboratory studies on decomposed peat up to 3 m deep in northern Minnesota suggest the opposite — that the K of decomposed peat can range between 0.1 and 8 m d^{-1} (Chason and Siegel 1986). This range in hydraulic conductivity of decomposed peat is probably caused by discontinuous zones of buried wood and other structural features in the peat, which can either obstruct or enhance water flow (Chason and Siegel 1986). For example, groundwater moves rapidly in pipelike channels under blanket bogs (Ingram 1982). These discontinuous "pipes" have K hundreds to thousands of times greater than the K of small samples of catotelmic decomposed peat. The K of geologic materials of low permeability is scale-dependent: the greater the size of the flow system, the greater the permeability (Neuzil 1986). The same scale dependency of hydraulic conductivity should also apply to thick organic soils.

A small value for hydraulic conductivity of decomposed peat, 0.01 m d^{-1}, and vertical hydraulic gradients from the cross-sectional models were used by Siegel (1983) to calculate the minimum annual amount of groundwater that could theoretically discharge from local flow systems under raised bogs to adjacent fens in the Red Lake peatlands. The calculated amount, $10,000 \text{ m}^3 \text{ yr}^{-1} \text{ km}^{-2}$ is about 6% of the annual amount of runoff from the peatlands. This small amount of groundwater may be sufficient to neutralize large amounts of acidity from bog runoff and cause the pronounced water-quality differences between bog and fen (Siegel 1983). Using a larger (and perhaps more realistic) figure for K would result in greater calculated amounts of groundwater and even more pronounced neutralization.

Implications of Groundwater Flow Patterns for Vegetation Patterns and Surface-Water Chemistry

Both theoretical results and data from field studies indicate that groundwater flow in the patterned peatlands is complex. Local flow systems caused by water-table mounds under raised bogs probably penetrate beneath the peat into the underlying mineral sediments. These flow systems force groundwater in underlying mineral soil to seep upward to bog margins and adjacent fens, where it mixes with acidic dilute water draining off the bog crest (Fig. 11.10a).

Chemical calculations indicate that less than 5% groundwater (of the total volume) is needed to neutralize the acid and raise the pH in the mixing zone to above 5.0 (Siegel 1983). This amount of groundwater agrees with the annual calculated contribution of groundwater discharge to the water budget of the peatlands. *Sphagnum* is generally intolerant of pH greater than 4.0, above which it is replaced by fen vegetation. Thus the lateral growth of *Sphagnum* may be self-regulated by the development and persistence of water-table mounds under the raised bogs.

Complicating this model for local groundwater flow in the peatlands are the effects of regional and intermediate groundwater flow systems. For example, changes in hydraulic head and water chemistry with depth at the Lost River peatland are consistent with the hypothesis that the peatland is located in the discharge area of a

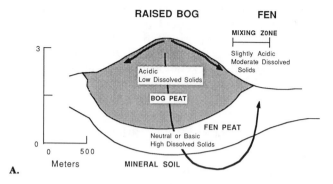

A.

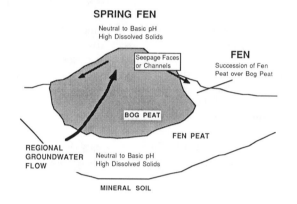

B.

Fig. 11.10. A. Chemical and hydraulic characteristics of local groundwater flow systems generated under groundwater mounds at raised bogs. Note that surface water at the edge of the bogs is partially neutralized by groundwater discharge in a mixing zone. **B.** Hypothetical succession of fen vegetation over bog vegetation because of groundwater discharge from intermediate or regional flow systems. Note that groundwater with high pH and high concentration of total dissolved solids (TDS) is found immediately below the acrotelm. Thus, subtle changes in the amount of groundwater discharge could raise the pH in the acrotelm beyond the tolerance of bog plants. With greater volumes of groundwater discharge at bog margins, fen vegetation would cover bog vegetation.

flow system of intermediate size, recharged at water-table mounds found on a beach ridge north of the peatland. Computer simulations suggest that the water levels and hydraulic gradients in the peatland are very sensitive to water-table elevations at the beach ridge. The discharge has converted a raised bog to a spring fen with minerotrophic channels that dissect the peat mound into islandlike forms (Fig. 11.10b).

The vegetational history of the spring fen at Lost River shows that the groundwater discharge is a geologically recent event. Most of the peat column in the channels consists of decomposed *Sphagnum* moss and other plants associated with raised bogs (P. H. Glaser, personal communication). Fen peat occurs only in the upper 70 cm (out of 218 cm total thickness) and at the base of the peat. No fen peat is found in the upper part of the bog or water-track peat column. By analogy with the spring fen, the raised bog may begin to change to fen if more minerotrophic water reaches the root zone of bog plants. The juxtaposition of regional flow systems with local flow systems may thus significantly control the distribution of the larger peat islands and other landforms in the patterned peatlands of Glacial Lake Agassiz and other mires (Glaser *et al.* 1986). Incipient water tracks that form on the flanks of large raised bogs may, in fact, be seepage faces for groundwater discharge from intermediate or regional flow systems.

Future studies to test this hydrogeologic model of peatland development will be challenging on many levels. Innovative field methods must be devised to obtain temporal data on water quality and hydraulic parameters at remote field sites to evaluate the transient nature and locations of reversals in hydraulic head. Computer simulations must be calibrated and verified against additional information on the subsurface geology. Finally, the effects of organic solutes on pH and other water-quality parameters must be determined to describe completely the geochemical effects of the mixing of groundwater and surface water. These and other avenues of research will ultimately establish the extent to which groundwater in the patterned peatlands of northern Minnesota can be considered part of a unique hydrobiologic system.

Literature Cited

Almendinger, J. C., J. E. Almendinger, and P. H. Glaser. 1986. Topographic changes on a spring fen and adjacent bog. Journal of Ecology 74:393-401.

Bear, J. 1972. Dynamics of fluids in porous media. Elsevier, New York, New York, USA.

Bidwell, L. E., T. C. Winter, and R. W. Maclay. 1970. Water resources of the Red River watershed, northwestern Minnesota. U.S. Geological Survey Water Resources Investigations Hydrologic Atlas HA-346.

Boelter, D. H. 1969. Physical properties of peats related to degree of decomposition. Proceedings of the Soil Science Society of America 33:606-609.

——— and E. S. Verry. 1977. Peatland and water in the northern lake states. U.S. Department of Agriculture Forest Service General Technical Report NC-31.

Boldt, D. R. 1986. Computer simulations of groundwater flow in a raised bog system, Glacial Lake Agassiz peatlands, northern Minnesota. M.S. thesis. Syracuse University, Syracuse, New York, USA.

——— and D. I. Siegel. 1985. Numerical simulation of groundwater flow through peat (Abstract). EOS 66:264.

Chason, D. B., and D. I. Siegel. 1986. Hydraulic conductivity and related physical properties of peat, Lost River peatland, northern Minnesota. Soil Science 42:91-99.

Domenico, P. 1972. Concepts and models in groundwater hydrology. McGraw-Hill, New York, New York, USA.

Eng, M. T. 1980a. An evaluation of the surficial geology and bog patterns of the Red Lake peatlands, Beltrami and Lake of the Woods Counties, Minnesota. Minnesota Department of Natural Resources, St. Paul, Minnesota, USA.

———. 1980b. Surficial geology of Koochiching County, Minnesota. Minnesota Department of Natural Resources, St. Paul, Minnesota, USA.

Fenton, M. M., S. R. Moran, J. T. Teller and C. Lee. 1983. Quaternary stratigraphy and history of the southern part of the Lake Agassiz basin. Pages 262-290 *in* J. T. Teller and C. Lee, editors. Glacial Lake Agassiz. Geological Association of Canada Special Paper 26.

Fetter, C. W., Jr. 1980. Applied hydrogeology. Merrill, Columbus, Ohio, USA.

Fox, R., T. Malterer, and R. Zarth. 1977. Inventory of peat resources in Minnesota: A progress report. Minnesota Department of Natural Resources, Division of Minerals, St. Paul, Minnesota, USA.

Freeze, R. A. and J. A. Cherry. 1979. Ground water. Prentice-Hall, Englewood Cliffs, New Jersey, USA.

——— and P. A. Witherspoon. 1966. Theoretical analysis of regional groundwater flow: 1. Analytical and numerical solutions to the mathematical model. Water Resources Research.

——— and P. A. Witherspoon. 1967. Theoretical analysis of regional groundwater flow: 2. Effect of water-table configuration and subsurface permeability variation. Water Resources Research 3:623-634.

Glaser, P. H., G. A. Wheeler, E. Gorham, and H. E. Wright, Jr. 1981. The patterned mires of Red Lake peatlands, northern Minnesota: Vegetation, water chemistry, and land forms. Journal of Ecology 69:575-599.

———, D. I. Siegel, and J. Janssens. 1986. Groundwater and nutrient cycling in peatlands (Abstract). Page 159 *in* Proceedings, Fourth International Congress of Ecology, Syracuse, New York, USA.

Gorham, E. 1957. The development of peatlands. Quarterly Review of Biology 32:145-166.

———, S. J. Eisenreich, J. Ford, and M. V. Santelmann. 1985. The chemistry of bog waters. Pages 339-361 *in* W. Stumm, editor. Chemical processes in lakes. Wiley, New York, New York, USA.

Heinselman, M. L. 1963. Forest sites, bog processes, and peatland types in the Glacial Lake Agassiz region, Minnesota. Ecological Monographs 33:327-372.

———. 1970. Landscape evolution, peatland types, and the environment in the Lake Agassiz Peatland Natural Area, Minnesota. Ecological Monographs 40:235-261.

Hegelson, J. O., G. F. Lindholm, and D. W. Ericson. 1975. Water resources of the Lake of the Woods watershed, north-central Minnesota. U.S. Geological Survey Hydrologic Investigations Atlas HA-544.

Hemmond, H. F., and J. C. Goldman. 1985. On non-Darcian water flow in peat. Journal of Ecology 73:579-584.

Hubbert, M. K. 1940. The theory of groundwater motion. Journal of Geology 48:785-944.

Ingram, H. A. P. 1982. Size and shape in raised mire ecosystems: A geophysical model. Nature 297:300-303.

———. 1967. Problems of hydrology and plant distribution in mires. Journal of Ecology 55:711-724.

———. 1983. Hydrology. Pages 67-168 *in* A. J. P. Gore, editor. Ecosystems of the world — Mires: Swamp, bog, fen and moor. General Studies. Elsevier, New York, New York, USA.

Ivanov, K. E. 1981. Water movement in mirelands. Academic Press, London, England.

Lindholm, G. F., J. O. Helgeson, and D. W. Ericson. 1976. Water resources of the Big Fork River watershed, north-central Minnesota. U.S. Geological Survey Hydrologic Investigations Atlas HA-549.

Neuzil, C. 1986. Groundwater flow in low-permeability environments. Water Resources Research 22:1163-1197.

Nicols, W. J., Jr. 1983. Hydrologic data for the Great and Denbow Heaths in eastern Maine, October 1981 through October 1982. U.S. Geological Survey Open-File Report 83-865.

Siegel, D. I. 1981. Hydrologic setting of the Glacial Lake Agassiz peatlands, northern Minnesota. U.S. Geological Survey Investigations 81-24.

─────. 1983. Groundwater and the evolution of patterned mires, Glacial Lake Agassiz peatlands, northern Minnesota. Journal of Ecology 71:913-921.

───── and P. H. Glaser. 1987. Groundwater flow in a boy/fen complex, Lost River peatland, northern Minnesota. Journal of Ecology 75:743-754.

Sims, P. K. 1970. Geologic map of Minnesota. Minnesota Geological Survey Miscellaneous Map Series, Map M-14.

Sjörs, H. 1963. Bogs and fens on the Attawapiskat River, northern Ontario. Museum of Canada Bulletin, Contributions to Botany 186:43-133.

Thurmond, E. M. 1985. Organic geochemistry of natural waters. Martinus Nijoff/Dr. W. Junk Publishers, Boston, Massachusetts, USA.

Toth, J. 1963. A theoretical analysis of groundwater flow in small drainage basins. Journal of Geophysical Research 68:4795-4811.

Verry, E. S., and D. R. Timmons. 1982. Waterborne nutrient flow through an upland-peatland watershed in Minnesota. Ecology 63: 1456-1467.

Wang, H. F., and M. P. Anderson. 1982. Introduction of groundwater modeling. Freeman, San Francisco, California, USA.

Wilcox, D. A., R. J. Shedlock, and W. R. Henrickson. 1986. Hydrology, water chemistry, and ecological relations in the raised mound of Cowles Bog. Journal of Ecology 74:1103-1117.

Wright, H. E., Jr. 1972. Quaternary history of Minnesota. Pages 515-548 *in* P. K. Sims and G. B. Morey, editors. Geology of Minnesota: A centennial volume. Minnesota Geological Survey, St. Paul, Minnesota, USA.

CHAPTER 12

Impact of Ditching and Road Construction on Red Lake Peatland

Kristine L. Bradof

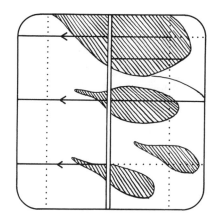

Many studies have examined the relationship between peatland water levels and the type of vegetation present. Heinselman (1963) originally called attention to the alignment of the major landforms in Minnesota's patterned peatlands with respect to the surface drainage. Glaser *et al.* (1981) observed "significant local alterations in the vegetation and landforms" associated with the system of drainage ditches in Red Lake peatland. Minerotrophic water tracks showed greater changes than ombrotrophic landforms. Both strings and flarks were drier in ditched areas, the patterns in some places becoming indistinct and merging into extensive shrublands. The vegetative composition shifted toward species better suited to the drier conditions and included more exotics. The most striking change in the patterned water tracks occurred downslope from Minnesota Highway 72, along which westward-flowing water is diverted to the north by ditches.

Peatland drainage has also been shown to stimulate the growth of trees. A study of northern Minnesota's peatlands drained between 1910 and 1920 indicated that black spruce trees close to ditches were indeed growing faster than those farther away (Averell and McGrew 1929).

These observed effects of drainage on vegetation were related to hydrologic changes in the peatland that could not be measured directly because data were not collected prior to ditch and road construction. Some inferences can be made, however, if water-table drawdown adjacent to the ditches and subsidence of the peat surface after drainage are compared to similar results from peatland areas for which pre-drainage data are available. In addition, the impact of drainage can be gauged by the comparison of growth rates for black spruce trees adjacent to a ditch and those in a control area a mile from the nearest ditch.

Fieldwork was concentrated in the most highly disturbed part of the peatland, adjacent to a six-mile stretch of Highway 72 and extending a mile west and two miles east. The study area includes parts of three ditch systems, Judicial Ditches (J.D.) 20, 30, and 36, which drain a major portion of Red Lake peatland (Fig. 12.1). Pre-drainage peat depths in the area (Soper 1919) ranged from 1 to 15 feet. By December 1914 construction of J.D. 20 along the R.30-31 W. range line from T.155-158 N. was completed, and the peat road on its west side, the forerunner of Highway 72, was scheduled to be ready for travel the following June (*Bemidji Pioneer*, December 10, 1914). Plans called for a maximum cut (ditch depth) of 6.2 feet and an average cut of 4.2 feet; one 30″ x 30′ and four 24″ x 30′ Portsmouth iron galvanized culverts were ordered, but their locations along the 22 miles of ditch were not specified (*Bemidji Pioneer*, January 9, 1914; contract for culvert pipe at Beltrami County Highway Department [BCHD]).

J.D. 30 and J.D. 36 were excavated in 1916. J.D. 30, which lies west of Highway 72, totals 165 miles in T.155-158 N., R.31-32 W.; maximum and average cuts are 8 feet and 5.5 feet respectively (*Bemidji Pioneer*, September 2, 1915). J.D. 20 forms the western edge of J.D. 36, 127 miles of ditch in T.155-158 N., R.29-30 W., with a maximum cut of 15.9 feet and an average cut of 5.0 feet (*Bemidji Pioneer*, November 18, 1915). The spacing between the different segments of J.D. 30 is one mile north to south and two miles east to west. For J.D. 36 the spacing is two miles in both directions. Ditch profiles at BCHD show that the bases of the ditches do not generally intersect mineral soil, except where the peat is relatively thin. The ditch gradient ranges from 0.02 to 0.04%.

The earliest road across Red Lake peatland was con-

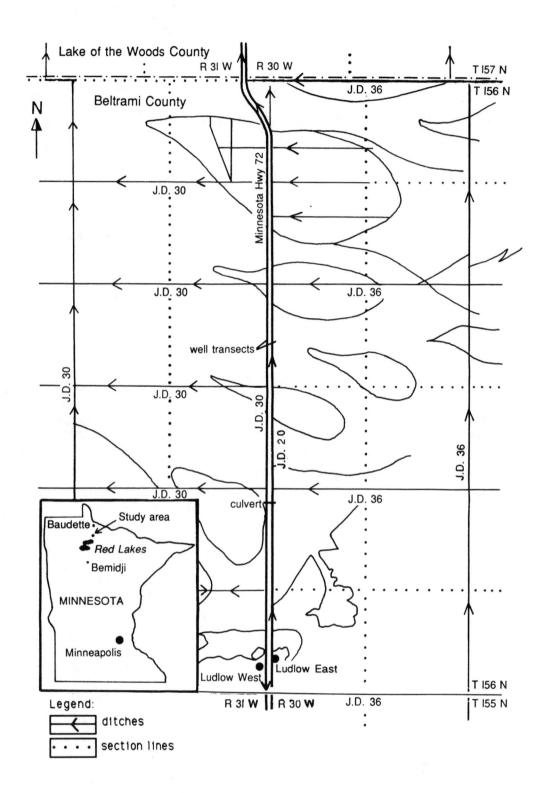

Fig. 12.1. The study area, a portion of Red Lake peatland crossed by Minnesota Highway 72, Beltrami County, Minnesota. *Solid lines with arrows* are ditch segments; *dotted lines* are section lines. From Bradof 1988.

Fig. 12.2. "A typical spruce swamp, looking along newly constructed State Road No. 6, Beltrami County." Photo by E. J. Bourgeois. (Soper 1919)

structed from the peat excavated from J.D. 20 and piled atop the peat along the west side of the ditch. Fig. 12.2 shows a typical peat road in Beltrami County around 1915. Gravel was first placed on the road early in 1922 at the same time that part of J.D. 36 was being redug (Ernest Blanchard, interview; *Bemidji Sentinel*, February 24, 1922).

The first grading project in 1931 was followed by gravel surfacing in 1932. The soil (meaning peat) subgrade indicates that the underlying peat was not removed. At the time of the first paving in 1937, a gravel base 5 inches thick by 24 feet wide was topped by a mixed bituminous (asphalt) surface 1.25 inches thick. The culvert 4.15 miles south of the county line was in place in 1953 but may have been installed as early as the first grading project in 1931 (Mark Gieseke, Minnesota Department of Transportation, Bemidji, 1982 letter).

Water-Table Drawdown Near Ditches

Methods

Fourteen water-table observation wells were installed in an east-west transect along the water track 2.6 miles south of the Lake of the Woods County line as measured along Highway 72 (Fig. 12.1). Each well consisted of polyvinyl chloride pipe 2.5 cm (1 inch) in diameter and 1.5 m (5 feet) long, with perforations drilled every 5 cm along its length to allow for equilibration of water levels inside and outside. A cap was placed on the bottom of each pipe to prevent peat from entering the well during installation. A cap on the top of each well could be removed when the water level was read.

Wells were pushed into the peat at distances of 1, 2, 5, 10, 20, 35, and 50 m (3.3 to 164 feet) from the east edge of the ditch on the east side of the road, and at the same distances from the west edge of the west-side ditch. The elevation of each well top relative to a datum on the shoulder of the road was determined with a transit and stadia rod. After equilibration the water level below the top of each well was measured. Air was blown through 1-cm-diameter plastic tubing as it was lowered into the well. The level at which bubbling was first heard was the water-table elevation, calculated as the well-top elevation minus the length of tubing below the well top.

Results and Discussion

Drawdown of the water table within 50 m of the ditches on either side of Highway 72 was measured along the observation-well transect in the water track several times during the summer and fall of 1984 and once in the spring of 1985 (Fig. 12.3). In wells 10 m or more

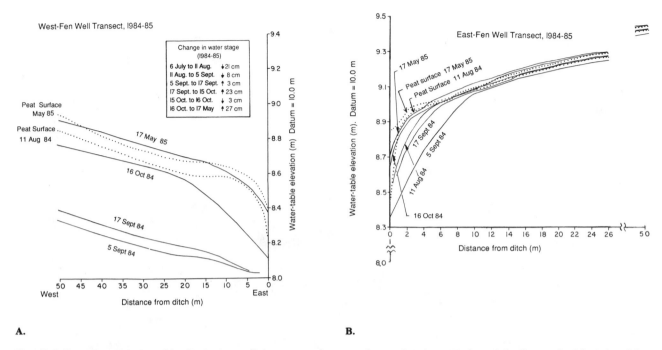

Fig. 12.3. Drawdown in water-table observation wells in water tracks east and west of Highway 72 about 2.5 miles south of the Lake of the Woods County line, 1984-85. Peat-surface and well-top elevations were measured with transit and stadia rod. Water levels on the sampling dates were subtracted from each well-top elevation. **A.** Water track west of Highway 72. **B.** Water track east of Highway 72; note that horizontal scale differs from that of A. From Bradof 1988.

from the ditch on the east (upgradient) side of the road, the water table remained at or only a few centimeters below the surface throughout the sampling period, which included an unusually dry August and September. During the dry period, the depth of the water table below the peat surface increased with decreasing distance from the ditch in wells within 10 m of the ditch. The ditch thus appears to have little effect on the water table at distances greater than 10 m.

The natural flow gradient before ditch and road construction was from east to west, but the present gradient west of the road is reversed: the water table slopes to the east, at least within 50 m of the ditch. Beyond the influence of the ditch, the water table presumably levels off and resumes a westward or northwestward slope in accordance with the topography (Fig. 12.4). This flow divide did not exist in the water track prior to drainage of the area.

Diversion of westward-flowing water to the north by the east roadside ditch has caused a lowering of the water table west of the road. For example, the water level in the well 1 m from the east-side ditch is 45 to 60 cm higher than that in the well 1 m from the west-side ditch on a given sampling day. If the east- and west-side well transects were separated by a single ditch instead of two ditches and a road, the water-table elevations 1 m distant on either side would be approximately equal. Along the length of the west-side well transect, the water table was 40 to 50 cm below the peat surface in September 1984. The water table had risen by

mid-October in response to rainfall and reduced evapotranspiration but still remained below the surface by 3 to 7 cm in wells 20 to 50 m from the ditch and 12 to 25 cm in wells within 10 m of the ditch. The water table was at or just below the surface at all wells east of the road at the same time. On May 17, 1985, during a very wet spring, the west-side water table was no more than 4 cm below the surface within 5 m of the ditch and was at or above the surface along the rest of the transect.

On two occasions after periods of rainfall, water levels were measured 24 hours apart (October 15-16, 1984, and May 17-18, 1985) to determine how quickly the water level dropped when the rain ended. Levels in the east-side wells declined 1 cm or less after 24 hours (no change from May 17 to 18), whereas those on the west side declined generally 1 to 2 cm, in some cases 4 to 8 cm.

Blockage of normal water flow by roads and pipelines across wetlands in the northern United States and Canada can cause a rise in the water table and reduced growth and increased mortality in timber stands upslope (Stoeckeler 1965, 1967a, b; Boelter and Close 1974). Glaser *et al.* (1981) suggest that blockage of flow by Highway 72 has caused stagnation in the fen water track to the east (upgradient). Compacted peat beneath the road probably obstructs westward flow through the deeper layers of peat, but Boelter (1972) found that the hydraulic gradient adjacent to a ditch produced flow from the deep layers of moderately decomposed (hemic) peat into the ditch. The observation-well re-

Transect showing location of water-table wells

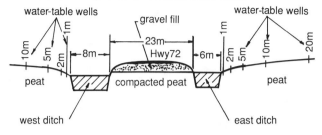

a. Water flow before road construction

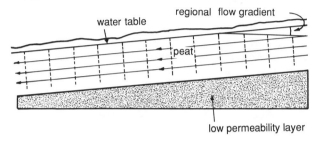

b. Water flow after road construction

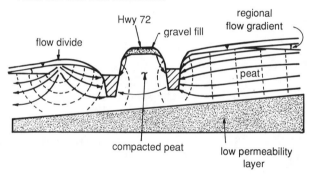

Fig. 12.4. Water-table observation well transects in the water track crossed by Minnesota Highway 72 about 2.5 miles south of the Lake of the Woods County line. Water flow through the peat (**A**) before and (**B**) after drainage. *Solid lines with arrows* represent water flow. From Bradof 1988.

sults indicate that water flows freely into the east roadside ditch from the upper layers of peat. There is no evidence of a rise in water table resulting from blocked drainage.

Studies of peatland road construction in Europe and pipeline construction in the Chippewa National Forest in Minnesota have shown that properly placed ditches and culverts provide cross-drainage sufficient to prevent a rise in water table upslope (Stoeckeler 1965; Boelter and Close 1974). In Red Lake peatland, where the only direct flow of water from east to west across Highway 72 is through a single culvert, the fen water track west of the road shows more evidence of stagnation — in terms of a lowered water table and a reversal of flow direction — than the water track east of the road.

Other studies in different types of peatlands have looked at water-table drawdown adjacent to a ditch. Boelter's (1972) work in northern Minnesota showed little drawdown more than 5 m from a ditch in moderately decomposed hemic peat. A ditch in porous, relatively undecomposed (fibric) peat, however, produced drawdown over the entire 50 m well transect. Lowering the ditch water table by 0.6 m did not affect the water table much beyond 10 m from the ditch, where the additional drawdown was only 1 cm.

Burke (1963) observed very little lowering of the water table at distances greater than 1.8 m from open ditches in a highly decomposed blanket peat in Ireland. Any drawdown beyond 1.8 m was influenced largely by evapotranspiration rather than ditch spacing. In Wicken Fen near Cambridge, England, Godwin and Bharucha (1932) found that a ditch influenced the water table up to 50 m away. Beyond that distance the water table was controlled by the balance between rainfall and evapotranspiration.

Godwin (1931) attributed a seasonal water-level decline from June to August or early September in Wicken Fen to an excess of transpiration over precipitation. He also noted diurnal fluctuations with an average drop of 3 cm during the day from transpiration and an average rise of 1 cm at night from inflow to the fen. Manson and Miller (1955) observed a similar seasonal water-level trend in seven northern Minnesota peatlands, ditched and unditched, forested and open: a rapid rise in water table after spring break-up, followed by a rapid recession during the summer and persistent low levels during the winter.

The water table in the transect west of Highway 72 fluctuates through a greater range and drops more rapidly after the cessation of rainfall than the water table east of the road. This difference suggests that the east side receives steady inflow from a source other than precipitation, whereas the west-side water level depends more on the balance between precipitation and evapotranspiration, having been cut off from westward flow. Bay's (1970) comparative study of water-table fluctuations in a perched bog and a groundwater fen in northern Minnesota showed that groundwater inflow to the fen maintained a higher, more stable water level than that of the nearby bog. Similar findings from an Ontario peatland were reported by Dai *et al*. (1974).

Mooratmung

The seasonal peat surface fluctuations known in Germany as *Mooratmung* (mire breathing) were not specifically studied in Red Lake peatland, but releveling of the water-table observation wells provided evidence that such surface-elevation changes occurred. Well-top elevations on the west side rose 5 to 8 cm between August 1984 and May 1985. The water table was below the surface by as much as 45 to 50 cm during September 1984, but it had risen from 4 cm below to 5 cm above the surface at the west-side wells on May 17, 1985.

On the east side, the peat surface rose 4 to 6 cm within 5 m of the ditch over the same time period. At distances of 10 m or more from the east-side ditch, where the water table remained at the peat surface throughout the monitoring period, the peat surface elevations were essentially constant at each well. Changes in peat surface elevation can thus be related to changes in water level. Frost heave was considered as a factor in changing well elevation, but it is unlikely that it could account for a relatively uniform increase in surface elevation.

Seasonal surface-elevation fluctuations related to water-level changes in peatlands are widely reported. Annual elevation changes ranging from 1.8 to 4.5 cm were observed in virgin Scots pine mires in eastern Finland by Kurimo (1983) and in raised-bog meadows in northwestern Germany by Baden and Eggelsmann (1964). Greater annual surface fluctuation occurred in other areas: 5 cm or more in the East Anglian peatlands (Hutchinson 1980), 15 cm in both drained and undrained areas of an open bog in southeastern Finland (Mustonen and Seuna 1971), and 20 cm in an East Carpathian raised bog (Kulczynski 1949).

In the Lost River peatland of northern Minnesota, changes in surface elevation of 7 to 12 cm on a raised bog, 4 to 7 cm on a spring fen, and 1 to 3 cm in a fen occurred over one year (Almendinger *et al.* 1986). The different amounts of elevation change were believed to be the result of variations among the landforms in peat thickness and peat type, which produced different responses to changes in fluid pressure generated by regional groundwater recharge. These surface fluctuations are of a magnitude similar to those observed at the well transects in Red Lake peatland not far away.

Peat Subsidence

Methods

Mooratmung is not the only process responsible for changes in surface elevation in drained peatlands. Paired difference analyses were used to determine if peat depths have decreased significantly in Red Lake peatland east and west of Minnesota Highway 72 since the construction of drainage ditches in 1916.

Soper (1919) published peat depths in Red Lake peatland measured along section lines by survey crews in 1915 before ditch construction began. The surveyors used a Davis peat sampler to measure the depth of the mineral soil contact below the peat surface every 500 feet (152 m) along the survey line. Most of the sites at which peat depths were determined in 1915 were in the path of excavation for the ditches and are thus not available for remeasurement. However, a reconnaissance-level survey of an area that included Red Lake peatland was conducted by the Minnesota Department of Natural Resources (DNR) as part of the Minnesota Peat Inventory Project (Minnesota DNR 1984). A Davis sampler was used to collect peat samples and to determine the depth of the peat. Many of the DNR sites are located close to those reported by Soper and thus can be paired with them for the purpose of a statistical comparison of pre- and post-drainage peat depths.

The DNR Minerals Division provided data from the field surveys of 1979-82. Because the peat depth at a given DNR site was not measured in 1915, it is impossible to know if the peat thickness was the same as at the nearest site reported by Soper. If both sites are on the same landform (e.g., both in a water track), however, the pre-drainage peat depths were probably similar. If the peat depths were different, the peat at the DNR site had an equal chance of being either thicker or thinner than the peat at the Soper site.

When a Soper site and the DNR site with which it was to be paired were located on different landforms, those sites were not included in the analysis. Similarly, if there was no Soper site near a given DNR site, no peat depth comparison could be made. When a DNR site was approximately equidistant from two Soper sites, the average of the peat depths at the latter sites was used for comparison; generally, the depths did not differ by more than 30 cm in such cases.

Results and Discussion

Each row in Table 12.1 presents peat depths for a pair of sites, one sampled by the DNR between 1979 and 1982, the other nearby site sampled by ditch surveyors in 1915 and reported by Soper (1919).

The paired difference analysis of peat depths east of Highway 72 (T.156 N., R.30 W.) shows a mean decrease in peat depth of 19 cm between 1915 and 1979-82 for the 24 pairs of sites sampled. This amounts to an average subsidence of about 3 mm per year since 1916. The 95% confidence interval around the mean is 3 to 36 cm.

For 22 pairs of sites west of Highway 72 (T.155-156 N., R.31-32 W.), a mean decrease in peat depth of 67 cm (95% confidence interval = 47 to 88 cm) occurred between 1915 and 1979-82. Subsidence thus has averaged approximately 10 mm per year since 1916.

The results of these analyses suggest that subsidence has been greater west of Highway 72 than east of the road. For the purpose of further analysis, the data from Table 12.1 were divided into three groups on the basis of site location. Highway 72 and the ditches along it, which follow the range line between R.30 W. and R.31 W., act as a barrier to the westward flow of water, as discussed in the section on water-table drawdown. The range line between R.31 W. and R.32 W. coincides with a natural hydrologic division between an area with water flow from the east and south (R.31 W.) and an area with water flow from the west and south (R.32 W.). Therefore, comparison of sites from R.30 W., R.31 W., and R.32 W. as hydrologically distinct groups is reasonable.

Table 12.1. Peat depths at pairs of localities in Red Lake peatland, northern Minnesota, 1915 and 1979-82

Peat depth (cm)			Peat depth (cm)		
Soper (1915)	DNR (1979-82)	Change in depth (cm)	Soper (1915)	DNR (1979-82)	Change in depth (cm)
T.156 N., R.30 W.			T.156 N., R.31 W.		
228	218	− 10	213	49	−164[a]
305	335	+ 30	244	150	− 94
183	180	− 3	213	150	− 63
213	147	− 66	213	150	− 63
213	198	− 15	274	210	− 64
213	203	− 10	427	330	− 97
183	150	− 33	366	270	− 96
305	256	− 49	289	195	− 94
244	290	+ 46	213	120	− 93
274	264	− 10	396	270	−126
335	244	− 91	366	270	− 96
335	330	− 5	427	300	−127
366	259	−107	274	300	+ 26
366	422	+ 56	T.156 N., R.32 W.		
366	292	− 74	198	180	−18
396	378	− 18	366	345	−21
274	270	− 4	259	225	−34
213	213	0	335	270	−65
213	221	+ 8	396	330	−66
244	213	− 31	427	390	−37
183	173	− 10	T.155 N., R.31 W.		
91	86	− 5	331	315	− 16
183	120	− 63	305	300	− 5
152	150	− 2	244	105	−139
T.155 N., R.32 W.					
228	105	−123			

[a]Site in small triangle of land bordered by ditches east of Highway 72, not included in statistical analyses.

Only sites from T.156 N. were included in the following Student's *t* tests.

The first hypothesis tested is that the mean decrease in peat thickness from 1915 to 1979-82 is greater west of Highway 72 in R.31 W. than it is east of the road in R.30 W. A Student's *t* test indicates that the mean decrease in peat depth for sites in R.31 W. is 63 cm greater than for R.30 W., with a 95% confidence interval of 35 to 90 cm. The mean decrease in peat depth from 1915 to 1979-82 is 42 cm greater for R.31 W., where water flow is from the *east* and south (disrupted by road), than for R.32 W., where flow is from the *west* and south. The 95% confidence interval is 5 to 80 cm. Sites in R.30 W. and R.32 W. have had essentially identical drainage history in that ditches are present in both areas, but the major sources of water flow, the eastern and western water tracks, respectively, have not been obstructed. A significant difference between these areas in the mean decrease in peat depth is therefore not expected, and none is indicated by the Student's *t* test at the 0.05 level of significance.

Studies have shown that subsidence generally increases with decreasing distance from a ditch (Burke 1963; Ilnicki 1983). A linear regression of change in peat depth from 1915 to 1979-82 versus distance from the nearest ditch for sites east of the road in T.156 N., R.30 W. did show a negative correlation, but not a very strong one ($r = -0.40$). The relationship was very poor for sites in R.31 W. ($r = 0.12$).

For each of the above statistical comparisons, equivalent nonparametric statistics (Wilcoxon signed rank or rank sum tests) were also calculated, with comparable results.

The data presented here may not represent the actual mean subsidence that has occurred in Red Lake peatland because the pairs of sites are all relatively close to the drainage ditches, where water-table drawdown leads to greater subsidence. The DNR sites east and west of Highway 72 average 47 and 128 m, respectively, from the nearest ditch. The greater subsidence in R.31 W. west of the road therefore is not the result of those sites being closer to the ditches than are sites east of the road.

Peat surface elevation changes of 8 cm or less occurred at the water-table well transect between the dry conditions of August 1984 and the very wet conditions of May 1985. Such seasonal or even year-to-year variations are too small to account for the measured subsidence. *Mooratmung* also cannot explain the significant difference in the amount of subsidence between sites in R.31 W. and those in R.30 W. or R.32 W.

According to Stephens *et al.* (1984), there are six causes of subsidence — (1) shrinkage due to desiccation, (2) loss of the buoyant force of groundwater, (3) compaction, (4) wind or water erosion, (5) burning, and (6) biochemical oxidation — all of which can be influenced by drainage. Subsidence rates are controlled by physical characteristics of the peat, depth to the water table, and temperature (Nesterenko 1976; Stephens *et al.* 1984). Of these factors, only depth to the water table is a function of drainage. Drought conditions and peat fires during the 1930s may account for much of the subsidence that has taken place since 1915. Subsidence occurs under such conditions whether ditches are present or not. However, the road and the ditches that run along it divert the natural westward flow of water to the north. This diversion of flow explains the greater subsidence in R.31 W. as compared to R.30 W. or R.32 W., where the natural flow of water is less disturbed.

Some very long records of peat subsidence exist in Europe. Schothorst (1977, cited in Stephens *et al.* 1984) recounts the history of low-moor (fen) peat drainage in the western Netherlands, which began between the ninth and thirteenth centuries. After 900 years of gravitational drainage, subsidence totalled 1 to 2 m, a rate of about 1.7 mm per year. The rate increased to 6 mm per year after steam pumping stations began controlling year-round water levels. An experimental field showed that a 0.4 m drop in ditch water level resulted in surface subsidence over a 20-year period of 23 cm, nearly half of which occurred in the first two years (50.6 mm per year). By the third year, the rate had declined to a constant 7 mm per year.

The most dramatic record of peat surface lowering exists in the East Anglian fenlands, where the Holme Post, a cast-iron column, was driven into the clay underlying a peatland that still approximated natural conditions in 1850 (Darby 1956; Hutchinson 1980). Subsidence was recorded along this fixed datum. Of the 6.7 m of peat present in 1848, 1.83 m was lost to subsidence in the first 12 years after drainage, and a total of 3.87 m was lost by 1978. Over this period, four stages of subsidence occurred, each associated with the operation of a pumping station that lowered the water level. During a 30- to 35-year period of static, relatively high water levels and no cultivation, subsidence appeared to cease.

In a study of four mostly open oligotrophic peatlands in southern Finland, each drained by ditches 20 m apart and 1.5 m deep, Sallantaus and Patila (1983) observed 18 cm of peat surface subsidence in the first 6 months after drainage. Also in southern Finland, Mustonen and Seuna (1975) conducted a paired watershed study with 22 years of calibration before drainage began in one of the basins, which consisted of mostly open bog and poor pine swamp. Main ditches 130 cm

deep that reached mineral soil were dug in 1958, leading to an average settling of the peat surface of 8 cm by 1960 (40 mm per year). An additional 12 cm of subsidence occurred by 1969 after small forest ditches 60 cm deep were dug in 1960; this period of subsidence thus averaged 13.3 mm per year. During the 1958-1969 period, the groundwater level dropped 30 cm and the peat surface settled about 20 cm.

Burke (1963) found an average subsidence of about 12 cm for a drain spacing of 7.6 m and about 5.7 cm for a drain spacing of 30.5 m in a highly decomposed blanket peat in Ireland five years after drainage. Close to the drain edge, 23 to 30.5 cm of subsidence occurred during the same period.

In a review of subsidence in the southern USSR and other countries, Nesterenko (1976) reports maximum subsidence of 7 to 15 cm within two to five years after drainage, followed by a decreased rate of 10 to 20 mm per year. Results from Scandinavia show subsidence of 7 to 42 mm per year (10 to 20 mm per year on average) between 10 and 65 years after drainage. Nesterenko makes a distinction between rapid post-construction subsidence within two to five years after drainage and the lower rate that follows.

Ilnicki (1983) found long-term subsidence rates for Polish low-bog (fen) peatlands to be about 10 mm per year on average and 13 to 14 mm per year or more for more intensive drainage. Subsidence was greater in looser, less decomposed low-ash peat, and it increased in rate with peat thickness and intensity of drainage. Higher rates were observed in raised bogs — 32 mm per year in one case. Drainage of one raised bog resulted in rapid subsidence of 3.0 m (mean subsidence of 176 mm per year). A second drainage in 1968 produced subsidence at a mean annual rate of 18 to 32 mm. Ilnicki attributes much of the observed peat subsidence to peat mineralization in the top 60 cm.

Studies of peat subsidence after drainage of low-moor (rich-fen) and high-bog (raised-bog) peats in northwestern Germany by Eggelsmann (1976) show that drainage influence (shrinkage and compaction) is greater in high bogs than in low moors, whereas oxidation is greater in low moors than in high bogs. Van der Molen (1975) reached similar conclusions in the Netherlands.

Eggelsmann found that the amount of subsidence due to oxidation is linear with respect to time, whereas that due to drainage levels off after about 50 to 60 years. Fig. 12.5 presents values observed by Eggelsmann for a typical low moor and high bog in northwestern Germany. Ten years after drainage, the raised bog has lost about 45 cm in surface elevation to drainage effect and about 10 cm to oxidation; the low moor has lost about 30 cm to drainage and about 28 cm to oxidation. Sixty years after drainage, the raised bog has lost 95 cm to drainage and 64 cm to oxidation, while the low moor has lost 64 cm to drainage and 130 cm to oxidation. Total subsidence after 10 years is about the same for low moor and raised bog, 55 to 58 cm, but after 60 years

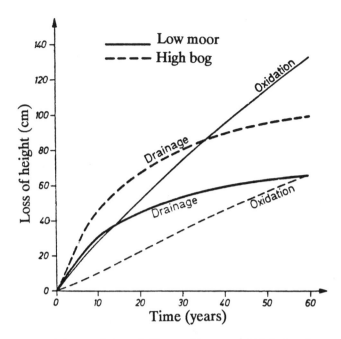

Fig. 12.5. Cumulative subsidence of low-moor (rich-fen) and high-moor (raised-bog) peat in northwestern Germany caused by drainage and oxidation. From Eggelsmann (1976, by permission).

the low moor has subsided 194 cm and the raised bog 169 cm.

Nearly all of the preceding studies present rates of subsidence greater than those inferred for Red Lake peatland. The higher rates observed elsewhere can be explained in part by the fact that many of the areas have been more intensively drained (by closely spaced ditches or by pumping) or have been utilized for agricultural purposes. Eggelsmann (1976) notes that the rate of oxidation is very low on peat grasslands and forests as compared to cultivated land. Differences in peat type from area to area also influence subsidence rates. At Red Lake peatland, the relatively low rate of subsidence of 3 to 10 mm per year since 1915 may simply reflect the failure of the drainage network to significantly lower the peatland water table overall, a conclusion that was quickly reached by prospective settlers before 1920. The mean subsidence rate probably would have been even lower were it not for the drought conditions of the 1930s.

Growth Response in Black Spruce to Drainage and Climatic Change

Methods

Two black spruce stands in Red Lake peatland, one adjacent to a drainage ditch and the other 1.5 km from the nearest ditch, were studied to determine the response of black spruce to drainage and to changes in climatic conditions. These trees have many shortcomings as subjects for climatological reconstructions.

Their annual growth rings are closely spaced, and their growth patterns vary from tree to tree and even within individual trees, making it difficult to discern trends. With a large enough sample, however, growth patterns can be correlated with climatic conditions. Any changes resulting from drainage will be superimposed on these climatic trends.

Selection of sites for study was complicated because much of Red Lake peatland has been burned since the drainage ditches were dug, and many areas have been logged to various extents. Cores taken by increment borer from 66 trees near the ditch and 16 trees in the control area were used for growth statistics. Radial growth in each core was measured by decades because individual annual growth increments often measured only a fraction of a millimeter, impossible to measure accurately with the aid of only a binocular microscope and metal rule.

The growth increment for each decade from 1900 to 1979 was taken as a percentage of the total 1900-79 measurement for that core as a means of standardizing the cores. The percentages for a given decade from all the cores were then averaged. A similar treatment was used by LeBarron and Neetzel (1942). Growth during the nine-year period before ditch construction (1908-16) and the nine-year period after drainage (1917-25) was also measured and taken as a percentage of the total 1900-79 growth for each core from the drained and control areas.

Long-term temperature and precipitation records, which date from 1820 and 1837, respectively, exist for the Twin Cities (Minneapolis-St. Paul) area, about 410 km (255 miles) southeast of the study area. These records were correlated with those from Red Lake Indian Agency at Redby, about 64 km (40 miles) southwest of the study area. The latter records were then correlated with records kept since 1963 at Baudette 21SSE (a private residence 21 miles south-southeast of Baudette) and Waskish Ranger Station, about 6.5 km (4 miles) north and 18 km (11 miles) south, respectively, of the study area. By such correlations the climatic record for Red Lake peatland could be extended back into the 1800s for comparison with tree growth since that time.

Results and Discussion

Black spruce growth generally has been much greater during the 1900s than during the 1800s. This is true even of trees whose natural growth rate would be expected to be slowing down as they approached the 200-year average lifespan for the species. Above-average temperature and below-average precipitation appear favorable to growth of black spruce on the peatland. The period of maximum growth for nearly all of the trees examined, some of which were more than 200 years old, occurred during the 1930s, when temperatures were at or near their highest recorded levels and precipitation approached its lowest recorded levels.

Hofstetter (1969) also observed peak growth during this period in black spruce and tamarack from the western water track of Red Lake peatland. He suggested that lower precipitation and increased wind-borne nutrient input from drought-stricken farms and from peat fires were likely contributors to this increased growth.

Tilton (1975) found similar growth trends in tamarack from four areas across northern Minnesota on both wet and dry sites. He argued that drier conditions and wind-borne nutrients alone could not explain better growth during the 1930s and 1940s. Suppression of larch sawfly populations by drought conditions was proposed as a possible cause of enhanced growth in tamaracks. This study did not address the increased growth of black spruce trees at the same time, but perhaps the harmful effects of dwarf mistletoe or spruce budworm infestations were also suppressed by drier conditions.

The present study shows a correlation between black spruce growth and temperature changes; precipitation trends appear to be of lesser importance (Table 12.2).

In a paired difference analysis, the spruce trees within 50 m of the drainage ditch showed a significant ($p > .995$) increase in growth following drainage, whereas those in the control area did not. Although drainage appears to have stimulated growth in the black spruce near the ditch, the small size of the control sample limits the interpretation of these results. It is clear, however, that any increase in radial growth attributable to drainage of this area is very small compared to the changes in growth rate that correspond to changes in climatic conditions. This may mean that the ditch had relatively little effect on the water table of the ovoid bog island that it crossed or that lowering of the water table was not the most important factor affecting growth in these black spruce.

A number of studies (LeBarron and Neetzel 1942; Manson and Miller 1955; Payendeh 1973; Laine and Seppala 1977; Heikurainen *et al.* 1978) have shown increased tree growth after drainage in Michigan, Minnesota, Ontario, and Finland. Their results would be more convincing, however, had the correlation between tree growth and climatic change also been examined.

One of the most comprehensive and frequently cited investigations was conducted by Averell and McGrew (1929) across northern Minnesota in a wide variety of peatlands that had been drained between 1910 and 1920. Their survey included 26 study areas (76 plots); 1,400 increment cores were measured, 400 trees cut and measured for volume relative to radius, and 11,812 trees measured for diameter. Two of their study areas fall within the field area of the present study, but they have since been burned over, apparently during the 1930s.

Averell and McGrew did not have separate control areas in undrained peatland, but each study area included a plot near the ditch and one 61.5 to 183 m distant. At the near plot, the average annual radial growth rate before drainage was compared to that over

Table 12.2. Changes in radial growth rate for black spruce (*Picea mariana*) and concurrent changes in climatic conditions from 1910 to 1979, Red Lake peatland, Minnesota.

Period	Change in radial tree growth		Change in climatic conditions	
	Drained	Control	Precipitation	Temperature
1910-19 to 1920-29	increase	slight increase	decrease	stable
1920-29 to 1930-39	increase	increase	increase	increase
1930-39 to 1940-49	decrease	decrease	increase	decrease
1940-49 to 1950-59	slight decrease	no change	decrease	decrease
1950-59 to 1960-69	decrease	decrease	increase	slight decrease
1960-69 to 1970-79	slight decrease	slight decrease	increase	slight increase

Summary of Results

Conditions for significant increase in growth:

(1) increase in temperature to well above average; increase in precipitation to average

(2) temperature stable; decrease in precipitation to lowest level

Conditions for significant decrease in growth:

(1) decrease in temperature from well above average to average; increase in precipitation from average to highest level

(2) very slight decrease in temperature; increase in precipitation from just below to just above average (decrease in temperature and increase in precipitation greater for growing season [May-September] than for annual)

Conditions for slight decrease (drained area) or no change (control):

(1) decrease in temperature from just above to below average (lowest value); decrease in precipitation from highest value to below average

(2) very slight increase in temperature; increase in precipitation

the same number of years after drainage. The rate of radial growth due to drainage (the difference between the after- and before-drainage growth rates) decreased with increasing distance from the ditch. Where the rate of radial growth no longer changed with distance from the ditch, it was assumed to be the natural growth rate for the stand beyond the influence of the ditch. The distant plot was located in this area. The following is a summary of their findings:

(1) Cedar (*Thuja occidentalis*) showed the greatest increase in growth rate due to drainage, tamarack (*Larix laricina*) the least; black spruce (*Picea mariana*) was intermediate in response.

(2) The greatest growth response to drainage occurred on sites of intermediate quality rather than on either the excellent or the poor sites.

(3) Spruce stands on sedge peat showed greater response to drainage than those on *Sphagnum* peat.

(4) The effect of drainage on tree growth fell off rapidly with distance, extending an average of 101 m from the ditch; the range was from 0 to more than 287 m and the effect was rarely equal on both sides of the ditch.

(5) There was a time lag, ranging from one to eight years, in the response to drainage as reflected by increased growth. This time lag increased with distance from the ditch.

(6) Stands of younger trees with better-developed crowns showed greater favorable response to drainage than older stands with poor crowns.

(7) Drainage was usually more effective on the side of the ditch that was downgradient in terms of natural water flow. The average effective draining width for the 26 areas was 77 m upslope and 135 m downslope of the ditch. Where the ditch was approximately at right angles to the natural water flow and the surface slope was greatest, the drainage effect extended farthest from the ditch.

(8) Drainage effectiveness was not significantly correlated with peat depth but was greater where the ditch reached mineral soil.

(9) Ditches could be rendered ineffective by beaver dams or by lack of maintenance.

Other Hydrologic Effects of Drainage

Two other effects of peatland drainage not directly addressed by this study deserve mention: changes in runoff pattern and increases in beaver population. No data are available for the comparison of pre- and post-drainage runoff from Red Lake peatland, but numerous studies have been done in Finland, the USSR, and other areas of Europe. Literature reviews by Clausen and Brooks (1980) and Starr and Paivanen (1981) conclude that drainage generally increases the quantity of runoff from peatlands, but the specific changes that take place depend on the type of drainage, drain spacing and orientation, time since drainage began, condition of the drainage system, peatland type, vegetation, and climatic factors. Both reviews divide the runoff results into two categories. One model shows decreased duration of peak flows, greater peak-flow rates, and reduced base flow after drainage; the other finds increased duration of peak flows, reduced peak-flow levels, and increased base flow as the result of drainage. Overall, the latter has more uniform, less "flashy" flow than the former. A third model combining elements of the other two, in which both maximum and minimum (base) flow levels are increased by drainage, is also discussed by Starr and Paivanen.

Huikari (1968) observed three times more runoff from an open (sedge) bog and two times more runoff from a forested (pine) bog in Finland with ditch spacings of 5 m than in the same peat types with ditch spacings of 60 m. Open ditches produced higher peak flows than covered drains or narrow vertical-walled ditches (Paivanen 1976). Burke (1963) found that for blanket peat in Ireland, precipitation generated runoff faster from undrained areas than from drained areas because of the greater water storage capacity in the latter. Runoff peaks were also higher in the undrained peat, but overall runoff from the two areas was about the same. A later study in the same area showed that overall runoff from the drained peat was 60% greater than that from the undrained peat (Burke 1975). Eggelsmann (1975) also reported higher peak flows from undrained peat than from well-drained peat grassland.

Studies in drained and undrained raised bogs in Germany showed that the water table remained close to the peat surface during a greater proportion of the year in undrained than in drained areas, suggesting to Ingram (1983) that drains remove excess water when precipitation exceeds evapotranspiration. An after-drainage increase in runoff from four peatland watersheds in southern Finland was attributed to a depletion of water storage resulting from subsidence (Sallantaus and Patila 1983). A paired watershed study in southeastern Finland documented significant post-drainage increases in mean annual runoff, spring and summer maximum runoff, and minimum runoff, but these effects diminished with time, approaching pre-drainage levels after 15 to 20 years (Seuna 1981). Impairment of the ditches and increased interception and evapotranspiration resulting from tree growth were suggested as explanations for the decreasing difference between drained- and undrained-area runoff over time.

During the 1940s, beaver activity killed or damaged significant amounts of timber in the "Big Bog" country of Koochiching, Lake of the Woods, and Beltrami counties (Vesall *et al.* 1947). The series of events leading to this situation was traced back to about 1915, the time of the large drainage projects. During the 1930s, several hundred control dams were installed in the ditches to maintain water levels sufficient to reduce fire hazards and improve wildlife habitat. Beavers built their own dams atop the human-made structures and elsewhere in the ditches, flooding large areas of the very flat peatland. Beaver damage to timber along drainage ditches was first reported in 1937.

Control measures — including the dynamiting of beaver dams and the legalization of trapping — began in 1939. A study in part of Koochiching County in 1946 by Vesall *et al.* (1947) found that the number of active beaver dams per mile of ditch ranged from 0 to 10, with an average of 2.5. Beaver activity also inundated and undermined roads used for logging, patrols, and firefighting.

A drainage survey of the peatlands north of Red Lake conducted between 1906 and 1908 considered beaver dams to be an "important factor in the formation of swamps" along with recessional moraines that blocked drainage (U.S. Department of the Interior 1909). Clearly, beavers were present in the area before the drainage ditches were dug. It is undoubtedly true, however, that the creation of a network of open channels through the peatlands enabled the beavers to colonize the entire area. Signs of beaver activity — dams, lodges, canals (some of which have undermined the shoulders along Highway 72), and gnawed aspen stumps — were readily apparent in the field study area between 1983 and 1985.

Another wildlife species may have been adversely affected by drainage-induced habitat changes. Carlos Avery of the Minnesota Game and Fish Commission reported thousands of caribou tracks and fresh markings north and northeast of Upper Red Lake prior to drainage (*Bemidji Pioneer*, September 16, 1909). Settlers also reported sighting caribou. Avery commented that he was "satisfied that these animals will remain there for a long time, or at least until that country is drained, as it is a natural feeding ground for them. There is a large quantity of what is known as 'reindeer moss,' which is the one food they like." Today caribou are absent from Minnesota, probably a casualty of hunting and of habitat changes more favorable to deer, which can be carriers of the "brainworm" parasite often fatal to moose and caribou.

Conclusions

Two major hydrologic implications of the drainage of Red Lake peatland emerge from this study. First, the blockage and northward diversion of westward-flowing water by the road has produced water-table drawdown toward the ditches on either side; this drawdown represents a reversal of the natural flow direction in the water track west of the road. Second, a decrease in peat thickness has occurred since drainage began. Although the subsidence of the peat surface may have been augmented by drought in the 1930s, differential subsidence east and west of Highway 72 is the result of drainage. Subsidence since 1915 is significantly greater in R.31 W. immediately downgradient from the road than it is in either R.30 W. or R.32 W., where the natural flow of water is relatively undisturbed.

Growth rates for black spruce trees on the peatland may have increased as the result of drainage, at least for trees relatively close to a ditch. Much more dramatic changes in growth rate can be correlated with trends in temperature and precipitation.

Other hydrologic effects of drainage not investigated at Red Lake peatland but likely to have occurred there are suggested by results from other peatlands. These effects include changes in runoff pattern, an increase in total runoff, and an increase in beaver population leading to undermining of roads and damage to timber from flooding.

One drainage-related question of considerable historical interest remains unanswered: Could Red Lake peatland have been successfully drained for agriculture, without incurring prohibitive costs, if the drainage system had been designed differently, or is the peatland by its nature virtually impossible to drain? Results from other areas provide some insight into the relationship of water-table lowering to ditch spacing and depth, peat type, and nature of the substrate beneath the peat.

The effect of ditch spacing and depth on water levels and forage crop yields was examined in a Newfoundland sedge bog (*Scirpus cespitosus* with *Sphagnum* and heath plants) by Rayment and Cooper (1968). Comparisons were made of results from virgin peat and drained areas with combinations of three ditch depths (0.6, 0.9, and 1.2 m) and four ditch spacings (15.2, 22.9, 30.5, and 45.6 m). Water-table depth was related most closely to ditch spacing. Only at a spacing of 15.2 m was the water table lowered throughout the plot during a wet year. A spacing of 22.9 m produced appreciable lowering of the water table at the center of the plot in an average year, while greater spacings did not. Deepening the ditch from 0.6 to 1.2 m had no significant effect.

Experimental drainage of a low-bog sedge peat in the USSR showed that deep canals constructed perpendicular to the natural flow direction at spacings of 700 to 900 m could successfully drain the area, provided that the canals cut into sand below the peat (Dzektser 1962). Thin layers of low permeability within the peat reduced the effectiveness of the ditches. In other areas underlain by a layer of highly decomposed organic matter (sapropel) 0.8 to 1.0 m thick instead of sand, the water table was not lowered by the ditches; it varied only according to precipitation and evaporation.

Correlation and multiple regression analyses based on data from deep peat in northern Norway (Braekke 1983) indicate that the mean growing-season water-table level midway between ditches can be predicted with good precision by ditch depth, ditch spacing, rainfall, peat permeability, and initial water level on June 1 (autocorrelation factor). An effective peat depth of 0.8 m and ditch distances of 9 to 25 m produced adequate drainage for establishing a forest. To achieve that effective ditch depth 5 years or so after ditching, it was necessary to dig drains 1.2 to 1.3 m deep in less-humified peat to allow for subsidence. Little subsidence occurs in well-humified peat, so initial depths of 0.8 to 1.0 m were sufficient. In addition, Tomberg (1957) pointed out that subsidence after drainage reduces the filtration coefficient (hydraulic conductivity) of peat, causing a rise in the average water table between ditches and a decrease in the overall effectiveness of drainage over time.

Ditch spacings on the order of 10 to 30 m obviously would be both impractical and prohibitively expensive for an area the size of Red Lake peatland. The spacings of 700 to 900 m discussed by Dzektser (1962) — the widest found to produce effective drainage by any study examined — were adequate only when the peat was underlain by a thick deposit of fine- to medium-grained sand. Ditch spacings in Red Lake peatland are much greater, 1,609 or 3,219 m, and the deposits below the peat are assumed to be lake clay or silt and till with some lenses of sand (Siegel 1981). Adding to this, the fact that the flatness of the topography did not provide sufficient outlets for runoff carried by the ditches shortly after their construction suggests that conversion of Red Lake peatland to agricultural land could not be accomplished in any reasonable manner unless a shift to warmer, drier climatic conditions were to occur.

Literature Cited

Almendinger, J. C., J. E. Almendinger, and P. H. Glaser. 1986. Topographic fluctuations across a spring fen and raised bog in the Lost River peatland, northern Minnesota. Journal of Ecology 74:393-401.

Averell, J. L., and P. C. McGrew. 1929. The reaction of swamp forests to drainage in northern Minnesota. Minnesota Department of Drainage and Waters, St. Paul, Minnesota, USA.

Baden, W., and R. Eggelsmann. 1964. Der Wasserkreislauf eines nordwestdeutschen Hochmoores. With English summary. Verlag Wasser und Boden, Hamburg, Germany.

Bay, R. R. 1970. The hydrology of several peat deposits in northern Minnesota, USA. Pages 212-218 in Proceedings of the Third International Peat Congress, Quebec, Canada, August 19-23, 1968.

Boelter, D. H. 1972. Water table drawdown around an open ditch in organic soils. Journal of Hydrology 15:329-340.

———— and G. E. Close. 1974. Pipelines in forested wetlands: Cross drainage needed to prevent timber damage. Journal of Forestry 72:561-563.

Bradof, K. L. 1988. Environmental impacts of drainage-ditch and road construction on Red Lake peatland, northern Minnesota: Drainage history, hydrology, water chemistry, and tree growth. M.S. thesis. University of Minnesota, Minneapolis, Minnesota, USA.

Braekke, F. H. 1983. Water table levels at different drainage intensities on deep peat in northern Norway. Forest Ecology and Management 5:169-192.

Burke, W. 1963. Drainage of blanket peat at Glenamoy. Pages 809-817 in Proceedings of the Second International Peat Congress, Leningrad, USSR.

————. 1975. Aspects of the hydrology of blanket peat in Ireland. Pages 171-182 in Hydrology of marsh-ridden areas. Proceedings of the Minsk Symposium, June 1972. Unesco Press, Paris, France.

Clausen, J. C., and K. N. Brooks. 1980. The water resources of peatlands: A literature review. Minnesota Department of Natural Resources, St. Paul, Minnesota, USA.

Dai, T. S., V. F. Haavisto, and J. H. Sparling. 1974. Water level fluctuation in a northeastern Ontario peatland. Canadian Journal of Forestry Research 4:76-81.

Darby, H. C. 1956. The drainage of the fens. Second edition. Cambridge University Press, Oxford, U.K.

Dzektser, E. S. 1962. Experiment in draining a bog by widely spaced deep canals. Gidrotekhnika i Melioratsiya 3:23-28. Translated from Russian.

Eggelsmann, R. 1975. The water balance of lowland areas in northwestern regions of the FRG [Federal Republic of Germany]. Pages 355-367 in Hydrology of marsh-ridden areas. Proceedings of the Minsk Symposium, June 1972. Unesco Press, Paris, France.

————. 1976. Peat consumption under influence of climate, soil condition, and utilization. Pages 233-247 in Peat and peatlands in the natural environment protection. Proceedings of the Fifth International Peat Congress, Poznan, Poland, September 21-25, 1976.

Glaser, P. H., G. A. Wheeler, E. Gorham, and H. E. Wright, Jr. 1981. The patterned mires of the Red Lake peatland, northern Minnesota:

Vegetation, water chemistry, and landforms. Journal of Ecology 69:575-599.

Godwin, H. 1931. Studies in the ecology of Wicken Fen. I. The ground water level of the fen. Journal of Ecology 19:449-472.

———— and F. R. Bharucha. 1932. Studies in the ecology of Wicken Fen. II. The fen water table and its control of plant communities. Journal of Ecology 20:157-191.

Heikurainen, L., K. Kenttamies, and J. Laine. 1978. The environmental effects of forest drainage. Suo 29:49-58.

Heinselman, M. L. 1963. Forest site bog processes and peatland types in the Glacial Lake Agassiz region, Minnesota. Ecological Monographs 33:327-374.

Hofstetter, R. H. 1969. Vegetational, hydrological, chemical, and evolutionary studies of ribbed fen, teardrop-shaped "islands," and bogs in the Red Lake peatland. Part A of Floristic and ecological studies of wetlands in Minnesota. Ph.D. thesis. University of Minnesota, Minneapolis, Minnesota, USA.

Huikari, O. 1968. Effect of distance between drains on the water economy and surface runoff of Sphagnum bogs. Pages 739-742 in Transactions, Second International Peat Congress, Leningrad, USSR, 1963. Her Majesty's Stationery Office, Edinburgh, U.K.

Hutchinson, J. N. 1980. The record of peat wastage in the East Anglian fenlands at Holme Post, 1848-1978. Journal of Ecology 68:229-249.

Ilnicki, P. 1983. Bog transformation resulting from drainage. Pages 13-25 in C. H. Fuchsman and S. A. Spigarelli, editors. Proceedings of the International Symposium on Peat Utilization, October 10-13, 1983. Bemidji State University, Bemidji, Minnesota, USA.

Ingram, H. A. P. 1983. Hydrology. Pages 67-158 in A. J. P. Gore, editor. Ecosystems of the World 4A, Mires: Swamp, bog, fen and moor. Elsevier, Amsterdam, Netherlands.

Kulczynski, S. 1949. Peat bogs of Polesie. Memoires de l'Academie Polonaise des Sciences et des Lettres. Classe de Sciences Mathematiques et Naturelles. Serie B: Sciences Naturelles. No. 15.

Kurimo, H. 1983. Surface fluctuation in three virgin pine mires in eastern Finland. Silva Fennica 17:45-64.

Laine, J., and K. Seppala. 1977. Development of radial growth in mineral soil stands bordering drained peatlands. Suo 28:67-74.

LeBarron, R. K., and J. R. Neetzel. 1942. Drainage of forested swamps. Ecology 23:457-465.

Manson, P. W., and D. G. Miller. 1955. Groundwater fluctuations in certain open and forested bogs of northern Minnesota. With notes on the effects of open drainage ditches on swamp forest growth. University of Minnesota Agricultural Experiment Station Technical Bulletin 217.

Minnesota Department of Natural Resources. 1984. Inventory of peat resources: An area of Beltrami and Lake of the Woods counties, Minnesota. Minnesota Department of Natural Resources, Division of Minerals, Peat Inventory Project. Hibbing, Minnesota, USA.

Mustonen, S. E., and P. Seuna. 1971. Metsaojituksen vaikutuksesta suon hydrologiaan. Summary: Influence of forest draining on the hydrology of peatlands. Pages 1-63 in Publication 2, National Board of Waters, Finland, Water Research Institute.

———— and P. Seuna. 1975. Influence of forest drainage on the hydrology of an open bog in Finland. Pages 519-523 in Hydrology of marsh-ridden areas. Proceedings of the Minsk Symposium, June 1972. Unesco Press, Paris, France.

Nesterenko, I. M. 1976. Subsidence and wearing out of peat soils as a result of reclamation and agricultural utilization of marshlands. Pages 218-232 in Peat and peatlands in the natural environment protection. Proceedings of the Fifth International Peat Congress, Poznan, Poland, September 21-25, 1976.

Paivanen, J. 1976. Effect of different types of contour ditches on the hydrology of an open bog. Pages 93-106 in Peat and peatlands in the natural environment protection. Proceedings of the Fifth

International Peat Congress, Poznan, Poland, September 21-25, 1976.

Payendeh, B. 1973. Analyses of a forest drainage experiment in northern Ontario. I: Growth analysis. Canadian Journal of Forestry Research 3:387-398.

Rayment, A. F., and D. J. Cooper. 1968. Drainage of Newfoundland peat soils for agricultural purposes. Pages 345-349 *in* Proceedings of the Third International Peat Congress, Quebec, Canada.

Sallantaus, T., and A. Patila. 1983. Runoff and water quality in peatland drainage areas. Pages 183-202 *in* International Peat Society, Commission III. Proceedings of the International Symposium on Forest Drainage, Tallinn, USSR, September 19-23, 1983.

Schothorst, C. J. 1977. Subsidence of low moor peat soil in the western Netherlands. Pages 265-291 *in* Technical Bulletin 102. Institute for Land and Water Management Research, Wageningen, Netherlands.

Seuna, P. 1981. Long-term influence of forestry drainage on the hydrology of an open bog in Finland. Publication 43, National Board of Waters, Finland, Water Research Institute.

Siegel, D. I. 1981. Hydrogeologic setting of the Glacial Lake Agassiz peatlands, northern Minnesota. U.S. Geological Survey Water Resources Investigations 81-24.

Soper, E. K. 1919. The peat deposits of Minnesota. Minnesota Geological Survey Bulletin 16.

Starr, M. R., and J. Paivanen. 1981. The influence of peatland forest drainage on runoff peak flows. Suo 32:79-84.

Stephens, J. C., L. H. Allen, Jr., and E. Chen. 1984. Organic soil subsidence. Pages 107-122 *in* T. L. Holzer, editor. Man-induced land subsidence. Reviews in Engineering Geology. Vol. 6. Geological Society of America, Boulder, Colorado, USA.

Stoeckeler, J. H. 1965. Drainage along swamp forest roads: Lessons from northern Europe. Journal of Forestry 63:772-776.

———. 1967a. Wetland road crossings: Drainage problems and timber damage. U.S. Forest Service Research Note NC-27. North Central Forest Experiment Station, St. Paul, Minnesota, USA.

———. 1967b. Size and placement of metal culverts critical on peatland woods roads. U.S. Forest Service Research Note NC-37. North Central Forest Experiment Station, St. Paul, Minnesota, USA.

Tilton, D. L. 1975. The growth and nutrition of tamarack (*Larix laricina* (du Roi) K. Koch). Ph.D. thesis. University of Minnesota, Minneapolis, Minnesota, USA.

Tomberg, U. 1957. Certain hydrophysical properties of peat soils and their alteration on reclamation under conditions prevailing in the Estonian SSR. (Translated from Russian.) Izvestiya Akademii Nauk Estonskoi SSR, Ser. Biol. 6(3):225-233.

U.S. Department of the Interior. 1909. A detailed report of a drainage survey of certain wet, overflowed, or swampy lands ceded by the Chippewa Indians in Minnesota. 61st Congress, 1st Session. U.S. House of Representatives Document no. 27.

Van der Molen, W. H. 1975. Subsidence of peat soils after drainage. Pages 183-186 *in* Hydrology of marsh-ridden areas. Proceedings of the Minsk Symposium, June 1972. Unesco Press, Paris, France.

Vesall, D., R. H. Gensch, and R. Nyman. 1947. Beaver-timber problem in Minnesota's "Big Bog." Conservation Volunteer 10(57):45-50.

PART IV
Two Studies of Ecological Development

CHAPTER 13

Development of a Raised-Bog Complex

Jan A. Janssens,
Barbara C. S. Hansen,
Paul H. Glaser, and Cathy Whitlock

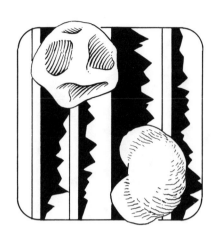

This chapter outlines the detailed development of the west-central bog complex of the Red Lake peatland, an example of a midcontinental forested bog complex, which was studied by means of fossil bryophyte, pollen, and radiocarbon analysis. The Red Lake peatland in northern Minnesota, with an area of 140 km² (55 square miles), contains the largest bog complexes in the United States outside Alaska. The intriguing differentiation in landforms and their pattern are described in detail by Heinselman (1963) and Glaser *et al.* (1981).

As early as 1928, Auer suggested that paludification (swamping) might be extensive in North America and mentioned Minnesota as an example (Heinselman 1963). After extensive surveys, Heinselman (1963, 1970) concluded that most of the Lake Agassiz peatlands have developed on gently sloping substrata through paludification rather than lake infilling. This chapter is a contribution to the task Heinselman defined as understanding the actual developmental history by elucidating the relationship among surface vegetation and landforms with topography and peat stratigraphy.

Knowledge of the environmental requirements of the bryophyte flora (Gorham and Janssens in press; Janssens 1989; Janssens and Glaser 1986; Vitt and Slack 1984) and the vascular plants and their pollen assemblages (Wheeler *et al.* 1983; Janssen 1984; Griffin 1975, 1977) provide the foundation for interpreting the macro- and microfossil assemblages identified in the peat cores, and thus the developmental history of the bog complexes. The conclusions can be applied across a much broader region of midcontinental lowlands of North America through aerial-photo interpretation.

Methods

Twelve peat cores 10 cm in diameter and 278-451 cm long were collected along two transects, one about 5 km long oriented parallel (Fig. 13.1, sites A to H) and one about 0.7 km long perpendicular to the slope (Fig. 13.1, sites I to L) (Wright *et al.* 1984). The topography of the surface and the mineral subsurface was initially interpreted from previous surveys (Farnham and Grubich 1966; Heinselman 1963; Soper 1919), and an additional survey was made with an electronic laser level to determine peat surface elevation (Almendinger *et al.* 1986) and with a Davis soil sampler to measure peat depths.

Pollen and macrofossil analysis strongly complement each other (Rybnicek 1973; Birks 1976; Birks and Mathewes 1978), and identification of pollen, wood, seeds, and needles and the quantitative analysis of bryophyte remains in the cores made it possible to reconstruct the succession of the peat-generating communities at each coring site (Janssens 1983a, 1989). The chronology of these changes was determined by 46 radiocarbon dates (Table 13.1) and by correlation of pollen zones. All dates have been converted from radiocarbon years B.P. (before 1950) into calendar years before 1981, the sampling year of most of the cores (yr B.S., years before sampling) by the calibration program CALIB of Stuiver and Reimer (1986).

The common peat-forming bryophytes exhibit clearly circumscribed habitat requirements, and their positions along moisture, trophic, and shade gradients are well known (chapter 4; Gorham and Janssens in press; Janssens and Glaser 1986; Vitt and Slack 1984; Vitt and Bayley 1984; Horton *et al.* 1979; Sjörs 1952). Quantitative bryophyte analysis makes estimates of pH and water level possible, and paleoenvironmental

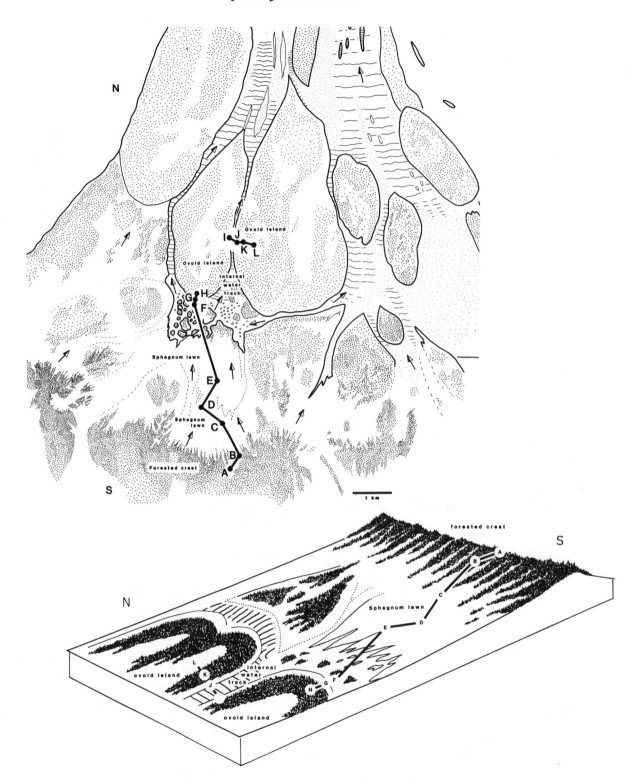

Fig. 13.1. Landforms in the west-central bog complex of the Red Lake peatland (48°15′ to 48°22′N, 94°34′ to 94°49′W). The water drains (*arrows*) from the forested bog crest northward across nonforested *Sphagnum* lawn into waterlogged, moderately minerotrophic areas with circular forested islands. Runoff from the *Sphagnum* lawns is then channeled between large lobes of raised bog, the ovoid islands, with a rim of forest and a nonforested center. Two transects of cores have been completed to reconstruct the development of the complex. The south/north transect (5325 m long) extends from site *A* near the bog crest downslope across the poor-fen *Sphagnum* lawn and adjacent water track to the head of an ovoid island (site *H*). An east/west transect (670 m long) crosses a water track between two ovoid islands (sites *I* to *L*).

Table 13.1. Accepted radiocarbon-date samples for the 10 Red Lake peatland cores of the west-central catchment

Site (see Fig. 13.1) Lab. #	Increment (cm)	^{14}C date and error (yr B.P.)
RLP8201 (A)		
SI-5985	62- 66	1255 ± 85
SI-6191	102-106	1755 ± 50
SI-6647	200-205	2060 ± 60
SI-6648	302-310	3545 ± 55
SI-5990	341-345	3780 ± 95
RLP8101 (B)		
SI-5962	119-121	1200 ± 60
SI-5963	242-245	2060 ± 85
SI-5964	318-324	2550 ± 80
SI-5965	339-341	2920 ± 90
SI-5966	380-382	3130 ± 105
SI-6641	438-443	4465 ± 55
SI-5967	447-450	6530 ± 270
RLP8304 (C)		
SI-6651	190-200	2010 ± 105
SI-6652	295-302	2635 ± 75
SI-6189	378-381	4205 ± 120
RLP8109 (D)		
SI-5599	167-172	1375 ± 45
SI-5600	229	2430 ± 50
SI-5601	312	2865 ± 55
SI-5602	416-420	4820 ± 50
RLP8305 (E)		
SI-6653	93-103	1195 ± 60
SI-6654	199-209	1970 ± 85
SI-6188	278-281	2540 ± 65
RLP8103 (F)		
SI-5439	87- 90	765 ± 65
SI-6478	150-155	1900 ± 60
SI-5441	223-227	2320 ± 70
SI-6643	268-278	2960 ± 40
SI-5442	300-305	3905 ± 55
RLP8104 (G)		
SI-6479	100-105	1185 ± 60
SI-5977	298-300	3060 ± 50
SI-5675	308-313	3080 ± 95
RLP7812 (I)		
SI-5677	95-105	1370 ± 70
SI-5678	235-245	2475 ± 75
SI-5679	365-375	4590 ± 60
RLP8102 (J)		
SI-5968	118-122	1405 ± 50
SI-6642	146-154	1840 ± 50
SI-5438	288-292	2655 ± 60
SI-5969	320-322	3635 ± 75
SI-5698	335-339	5250 ± 65
RLP8111 (K)		
SI-6644	197-202	2635 ± 95
SI-5981	336-338	3070 ± 75
SI-6645	350-353	3785 ± 95
SI-5435	380-388	3965 ± 60
RLP8112 (L)		
SI-6480	120-125	1420 ± 55
SI-5431	253-258	2040 ± 50
SI-5984	322-325	2685 ± 50
SI-5432	393-398	3320 ± 50

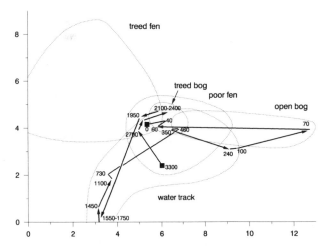

Fig. 13.2. DCA ordination of Minnesota bryophyte populations, classified by habitat, superimposed by the trace of core 8104 fossil assemblages as passive samples (see explanation in Methods section).

conditions can be reconstructed in detail from these cores (Janssens 1983a, 1989). Fig. 13.2 illustrates the paleoenvironmental reconstruction for the fossil assemblages of core 8104 (site G, discussed in detail below). The reconstruction is traced upon a detrended correspondence analysis (DCA) ordination (see chapter 4) produced from the modern data-set of Minnesota bryophyte populations. The DCA (active samples or reference data-set) and trace (passive samples) are produced with the program CANOCO (ter Braak 1987), and more details are provided in chapter 4. The trace illustrates in a qualitative way — in contrast to the quantitative inference curves presented below — the development of the peatland at site G from one bryophyte community to another. It also shows graphically the reversal from bog to fen habitat characteristic during the development of the peatland at this site. This figure provides the visual link between the autecological data-set on bryophytes discussed in chapter 4 and the paleoenvironmental reconstruction based on fossil mosses presented in this chapter.

Physical Variables: A number of physical variables, such as bulk density (dry mass per wet volume as determined in the field), ash content (ash mass as percent of total dry mass), and absorbance of a pyrophosphate extract (light extinction at 350 nm in 1-cm cells) were determined (Janssens 1989). These and microscopic peat-component analyses are routinely included because they sometimes indicate major biostratigraphic zones in the peat profile and because they also suggest the sampling interval needed for the more time-consuming analyses of biological variables (pollen, bryophytes, and other macrofossils) in long core sections. Bulk density, ash content, and absorbance are also partly correlated with the degree of humification of peat. Correlations among peat types, humification degree, and peat accumulation rate are poor, however. Different ages and secondary processes such as autocompaction, variable acrotelmic aerobic decomposition, and anaerobic decomposition in the catotelm (Aaby and Tauber 1975) will reduce any direct correlation between the original growth rates and the productivity of peat-producing communities and physical peat characteristics of the resulting peat (R. S. Clymo, personal communication). Preliminary estimates of the rate in which mass enters the catotelm (p_c) and of anaerobic decay (α_c; Clymo 1984) are very similar for most peat types (Gorham *et al.*, in manuscript).

Pollen: Extensive information on the differentiation of local from regional pollen is now available for north-central Minnesota (Janssen 1967a, 1984; Griffin 1975), and detailed pollen diagrams from five coring sites (Fig. 13.1: B, G, I, K, and L) in the west-central catchment of the Red Lake peatland are used in the interpretation presented here. The tough fibrous matrix of undecomposed peat prevented the use of the usual technique for subsampling the core for pollen analysis (Birks 1976). Instead, an accurate volume of a microfossil subsample was estimated from field measurements of total peat volume of the monolith or core section and then determined in the laboratory from the wet weight of the subsample (Janssens 1989). Samples for pollen analysis were processed by the acetolysis method (Faegri and Iversen 1975) as modified by Cushing (1963). Samples were mounted in silicone oil, and counts were made until a pollen sum of 400 to 600 local and regional pollen grains was reached. Pollen percentages for both upland and lowland taxa are based on the upland pollen sum alone, as determined by Janssen (1967b, 1984). Poaceae (grasses) and *Betula* are not included in the pollen sum, and they are plotted on the diagrams as ecologically indeterminable, for Poaceae could include *Zizania* (wild rice) as well as upland grasses, and *Betula* could include both the lowland dwarf birch (*B. pumila*) and the upland white birch (*B. papyrifera*). The pollen of *Picea mariana* and *P. glauca* (black and white spruce) were separated for cores analyzed by Hansen according to morphological criteria outlined in Hansen and Engstrom (1985). Pollen of *Fraxinus nigra* (black ash), *F. pennsylvanica* (green ash), *Alnus rugosa* (speckled alder), and *A. crispa* (green alder) was separated on the criteria of Cushing (1963).

Bryophytes and Other Macrofossils: High concentrations of resistant lignified vascular-plant structures, such as seeds, are found only in the extreme basal peat, where they are significant for the reconstruction of paleoenvironment (Griffin 1977). Bryophytes, however, are extremely abundant throughout the rest of each Red Lake core. Moss remains form the major component of the fen and bog peats; the rest is unidentifiable rootlets and fungal hyphae (Janssens 1983a).

After the volumetric contribution of the diverse peat components — *Sphagnum*, mosses, wood, herbaceous remains, rootlets, and detritus — in a dispersed coarse fraction was estimated, the macrofossils and *Sphagnum* and moss remains were extracted for identification. Stem fragments of *Sphagnum* plants with attached fascicles and leaves were picked out by forceps and prepared with the techniques described for living or herbarium material (Janssens 1989). Most frequently, however, fragmented material of *Sphagnum* was extracted with a syringe (without needle) from the dispersed screened subsample in the macrofossil tray. Detached branch and stem leaves were recovered from the water by screening the content of the syringe through a small 300 μm mesh sieve that retains all identifiable fragments. Fragments were stained by immersion of the entire screen in gentian violet for a few seconds, followed by repeated rinsing. A subsample of the stained leaves was tallied under the high power of a stereoscopic microscope or mounted on a microscope slide and studied under a compound microscope. Branch leaves belonging to each of the *Sphagnum* sections were tallied. Stem leaves included in the preparation or extracted separately from the batch of stained leaves were then used to confirm or identify species belonging to each of the *Sphagnum* sections (Janssens 1989).

Most *Sphagnum* remains were identified to species. Exceptions are the members of the *S. recurvum* aggregate (*S. angustifolium*, *S. fallax*, and *S. flexuosum*). With fragmented fossil material it was possible to differentiate among these species of the aggregate only when stem leaves are preserved. A maximal pore-size differentiation in the branch leaves of the spreading branches exists (Klinger 1968), but the position that a detached fossil leaf occupied along a branch is not known, so this character cannot be used.

Well-preserved moss fragments other than *Sphagnum* were temporarily stored in glycerine and prepared with techniques similar to those used for herbarium material. Voucher slides were prepared after enclosing the material in Hoyer's mounting medium (Schuster 1966) and are deposited in the senior author's personal herbarium. A review of identification literature is given in Dickson (1986), but an extensive reference collection of herbarium material and microslides is indispensable.

The proportion of Bryopsida and *Sphagnum* species identified was calculated using the volumetric estimates of both components and the counts of individual fragments (see Janssens 1989: Table 3.2 for details). This technique of quantification partly circumvented the problems discussed by Watts and Winter (1966), Rybnicek and Rybnicková (1968), and Rybnicek (1973) of quantification of macrofossils with dissimilar structures (seeds versus leafy fragments versus detached leaves). Differential fragmentation and size of fragments among different *Sphagnum* species (such as *S. magellanicum* versus *S. recurvum* agg., Rybnicek and Rybnicková 1968) is, however, still ignored, although the proportion of detached leaves is used to calculate relative abundance. Selective decomposition (Clymo 1965) is also ignored.

Quantification of bryophyte remains and other macrofossils is also discussed by Janssens (1989, 1983a), Barber (1981), and Watts (1978). A reliability index for fossil bryophyte fragments is explained by Janssens (1989), and this manual also includes dichotomous keys for the identification of fossil fragments of *Sphagnum* and *Drepanocladus*.

Inference: Quantitative paleoenvironmental reconstruction (Birks and Birks 1980; Birks 1985, 1987; Birks *et al.* 1990) consists of the reconstruction (inference) of environmental variables, such as pH or HMWT (height [of the peat surface] above the mean water table), from fossil assemblages. Initially, a modern reference data-set of living populations associated with environmental variables is needed (chapter 4; Janssens 1989; Janssens and Glaser 1986). These results can then be used to predict values for the environmental variables from the fossil assemblages. In the following the reference data-set on North American bryophyte populations as it presently exists is described and the inference techniques and some results are summarized.

The major environmental variables explaining most of the biologically relevant variability in the North American bryophyte reference data-sets are pH and HMWT. The relationship between these two major peatland gradients is illustrated in Janssens (1989) and — for the Minnesota data-set — in chapter 4.

For the inferences presented in this chapter, the combined North American data-sets have been used to derive the transfer functions for pH and HMWT. Transfer functions derived from individual regional data-sets compare very well with the combined North American data-sets.

Birks (1985, 1987) reviews the numerical procedures to obtain the inferences, and Birks *et al.* (1990) conclude that the statistical calibration technique of weighted averaging (WA) gives superior results in the prediction of pH. Line and Birks (1989) and Birks *et al.* (1990) discuss the underlying theory of WA regression and calibration and compare its performance with computationally more difficult techniques such as maximum likelihood regression and calibration. The following is a summary of the major ecological assumptions underlying WA, used for quantitative paleoenvironmental reconstruction, from Birks *et al.* (1990), discussed with fossil bryophytes in mind.

(1) The environmental variables to be reconstructed, pH and HMWT, have to be ecologically the most important variables or the best parameters to reconstruct the gradients of interest in peatlands. This is supported by the literature review in Janssens (1989) and Gorham and Janssens (in press). It was also illustrated in chapter 4 by the results of the CCA (canonical correspondence analysis) ordination — constrained by pH, HMWT, and shade — of the Minnesota data-set.

(2) The bryophyte species in the reference data-set have to be the same as in the fossil assemblages, and their ecological response must not have changed significantly over the late-Holocene time span represented by the peatland profiles. Paleoenvironmental reconstruction is discussed by Janssens (1990), who explains the rare occurrence of ecotypes and nonanalogous fossil assemblages. It suffices to indicate here that no species are recorded as fossils in the Holocene peat deposits of North America that are not found as living populations. In addition, few fossil assemblages have no representatives among the plots of the reference data-set (Fig. 13.2).

(3) The mathematical method used in WA should adequately model the biological response to the environmental variables of interest. Gorham and Janssens (in press) clearly indicate that most bryophyte species have a well-defined optimum and a species-specific tolerance along the pH gradient. The WA estimate of the optimum (= abundance weighted mean) of species k is:

$$\hat{u}_k = \sum_{i=1}^{n} y_{ik} x_i \Big/ \sum_{i=1}^{n} y_{ik}$$

and its tolerance (weighted standard deviation) is:

$$\hat{t}_k = \left[\sum_{i=1}^{n} y_{ik} (x_i - \hat{u}_k)^2 \Big/ \sum_{i=1}^{n} y_{ik} \right]^{1/2}$$

where

x = the environmental variable (pH or HMWT)
x_i = the value of x in sample i (a plot)
y_{ik} = the abundance of species k in plot i (i = 1, ..., n plots; k = 1, ..., m moss species).

Table 13.2 lists the WA optima and tolerances for pH and HMWT of the most important fossil species present in the Red Lake peatland cores. These values are based on the combined North American data-sets, in contrast to those in chapter 4, which presents only the Minnesota data-set.

Inference results for five of the cores of the Red Lake peatland transect are presented in the figures of the Red Lake Peatland Development section that follows.

Red Lake Peatland Development

Twelve cores along the Red Lake west-central catchment were analyzed for fossil bryophytes and physical variables. In addition to the analysis of physical variables, bryophytes, and macrofossils, pollen was also analyzed for five cores along the transect. The paleoenvironmental results of all the analyses for these five cores are discussed later in the Appendices.

A comparison of the upland and lowland pollen sequences from the west-central catchment suggests that

Table 13.2. Optima, tolerances, and occurrences for pH and HMWT calculated with WACALIB 2.1 for the most common bryophyte species of the combined North American reference data-sets (total n = 431, total m = 271)

species	pH			HMWT		
	$\hat{u}_k$	$\hat{t}_k$	n	$\hat{u}_k$	$\hat{t}_k$	m
Sphagnaceae						
S. fallax	4.27	0.68	32	12.5	10.2	22
S. fuscum	4.35	0.73	46	27.5	9.5	25
S. capillifolium	4.33	0.71	83	19.3	7.4	41
S. majus	4.40	0.44	20	6.6	4.7	18
S. angustifolium	4.69	0.89	99	19.5	8.9	58
S. papillosum	4.33	0.38	59	12.0	6.8	41
S. magellanicum	4.39	0.65	103	19.4	8.3	65
S. centrale	5.68	0.99	27	19.9	6.9	10
S. teres	5.78	0.88	24	12.7	8.5	6
S. subsecundum	5.29	0.43	31	10.0	4.9	22
S. warnstorfii	6.77	0.47	33	19.2	8.7	9
Amblystegiaceae						
Drepanocladus fluitans	4.66	0.70	12	3.7	7.0	8
Calliergon stramineum	5.03	0.72	46	11.9	8.2	25
C. cordifolium	6.17	0.62	15	7.4	7.5	4
Calliergonella cuspidata	6.19	1.05	14	–	–	0
Calliergon giganteum[a]	6.91	0.35	11	6.0	3.1	7
Campylium stellatum	6.56	0.78	29	10.8	5.9	2
Scorpidium scorpioides[a]	6.65	0.71	16	3.3	2.4	8
Drepanocladus revolvens[a]	6.88	0.54	10	8.7	4.2	6
Amblystegium riparium	6.79	0.44	17	3.6	–	1
Others						
Mylia anomala	4.11	0.26	25	20.7	8.4	12
Pleurozium schreberi	4.95	1.34	73	26.9	9.7	37
Polytrichum strictum	4.71	1.01	36	29.7	9.2	18
Dicranum undulatum	4.91	1.31	22	24.3	8.2	8
Cladopodiella fluitans	4.39	0.30	32	6.5	3.6	22
Aulacomnium palustre	5.47	1.15	86	19.5	8.5	40
Helodium blandowii	6.96	0.41	26	11.2	11.7	3
Plagiomnium ellipticum	6.88	0.61	22	13.0	11.4	6
Hylocomium splendens	5.90	1.38	21	19.4	13.4	11

[a]The height optima and tolerances for these three species is above the *local* water table, because no plots with these species were available in the *Sphagnum angustifolium* standardized *mean* water table dataset.

Source: Janssens 1989.

major fluctuations in the wetland vegetation through time are closely related to regional vegetation succession. Regional vegetation fluctuations are related primarily to climatic change and more recently to human influence. In the peatlands, besides the climatic impetus, there is also the local vegetation dynamics, propelled by changes in regional hydrology, water chemistry, and the autochthonous (*in situ*) growth of the peat itself. Three major pollen zones are delineated on the basis of regional pollen assemblages (Fig. 13.3). Local pollen succession is more complex, so local zones have been further delineated in the individual pollen diagrams (Appendixes 1B to 5B). The local pollen succession complements the bryophyte data and supports the following paleoenvironmental reconstruction.

When the peatlands started to develop in the Lake Agassiz lowlands (8,000? to 3,000 years ago; Heinselman 1963, 1970; Griffin 1977), the upland landscape in the Red Lake area was still covered by prairie and oak savanna, as inferred from high pollen percentages of grass, *Artemisia*, *Ambrosia* type, oak, and Chenopodiineae (Fig. 13.3, Regional Zone 3, corresponding to regional zones in chapter 14; Griffin 1977). Prairie pothole peatlands were probably present before large-scale paludification, and there is no record of earlier extreme peatland cover, which must have decayed or burned away during the hypsithermal interval, the mid-Holocene optimum (Heinselman 1963, 1965; Griffin 1977). Nor is there evidence along the Red Lake transects of extensive forest cover before paludification. Wood remains are common only after the initial sedge-meadow phase in most cores (Fig. 13.4; see also Heinselman [1963, 1965] and Griffin [1977] about the interpretation of buried wood in paludified peatlands). Basal wood remains appear more common toward the east of the Red Lake west-central catchment (Heinselman 1963) but are absent in the patterned peatland near Seney, in upper Michigan (Heinselman 1965). Pine and deciduous trees other than oak were not important components of the uplands. Sedge meadows were interspersed with cattail swales, and small intermittent pools held aquatics such as *Potamogeton*, *Utricularia*, and *Myriophyllum*. These aquatic plants flourish under solute-rich conditions. Other pollen sites in northern Minnesota repeat this regional vegetation picture with individual variations in the local pollen flora (Cushing 1963; Shay 1967; McAndrews 1966; Janssen 1967a, 1968; Griffin 1977).

The sedge-meadow assemblage in these basal peats contains macroremains of *Typha*, *Carex*, *Scirpus*, *Cladium mariscoides*, *Rumex maritimus*, *Rhynchospora*, *Potamogeton*, and *Myriophyllum*. Bryophytes are either absent or represented in low quantities. This basal peat is overlain by rich-fen moss/sedge peat: the dominant mosses *Scorpidium scorpioides* and *Calliergon trifarium*, associated with *Campylium stellatum*, *Bryum pseudotriquetrum*, *Drepanocladus revolvens*, and *Meesia triquetra*, replaced the sedge-dominated communities in some cores as early as 5500 yr B.S. At some sites, however (Fig. 13.4), this primary peat (*sensu* Sjörs 1983) was possibly initiated more than 5,000 years ago, but the dates are uncertain because the contact of autochthonous peat with mineral sediment is poorly defined, and 'old' carbon contained in fragments of Cretaceous lignite may have been present in the underlying mineral soil (H. E. Wright, Jr., personal communication). No Holocene lake sediments of any significant extent have been probed in the west-central catchment, and the developmental sequence is entirely by paludification (Sjörs 1983) rather than partly by infilling, as is evident in the Myrtle Lake peatland (chapter 14).

By 3500 yr B.S., the climate began to shift toward a moister regime over much of northern Minnesota, and

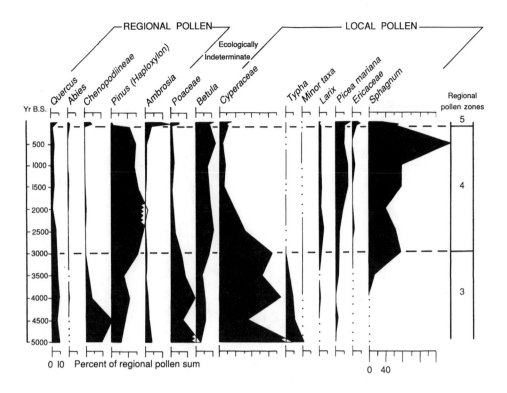

Fig. 13.3. Summary percentage pollen profiles versus calibrated age in years before sampling (B.S.). The values plotted are five-core averages at each 500-year interval for the selected regional and local pollen types. The regional pollen sum is the same as the upland pollen sum in Appendices 1B to 5B and discussed in the Methods section. The regional pollen zones correspond with Janssen's (chapter 14) and are discussed in the text (Summary of Stratigraphy).

in response the regional vegetation shifted from an oak savanna/prairie to a mixed conifer/deciduous forest, as noted in the transition to a regional pollen assemblage dominated by pine pollen (predominantly haploxylon type *Pinus strobus*), together with the occurrence of *Picea glauca, Abies,* and *Ostrya/Carpinus,* and *Acer* pollen (Fig. 13.3, Regional Zone 4). At Stevens Pond the succession from an oak savanna/prairie-dominated upland to a mixed deciduous/pine forest is credited with changing the inflow of nutrients, allowing the introduction of more acidophilous species (Janssen 1967b). The change to a more acidic and less minerotrophic environment in the Red Lake peatlands is corroborated both by acidophilous bryophyte species as well as the lowland pollen types *Picea mariana, Larix,* Ericaceae, and *Sphagnum.*

In the central part of the catchment (Fig. 13.4A: sites F-H), bryophyte-rich swamp forest initially was prevalent, and its peat is characterized by the mosses *Calliergon aftonianum* (cf. *C. richardsonii*), *C. cordifolium, Tomenthypnum nitens, Sphagnum warnstorfii,* and *S. teres.* In the southern part (Fig. 13.4A: sites A to D) the fens were abruptly supplanted by poor-fen and bog forest (3,000-2,000 years ago, except at site A, where oligotrophic peats originated prior to 4000 yr B.S.), but northward (Fig. 13.4B: sites I to

L) they were more gradually replaced by *Sphagnum*-dominated intermediate fens (*Sphagnum contortum, S. subsecundum,* and *S. platyphyllum*: 2,500-1,500 years ago).

In the Red Lake peatlands, the major shift from a sedge-dominated fen with minerotrophic lowland plants follows the upland transition to the mixed conifer/deciduous forest in Regional Zone 4 (3000 yr B.S., Fig. 13.3). Between 3,000 and 2,000 years ago, the minerotrophic species declined first in the area of the present-day crest and gradually toward the ovoid islands (site B to L, Figs. 13.4, 13.5) and oligotrophic mire became established. The bog/poor-fen forest extended downslope and northward from the location of the bog crest (site A) to ovoid island sites (site L). The farther northward extent of this oligotrophic mire is uncertain, however, and more closely spaced coring sites in the ovoid-island area are needed to establish the continuity of the early bog forest. The bog forest was gradually replaced after 2,000 years ago by a nonforested (open) bog and poor-fen *Sphagnum* lawn downslope from the forested crest (Figs. 13.4A, 13.5). This reconstruction is consistent with the hypothesis in Glaser and Wheeler (1980) and Glaser *et al.* (1981) that the bog drains expand and coalesce downslope to form broad *Sphagnum* lawns on the lower bog flanks. The

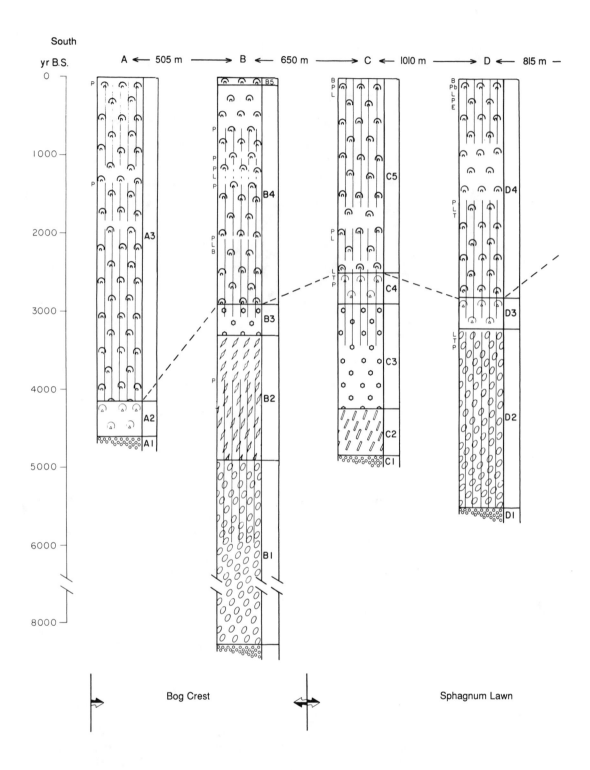

A.

Fig. 13.4. Correlation diagram of the 12 cores along the north-south (**A**) and east-west (**B**) transects through the west-central bog complex at Red Lake. The vertical axis of the columns is in calendar years before sampling date (yr B.S.). The distance in meters between each core pair is listed at the top of the diagram, including the landforms presently at the surface of the transects. The columns summarize the paleoenvironmental reconstruction for all cores based on information as presented in

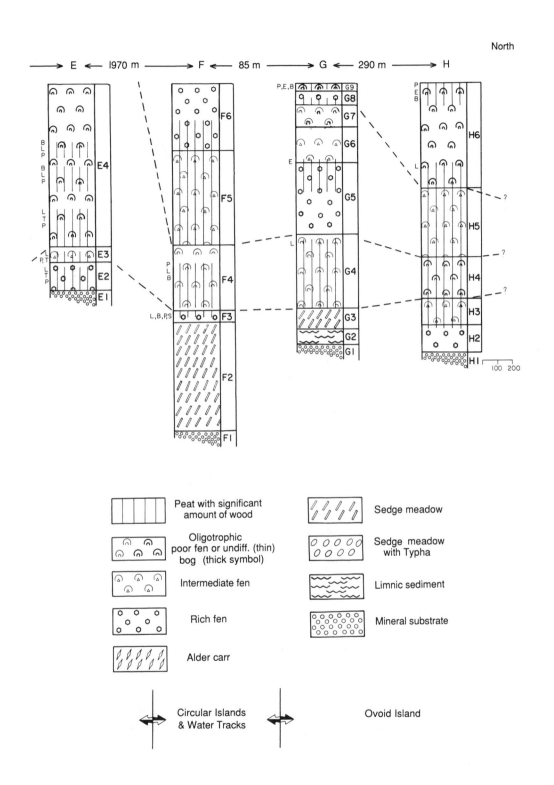

Appendices 1 to 5 for sites B, G, J, L, and K. The original vegetation is classified as (1) oligotrophic mire (bog, poor fen, or undifferentiated bog/poor fen), (2) intermediate fen (minerotrophic *Sphagnum* species are a major component), (3) rich fen (Amblystegiaceae or brown mosses are significant), (4) alder carr, (5) sedge communities (sedge meadow and sedge-cattail), and (6) aquatic community, depositing limnic sediment. The oligotrophic-minerotrophic boundary of each column is connected by the *broken*

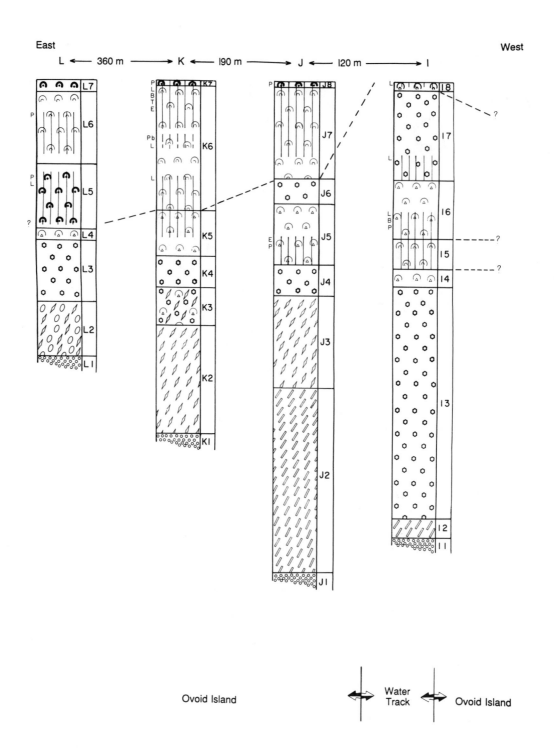

line and discussed in the text. The presence of wood remains is superimposed on the vegetation-reconstruction symbol by *vertical hatching*. Wood identifications are listed near the left upper corner of each discrete wood zone (P = *Picea*, L = *Larix*, T = *Thuja*, B = *Betula*, Pb = *Pinus banksiana*, E = Ericaceous wood remains, S = *Salix*).

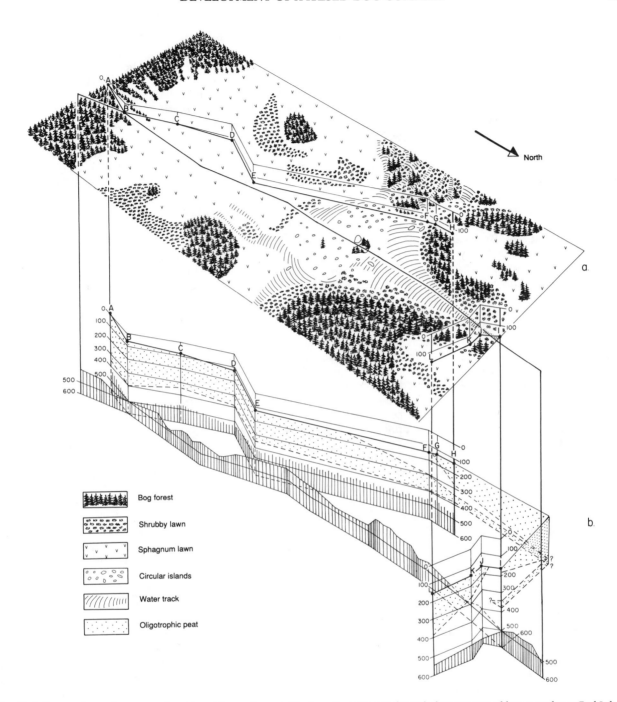

Fig. 13.5. Present-day surface vegetation (a) and fence diagram (b) along the transects through the west-central bog complex at Red Lake. The present-day landform pattern in (a) is an interpretation from air photos and topographic map (Chase Brook SW quadrangle). The area reconstructed is 2.4 km wide (along the east-west axis) and 6.0 km long (north-south axis). In (b), vertical relief is exaggerated by a factor of approximately 150. Three fences are shown: (1) the subsurface topography (*vertical hatching delimited by horizontal line*) is a reconstruction by laser leveling and peat-depth measurements along a north-south direction by J. C. and J. E. Almendinger in March 1984. Absolute elevation above sea level is recalculated to a 0-m datum at the surface of core site A (for practical purposes the crest of the radiating bog complex). The influence of subsurface relief on the peatland-landform development is discussed in the text. (2) The south-north core transect from sites A to H indicates the slope of the bog complex along the transect. This is also reproduced in (a) (140 cm below the 0-m datum at core site H). The contact of the oligotrophic-minerotrophic peat and the mineral subsurface contact is indicated along the fence, although the mineral subsurface has a far lower resolution than the one along the leveling transect and is hatched vertically without a delimiting horizontal line. (3) The west-east fence indicates the mineral subsurface and present-day peat-surface relief along the transect of core sites I to L. The oligotrophic-minerotrophic boundary is also indicated, and oligotrophic peat is indicated by the dotted pattern.

stratigraphy also indicates that the internal water tracks in this bog complex are not remnants of the original sedge fen but apparently developed by replacing more oligotrophic communities and became narrower and better defined by subsequent enlargement of the bog islands. These bog islands have expanded over highly minerotrophic drains from remnants of oligotrophic communities not completely obliterated by the minerotrophic channels. This is in contrast to Hofstetter's hypothesis (1969), which suggested that the initial bog communities developed in depressions in a large fen expanse over a thicker, more insulating peat layer.

The poor-fen *Sphagnum* lawn and nascent ovoid islands (their earliest basal peat dated at 2,100 years ago at site L, Fig. 13.4B) became separated by developing internal water tracks (sites G and H, Fig. 13.4A) that channeled increased surface flow from the lawn and bog drains. The subsequent expansion of the ovoid islands gradually restricted the water tracks to their present position and width. Highly minerotrophic communities (sites G and I) initially occupied the widened water tracks, consistent with the discharge of highly minerotrophic waters driven by the hydraulic-flow cells that penetrate deep into the underlying mineral till and are generated by the bogs (Siegel 1983). *Scorpidium scorpioides* rich-fen peats underlying the present-day bog peat of the ovoid-island margins (sites G and H) or underlying the intermediate-fen peat of strings in the present water track (sites I and J) contain abundant remains of such moss species as *Calliergon trifarium* and *Drepanocladus lapponicus*, which today do not occur in the bog complex (Janssens and Glaser 1986; Janssens 1983a) but are characteristic of the rare, highly minerotrophic spring fens in northern Minnesota. Spring fens are domed and forested rich-fen mounds with drains formed by discharge of upwelling minerotrophic groundwater (Glaser *et al.* 1990). Today, however, the water chemistry and communities in the water tracks are less minerotrophic (Glaser and Janssens 1986; Janssens 1983a; Glaser *et al.* 1981). and the poor-fen strings probably developed quite recently by coalescence of hummocks in response to increased water flow through the narrowing water tracks (Foster and King 1984). This decrease in minerotrophy during the past 500 years is apparently caused by (1) the blanketing of the vast landscape of the Lake Agassiz plain by thicker and more oligotrophic peat, resulting in an increase of the output of oligotrophic water over that of mineral groundwater, and (2) the growth of the ovoid bog islands, decreasing the area of the water-track drainage.

By the late 1800s and early 1900s, logging and farming were changing the regional picture. A precipitous drop in haploxylon pine pollen together with the rise of *Ambrosia* type pollen (Fig. 13.3, Regional Zone 5) reflects the logging of white pine and the forest clearance in new agricultural areas to the west. This event is seen in the lowland pollen assemblages, for black spruce and Ericaceae together with sedges and cattails create a pollen assemblage characteristic of both bog and poor-fen conditions. The ovoid islands also expanded slightly after the abortive ditching efforts of 1913-1920 (Wright 1984), as is evident from the interpretation of aerial photographs, stratigraphic analysis, and tree-ring studies (chapter 12; Fig. 13.4A, site G, and Fig. 13.4B, site I). Ditching or a drier climate would reduce the recharge function of the local flow cells and therefore reduce the input of minerotrophic water in the water tracks (Siegel 1983). The resulting drop in HCO_3 concentration of the mixture of basic groundwater with acidic runoff from the bog allows *Sphagnum*, intolerant of high cation concentrations and pH (Clymo 1973), to expand and restrict the water tracks further. Detailed stratigraphical analysis of the ovoid island margins should make it possible to time previous hydrological changes in the bog/fen complex and by inference evapotranspiration and human influences.

Discussion

The Fossil Bryophyte Record

Table 13.3 compares the fossil record of mosses in the 12 cores of the west-central catchment with the present-day occurrence of these bryophytes as living populations in the catchment. Thirty-one species and one aggregate taxon — *Sphagnum recurvum* (= *S. angustifolium* + *S. fallax* + *S. flexuosum*) — have a fossil record. They are ranked in Table 13.3 descending from highest frequency of occurrence (total of 650 fossil assemblages). The mean-abundance class of species that have a frequency greater than 1% is calculated on a five-point scale assigned to the percent representation of the fossil taxon in the assemblage, similar to the coverclass assigned to a taxon in a plot (see Methods section). *Sphagnum magellanicum* has the highest frequency of occurrence in fossil assemblages (19% of a total of 650) and is also a major component of most peat samples in which it occurs (mean abundance of 4.1 on a 5-point scale). Other species that usually contribute approximately half of fossil fragments in the assemblages in which they occur (abundance class > 4) are *Scorpidium scorpioides* and *Sphagnum platyphyllum*. In contrast, *Sphagnum centrale* (which is a very frequent fossil), *Campylium stellatum*, and *Polytrichum strictum* contribute only with low abundances (< 3) to the fossil assemblages in which they occur. Highly minerotrophic taxa such as *Calliergon trifarium* and *Scorpidium scorpioides*, but also oligotrophic bog/poor-fen dominants such as *Sphagnum magellanicum* are well represented in the fossil record. *Sphagnum magellanicum* fossils indicate that the minerotrophic groundwater influence has dropped consistently below a certain threshold, and further peat accumulation will be accomplished mainly by oligotrophic mire communities (Rybnicek and Ryb-

Table 13.3. Frequency of occurrence and mean abundance of fossil moss species in the 12 cores from the west-central catchment at Red Lake peatland compared with total record of present-day bryophyte populations collected in the catchment.

Species	Frequency of total fossil occurrence (total n = 650)	Mean abundance (5 point-) scale
Sphagnum magellanicum (8%)[a]	123 (19%)	4.1
Calliergon trifarium	73 (11%)	3.6
Scorpidium scorpioides (2%)[a]	63 (10%)	4.1
Sphagnum recurvum agg. (19%)[a]	48 (7%)	3.0
Sphagnum contortum	44 (7%)	3.5
Sphagnum centrale (1%)[a]	42 (6%)	2.7
Sphagnum papillosum (6%)[a]	37 (6%)	3.3
Sphagnum capillifolium (12%)[a]	25 (4%)	3.9
Sphagnum subsecundum (2%)[a]	23 (4%)	3.3
Campylium stellatum (3%)[a]	21 (3%)	2.7
Sphagnum fuscum (3%)[a]	19 (3%)	3.6
Meesia triquetra	16 (2%)	3.1
Sphagnum majus (<1%)[a]	16 (2%)	3.2
Sphagnum platyphyllum	10 (2%)	4.3
Calliergon stramineum	9 (1%)	3.2
Drepanocladus lapponicus	7 (1%)	3.6
Polytrichum strictum (4%)[a]	7 (1%)	2.7
Sphagnum teres	7 (1%)	3.9

Additional taxa:

Aulacomnium palustre (6%)[a], *Amblystegium riparium*, *Bryum pseudotriquetrum*, *Calliergon aftonianum*, *C. cordifolium*[a] (1%), *C. giganteum*, *Campylium polygamum*[a] (<1%), *Dicranum undulatum*[a] (2%), *Drepanocladus exannulatus*, *D. fluitans*, *D. revolvens*, *Plagiomnium ellipticum*, *Sphagnum warnstorfii*, and *Tomenthypnum nitens* each have less than 1% of the total number of records (650).

[a](x%): taxa presently found growing in the Red Lake peatland west-central watershed at 'x'% of total frequency (total of 238 collections). Taxa absent in the fossil record are *Callicladium haldanianum* (<1%), *Cladopodiella fluitans* (1%), *Dicranum flagellare* (<1%), *Dicranum ontariense* (2%), *D. polysetum* (3%), *Drepanocladus uncinatus* (<1%), *Hypnum pratense* (<1%), *Lophocolea heterophylla* (<1%), *L. minor* (<1%), *Mylia anomala* (1%), *Oncophorus virens* (<1%), *Plagiothecium laetum* (<1%), *Pohlia nutans* (1%), *Ptilidium pulcherrimum* (2%), *Pylaisiella polyantha* (1%), *Pleurozium schreberi* (8%), *Scapania irrigua* (1%), *Sphagnum pulchrum* (<1%), and *S. russowii* (<1%).

nicková 1968), except in rare occurrences of major hydrological change, such as at site G (Appendix 2A).

Table 13.2 also includes all bryophytes collected in the west-central catchment of Red Lake peatland from living populations. Nine localities and 238 collections are represented, covering all major present-day landforms. Thirty-three species and *S. recurvum* agg. (all three species) are recorded presently living in the catchment. The most common taxa are *S. recurvum* agg. (19%), *S. capillifolium* (12%), *S. magellanicum* (8%), and *Pleurozium schreberi* (8%). All four taxa commonly have high cover-class values on plots (Janssens and Glaser 1986).

Comparison of the present-day bryophyte flora of the Red Lake peatland west-central catchment with the fossil moss record reveals several interesting points. (1) Highly minerotrophic taxa are either absent (*Calliergon trifarium*) or poorly presented (*Scorpidium scorpioides*) among the present bryophyte populations, but are often well represented in fossil assemblages: a total of 16 taxa (half of the fossil record) are not found in the catchment any more, and 11 of them are rich-fen indicators. (2) Most extant species in the catchment that are not represented in the fossil assemblages (19 of 34 taxa) are rare or, if more commonly found, are high-hummock-top growers or epiphytes (e.g., *Pleurozium schreberi*, *Dicranum polysetum*, and *Ptilidium pulcherrimum*) corroborating the expected underrepresentation of hummock-growing taxa in peat. The only *Drepanocladus* sp. absent from the fossil assemblages is *D. uncinatus*, the most mesic of all taxa (Janssens 1983b). (3) None of the six liverworts recorded among the living bryophytes is recovered as a fossil, even the semiaquatic *Cladopodiella fluitans*. (4) The a priori exclusion of *Sphagnum* species during identification of fossil fragments on grounds of rarity or ecological incompatibility is indeed dangerous practice (Barber 1981; Wright and Janssens 1982) as Table 13.3 indicates: *S. contortum*, *S. platyphyllum*, *S. teres*, and *S. warnstorfii* are all represented as fossils but are not found in the catchment now. They are, however, represented in the large fen water track just westward from the west-central catchment.

The presence of *Scorpidium scorpioides*, usually considered a rich-fen indicator (Sjörs 1959, 1961, 1963; Miller 1980), in the present water tracks of the west-central catchment (Janssens and Glaser 1986) without its usual associates (Jasnowski 1957; Rybnicek 1966) illustrates two points of general ecological and paleoecological relevance: (1) even in mosses — plants without any deep-rooted structures — depauperate communities can continue for a long time after the conditions suitable for their establishment have ceased (biological inertia: Gorham 1957) and (2) individual taxa are poor indicators of macroclimatic or edaphic change (Miller 1980) or could have had past populations consisting of different ecotypes (Green 1968). Paleoenvironmental reconstruction based on a comparative approach involving as many taxa as possible is preferable to the 'indicator' species approach (Birks and Birks 1980). Practically, this is accomplished by computer programs for paleoenvironmental ordination and calibration such as CANOCO and WACALIB.

Peatland Development

This study supports the concept that bog development on a large scale (catchment and landform pattern: ±5 km) and in a coarse time frame (±500 years) is a climate response, but that over short time periods and on a small scale (i.e., a water-track string/flark pattern), local hydrological change influences water chemistry and clearly dominates the direction of development. The following discussion of the Red Lake peatland development takes into account these concepts and previous studies on peatland development.

Direct climate-correlated development is evident only in the nearly synchronous initiation (around 3000 yr B.S.) of less minerotrophic peats over a large part of the west-central catchment at Red Lake peatland (Figs. 13.4, 13.5) and in Myrtle Lake peatland (chapter 14; Heinselman 1970) and coincides with a major shift in the regional pollen spectra (chapter 14; Fig. 13.3; Ritchie 1983). Core sites where paludification started later, and where a shallower layer of peat had accumulated, show a slightly delayed conversion to oligotrophic mire (e.g., site E, Fig. 13.4A, situated above a topographic high, Fig. 13.5).

There is no evidence of any other nearly synchronous changes, certainly not in the timing of the initial paludification of northern Minnesota peatlands (8000? to 2500 yr B.S.) (see also Heinselman 1963, 1970; Griffin 1977): apparently local subsurface topography, associated hydrology, and the texture of the mineral substrate are the most important factors here. Similarly widely spread timing of individual lake infilling and paludification initiation is documented by Futyma and Miller (1986) and Miller and Futyma (1987). The initial deposition in lake basins in northern Michigan began during the period 6500-2800 yr B.P., triggered by a rise in lake levels that was itself caused by a general trend to cooler and wetter climate. Individual chronologies appear to relate rather to basin characteristics and autogenic processes than to the onset of climate change. Other allogenic factors at a lower level than climate change — beaver activity (Sjörs 1959; Futyma and Miller 1986) and deforestation (French and Moore 1986), for example — also play a role. Even after peat has started to accumulate, the sudden onset of natural acidification (part of the autogenic "hydrosere" development) is very different for individual sites: it can occur rapidly in localities poor in solutes, as demonstrated by the presence of *Sphagnum imbricatum* and *S. flavicomans* in late glacial sediments (Tolonen and Tolonen 1984).

Subsurface topography appears to have an effect on the position of landforms (see also the discussion on catchment divides in Heinselman 1970). The ovoid islands in the west-central catchment developed downslope of a rise in the mineral surface. Similarly, Janssen (chapter 14) shows that the raised bog in Myrtle Lake peatland developed downslope from a mineral rise. Heinselman (1965) discussed the position of forest islands in relation to fossil dunes near Seney, in upper Michigan. The effect of the topography of the mineral surface is also implicated in the present position of the bog margin in a domed bog with linear-convex crest near Cochrane, Ontario (Zoltai et al. 1988). Survey of peat depths and gross peat stratigraphy indicates the possible influence of a rise in the topography of the underlying mineral substrate. Just as in the Red Lake peatland west-central catchment, the entire basin was

initially filled by an open fen characterized by brown moss-sedge peat. Afterwards, sedge peat (possibly including *Sphagnum* section *Subsecunda* taxa characteristic of intermediate fen) was formed over the entire basin. In the smaller part of the basin behind the topographic rise of the mineral substrate, more oligotrophic bog/poor fen peat developed, eventually forming the present-day domed bog. At the present time, however, the fen part of the basin is forming brown moss-sedge peat similar to the initial peat deposited before the less-minerotrophic sedge peat deposition. Peat depths and presumably accumulation rates are similar to those in the Red Lake west-central catchment. No radiocarbon dates are available, so it is not possible to correlate the poor-fen extension at Red Lake with the sedge peat development at Cochrane, Ontario.

In contrast, Heinselman (1963) could not find any correlation with peat depth and "teardrop" island position in the western water track in Red Lake (see Glaser et al. 1981). Rather, peat depth was slightly greater under the islands because of their buildup above the surrounding water track. Griffin's work (1977) established the recent development of the minerotrophic teardrop islands on this nonforested water track, coinciding with the initiation of the ovoid bog islands in the west-central catchment.

Zoltai et al. (1988) describe the ovoid islands and other teardrop-shaped treed bogs in large expanses of fens under the classification of "northern plateau bogs." They are formed away from the main water seepage stream. Gross stratigraphy indicates three main layers: (1) a basal woody forest peat or well-decomposed fen or aquatic peat, (2) a middle layer of more or less oligotrophic forest or fen peat, and (3) a surficial layer of highly acidic, fibric *Sphagnum* peat. No time frame for the initiation and expansion of the plateau bogs is suggested, but the concept of a small initial nucleus and later expansion is also supported by Zoltai et. al. (1988) in their description of the boat-shaped bog-peat body for a plateau bog near Riverton, Manitoba.

The complex development of landform patterns, influenced indirectly by Holocene climate changes and directly correlated with local hydrological changes (such as the development of a "raised-bog lake" and bog-bursts), is also presented in detail by Casparie (1972). Similarly, Barber (1981) introduces the "phasic theory" of bog development, in which climate controls raised-bog growth. Phase-shift in peat (*Sphagnum* "succession," or sudden changes from major assemblage type to type) is the result of climate shifts. Individual bogs have different threshold values for phase shifts depending on their particular hydrology and vegetation type. This latter conclusion from Barber's work also can be illustrated by the not-completely-synchronous development of the oligotrophic mire system (3250-2750 yr B.S.) over the west-central catchment of Red Lake peatland. However, minor changes in abundances among species in fossil assemblages that are recorded in the Red Lake peatland cores are better explained in terms of competition and Hutchinsonian niche theory (Bennett 1988) linked directly to local hydrological, chemical, and some autogenic successional changes (Ratcliffe and Walker 1958; Rybnicek 1973) than by directly applying higher-level explanations such as climate change.

Future development of the Red Lake peatland will be directed mainly by regional changes induced by the present climatic warming, likely to be exacerbated by the "greenhouse effect." *Sphagnum* growth is still active now but will be arrested (cf. Gorham 1988), and it is very unlikely that the present bogs will further develop toward nonforested bog plains with distinct pools, as suggested in Glaser and Janssens (1986) and presently observed toward the north and east in Hudson Bay lowlands and in the Maritime Provinces. Expansion of catchments (Darlington 1943; Heinselman 1970; Sjörs 1963) and more frequent fires (Sjörs 1963) will reduce peatland extent. The Holocene trend toward landscape diversification will not be stopped, but peatlands will probably occupy a smaller area.

The landform patterns at Red Lake occur also in the other large bog complexes on paludified lowland in Michigan, Wisconsin, Minnesota, Manitoba, and northern Ontario (Sjörs 1961, 1963; Heinselman 1963, 1965, 1970; Glaser 1983; Wright 1984; Glaser and Janssens 1986; Zoltai *et al.* 1988). Farther north in the Hudson Bay lowlands, the linear forested bog crest is replaced by a nonforested bog plain, but the system of water tracks and ovoid islands is similar to that formed farther south. The recurrence of the bog pattern across such a vast region in the midcontinental lowlands suggests that the development sequence described here by stratigraphic bryophyte analysis may also be applicable to other large bog complexes over this broad region. Climatic warming may cause dramatic changes in these large ecosystems, and its effects will probably be recorded first in peripheral areas such as the Red Lake peatland.

Acknowledgments

Most of the funds for this study were provided by contracts with the Peat Program of the Minnesota Department of Natural Resources and with the U.S. Department of Energy (DOE AC01-80EV10414) through H. E. Wright (stratigraphical work and logistics), and by grants from the National Science Foundation (DEB-7922142) and the Andrew W. Mellon Foundation through E. Gorham (reference data-sets).

We appreciate the essential skills of John and James Almendinger in the fieldwork and the overall field assistance offered by members of the Limnological Research Center and the Department of Natural Resources. We thank Erna Janssens-Verbelen and Rebne Karchevsky respectively for typing and drafting numerous versions of the manuscript and figures.

We thank H. J. B. Birks, R. S. Clymo, E. Gorham, M. L. Heinselman, S. Juggins, and N. G. Miller for critical reviews of the manuscript and numerous other persons of the paleoecological community for helpful comments and hints.

APPENDICES: Chapter 13

Appendix 1A and 1B

CORE 8101 (site B)

The core was taken near the crest of the forested bog of the west-central catchment. The site is in an incipient drain in the bog forest. Stunted black spruce trees surround the immediate coring site, and better-developed bog forest is several hundred meters from the site. The following local zone descriptions are based on Appendix 1.

Zone B1: cattail-sedge marsh (4,900 ± 225 yr B.S. and older; 455-433 cm below surface). The extremely old basal date (6530 ± 270 yr B.P.: Table 13.1) could be the result of very slow accumulation in a rare wet depression, starting long before the other coring sites became paludified (Heinselman 1963, 1970). Another explanation for the old date is possible contamination with lignite fragments from the underlying Glacial Lake Agassiz sediments (H. E. Wright, Jr., personal communication). Extrapolation of the 2 σ error range on the basis of the vertical accumulation rate calculated from the basal date and the next higher radiocarbon date places the age of the mineral contact at 455 cm between 7650 and 8950 yr B.S. In contrast, the regional pollen data for this zone (pollen samples at 440, 448, and 455 cm depth) suggest a more recent deposition time, such as the end of the prairie period at 6000-5000 yr B.S., for these basal sediments, corresponding more with the other cores along the transect (see also Fig. 13.4).

A highly organic, black sandy silt forms the mineral contact in this core segment. Seeds and achenes of *Potamogeton* (pondweed), *Scirpus* (bulrush), and *Carex* (sedge) are present. The lowland pollen assemblage is dominated by species typical of highly minerotrophic sites: sedges (Cyperaceae), cattail (*Typha*), alder (*Alnus*), and willow (*Salix*) (Janssen 1968). *Equisetum* (horsetail) and *Potamogeton* pollen are represented by low pollen percentages; the latter is significant because its presence is generally associated with standing water (Griffin 1977).

The lower part of the highly humified peat above the mineral contact is nonfibrous, consisting of organic matter, possibly of detrital origin, and some reworked lignite. The upper part of this zone is composed of more-fibrous organic detritus, the fibers mostly derived from herbaceous plants (bulk density declines to less than 150 g L^{-1}, and absorbance declines slightly).

Relatively high pollen percentages of oak (*Quercus*), Chenopodiineae (goosefoot and amaranth families), *Artemisia* (wormwood), and grasses (Poaceae) suggest an oak savanna in the uplands. Griffin (1977) found no recognizable macrofossils of *Zizania aquatica* (wild rice) or *Phragmites* sp. (common reed) at Camp Island, Red Lake Bog, leading her to suggest that the grasses represented may be regional prairie types rather than

aquatic types. *Betula* is relatively well represented in zone B1. Pollen percentages of haploxylon pine (white pine) are low, and diploxylon pine pollen (red or jack pine) increases from 20% to 40% at the top of the zone.

Zone B2: sedge-*Equisetum* meadow (*ca.* 4900 yr B.S. to 3300 ± 300 yr B.S.; 433-375 cm below surface). Half of the peat is highly humified, and the rest consists of herbaceous remains. Seeds and achenes of *Potamogeton*, *Scirpus*, and *Carex* are found in this zone as well as in B1. No bryophyte remains were recovered except for a clearly reworked assemblage of *Sphagnum magellanicum*, *S. majus*, *S. capillifolium*, *S. warnstorfii*, and *S.* section *Cuspidata* spp. at 399 cm depth. This assemblage is not indicated on the profiles in Appendix 1A because of its possible detrital origin, and no inferences are calculated. An abrupt reduction in pollen of *Typha* (25% to less than 2%) and Cyperaceae (95% to 30% characterizes this zone, along with minor representation of *Salix*, *Sarracenia* (pitcher plant), and *Scheuchzeria* pollen and *Equisetum* spores.

Upland pollen is dominated by diploxylon pine pollen, with slightly decreased amounts of *Quercus*, *Artemisia*, and *Ambrosia* type (ragweed). An abrupt decrease in Poaceae and Chenopodiineae pollen further characterizes the pollen assemblages in this zone.

Zone B3: rich fen (3300 ± 300 to 2900 ± 250 yr B.S.; 375-330 cm below surface). This zone is differentiated from B2 by the first occurrence of numerous autochthonous moss remains in the peat. Remains of highly minerotrophic taxa were recovered (Amblystegiaceae or brown mosses: *Calliergon trifarium*, *Scorpidium scorpioides*, *Campylium stellatum*; *Sphagnum* section *Subsecunda*: *S. contortum* and *S. subsecundum*; *Bryum pseudotriquetrum*). The inferred pH is initially circumneutral, and HMWT inference indicates a nearly aquatic habitat. Some pattern development in the rich fen is suggested by the presence of *S. papillosum* and *S. centrale* (the latter not indicated on Appendix 1A and the development of *Larix*. Both *Sphagnum* species are lawn formers in minerotrophic fens and indicate the succession to a more oligotrophic habitat, developed fully in the next zone. The rich-fen characterization of the peat in zone B3 is further supported by the occurrence of *Utricularia* (bladderwort), *Menyanthes* (buckbean), *Rhynchospora*, and *Sarracenia* pollen. *Fraxinus nigra* and *Salix* are present, and *Larix* pollen increases to 3% at the transition to zone B4. Cyperaceae pollen decreases strongly from the base of this zone to the top.

In the regional picture, jack and red pine (diploxylon pine pollen) decrease, while white pine (haploxylon type pollen) increases in this zone. A slight increase in elm (*Ulmus*) pollen and the first continuous representation of fir (*Abies*) pollen are noteworthy. Poaceae, *Quercus*, and other upland herb pollen does not change in representation appreciably from zone B2.

Zone B4: oligotrophic mire (2900 to 86 yr B.S.; 330-42 cm below surface). This zone is composed mainly of *Sphagnum* peat. Bulk density, absorbance, and organic-detritus content decline except for a high-ash event (loss-on-ignition ash) around 1000 yr B.S. Initial high peat accumulation rates of 1.5 mm yr^{-1} and 130 g m^{-2} yr^{-1} decline gradually to values of 0.7 mm yr^{-1} and 50 g m^{-2} yr^{-1} upward to the cultural horizon. The bryophyte composition is mostly *Sphagnum magellanicum* and *S. capillifolium*, the latter a hummock-forming species. Other hummock-forming species, not indicated on Appendix 1A, are *Aulacomnium palustre* and *Sphagnum fuscum*, recovered between 300-330 cm depth. *Sphagnum recurvum* agg. spp. were also a minor component in some of the peats in this zone. They occur in depressions among hummocks. Numerous needles and wood remains of *Picea* and *Larix* reinforce the poor-fen/bog reconstruction. Synchronous with the high-ash event around 1000 yr B.S. are slightly more minerotrophic and more aquatic bryophyte taxa (*S. majus, S. papillosum*), as illustrated by the inference curves. This high-ash event — in some cores there are two distinct peaks — is found at many other sites throughout Minnesota (Janssens, report to the Minnesota Department of Natural Resources) but is not yet dated accurately enough in most cores to postulate about causal factors. The indication of a more aquatic habitat inferred from the fossil mosses could have induced higher humification and an increased concentration of plant ash.

The pollen record supports the characterization of a poor fen with high percentages of *Picea mariana, Arceuthobium* (dwarf mistletoe), and Ericaceae (heaths). *Sphagnum* replaces Cyperaceae as the major wetland plant, reflecting a trend toward an oligotrophic mire. *Equisetum* spores decrease in this zone.

Among the upland types, pine pollen continues to dominate the section, and Poaceae, *Quercus*, and other upland herb types reach their lowest pollen percentages for the core.

Zone B5: hummocky oligotrophic mire (86 yr B.S. to present; 42 cm to surface). At 42 cm an abrupt increase in *Ambrosia* pollen and concurrent decrease in *Pinus strobus* (Jacobson 1979; Griffin 1977) indicate the settlement horizon (approximately 1895 A.D.). Cyperaceae and *Sphagnum* decrease in pollen percentages just prior to this level, and *Picea mariana* and *Betula* both increase within this zone. *Alnus rugosa* pollen reaches a maximum in the uppermost sample. The extralocal pollen assemblage suggests a poor fen, possibly slightly more minerotrophic than the preceding zone B4, indicating further development of the nearby bog drain. No significant changes are observed within the bryophyte populations (all hummock taxa) during the cultural period — see inference curves — but depth accumulation rate and dry-mass accumulation rate show sharp increases, commonly observed in acrotelm peats, at the top of the core.

Appendix 2A and 2B

CORE 8104 (site G)

This site is situated at the southernmost tip of a well-developed ovoid island. The coring site is in the ecotonal area between the horseshoe-shaped bog forest on the ovoid island and the waterlogged fen area with small, circular forested islands (Fig. 13.1). Waterflow diverges around the ovoid island in both directions at the site of core 8104. The local zone descriptions (G1 to G9) are based on interpretation of Appendix 2. Compared with core 8101, a more complex stratigraphy is evident. The paleoenvironmental reconstruction detailed below can also be followed along the ordination trace of this core on Fig. 13.2

Zone G1: below the mineral contact (3300 ± 300 yr B.S. and older; 315-312 cm below surface). A dark-gray fine sandy silt with some gravel forms the mineral contact in this core. No samples for micro- or macrofossil analyses were taken from this material.

Zone G2: minerotrophic aquatic community (3300 to 3100 ± 170 yr B.S.; 312-285 cm below surface). The highly decomposed *Limus detrituosus* contains abundant remains of aquatic modifications of the mosses *Drepanocladus fluitans, Amblystegium riparium*, and *Sphagnum platyphyllum*, reflected in the low HMWT inference. The acidophilous nature of *Drepanocladus fluitans* is indicated by the initial low pH inference values. Lowland pollen is poorly represented. The upland pollen component of the single pollen sample from this zone is dominated by *Pinus* diploxylon, with significant amounts of Tubuliflorae and Poaceae pollen.

Zone G3: sedge meadow (3100 to 2850 ± 150 yr B.S.; 285-260 cm below surface). Bryophyte remains of highly minerotrophic taxa are still evident — indicated by the circumneutral inferred pH — but with a more-emergent growth form. The lowland pollen is dominated by Cyperaceae, with minor amounts of *Typha latifolia* and *Rhynchospora*, a pollen assemblage typical of fens with hollows and flarks. Haploxylon pine pollen increases in this zone as diploxylon pine pollen decreases. *Quercus* and *Artemisia* pollen are present.

Zone G4: *Sphagnum*-dominated intermediate and poor fen (2850 to 1900 ± 130 yr B.S.; 260-180 cm below surface). During this long period, the peatland vegetation became gradually less minerotrophic — indicated by the declining pH and increased HMWT inferences — developing from a circumneutral and very wet intermediate fen with *Sphagnum* section *Subsecunda* (*Sphagnum contortum*) species to a less minerotrophic, more raised *Sphagnum* poor fen with *S. magellanicum* and *S. centrale*. This trend toward a more acidic fen is corroborated by the pollen record, which is dominated by *Sphagnum* spores. *Larix, Alnus rugosa*, and *Betula* pollen maxima suggest a poor-fen environment with hummocks (*Aulacomnium palustre*) and islands. Haploxylon pine is the major component of the upland pollen types, with a minor oak peak in the lower part of zone G4. *Abies* pollen is important as a regional zone marker even though the percentages are low.

Zone G5: brown-moss, rich-fen flark (1900 to 1000 ± 100 yr B.S.; 180-95 cm below surface). *Sphagnum* spores decline precipitously in this zone. The bryophyte assemblage contains the most highly minerotrophic, aquatic fen species (Amblystegiaceae: *Scorpidium scorpioides, Calliergon trifarium, Campylium stellatum,* and *C. polygamum*). All taxa contribute to the low HMWT inferences, including the only remaining *Sphagnum* sp., *S. contortum*. A major decrease in the occurrence of *Larix, Alnus,* and *Betula* pollen, accompanied by change in *Picea mariana* (up to 10%, then down to 5% or less) seems suggestive of local substrate change. A slight increase in pollen of *Typha,* Cyperaceae, and *Myrica* (waxmyrtle) and the decline in *Sphagnum* spores support the characterization of rich-fen flark. Upland pollen shows little change except for a minor peak in *Picea glauca* pollen, which may be due to the diminished amount of the lowland trees and shrubs.

Zone G6: Sphagnum contortum flark (1000 to 500 ± 40 yr B.S.; 95-60 cm below surface). Abundant remains of *Sphagnum contortum* occur in this zone, together with increased spores of *Sphagnum*. Of all *Sphagnum* species, *S. contortum* is one of the most aquatic, barely emergent or carpet-forming, but highly minerotrophic, in contrast with *Sphagnum cuspidatum*. Regional pollen does not change appreciably from zone G5 through zone G7.

Zone G7: Sphagnum lawn (500 to 270 ± 25 yr B.S.; 60-45 cm below surface). The aquatic community dominated by *Sphagnum contortum* in the previous zone is replaced by the more elevated, lawn-forming species *S. centrale* and *S. papillosum,* both still indicating slight minerotrophic influence (inferred pH of 4.5). The second maximum of *Sphagnum* spores in this core and the slight decrease in Cyperaceae supports also the trend toward a poor fen. This is the first indication of the encroachment of the ovoid island toward this coring site.

Zone G8: wooded *Sphagnum* and brown-moss rich fen (270 to 130 ± 5 yr B.S.; 45-35 cm below surface). The *Sphagnum* species give way again to an assemblage dominated by *Drepanocladus lapponicus* in this zone. The latter is a rare species, recorded live in Minnesota only in the adjacent Lost River peatland (Janssens 1983a, b, 1984; Janssens and Glaser 1986). It indicates a circumneutral and aquatic habitat, supported by its associated taxa *Meesia triquetra, Sphagnum majus, S. subsecundum,* and *S. platyphyllum*.

The woodiness of the peat is supported in the pollen record by increasing percentages of *Picea mariana* pollen, together with *Alnus rugosa* and *Betula* (probably *B. pumila,* shrub birch), and Ericaceae. *Menyanthes* pollen present in this zone indicates distinct influence of minerotrophic water chemistry. *Sphagnum* spores remain well represented. Among the upland pollen

types, *Artemisia* and *Quercus* exhibit increasing pollen values.

Zone G9: Sphagnum poor fen (130 to 0 yr B.S.; 35 cm to surface). This uppermost zone is dominated by *Sphagnum* section *Cuspidata* (*S. recurvum* agg. and *S. majus*) remains; the topmost samples of *S. recurvum* agg. can be identified as *S. angustifolium* by the presence of stem leaves. Slightly minerotrophic indicators (*S. centrale, S. papillosum, S. majus,* and *Calliergon stramineum*) are still present throughout. Wood remains are present, and dry hummock mosses (*Aulacomnium palustre*) indicate the second encroachment of the nearby ovoid island bog forest. Increasing black spruce and maxima for alder, birch, and Ericaceae confirm bog-forest succession.

High percentages of *Ambrosia* type and Chenopodiineae pollen signal the advent of cultural disturbance halfway up in this zone, as does the decrease in white pine pollen in this zone.

Appendix 3A and 3B

CORE 8102 (site J)

This core came from the ecotonal area separating the two major ovoid islands of the west-central catchment (Fig. 13.1). The local zone descriptions (J1 to J8) are based on the interpretation of Appendix 3.

Zone J1: below mineral contact (6400 ± 200 yr B.S. and older; 354-339 cm below surface). A very dark gray, sandy silt forms the mineral substrate in this zone. *Potamogeton* and *Scirpus* achenes were recovered from this core segment. Maximum percentages of *Typha* and Cyperaceae pollen suggest a sedge meadow with hollows. Upland pollen is dominated by Chenopodiineae and grasses (Poaceae). Tree pollen is negligible.

Zone J2: open sedge-meadow community (6400 to 4000 ± 260 yr B.S.; 339-321 cm below surface). This highly decomposed *Limus detrituosus* contains only fibrous remains of herbaceous plants. The wetland pollen assemblage is dominated by Cyperaceae, confirming the sedge-meadow aspect of the area. *Typha* pollen is strongly reduced, perhaps reflecting the drying up of hollows. Pine, oak, and *Artemisia* are the major upland pollen types.

Zone J3: shrubby sedge-meadow community (4000 to 2800 ± 150 yr B.S.; 321-290 cm below surface). Bryophyte remains of highly minerotrophic and aquatic brown mosses are evident in the samples, but still in low concentration. Cyperaceae is still the dominant wetland pollen type, although it decreases to the top of this zone. *Alnus rugosa, Betula,* and *Salix* exhibit maxima accompanied by pollen of *Fraxinus nigra,* corroborating the shrubby nature of the peat. Upland pollen remains basically the same as in the preceding zone.

Zone J4: brown-moss rich fen (flark?) (2800 to 2400 ± 150 yr B.S.; 290-235 cm below surface). The lower part of this zone is pure brown-moss peat. The Amblystegiaceae represented are the highly minerotrophic *Calliergon trifarium* (now very rare in Red Lake peatland), *Scorpidium scorpioides*, and later in the zone *Campylium stellatum*. *Campylium* indicates the formation of an emergent bryophyte carpet community, illustrated by the increase in inferred HMWT. *Larix* pollen assumes a slightly more distinct profile beginning in this zone, and *Sarracenia* pollen and *Equisetum* spores are present. Cyperaceae pollen still dominates the lowland assemblage, which is compatible with the rich-fen characterization of the peat. Notable among the upland pollen types are the first significant occurrences of *Abies* and *Ulmus* and increasing values of *Populus* pollen. Haploxylon pine shows a slight increase, while diploxylon type decreases.

Zone J5: Sphagnum intermediate fen (string/flark alternation) (2400 to 1600 ± 100 yr B.S.; 235-135 cm below surface). This meter-thick peat layer contains high concentrations of aquatic, minerotrophic *Sphagnum* species, but also more acidophilous lawn-forming species. This is illustrated by the fluctuating pH and HMWT inference curves. The lower part also has a significant woody Ericaceae component. Both zone J4 and zone J5 have high peat-accumulation rates. Increasing pollen percentages of *Picea*, Ericaceae, and *Sphagnum*, together with fluctuating percentages of Cyperaceae pollen, suggest a poor or intermediate fen, whereas percentages of *Utricularia* indicate a rich-fen flark. The upland pollen record exhibits maximum percentages of haploxylon pine and decreasing oak.

Zone J6: Calliergon trifarium rich-fen flark (1600 to 1275 ± 70 yr B.S.; 135-118 cm below surface). Apparently the local environment shifted back to a well-developed flark at the coring site. Zones J4, J5, and J6 indicate the development of the string-flark pattern in the water track that confines the runoff water between the two nascent ovoid islands (see cores 8111 and 8112). This may be reflected in the reciprocally oscillating pollen curves for *Sphagnum* and Cyperaceae. The peat of the zone is mainly brown-moss peat with Cyperaceae remains. A submerged environment is indicated by the aquatic brown-moss and *Sphagnum* modifications. *Sphagnum* spores decline precipitously at the transition from J5 to J6, as does Ericaceae and Cyperaceae pollen. Spruce pollen starts to increase in this zone, suggesting spruce-bog development and local paludification. Most lowland pollen types except *Sparganium* decrease in representation in this zone. Of the upland and ecologically indeterminant elements, haploxylon pine type decreases substantially, and Poaceae reaches a minimum for this site.

Zone J7: Sphagnum poor fen (1275 to 72 yr B.S.; 118-45 cm below surface). The *Sphagnum* peat deposited during this period became highly humified (see bulk-density, absorbance, and organic detritus profiles). The HMWT inference indicates a lawn, low-hummock environment. *Sphagnum* spores and *Picea* and Ericaceae pollen again dominate the wetland microfossil assemblage. Little change is noticeable among the upland pollen types. The high percentages of spruce and *Betula* (assumed to be *B. pumila*) suggest the encroachment of the western ovoid island over the coring site, which now is located in the ecotonal zone between ovoid island and water track.

Zone J8: Sphagnum poor-fen or bog (72 to 0 yr B.S.; 45 cm below surface). This upper zone is differentiated from J7 only by its undecomposed acrotelm peat and the presence of *Sphagnum capillifolium*. Higher percentages of Cyperaceae pollen, together with continued fluctuations and high pollen percentages of spruce and *Sphagnum* spores, may reflect short-term fluctuation of the nearby bog margin. A pronounced peak in *Ambrosia* type pollen and an equally pronounced reduction in haploxylon pine indicate the cultural horizon.

Appendix 4A and 4B

CORE 8111 (site K)

This site is situated in the horseshoe-shaped bog forest of the large eastern ovoid island of the west-central catchment. The local zone descriptions (K1 to K7) are based on the interpretation of Appendix 4.

Zone K1: below mineral contact (4600 ± 250 yr B.S. and older; 412-389 cm below surface). A black or dark-gray sandy silt with high organic content and vivianite forms the mineral contact on this core. Two pollen samples were collected in this zone. The lower sample is dominated by *Pinus* pollen, with high percentages of lowland *Typha, Myriophyllum*, and *Sphagnum*. The abundance of *Sphagnum* spores and macrofossil remains of *Calliergon cordifolium, S. magellanicum*, and *S. centrale* (not plotted in Appendix 4A) indicates a mixed fossil assemblage possibly with some reworked organics deposited under high-energy conditions with the inorganic sediments. A shaded lagg is suggested, supported by the presence of the aquatic *Myriophyllum* in the lowland pollen assemblage.

The upper pollen sample at the mineral contact is strongly differentiated from the one at 410 cm in both upland and lowland spectra. The upland pollen is now dominated by diploxylon *Pinus*, Chenopodiineae, *Quercus*, and *Artemisia*, together with Poaceae (ecologically indeterminate). The lowland pollen curves exhibit maxima of Cyperaceae and *Typha*, characteristic of a rich sedge fen with open pools or hollows.

Zone K2: Alnus carr (4600 to 3200 ± 225 yr B.S.l; 389-328 cm below surface). This highly humified peat contains only recognizable brown-moss and *Bryum pseudotriquetrum* remains, indicating a highly minerotrophic status with circumneutral pH and aquatic HMWT inferences. The lowland pollen assemblage is co-dominated by alder and sedge pollen, with significant representation of *Salix* and *Thuja* (cedar) type pollen, indicative of rich fens and alder thickets (Griffin 1977). A slight increase in *Ambrosia* and the appearance of *Ostrya/Carpinus* pollen are the major differences in the regional pollen flora.

Zone K3: minerotrophic *Sphagnum* and brown-moss alder carr (3200 to 2700 ± 275 yr B.S.; 328-200 cm below surface). This zone is similar to K2 but shows a far higher accumulation rate and a greater diversity of bryophytes. High concentrations of minerotrophic *Sphagnum* species occur at the top of this zone, along with high percentages of *Sphagnum* spores in the pollen diagram. *Larix* and *Rhynchospora* occur early in this zone, followed by increasing *Picea mariana* and *Fraxinus nigra* pollen. Diversity of taxa is also evident in the lowland pollen assemblage, with *Drosera, Sarracenia, Menyanthes,* and *Myriophyllum* all present, suggestive of a rich fen (Glaser *et al.* 1981; Griffin 1977). Cyperaceae and *Alnus* pollen gradually decrease upward in this zone as black spruce and black ash become more important. The major upland pollen type, pine, changes little from the previous zone. *Abies, Acer,* and some other minor types have their first consistent representation in K3.

Zone K4: rich-fen flark (2700 to 2300 ± 250 yr B.S.; 200-170 cm below surface). The peat in this zone is composed mainly of *Calliergon trifarium* and *Scorpidium scorpioides* remains. Both taxa are aquatics and indicators of circumneutral pH. *Sphagnum* spores in the pollen diagram decrease abruptly. The lowland pollen is dominated by Cyperaceae pollen, suggesting rich fen with increasing percentages of *Larix* and Ericaceae. *Betula* pollen peaks at the base of this zone. Many of the typical rich-fen pollen types (*Sarracenia, Drosera, Menyanthes,* and *Myriophyllum*) drop out of the diagram at the end of this zone.

Zone K5: *Sphagnum* intermediate-fen/poor-fen transition (2300 to 1700 ± 225 yr B.S.; 170-135 cm below surface). The major peat component in this zone is now *Sphagnum* remains. Initially, minerotrophic indicators are present, completely replaced by *Sphagnum magellanicum* in the upper part of the zone. This species has a large tolerance for minerotrophic conditions, and a large error should be associated with the pH inferences based on a monospecific *S. magellanicum* assemblage. Black spruce, Ericaceae, and *Sphagnum* increase as Cyperaceae, *Betula,* and *Fraxinus nigra* pollen decline. *Thuja occidentalis*-type pollen is a slightly more obvious contributor to the pollen rain. This lowland pollen assemblage suggests an evolving ovoid island nearby. The upland pollen exhibits very little change from zone K4.

The main reason for positioning the K5/K6 boundary at 1700 yr B.S. is the precipitous drop in the bulk density of the peat at this level. When placed at 2000 yr B.S., the development of the ovoid bog island would be synchronous at both coring sites 8111 and 8112.

Zone K6: forested poor-fen or bog community (1700 to 86 yr B.S.; 135-30 cm below surface). This oligotrophic *Sphagnum* peat is composed mostly of *S. magellanicum.* The humification measurements increase toward the top of this zone. The local pollen and spore counts are dominated by *Sphagnum* spores, with *Picea mariana, Arceuthobium,* Ericaceae, and *Alnus rugosa* assuming major importance within the lowland pollen types. The upland pollen exhibits little change aside from a slight increase in *Picea glauca* pollen percentages.

Zone K7: forested-bog community (86 yr B.S. to present; 30 cm to surface). This zone is differentiated from K6 by higher percentages of Ericaceae, *Alnus,* and *Picea mariana* in the wetland pollen assemblage, indicating further development of the horseshoe-shaped forest, and an abrupt decrease in *Sphagnum* spores. High pollen percentages of *Ambrosia* and Chenopodiineae and a significant reduction in haploxylon pine signal the cultural horizon at the base of this zone.

Appendix 5A and 5B

CORE 8112 (site L)

This site is in the unforested center of the large eastern ovoid island of the west-central catchment (Fig. 13.1). The local zone descriptions (L1 to L7) are based on interpretation of Appendix 5.

Zone L1: below mineral contact (3600 ± 125 yr B.S. and older; 445-398 cm below surface). A black or dark-gray silty sand with marly concretions and pebbles from the mineral contact in this core. Reworked remains of mosses (*Sphagnum papillosum, Sphagnum* section *Acutifolia,* and *S. magellanicum*) and sedge seeds are found at the mineral contact. Four pollen samples were collected in this zone. *Typha* and sedge pollen are the dominant lowland bog types, together with *Fraxinus nigra, Salix, Picea, Arceuthobium,* and *Myriophyllum,* reflecting a rich-fen sedge meadow. The upland pollen is dominated by Tubuliflorae, Chenopodiineae, and *Ambrosia* type pollen, together with a small peak of *Pteridium,* an upland fern. Grass pollen (Poaceae), constitutes 70% to 160% of the upland pollen sum.

Zone L2: sedge meadow and *Alnus* carr development (3600 to 2900 ± 100 yr B.S.; 398-325 cm below surface). Initially sedge and *Typha* dominate the lowland pollen picture, followed by a decrease in *Typha* and the disappearance of *Myriophyllum* pollen. *Alnus rugosa* increases along with *Salix,* as Cyperaceae abruptly decline in the middle part of the zone. Sedge and *Salix* become the major pollen types in the upper part of this zone. Among the upland taxa, pine is the most significant component; oak and elm are well represented. The upland herbs *Ambrosia* type, Chenopodiineae, and *Artemisia* decrease in importance in this zone. Highly minerotrophic and aquatic brown mosses are present in this latter stage.

Zone L3: brown-moss rich-fen (2900 to 2100 ± 125 yr B.S.; 325-258 cm below surface). The peat consists mostly of the remains of *Scorpidium scorpioides, Calliergon trifarium,* and *Drepanocladus revolvens.* These three species are the most common Amblystegiaceae in the circumneutral, aquatic brown-moss assemblage. The increase of *Campylium stellatum* near the end of the zone indicates the emergence of the moss mat, illustrated by the higher HMWT inference value. The rich fen characterized by the mosses is duplicated in the lowland pollen by continued high percentages of Cyperaceae, increased *Fraxinus nigra*, and a maximum representation of *Sarracenia* pollen. A decrease in haploxylon pine and Poaceae pollen and the first notable occurrence of *Celtis* and *Corylus* pollen are the only distinctive changes in the upland pollen.

Zone L4: Sphagnum intermediate fen (2100 to 1950 ± 140 yr B.S.; 258-236 cm below surface). During this brief period, the Amblystegiaceae community is abruptly succeeded by moderately minerotrophic *Sphagnum* species. The peat is highly humified, and few of the *Sphagnum* remains are well preserved. An abrupt decrease in sedge pollen and increasing black spruce and *Sphagnum* percentages also suggest a less minerotrophic environment. The upland pollen remains static.

Zone L5: wooded *Sphagnum* poor fen or bog (1950 to 1000 ± yr B.S.; 236-105 cm below surface). *Sphagnum* and wood remains form the bulk of the peat during this long period. The bog/poor-fen assemblage developed rapidly into a forested community, as suggested by the high *Picea*, *Alnus rugosa*, and *Fraxinus nigra* pollen contributions, along with increased Ericaceae. *Sphagnum* replaces Cyperaceae as the dominant contributor to the wetland pollen assemblage, and *Sarracenia* drops out of the pollen picture. Indicators of minerotrophic influence are still present among the bryophyte components of this zone (*S. papillosum, S. majus*) and the succeeding zone, L6, indicated by the slightly elevated pH inference.

Zone L6: Sphagnum poor fen (1000 to 160 yr B.S.; 105-45 cm below surface). There is no significant change in the *Sphagnum* composition of this zone, although *Sphagnum*-spore counts drop precipitously. The peat is composed of well-preserved *Sphagnum* remains and less wood than in the previous zone. *Picea* and Ericaceae pollen decreases. *Larix* continues to be relatively well represented, and *Typha* shows a slight increase. Although tree pollen is still a major contributor to the lowland pollen assemblage, the absence of woody remains and the increase in sedge pollen near the top of this zone indicate that the bog forest is opening up (formation of the horseshoe-shaped ring of forest around the island rim). Upland pollen manifests little change from the previous zone.

Zone L7: open bog community (160 yr B.S. to the present; 45 cm to surface). This zone is differentiated from L6 by the absence of *Sphagnum majus*, the slightly minerotrophic indicator in the bryophyte assemblages of the preceding zone. It is replaced by hummock-forming *Sphagnum fuscum* and *Polytrichum strictum*. A drier and ombrotrophic habitat is present, as illustrated by the inference curves. This is supported by an increase in Ericaceae pollen and a decrease in Cyperaceae. *Sphagnum* spores reach high percentages again

as in zone L5. High percentages of *Ambrosia* type pollen along with decreased percentages of white pine pollen again signify the cultural horizon in the upland pollen assemblage.

Figure description

The figures that follow are profiles for the peat cores mentioned in the appendix and noted in Figure 13.1 (cores 8101-site B, 8104-site G, 8102-site J, 8111-site K, and 8112-site L). The A figures show selected profiles for analyses from the individual peat cores. The vertical scale is age (in calendar years before sampling), not the usual linear depth. The age scale is not constant along its entire length because of extremely slow accumulation in the basal part of the core. The paleoenvironmental reconstruction is discussed in the text. Some notes on individual variables: (1) the ^{14}C scale illustrates the thickness of each sample submitted for radiocarbon determination, in addition to the radiocarbon date in yr B.P. (Table 13.1). The age of the top and bottom value of the increment submitted for radiocarbon analysis along the yr B.S. axis (years before sampling, for all cores illustrated here 1981) is calculated from their depth level using the vertical accumulation rate. The error of the calibrated date, which is based on the statistical error of the radiocarbon date and the quality of the calibration (Stuiver and Reimer 1986) is listed in the text as the 2 σ error range. (2) The profiles for bulk density and absorbance (and in some of the figures also for fines) are in the form of continuous-bar histograms. Bulk density and absorbance are two partial measures of humification and correlate well with each other (r = 0.76, P < 0.001, n = 419; bulk density ranged from 7 to 250 g L^{-1}, absorbance from 0.12 to 1.249). (3) Peat-component profiles are plotted on a quartile scale in continuous-bar histograms. Gaps between core sections are illustrated in these profiles. (4) The bryophyte profiles are discontinuous bar histograms at the level of each sample along a five-point estimated volume scale (see Janssens 1989 for details). Additional bryophytes not plotted are listed in the text. (5) The inferred pH is reconstructed for each fossil bryophyte assemblage using WACALIB (see Methods section).

The B figures show selected percentage pollen profiles versus the individual peat cores. Pollen percentages for both upland and lowland taxa are based on the upland pollen sum alone (see Methods section). Poaceae and *Betula* are not included in the pollen sum and are plotted as ecologically indeterminate. Types plotted to the left of the pollen sum are included in the sum, but not all the included types are shown. Types to the right are excluded from the sum. Complete pollen count, percentage, and concentration-value tables are available from the senior author of this chapter.

CORE 8101

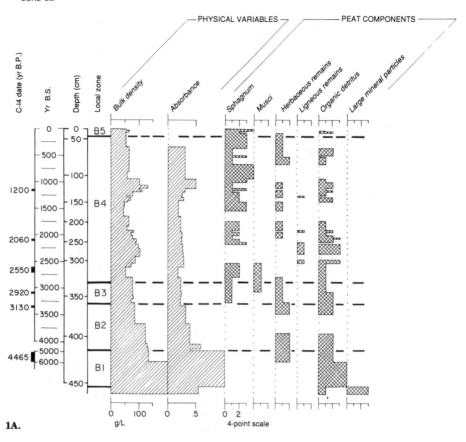

1A.

UPLAND POLLEN

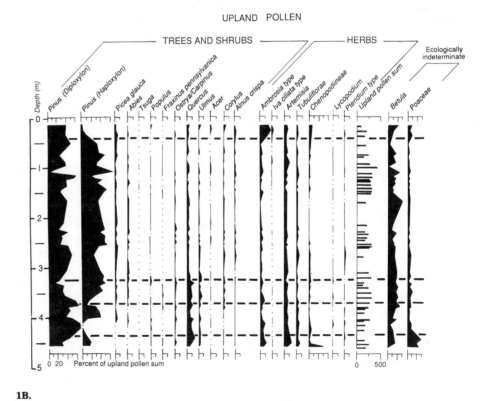

1B.

Appendix 1A: Selected profiles for analyses from peat core 8101 (Fig. 13.1, site B). See p. 209 for construction details.

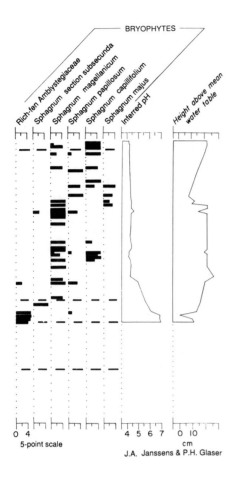

J.A. Janssens & P.H. Glaser

LOWLAND POLLEN

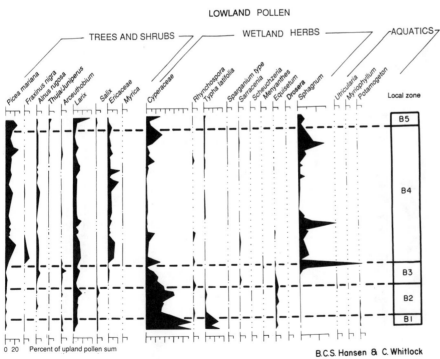

B.C.S. Hansen & C. Whitlock

1B: Selected percentage pollen profiles versus depth in m for peat core 8101 (site B). See p. 209 for construction details.

CORE 8104

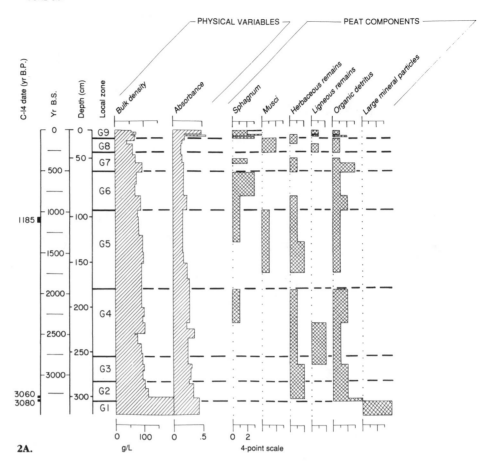

2A.

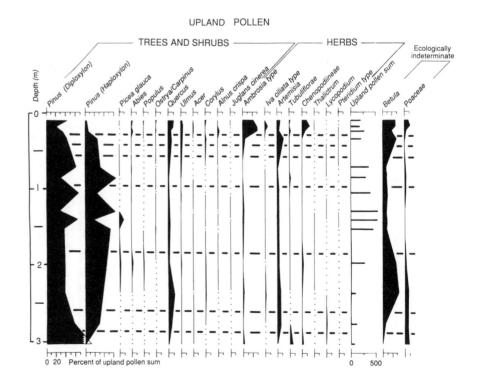

2B.

Appendix 2A. Selected profiles versus calendar-year age for analyses from peat core 8104 (Figs. 13.1, 13.2, site G). See p. 209 for construction details.

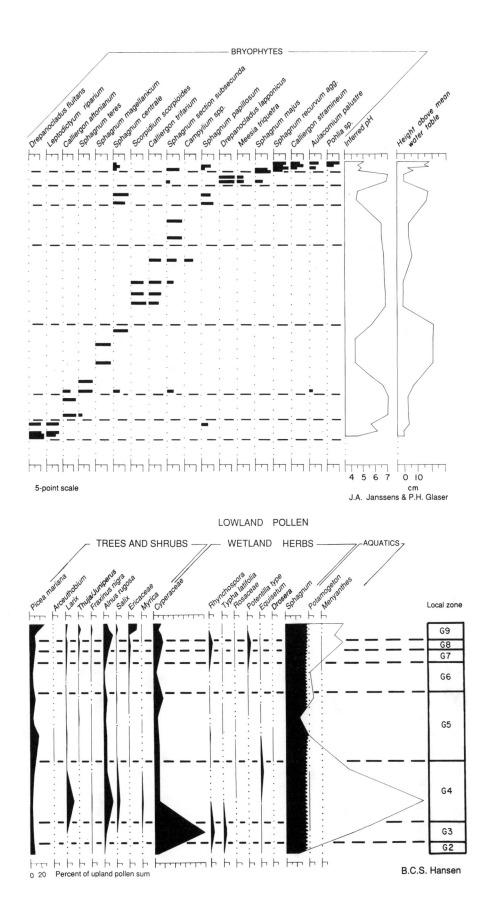

BRYOPHYTES

Drepanocladus fluitans · *Leptodictyum riparium* · *Calliergon riparium* · *Sphagnum aftonianum* · *Sphagnum teres* · *Sphagnum magellanicum* · *Sphagnum centrale* · *Scorpidium scorpioides* · *Calliergon trifarium* · *Sphagnum section subsecunda* · *Campylium spp.* · *Sphagnum papillosum* · *Drepanocladus lapponicus* · *Meesia triquetra* · *Sphagnum majus* · *Sphagnum recurvum agg.* · *Calliergon stramineum* · *Aulacomium palustre* · *Pohlia sp.* · Inferred pH · Height above mean water table

5-point scale

4 5 6 7   0 10
cm

J.A. Janssens & P.H. Glaser

LOWLAND POLLEN

TREES AND SHRUBS — WETLAND HERBS — AQUATICS

Picea mariana · *Arceuthobium* · *Larix* · *Thuja/Juniperus* · *Fraxinus nigra* · *Alnus rugosa* · *Salix* · *Ericaceae* · *Myrica* · *Cyperaceae* · *Rhynchospora* · *Typha latifolia* · *Rosaceae* · *Potentilla type* · *Equisetum* · *Drosera* · *Sphagnum* · *Potamogeton* · *Menyanthes*

Local zone

G9
G8
G7
G6
G5
G4
G3
G2

0 20   Percent of upland pollen sum

B.C.S. Hansen

B. Selected percentage pollen profiles versus depth in meters for peat core 8104 (site G). See p. 209 for details.

CORE 8102

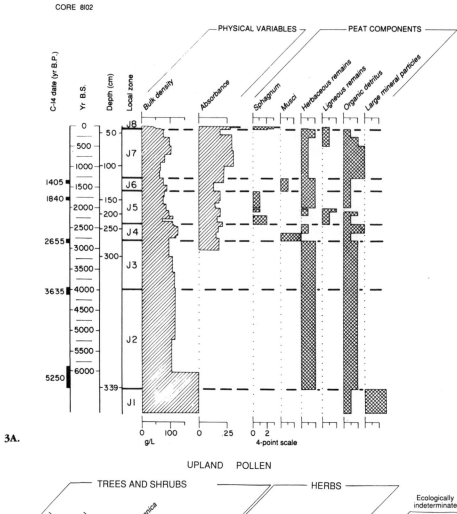

3A.

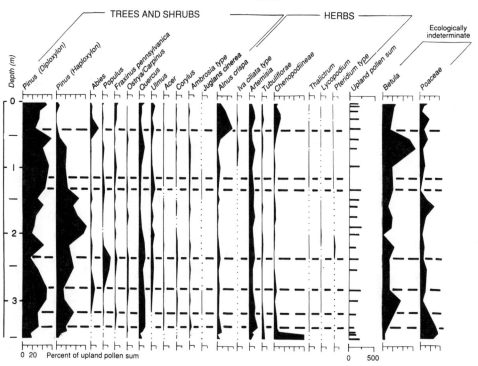

3B.

Appendix 3A. Selected profiles versus calendar-year age for analyses from peat core 8102 (Fig. 13.1, site J). See p. 209 for construction details.

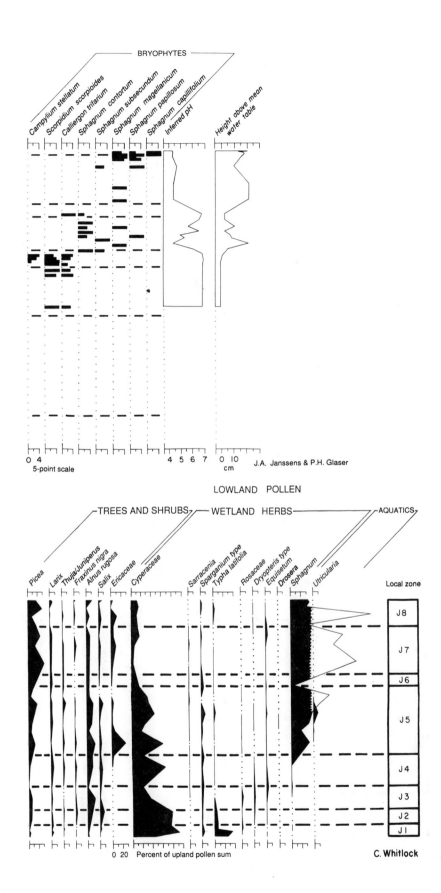

3B. Selected percentage pollen profiles versus depth in meters for peat core 8102 (site J). See p. 209 for details.

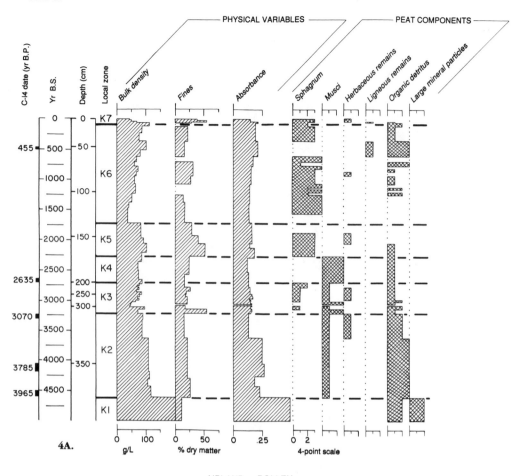

4A.

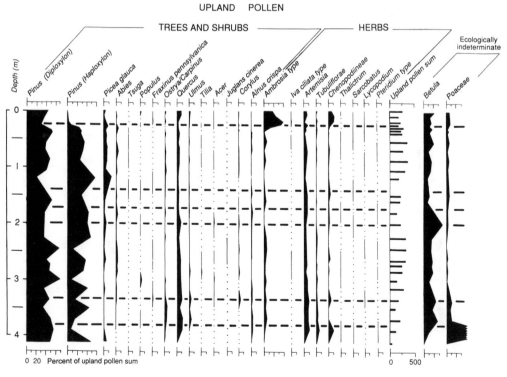

4B.

Appendix 4A. Selected profiles versus calendar-year age for analyses from peat core 8111 (Fig. 13.1, site K). See p. 209 for construction details.

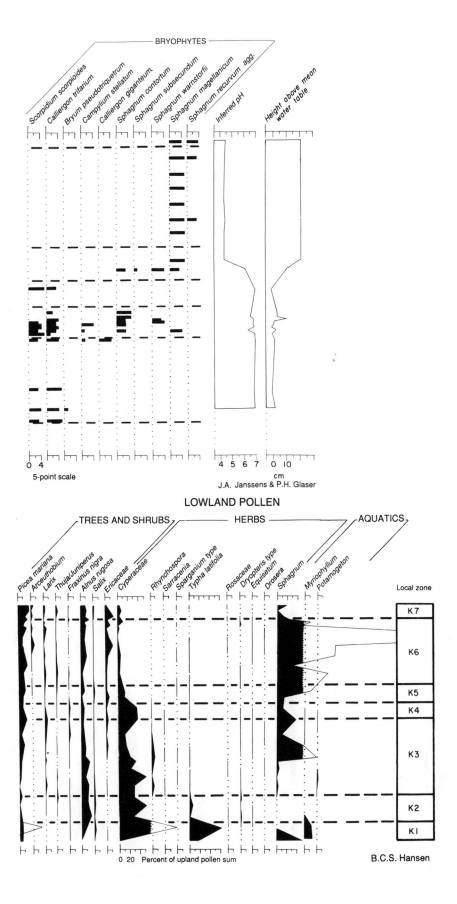

4B. Selected percentage pollen profiles versus depth in meters for peat core 8111 (site K). See p. 209 for details.

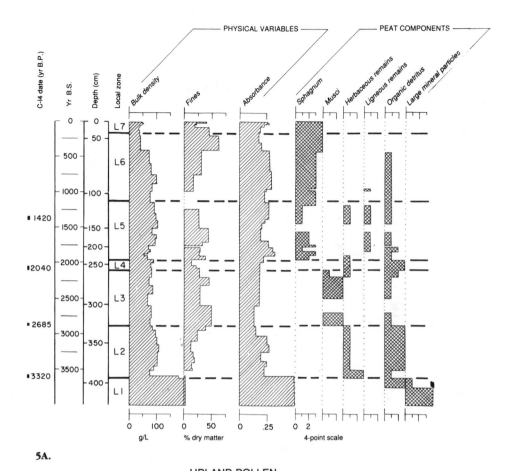

5A.

UPLAND POLLEN

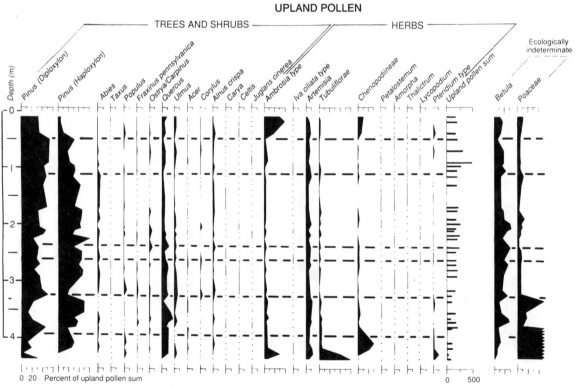

5B.

Appendix 5A. Selected profiles versus calendar-year age for analyses from peat core 8112 (Fig. 13.1, site L). See p. 209 for construction details.

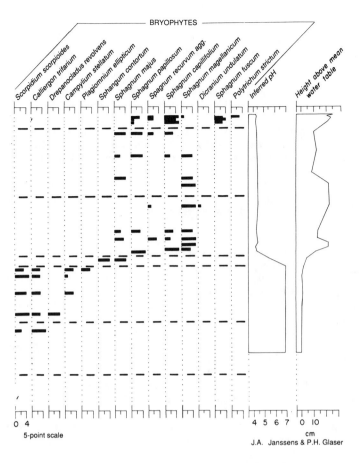

BRYOPHYTES

0 4

5-point scale

4 5 6 7   0 10

cm

J.A. Janssens & P.H. Glaser

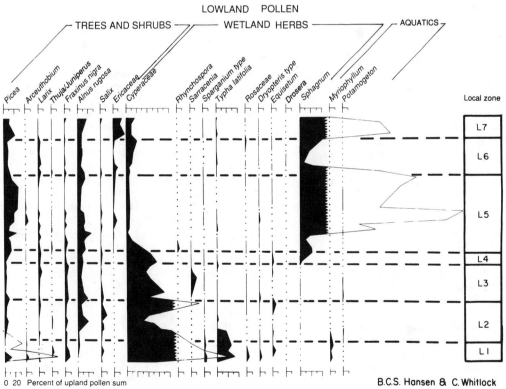

LOWLAND POLLEN

TREES AND SHRUBS — WETLAND HERBS — AQUATICS

0 20 Percent of upland pollen sum

B.C.S. Hansen & C. Whitlock

5B. Selected percentage pollen profiles versus depth in meters for peat core 8112 (site L). See p. 209 for details.

Literature Cited

Aaby, B., and H. Tauber. 1975. Rates of peat formation in relation to degree of humification and local environment, as shown by studies of a raised bog in Denmark. Boreas 4:1-17.

Almendinger, J. C., J. E. Almendinger, and P. H. Glaser. 1986. Topographic fluctuations across a spring fen and raised bog in the Lost River peatland, northern Minnesota. Journal of Ecology 74:393-401.

Barber, K. E. 1981. Peat stratigraphy and climatic change. Balkema, Rotterdam, Netherlands.

Bennett, K. D. 1988. Post-glacial vegetation history: Ecological considerations. Pages 699-724 in B. Huntley and T. Webb III, editors. Vegetation history. Kluwer Academic Publishers, Dordrecht, Netherlands.

Birks, H. H., and R. W. Mathewes. 1978. Studies in the vegetational history of Scotland. V: Late Devensian and early Flandrian pollen and macrofossil stratigraphy at Abernethy Forest, Inverness-shire. New Phytologist 80:455-484.

Birks, H. J. B. 1976. Late-Wisconsinan vegetational history at Wolf Creek, central Minnesota. Ecological Monographs 46:395-428.

———. 1985. Recent and possible future mathematical developments in quantitative paleoecology. Paleogeography, Paleoclimatology, Paleoecology 50:107-147.

———. 1987. Methods for pH-calibration and reconstruction from paleolimnological data: Procedures, problems, potential techniques. Pages 370-380 in Proceedings of Surface Water Acidification Programme (SWAP). Mid-Term Review Conference, June 22-26, 1987, Bergen, Norway.

——— and H. H. Birks. 1980. Quaternary palaeoecology. Edward Arnold, London, U.K.

———, J. M. Line, S. Juggins, A. C. Stevenson, and C. J. F. ter Braak. 1990. Diatoms and pH reconstruction. Philosophical Transactions of the Royal Society of London B: 327:263-278.

Casparie, W. A. 1972. Bog development in southwestern Drenthe (The Netherlands). Junk, The Hague, Netherlands.

Clymo, R. S. 1965. Experiments on breakdown of Sphagnum in two bogs. Journal of Ecology 53:747-758.

———. 1973. The growth of Sphagnum: Some effects of environment. Journal of Ecology 61:849-869.

———. 1984. The limits to peat bog growth. Philosophical Transactions of the Royal Society of London B: 303:605-654.

Cushing, E. J. 1963. Late-Wisconsin pollen stratigraphy in east-central Minnesota. Ph.D. dissertation. University of Minnesota, St. Paul, Minnesota, USA.

Darlington, H. C. 1943. Vegetation and substrata of Cranberry Glades, West Virginia. Botanical Gazette 104:371-393.

Dickson, J. H. 1986. Bryophyte analysis. Pages 627-643 in B. E. Berglund, editor. Handbook of Holocene paleoecology and paleohydrology. Wiley, New York, New York, USA.

Faegri, K., and J. Iversen. 1975. Textbook of pollen analysis. 3rd edition. Munksgaard, Copenhagen, Denmark.

Farnham, R. S., and D. N. Grubich. 1966. Peat resources of Minnesota. Report of Inventory no. 3. Red Lake Bog, Beltrami County, Minnesota. State of Minnesota Office of Iron Range Resources and Rehabilitation.

Foster, D. R., and G. A. King. 1984. Landscape features, vegetation, and developmental history of a patterned fen in southeastern Labrador, Canada. Journal of Ecology 72:115-143.

French, C. N., and P. D. Moore. 1986. Deforestation, Cannabis cultivation and Schwing moor formation at Cors Llyn (Llyn Mire), central Wales. New Phytologist 102:469-482.

Futyma, R. P., and N. G. Miller. 1986. Stratigraphy and genesis of the Lake Sixteen peatland, northern Michigan. Canadian Journal of Botany 64:3008-3019.

Glaser, P. H. 1983. Vegetation patterns in the North Black River peatland, northern Minnesota. Canadian Journal of Botany 61:2085-2104.

——— and J. A. Janssens. 1986. Raised bogs in eastern North America: Transition in surface patterns and stratigraphy. Canadian Journal of Botany 64:395-415

——— and G. A. Wheeler. 1980. The development of surface patterns in Red Lake peatland, northern Minnesota. Pages 31-45 in Proceedings of the Sixth International Peat Congress, Duluth, Minnesota, USA.

———, G. A. Wheeler, E. Gorham, and H. E. Wright, Jr. 1981. The patterned mires in the Red Lake peatland, northern Minnesota: Vegetation, water chemistry and landforms. Journal Ecology 69:575-599.

———, J. A. Janssens, and D. I Siegel. 1990. The response of vegetation to chemical and hydrological gradients in the Lost River peatland, northern Minnesota. Journal of Ecology 78:1021-1048.

Gorham, E. 1957. The development of peatlands. Quarterly Reviews of Biology 32:145-166.

———. 1988. Canada's peatlands: Their importance for the global carbon cycle and possible effects of "greenhouse" climatic warming. Transactions of the Royal Society of Canada, Series V, 3:21-23.

——— and J. A. Janssens. Concepts of fen and bog reexamined in relation to bryophyte cover and the acidity of surface waters. Acta Societatis Botanicorum Poloniae, in press.

Green, B. H. 1968. Factors influencing the spatial and temporal distribution of Sphagnum imbricatum Hornsch. ex Russ. in the British Isles. Journal of Ecology 56:47-58.

Griffin, K. O. 1975. Vegetation studies and modern pollen spectra from the Red Lake peatland, northern Minnesota. Ecology 56:531-546.

———. 1977. Paleoecological aspects of the Red Lake peatland, northern Minnesota. Canadian Journal of Botany 55:172-192.

Hansen, B. C. S., and D. R. Engstrom. 1985. A comparison of numerical and qualitative methods of separating pollen of black and white spruce. Canadian Journal of Botany 63:2159-2163.

Heinselman, M. L. 1963. Forest sites, bog processes, and peatland types in the Glacial Lake Agassiz region. Ecological Monographs 33:327-374.

———. 1965. Spring bogs and other patterned organic terrain near Seney, Upper Michigan. Ecology 46:185-188.

———. 1970. Landscape evolution, peatland types and the environment on the Lake Agassiz Peatlands Natural Area, Minnesota. Ecological Monographs 40:235-261.

Hemond, H. F., and J. C. Goldman. 1985. On non-Darcian water flow in peat. Journal of Ecology 73:579-584.

Hill, M. O. 1979. TWINSPAN — a FORTRAN program for arranging multivariate data in an ordered two-way table by classification of the individuals and attributes. Section of Ecology and Systematics, Cornell University, Ithaca, New York, USA.

Hinde, G. J. 1877. The glacial and interglacial strata of Scarboro' Heights, and other localities near Toronto. Canadian Journal of Science N.S. 15:388-413.

Hofstetter, R. H. 1969. Floristic and ecological studies of wetlands in Minnesota. Ph.D. dissertation. University of Minnesota, St. Paul, Minnesota, USA.

Horton, D. G., D. H. Vitt, and N. G. Slack. 1979. Habitats of circumboreal-subarctic Sphagna: I. A quantitative analysis and review of species of the Caribou Mountains, northern Alberta. Canadian Journal of Botany 57:2283-2317.

Jacobson, G. L., Jr. 1979. The paleoecology of white pine (Pinus strobus) in Minnesota. Journal of Ecology 67:697-726.

Janssen, C. R. 1967a. A floristic study of forest and bog vegetation, northwestern Minnesota. Ecology 48:751-765.

————. 1967b. A postglacial pollen diagram from a small *Typha* swamp in northwestern Minnesota, interpreted from pollen indicators and surface samples. Ecological Monographs 37:145-172.

————. 1968. Myrtle Lake: A late- and post-glacial pollen diagram from northern Minnesota. Canadian Journal of Botany 46:1398-1408.

————. 1984. Modern pollen assemblages and vegetation in the Myrtle Lake peatland, Minnesota. Ecological Monographs 54:213-252.

Janssens, J. A. 1983a. A quantitative method for stratigraphical analysis of bryophytes in Holocene peat. Journal of Ecology 71:189-196.

————. 1983b. Past and extant distribution of *Drepanocladus* in North America, with notes on the differentiation of fossil fragments. Journal of the Hattori Botanical Laboratory 54:251-298.

————. 1984. Quaternary fossil bryophytes in North America: New records. Lindbergia 9:137-151.

————. 1989. Ecology of peatland bryophytes and paleoenvironmental reconstruction of peatlands using fossil bryophytes. Methods manual, updated December 1989. Available from the author.

————. 1990. Methods in Quaternary ecology — Bryophytes. Geoscience Canada 17:13-24.

———— and P. H. Glaser. 1986. The bryophyte flora and major peatforming mosses at Red Lake peatland, Minnesota. Canadian Journal of Botany 64:427-442.

Jasnowski, M. 1957. *Calliergon trifarium* Kindb. Pages 701-718 *in* der Stratigraphie und Flora der holozänen Niedermoore Polens. Acta Societatis Botanicorum Poloniae XXVI (4). (In Polish, German summary.)

Klinger, P. U. 1968. Feinstratigraphische Untersuchungen an Hochmooren mit Hinweisen zur Bestimmung der wichtigsten Großreste in nordwestdeutschen Hochmoortorfen und einer gesonderten Bearbeitung der mittel-europäischen Sphagna cuspidata. 135 S., Dissertation Kiel 1968.

Line, J. M., and H. J. B. Birks. 1989. WACALIB 2.1 — a computer program to reconstruct environmental variables from fossil assemblages by weighted averaging. Available for US $50 from H. J. B. Birks, Botanical Institute, University of Bergen, Allegaten 41, N-5007 Bergen, Norway.

McAndrews, J. H. 1966. Postglacial history of prairie, savanna, and forest in northwestern Minnesota. Torrey Botanical Club Memoirs 22:1-72.

Miller, N. G. 1980. Mosses as paleoecological indicators of late glacial terrestrial environments: Some North American studies. Torrey Botanical Club, Bulletin 107:373-391.

———— and R. P. Futyma. 1987. Paleohydrological implications of Holocene peatland development in northern Michigan. Quaternary Research 27:297-311.

Ratcliffe, D. A., and D. Walker. 1958. The Silver Flowe, Galloway, Scotland. Journal of Ecology 46:407-445.

Ritchie, J. C. 1983. The paleoecology of the central and northern parts of the Glacial Lake Agassiz basin. Pages 157-170 *in* J. T. Teller and L. Clayton, editors. Glacial Lake Agassiz. Geological Association of Canada Special Paper 26.

Rybnicek, K. 1966. Glacial relics in the bryoflora of the highlands Ceskomoravská vichvina (Bohemian-Moravian Highlands); their habitat and cenotaxonomic value. Folia Geobotanica Phytotaxonomica Praha 1:101-119.

————. 1973. A comparison of the present and past mire communities of central Europe. Pages 238-261 *in* H. J. B. Birks and R. G. West, editors. Quaternary plant-ecology. 14th Symposium of the British Ecological Society. Blackwell Scientific Publications, Oxford, U.K.

———— and E. Rybnicková. 1968. The history of flora and vegetation on the Bláto Mire in southeastern Bohemia, Czechoslovakia. Foliá Geobotanica Phytotaxonomica Praha 3:117-142.

Schuster, R. M. 1966. The Hepaticae and Anthocerotae of North America, east of the hundredth meridian. Vol 1. Columbia University Press, New York, New York, USA.

Shay, C. T. 1967. Vegetation history of the southern Lake Agassiz basin during the past 12,000 years. Occasional Papers, Department of Anthropology, University Manitoba 1:231-252

Siegel, D. I. 1983. Ground water and the evolution of patterned mires, Glacial Lake Agassiz peatlands, northern Minnesota. Journal of Ecology 71:913-921.

Sjörs, H. 1952. On the relation between vegetation and electrolytes in north Swedish mire waters. Oikos 2 (1950):241-258.

————. 1959. Bogs and fens in the Hudson Bay lowlands. Arctic 12:2-19.

————. 1961. Forest and peatland at Hawley Lake, northern Ontario. National Museums of Canada. Bulletin 171:1-30.

————. 1963. Bogs and fens on Attawapiskat River, northern Ontario. Museum of Canada Bulletin, Contributions to Botany 186:45-133.

————. 1983. Mires in Sweden. Pages 69-94 *in* A. J. P. Gore, editor. Mires: Swamp, bog, fen and moor. Regional Studies. Ecosystems of the World 4B. Elsevier, Amsterdam, Netherlands.

Soper, E. K. 1919. The peat deposits of Minnesota. Minnesota Geological Survey, Bulletin 16:1-261.

Stuiver, M., and P. J. Reimer. 1986. A computer program for radiocarbon age calibration. Radiocarbon 28:1022-1030.

ter Braak, C. J. F. 1987. CANOCO — a FORTRAN program for canonical community ordination by [partial][detrended][canonical] correspondence analysis, principal components analysis and redundancy analysis (version 2.1). TNO Institute of Applied Computer Science, Statistics Department, Report 89 ITI A11:1-95. Wageningen, Netherlands.

Tolonen, K., and M. Tolonen. 1984. Late-glacial vegetational succession at four coastal sites in northeastern New England: Ecological and phytogeographical aspects. Annales Botanici Fennici 21:59-78.

Vitt, D. H., and S. Bayley. 1984. The vegetation and water chemistry of four oligotrophic basin mires in northwestern Ontario. Canadian Journal of Botany 62:1485-1500.

———— and N. G. Slack. 1984. Niche diversification of *Sphagnum* relative to environmental factors in northern Minnesota peatlands. Canadian Journal of Botany 62:1409-1430.

Watts, W. A. 1978. Plant macrofossils and Quaternary paleoecology. Pages 53-67 *in* D. Walker and J. C. Guppy. Biology and Quaternary environments. Australian Academy of Science.

———— and T. C. Winter. 1966. Plant macrofossils from Kirchner Marsh, Minnesota — a paleoecological study. Geological Society of America, Bulletin 77:1339-1360.

Wheeler, G. A., P. H. Glaser, E. Gorham, C. M. Wetmore, F. D. Bowers, and J. A. Janssens. 1983. Contributions to the flora of the Red Lake peatland, northern Minnesota, with special attention to *Carex*. American Midland Naturalist 110:66-96.

Wright, H. E., Jr. 1984. Red Lake peatland: Its past and patterns. James Ford Bell Museum, Imprint vol. I, summer 1984.

———— and J. A. Janssens. 1982. Book review of "Peat Stratigraphy and Climatic Change" by K. E. Barber. Science 216:616-617.

————, D. H. Mann, and P. H. Glaser. 1984. Piston corers for peat and lake sediment. Journal of Ecology 65:657-659.

Zoltai, S. C., S. Taylor, J. K. Jeglum, G. F. Mills, and J. D. Johnson. 1988. Wetlands of boreal Canada. Pages 97-154 *in* C. D. A. Rubec, editor. Wetlands of Canada. Polyscience Publications, Montreal, Quebec, Canada.

CHAPTER 14

The Myrtle Lake Peatland

C. R. Janssen

The Myrtle Lake peatland, as the second-finest peat-land in northern Minnesota, has been designated as an area to be protected from degradation by human activity (Minnesota Department of Natural Resources 1984). It is almost entirely surrounded by uplands, and does not receive any major input or water from other peatlands. With its watershed area, it is thus an ecological entity. It is large enough (23,000 acres) to contain most of the usual peatland features of northern Minnesota. The watershed area is only an additional 12,600 acres. Among the 17 peatlands proposed to be protected by the Minnesota Department of Natural Resources (1984), only one other peatland (Hole-in-the-Bog) has a watershed protection area smaller than the core area, and it is much smaller than the Myr-tle Lake peatland. The area that feeds the Myrtle Lake peatland is thus relatively small, and this unusual situation offers a fine opportunity to study the de-velopment of a peatland in direct connection with the development of the vegetation of the surrounding upland.

Present-day Geomorphology

Myrtle Lake peatland is situated in the southeastern corner of Koochiching County. It developed in the southeastern part of former Glacial Lake Agassiz on a gentle slope to the north (Heinselman 1963, 1970). The peatland can be divided roughly into four major units (Fig. 14.1). The southern part, separated from the rest of the peatland by a ridge completely over-grown by peat, contains Myrtle Lake itself in a deep pit in the substratum. An inflowing stream on the south is almost obliterated by peat growth. To the northwest lies unit two, a large wet fen dotted with a number of rock outcrops. One of these outcrops is the so-called *Abies* island, shown in Plate 14A and Fig. 14.2

This unit is partly patterned with linear flarks (pools) and strings (ridges) perpendicular to the drainage di-rection. It resembles the aapa mires in the middle boreal zone in Scandinavia, but they may have a dif-ferent origin (chapter 3). The main drainage occurs by way of this patterned area, called a water track, al-though moving water cannot be observed during most of the year. Water drains to tributaries of the Little Fork and Big Fork rivers, which are eroding into the peatland.

The third unit of the peatland is a large raised bog in the north. An aerial photograph (Plate 16) shows a radial pattern in the vegetation, indicating that wa-ter drains in all directions from a central ridge. Careful measurements by Heinselman (1970) do indeed indi-cate that the peat surface is convex, although the slopes are only 2 to 6 feet per mile. Despite the small differ-ences in elevation, the changes in the vegetation are obvious.

The fourth, rather complex, unit is in the eastern part of the peatland, north and south of a large upland peninsula.

Present-day Vegetation

Seen from the air, many peatlands display beautiful vegetation patterns. Water is usually in abundant sup-ply, and the overriding ecological factor causing these patterns is often the mineral content of the water. A number of vegetation types can be defined largely on the basis of the water chemistry. In Fig. 14.1, the major vegetation types are shown. In a raised bog, the peat accumulates in a domelike form, so that water with dis-solved minerals (nutrients) from other areas is diverted around it.

The vegetation is ombrotrophic, or fed by rain, de-pending for its nutrients on the rain and snow that

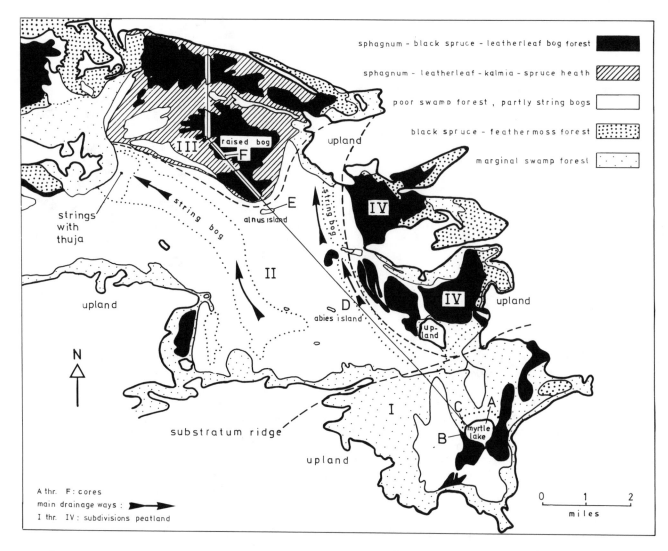

Fig. 14.1. The Myrtle Lake peatland: location of the major subdivisions of peatland vegetation types and miscellaneous topographic features described in the text. (From M. L. Heinselman, "Landscape Evolution, Peatland Types, and the Environment in the Lake Agassiz Peatlands Natural Area, Minnesota," *Ecological Monographs* 40 [1970]: 235-261. Copyright 1970 by the Ecological Society of America, by permission.)

under normal conditions are poor in minerals — at least before the recent pollution of the atmosphere by agricultural and industrial activities. These small amounts of minerals are increasingly incorporated in the accumulating peat. As a result, the available mineral content of peat is low in raised bogs, usually less than 1 mg/liter dissolved calcium (the so-called "mineral soil water limit"). In addition, hydrogen ions in the substrate are not neutralized. Under these conditions, the pH and the specific conductivity have low values. The vegetation is adapted to such an environment. In regions with an oceanic climate, raised bogs are treeless, except along the margins. In regions with a more continental climate, however, raised bogs have at least a partial cover of trees (spruce in North America, pine in continental Europe). In northern Minnesota this ombrotrophic vegetation is either a stunted black spruce-leatherleaf bog forest

or a heathland with varying amounts of black spruce (*Picea mariana*), bog laurel (*Kalmia polifolia*), bog rosemary (*Andromeda glaucophylla*), and leatherleaf (*Chamaedaphne calyculata*). The Myrtle Lake raised bog is thus a typical example of a continental raised bog.

In areas outside the raised bog, the mineral-rich water provides nutrients to the roots of the plants. Two rough vegetation types can be recognized: (1) a species-poor swamp forest in the center of the peatland, with widely dispersed tamarack (*Larix laricina*) and western white cedar (*Thuja occidentalis*), along with ericaceous shrubs such as bog rosemary, leatherleaf, and swamp dwarf birch (*Betula pumila* var. *glandulifera*) and (2) a species-rich swamp forest along the margins of the peatland and around upland islands, here called marginal swamp forest. Tamarack and cedar are much taller here than in the poor

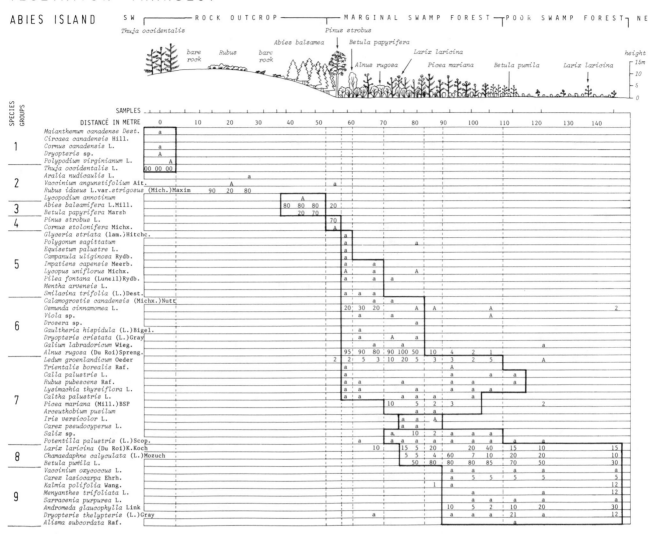

VEGETATION TRANSECT

Fig. 14.2. Vegetation transect through the marginal swamp forest around the *Abies* island and into the poor swamp forest. The plant species are arranged in groups according to occurrence along the transect. Groups 1-4 represent plant species on the rock outcrop. Groups 5-9 depict the differences in vegetation in response to diminishing mineral content in the substrate. Roughly, a dense tall alder shrub with some tamarack and black spruce is reduced and replaced by black spruce-tamarack swamp forest with an understory of bog birch, which finally grades into poor swamp forest with low, widely spaced tamarack and an understory of bog birch and sedges. Abundances of plant species are shown by numbers that indicate percentage cover of relevant plant species or by the letters *A* and *a*, referring to two categories of high (*A*) and low (*a*) cover value or abundance. *Vertical dashed lines* that divide the table delineate plant communities defined on the basis of species groups. Transects like these were used to study the change in the local pollen deposition in relation to the local vegetation along the transect. (From C. R. Janssen, "Modern Pollen Assemblages and Vegetation in the Myrtle Lake Peatland, Minnesota," *Ecological Monographs* 54 [1984]: 213-252. Copyright 1984 by the Ecological Society of America, by permission.)

swamp forest; black ash (*Fraxinus nigra*) is present, and the shrub and herb layer has a large species diversity.

The relation between vegetation and substrate can be studied in fine detail along environmental gradients from one vegetation type to another. Figure 14.2 is a vegetation transect extending a 150 meters (some 500 feet) from a rock outcrop (*Abies* island; Plate 1) into the

surrounding poor tamarack swamp forest. Rainwater falling on the island penetrates the soil and fissures in the rock, picking up minerals before the outflow to the surrounding peatland. Here this drainage from the island gives rise to a marginal swamp forest with a dense shrub layer of alder (*Alnus rugosa*), on the northwest side of the island, where water moves away from the mineral source. The influence of the mineral-rich water is visible downstream in the form of a long "tail" of marginal swamp forest.

On the opposite southern and southeastern side of the island, the water moves toward the island, and here the marginal swamp forest is very narrow; the gradient from island to poor swamp forest is so short and strong that the intermediate plant communities are not fully developed. At the northern side of the island, the marginal swamp forest grades into poor swamp forest over an intermediate distance, resulting in a neat alignment of plant communities according to the mineral content of the substrate. Plant species appear and disappear with increasing distance from the island, and the reduction in the mineral content in the substrate causes a drastic reduction in the height and spacing of the trees. Tree growth is increasingly inhibited, and competition among the shallow tree roots prevents growth of trees within short distances of each other.

Strong environmental gradients are also present around Myrtle Lake itself (Plates 14b and 15). The lake waters are rich in minerals and have a pH near neutral, as is testified by the presence of wild rice (*Zizania aquatica*) as well as water lilies (*Nymphaea odorata*) and yellow pond lilies (*Nuphar* sp.) floating on the water surface. It is not entirely clear where the minerals come from. Because upland areas that may serve as a source are not within close range of the lake, the minerals must come either from the incoming stream on the south or from springs that discharge mineral-rich groundwater into the lake. The stream is surrounded by ombrotrophic black spruce forest, resulting in a very strong ecological gradient at the margin of the lake. Typical for such a situation is sweet gale (*Myrica gale*). This shrub forms a wide belt on the south between the stream and the black spruce forest.

A different situation exists on the opposite (northern) side of the lake. Because the lake does not have an outlet on the north, the excess mineral-rich water is forced through the peat like a fan, and a very rich cedar swamp with a dense understory of cinnamon fern (*Osmunda cinnamomea*) — a veritable jungle, almost impassable — has developed. A nutrient gradient between this forest and the lake waters is not present because mineral-rich conditions exist both in the lake and in the substrate of the adjoining peatland: a belt of sweet gale is absent, and a cedar swamp forest is well developed right at the edge of the lake. The fan shape of the cedar swamp north of the lake is thus determined by the northward drainage of mineral-rich water through the peat. The swamp is bounded sharply by the lake and on both sides by ombrotrophic spruce forest (Plates 14b and 15). Between these two different types of forest a strong ecological gradient exists, best developed near the lake. The cedar swamp forest diminishes gradually northwards as the mineral content of the water is diluted in that direction, and as a result the ecological gradient is increasingly less steep away from the lake.

Past Vegetation and Peatland Development

Past peatland vegetation can be reconstructed from the kinds of peat in cores taken from various parts of the peatland: peat consists of the plant debris that in compacted form is what remains of past vegetation. An experienced eye can distinguish sedge peat from forest peat or lake sediments by the structure of the material. In the Myrtle Lake peatland these kinds of observations were carried out extensively by Heinselman (1970; Figs. 14.3 and 14.4). A more elaborate way to determine the sequence of peat types is quantitative analysis of the components — wood, leaves, seeds, and fruits of the plant species, commonly called macrofossils — of the peat matrix. This method offers a detailed picture of the peatland vegetation

The disadvantage of macrofossil analysis is that we gain an idea of the peatland vegetation only at specific sites. Macrofossils from other parts of the peatland or from the upland usually do not reach the coring sites. The various parts of the peatland are highly interrelated in development and depend in turn on the vegetation development of the upland, especially in the watershed area, which provides the source or the mineral-rich water that enters the peatland. It is therefore important to relate the local sequence of peatland development to the contemporaneous sequences elsewhere in the peatland. For that reason, a different kind of strategy — analysis of pollen grains and spores produced by the various plant species — was chosen to give information on the development both of the various parts of the peatland and of the upland vegetation. The pollen analyst can recognize numerous plant genera and even species from morphological differences in the pollen grains and spores that are well preserved in anaerobic environments such as peat bogs and lake sediments.

Figure 14.5 shows pollen of oak and black spruce and a spore of bracken, all three common in the past and present in the Myrtle Lake peatland. The pollen analyst separates the pollen grains into categories related to plant taxa. Pollen grains not only are well preserved in peat, after deposition they are not displaced vertically in the peat column, as may occur in soils by downwash. A column of peat is thus an archive of past pollen deposition and therefore of the history of the vegetation cover. Access to the archive is through a small amount of peat chemically treated to dissolve other organic material, enabling the scientist to identify and count the pollen grains under a high-power microscope.

Unlike most macrofossils, pollen grains are released from their sources — the anthers of flowers and the sporangia of mosses and ferns — and some are transported by the wind and dispersed evenly before they are deposited, creating what is called the regional component. Pollen grains thus reflect events in the vegetation development at localities other than the site of deposition, mostly in the upland vegetation around the peatland or from areas even hundreds of miles away.

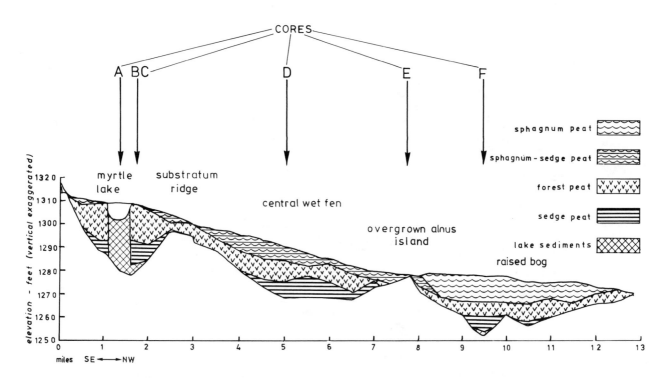

Fig. 14.3. A cross section through the peatland along the Heinselman Trail, depicting the main peat types, the substratum ridge, and the *Alnus* island (after Heinselman 1970). *A* through *F* indicate location of peat cores taken for analysis of pollen and macrofossils. (From M. L. Heinselman, "Landscape Evolution, Peatland Types, and the Environment in the Lake Agassiz Peatlands Natural Area, Minnesota," *Ecological Monographs* 40 [1970]: 235-261. Copyright 1970 by the Ecological Society of America, by permission.)

Most of the pollen grains, however, fall to the ground at or very near the point of release, creating what is called the local component. The pollen at any given point thus includes both local and regional components, recording the events both in the peatland and on the surrounding uplands.

Pollen production and dispersal are variable depending on the plant species involved, the time of day and year, wind direction and other atmospheric conditions, relief of the surface, and the type of vegetation cover. Separation of the local and regional components is therefore difficult but can be determined approximately by studying the final product of these processes — modern pollen assemblages — in relation to the surrounding vegetation. This approach has been applied extensively in the Myrtle Lake peatland (Janssen 1973, 1984) and elsewhere in Minnesota (Janssen 1966; Griffin 1975). In the Myrtle Lake peatland, pollen analyses have been carried out on moss samples collected on a transect along the so-called Heinselman Trail (see Fig. 14.4) across the peatland and also along short transects (e.g., Fig. 14.3). The main conclusion of these studies is that there is a rather even deposition of regional pollen types across the peatland. If this were also true in the past, then similar regional pollen assemblages from various cores must have deposited at the same time. The recognition of the regional component in the pollen deposition is

thus an important tool to determine synchronous levels in pollen sequences from various locations in the peatland.

Another important conclusion from studies of the relation between the modern pollen and recent vegetation is that peatland vegetation types can best be defined on the basis of the local pollen assemblage. Many of the plant species along the transect of Fig. 14.2 are represented only in the local pollen deposition. Pollen groups can be established to represent the species groups in the present-day vegetation, and the paleovegetation can be reconstructed by combining these pollen groups.

Peat Types

Heinselman (1970) determined the sequence of peat types along a number of transects across the peatland along the Heinselman trail. The cross section (Fig. 14.3) shows the two localities where the mineral substrate is close to the peat surface: the substratum ridge north of Myrtle Lake and an overgrown upland island, the so-called *Alnus* island just southeast of the raised bog dividing the peat body effectively into three parts. Heinselman recognized five types of peat by visual inspection. In a complete profile in the raised bog area, lake sediments are overlain in turn by sedge peat, forest peat, and *Sphagnum* peat. Figure 14.3 shows that

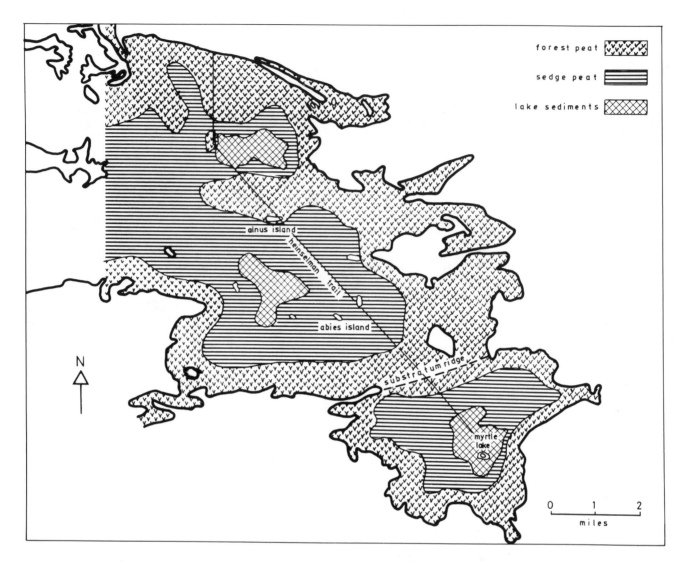

Fig. 14.4. Basal peat types in the Myrtle Lake peatland and location of substratum ridge (after Heinselman 1970). Lake sediments are known to exist at Myrtle Lake and below the raised bog and western fen areas. The remainder of the peatland is underlain by sedge peat and, along the perimeter of the peatland, by forest peat. (From M. L. Heinselman, "Landscape Evolution, Peatland Types, and the Environment in the Lake Agassiz Peatlands Natural Area, Minnesota," *Ecological Monographs* 40 [1970]: 235-261. Copyright 1970 by the Ecological Society of America, by permission.)

in three areas, lake sediments are present at the base of the peat: in the still-existing Myrtle Lake itself and in two areas in the middle and northern parts of the peatland. The latter two former lakes are now overgrown to such an extent that there is no indication at the surface that a shallow lake was ever present. These two lakes may have disappeared by the process of lake filling or terrestrialization.

Myrtle Lake is an anomaly in that it is still open, and it does not show the usual stages involving peat formation, starting with a sedge mat encroaching upon a lake. At Myrtle Lake there is no floating sedge mat; ombrotrophic bog forest extends right to the margin of the lake. It is not clear why the normal process of lake filling is absent in this shallow lake. Perhaps the lake serves

as a discharge for groundwater originating in the high ground south of the peatland, as postulated for other parts of the Lake Agassiz peatlands by Siegel (chapter 11). At any rate, ombrotrophic peat growth next to the lake is keeping pace with a rise in the lake level, sustaining the extraordinary gradient between the lake and the bog environment.

The large fen area in the middle of the Myrtle Lake peatland has a sequence in which *Sphagnum*-sedge peat overlies a basal forest peat. The cross section shows clearly that at many places the sequences are less complete. The sequences of peat types indicate that in these localities peat formation started by the process of swamping or paludification (chapter 3). In this process (in contrast to terrestrialization), an originally dry sur-

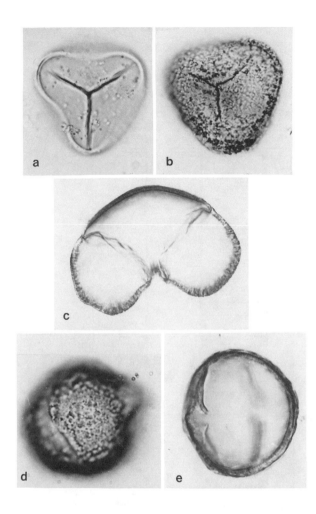

Fig. 14.5. Photomicrograph of pollen and spore types commonly found in the present-day pollen deposition and in the peats and lake sediments of the Myrtle Lake peatland. **a** and **b**: Spore of bracken (*Pteridium aquilinum*) without (**a**) and with (**b**) the outer layer, the so-called perisporium. Magnification x 1200. **c**: Pollen grain of black spruce (*Picea mariana*): a large grain with two bladders attached to the body of the grain. Magnification x 450. **d** and **e**: Pollen grain of red oak (*Quercus rubra*) in surface view (**d**) and in optical section (**e**). Surface coarsely, irregularly granular. Pollen grain provided with three (two visible) longitudinal furrows in the surface. Magnification x 1200.

face becomes so wet that the decay of organic matter from the vegetation is retarded, resulting in the accumulation of organic material. The Heinselman survey indicates that sedge peat typically occupies the basal layers of the peat beds in the center of the peatland and that forest peat is present at the bottom of the peat column at the perimeter of the peatland and at other localities where the mineral substrate is close to the surface. How do these lake sediments and peat beds fit into the general vegetation development of north-

ern Minnesota? To answer this question we can look at pollen diagrams from two sites in the peatland.

Pollen-analytical Studies

Data gathered from analyses of pollen are presented in a pollen diagram, a collection of curves representing the quantities of the pollen types found in the sediment plotted against depth. Quantities are generally expressed as percentages of the total of an established number of pollen types, usually the regional types. Quantities may also be expressed as concentrations of pollen grains per cubic centimeter or as the influx of pollen grains on a square centimeter per year. For the core in the lake sediments of Myrtle Lake, both percentage and concentration values are available.

Cores of peat were collected for pollen analysis from various parts of the peatland (A-F along the Heinselman Trail in Figs. 14.1 and 14.3). Pollen diagrams for cores from Myrtle Lake and from the raised bog in the north are presented here.

The Sequence at Myrtle Lake

The regional pollen curves for Myrtle Lake (Fig. 14.6) display the general pattern found in cores from the forested area of north-central Minnesota. Pollen zone 1 at the base reflects a spruce forest or spruce parkland that covered most of northern Minnesota before temperate trees migrated into the area. The sediment is silt, indicating that after the retreat of the water of Lake Agassiz the environment was still unstable, allowing transport of mineral particles into the lake. During the time covered by zone 2, the immigration of several elements such as pine (*Pinus resinosa* or *P. banksiana*), balsam fir (*Abies balsamea*), bracken (*Pteridium aquilinum*), and elm (*Ulmus* sp.) took place. These taxa were absent from northern Minnesota during the glacial period, but when climatic conditions improved they immigrated from refugia in the south and southeast. The spruce forest/parkland declined, and the sediment became organic, indicating that a closed vegetation cover stabilized the surroundings of Myrtle Lake.

The pollen assemblages of zone 3 reflect the postglacial warm/dry interval, the so-called prairie-period, recognizable in most of the pollen diagrams of north-central Minnesota. During this phase the prairie/forest border was located farther east than at the present time (McAndrews 1966), and the westward migration of white pine (*Pinus strobus*) was halted (Jacobson 1979). In the Myrtle Lake pollen diagram, this phase is characterized by increased pollen values of the prairie taxa wormwood (*Artemisia*), ragweed (*Ambrosia*), and Chenopodiaceae. Above the base of zone 3, white pine (*Pinus strobus*) became an element of the upland vegetation, initially mixed with deciduous elements such as ironwood (*Carpinus*), elm (*Ulmus*) and basswood (*Tilia*). Above zone 3, however, the upland forests were

MYRTLE LAKE

REGIONAL POLLEN TYPES

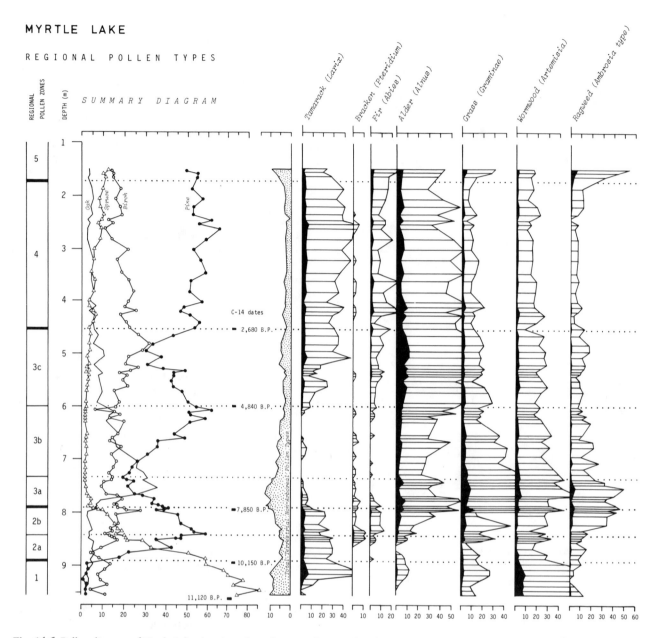

Fig. 14.6. Pollen diagram of Myrtle Lake showing selected curves of regional pollen types, that is, those produced by plant species that do not belong to the peatland vegetation. These are used to establish a regional relative chronology, which provides the basis for dating the levels of the cores in the peatland. Zones 3a and 3b represent the "prairie" period, a mid-Holocene warm phase during which prairie vegetation and deciduous forest vegetation expanded. Dates are in conventional uncalibrated radiocarbon years before the present. (Reproduced by permission of the National Research Council of Canada, from the *Canadian Journal of Botany* 46 [1968]: 1397-1408.)

dominated by pine, fir, and birch, and the prairie component diminished in the pollen diagram to the low values we find today. In the topmost zone, zone 5, is the usual increase of the values for *Ambrosia* and Chenopodiaceae, connected with the occupation of land by immigrant farmers late in the last century.

What kind of information on peatland development can be extracted from the pollen analysis of the Myrtle Lake core? A number of pollen curves from peatland taxa are shown in Fig. 14.7. The high pollen values of

spruce and tamarack in zones 1 and 2 indicate that these genera were quite common at that time. They were probably not part of the peatland vegetation, but they occupied mineral soils as they do today in the boreal forest of northeastern Minnesota. The warm/dry prairie period (zone 3) is decidedly unfavorable for ombrotrophic peat formation, as is the case in the present-day prairie. The poorly drained soils were covered by sedges, cattails (*Typha latifolia*), ferns, sweet gale, and perhaps alder.

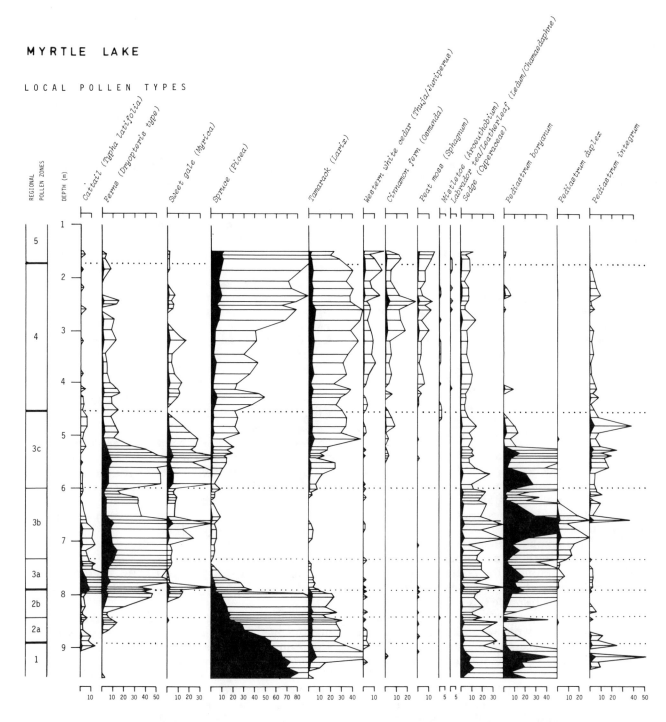

Fig. 14.7. Pollen diagram of Myrtle Lake, with selected curves of local peatland plant species. (Reproduced by permission of the National Research Council of Canada, from the *Canadian Journal of Botany* 46 [1968]: 1397-1408.)

The lake ecosystem responded to the dry interval by the introduction of the alga *Pediastrum duplex* var. *cohaerens* at the transition of zones 2b and 3a. Values for sweet gale increased and retracted in three waves, indicating that stands of sweet gale expanded and shrank depending on the existence of gradients between mineral-rich and mineral-poor substrates some-

where around the lake. Such an alternation could be the result of fluctuations in the climate, for changes in the precipitation and evaporation could cause changes in the input of mineral-rich water into the lake. The pollen concentration diagram indicates that the fluctuations in the values for sweet gale are connected with changes in the rate of sediment accumulation, another indication that the lake ecosystem changed as a response to climate fluctuations. Pollen concentration indicates that fluctuations in the values for sweet gale are connected with changes in the rate of sediment accumulation, reflecting changes in lake level caused by climatic changes.

The first indication that peat began to accumulate again around the lake is at the zone 3b/3c transition 4,000 years ago. Pollen of spruce and tamarack begin to increase in value, and spores of cinnamon fern (*Osmunda cinnamomea*) make their appearance, indicating that the rich swamp forest at the northern end of the lake came into being at this time. Conditions were different from present conditions, however, for the sweet gale pollen values were quite high, and alder was at a maximum. The ombrotrophic bog forest around the southern perimeter of the lake began more than 1,000 years later, as reflected by the continuous expansion of the spruce pollen values. The appearance of *Sphagnum* spores and of pollen of the bog Ericales, along with the demise of sweet gale and most of the *Pediastrum* species in the lake, support the change to more ombrotrophic conditions at the southern end of the lake. The spread of ombrotrophic bog forest is also confirmed by the appearance of pollen of dwarf mistletoe (*Arceuthobium pusillum*, a parasite that lives on spruce).

The Sequence at the Raised Bog

The relation between the present-day vegetation and the modern pollen deposition indicates that the area of pollen dispersal of peatland plant species is quite local. Clearly, we need pollen diagrams from other parts of the peatland to elucidate the vegetation developments elsewhere in the peatland.

Figure 14.8 is a simplified pollen diagram from a core of the raised bog in the north at a locality where organic lake sediment is present at the bottom. The regional pollen assemblages at the base belong to zones 1 and 2a, the phases in which spruce is replaced by pine. The ancient lake that was present here apparently started at the same time as Myrtle Lake. Both lakes can be considered as remnants of Glacial Lake Agassiz. The ancient lake at the raised bog site was soon overgrown by vegetation.

The peat at the levels of pollen zone 2a is charred and contains scarcely any pollen, except around a depth of 675 cm, where the pollen assemblage indicates that it belongs to zone 2b. Above that level the peat is amorphous, with much charcoal and almost no pollen. Peat is again present from 640 cm on. What is the time represented by this level? The regional pollen assemblages, especially the low values (10%) of oak, the dominance of pine, and the beginning of the fir curve, can be compared with those at the lower part of zone 3c at Myrtle Lake. The conclusion is that zones 3a and 3b of the Myrtle Lake diagram, representing the prairie period, are missing in the raised bog diagram, along with

most of zone 2b. This phase may have been too dry for peat formation, although it is more likely that the peat deposited at that time was burned at the end of the phase represented by zone 2b. A comparable absence of material from the prairie period was recorded at the small Stevens Pond in Itasca State Park (Janssen 1967). In critical situations — when the peat bog is small, for instance — the surface of the peat apparently becomes so dry that accumulation of peat stops and fire becomes possible. When peat accumulation began again we observe high local values of bracken (*Pteridium aquilinum*), which expands especially on dry peat surfaces. Conditions then became wetter and accumulation of peat accelerated.

The change from lake sediments to peat implies terrestrialization. The high values of bracken at the renewal of peat accumulation, however, indicate that the original lake had largely dried out and the formation of peat began by swamping or paludification of the area. In the pollen diagram of the raised bog, the process of paludification is marked by the appearance of tamarack pollen and by a huge peak in the *Sphagnum* curve just above the bracken maximum. This process started about 5,000 years ago, according to a radiocarbon date of the transition between zones 3b and 3c at Myrtle Lake.

Surprisingly little is known exactly about the time when peat accumulation began in the various areas of the large peatlands of northern Minnesota. Most scientists studying this field agree that this accumulation began 4,000-4,500 years ago (Wright and Glaser 1983). Clearly this event is connected with the end of the prairie period, when the precipitation-evaporation ratio shifted toward wetter conditions. It is also around this time that white pine (*Pinus strobus*) resumed its westward migration (Jacobson 1979).

Further Development of the Vegetation at the Raised Bog

The local pollen assemblages show interesting fluctuations in the succession leading to the raised bog. The main components are sedge (Cyperaceae), heaths (*Chamaedaphne, Kalmia, Andromeda*), tamarack, and black spruce. The pollen representation of tamarack today shows that the stunted trees in the poor swamp forest and in the raised bog have low pollen production, so peat samples below these stunted trees show almost no tamarack pollen. In rich swamp forests, however, tamarack grows taller and is able to produce more pollen. Tamarack produces rather heavy pollen grains; when they are released they fall to the ground within a short distance from the source. Thus low tamarack pollen values in peats indicate absence of tamarack or the presence of poor tamarack swamp forest, whereas higher values indicate the local presence of a richer type of swamp forest. The pollen diagram (Fig. 14.8) shows that tamarack has low pollen values in the local

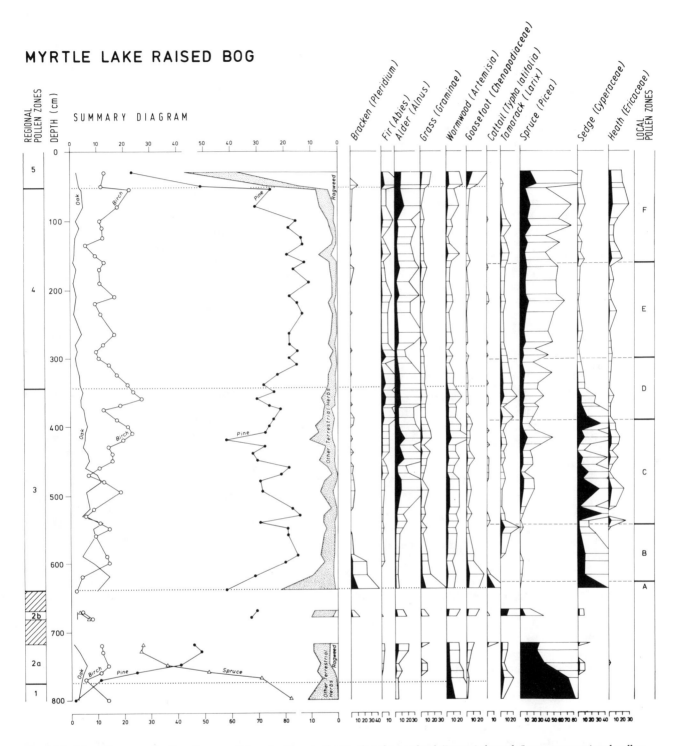

Fig. 14.8. Selected curves of pollen diagram of the raised bog in the Myrtle Lake peatland. Zones 1 through 5 represent regional pollen zones used for dating. The shaded part indicates charred peat with little pollen. Zones A through F represent local pollen zones depicting the succession in peatland vegetation.

pollen zones A, C, E, and F. In local pollen zones B and D, however, tamarack has (for tamarack) such high values that it must be assumed that taller, more productive trees were present than today.

Although a rich swamp forest never developed, the substrate was probably richer than in today's poor swamp forest. The pollen diagram indicates in the low regional pollen values of alder (*Alnus rugosa*) and black ash (*Fraxinus nigra*) (not shown in Fig. 14.8) that the substrate never became so rich in minerals that the succession led to marginal swamp forest. Higher values of tamarack are also present in the sediments of Myrtle Lake, which is distant from sites where most of tamarack grows. The lake core is special, however, for tamarack grows in the rich swamp forest at the water's edge, and pollen is easily transported by water currents to the middle of the lake where the core was taken.

Studies of the present-day pollen deposition of heath and sedge in relation to the pattern of vegetation proved to be equally useful for the reconstruction of the peatland vegetation. Heath and sedge have a higher pollen production than tamarack, but the dispersal capacities are also poor. The pollen representation of these three taxa reveal the local vegetation development. Six local pollen zones can be distinguished:

Local pollen zone A: just after the fires. Incipient paludification with sedges and cattail (*Typha latifolia*) in wet habitats and bracken in the drier spots.

Local pollen zone B: tamarack-sedge swamp, disappearance of cattail.

Local pollen zone C: ericaceous sedge fen.

Local pollen zone D: transition to raised bog, tamarack more productive in pollen, sedge gradually disappearing.

Local Pollen zone E: invasion of black spruce in the area of the present-day raised bog.

Local pollen zone F: present-day conditions, with expansion of black spruce swamp forest and leatherleaf heathland.

What would explain the presence of a rather productive swamp forest in sites where today only stunted, depauperate trees occur? For the lowest zones A and B, the answer is not too difficult: minerals may have been released in the substrate by fires, giving rise to a richer vegetation. Soon, however, in local zone C the declining values for tamarack indicate that the substrate became increasingly depleted in minerals. Judging from the high pollen values of sedge, conditions must have been quite wet. It is not clear, however, why tamarack became more productive during local zone D. Sedge retreated in the vegetation, perhaps indicating a temporarily drier environment suitable for tamarack.

From local pollen zone E upward, accumulation of organic matter raised the surface of the bog to such an extent that mineral-rich water from elsewhere was excluded. The present-day radiating pattern then may have come into existence, This level can be correlated in time by comparing the trends in the regional pollen curves — the end of the temporary rise in the terrestrial herbs, the beginning of stable high values of pine, and

the low values of oak — at Myrtle Lake and the raised bog. The radiocarbon date at Myrtle Lake is 2,680 years ago for this level.

Intriguing questions remain: How did the Myrtle Lake peatland look before the prairie period? Was most of the area covered by boreal upland forest, or was it already partly covered by marshes and swamps? To what extent were swamps present during the dry prairie period? Did fires burn away most of the peat, or were some parts of the area so wet that they escaped fires? To answer these questions, a number of long cores across the peatland, along with many short cores, preferably in a grid to cover the transition between the mineral substrate and the basal peat, would be necessary. The results from these cores would have to be compared with the results of the analyses of macrofossils, which often can be identified to a lower taxonomic level. For instance, seeds of sedge may give us a clue about the species that were part of the peatland vegetation during local pollen zones A-D at the raised-bog site. The integrated analysis of macrofossils, pollen, and possibly other organic elements in the peat must be carried out in other parts of the peatland, along with hydrological studies, to decipher the interrelation between the development of the various landscape elements in the Myrtle Lake peatland. Important questions to be addressed are the influence of upland vegetation development in the watershed area upon the peatland development and the importance of the substratum ridge northwest of Myrtle Lake and the now-submerged *Alnus* island southeast of the raised bog in obstructing or diverting the drainage flow across the peatland. The *Alnus* island may constitute a protective barrier behind which a raised bog could develop more easily than elsewhere. The Myrtle Lake peatland constitutes a unique opportunity to study these problems for a large peatland with a rather limited watershed area. The small size of its watershed area offers a realistic chance to protect both peatland and surrounding upland, safeguarding this beautiful peatland from contamination and finally from destruction from the outside world.

Literature Cited

Griffin, K. O. 1975. Vegetation studies and modern pollen spectra from the Red Lake peatland, northern Minnesota. Ecology 56:531-546.

Heinselman, M. L. 1963. Forest sites, bog processes, and peatland types in the Glacial Lake Agassiz region, Minnesota. Ecological Monographs 33:327-374.

———. 1970. Landscape evolution, peatland types, and the environment in the Lake Agassiz Peatlands Natural Area, Minnesota. Ecological Monographs 40:235-261.

Jacobson, G. L., Jr. 1979. The palaeoecology of white pine (*Pinus strobus*) in Minnesota. Journal of Ecology 67:697-726.

Janssen, C. R. 1966 Recent pollen spectra from the deciduous and coniferous-deciduous forests of northwestern Minnesota: A study in pollen dispersal. Ecology 47:804-825.

PART V
Human Influences

CHAPTER 15

The Archaeological and Ethnohistoric Evidence for Prehistoric Occupation

Mary K. Whelan

Archaeological material provides evidence of the human occupation of Minnesota from at least 12,000 years ago to the present. Unfortunately, many cultural practices are elusive, if not invisible, in the archaeological record. The use of peat by prehistoric peoples in Minnesota is an example. Whether archaeologists have not recognized the evidence of past peat use or whether prehistoric people did not in fact use peat remains to be determined. While there is no historic or archaeological evidence that historic Indian groups (Ojibwe and Dakota) burned peat for fuel or used dried peat blocks as building material, many of the individual plants and animals indigenous to the Minnesota peatlands were, and are, used by Indian people today, and it is likely that those materials were used in much the same way by prehistoric groups. On the basis of written records and ethnographic analogy, historically described peat use can be projected into the more-distant past with some certainty. For the historic period (A.D. 1492 to present), we have documentary evidence fully describing the cultural practices (including uses of peatland plants and animals) of the Indian societies that once occupied the state.

Because this discussion is based in large measure on ethnographic analogy, a connection must be established between the Indian people who were described by early explorers as inhabiting northern Minnesota during the 17th and 18th centuries and their prehistoric ancestors known only from archaeological investigation. Euroamerican colonization of the New World resulted in the widespread displacement of native people, so the distribution of Indian cultures presented on colonial maps often illustrates only the more re-

cent boundaries imposed by Euroamerican resettlement policies and warfare.

Historic description and Indian oral traditions report that the earliest inhabitants of northern Minnesota were the Dakota people, Siouan tribes of hunter-gatherer-horticulturalists (Warren 1885; Hickerson 1962; Walker 1983). This is consistent with 17th- and 18th-century cartographic representations of the region (Wedel 1974; Karpinksi 1977).

Sioux is a cultural designation that refers to a large association of related tribes who once inhabited the woodlands and plains from western Wisconsin to the Rocky Mountains. Rarely did the various members of this group act as a corporate body or define themselves as part of such a large group, but when this level of generality was needed the designation they used for all of the divisions was *Oceti Sakowin*, the "seven council fires" or "seven fireplaces" (Riggs 1893; Hodge 1912; Howard 1984; Powers 1975).

This larger polity was composed of three linguistic divisions: those who lived primarily in the forested regions of Minnesota and spoke the Dakota dialect; those who lived on the western border of Minnesota and along the eastern edges of North and South Dakota and spoke Nakota; and those who lived in the western prairie areas along the Missouri River and spoke Lakota.

In the historic period, other cultural groups — including the Ojibwe, Cheyenne, Iowa, Oto, and Omaha — are thought to have occupied different parts of Minnesota (Johnson 1985). Indian myth and oral tradition state that at some point in the 18th century the Dakota and the recently arrived Ojibwe engaged in a fierce war over the right to live in and use northern Minnesota lands. The Ojibwe were expanding westward from their previous homelands in Wisconsin and Michigan as the frontier of Euroamerican occu-

The author would like to thank Blane H. Nansel for his help in researching this paper.

pation pushed to the west. Outfitted with firearms and weapons obtained from Euroamerican traders, the Ojibwe began a period of hostile attacks on other Native American societies in Wisconsin and neighboring northern Minnesota. By virtue of their superior weaponry, the Ojibwe were able to push the Dakota out of northern Minnesota. According to Ojibwe and Dakota legends, the territorial war culminated in the Battle of Kathio, in which the Ojibwe succeeded in vanquishing their Dakota enemies. The Dakota were forced to escape southward into the lands around the Minnesota River (Warren 1885; Folwell 1921; Hickerson 1962). Suggested dates for the Battle of Kathio and the subsequent Dakota shift to the south range from 1720 to 1800 (Anderson 1980; Hickerson 1962, 1974; Warren 1885; Winchell 1911). Although there is some discussion of the veracity of the details of this scenario (Johnson 1985; Anderson 1980), it is generally accepted that the Ojibwe did replace the Dakota in northern Minnesota during the 18th century. Thus, both Ojibwe and Dakota peoples have adapted to the resources of northern Minnesota's peatlands. Historic and ethnographic information summarized in this chapter indicate that both cultural groups made extensive use of peatland plant and animal taxa for food, medicine, ritual substances, and a variety of domestic items.

Because it is likely that Dakota or Ojibwe people inhabited northern Minnesota prehistorically, it is tempting to project modern peatland exploitation practices into unknown prehistoric millennia. Large-scale population movements occurred prehistorically as well as historically, however, and it is difficult to assign specific labels (e.g., Dakota) to prehistoric cultures with any confidence. A series of archaeological sites near Mille Lacs Lake in central Minnesota provides some information on the prehistoric antecedents of the historic Dakota (Johnson 1985; Ossenberg 1974). Ossenberg examined skeletal material from a number of prehistoric burial locations and compared osteological traits (those known to be under genetic control) with traits in modern American Indian skeletons. She argued that the Dakota were probably the descendants of several distinct archaeological cultures. In particular, she concluded that the Laurel culture (100 B.C.-800 A.D.) and the Blackduck culture (800-1400 A.D.) were closely related to historic Dakota people and thus were probably ancestral to them (Ossenberg 1974).

Johnson argued that the late prehistoric groups known archaeologically from the Vineland Bay site (21-ML-7), the Cooper Village site (21-ML-9), the Cooper Mound site (21-ML-16), and the Wilford site (21-ML-12) in Mille Lacs County provided the most recent link to the Dakota. Using prehistoric artifacts and historically dated French trade materials, Johnson dated these sites between 600 and 1700 A.D. (1985). He concluded that the continuity of a number of cultural aspects (including subsistence pattern, seasonal movement patterns, artifact styles, and house forms)

indicated that the archaeological cultures referred to as Bradbury phase were ancestral to the historically known Dakota (Johnson 1985). Farther back in time — that is, before 100 B.C. — cultures that can be shown to be ancestral to the historic Dakota are unknown. Neither Ossenberg nor Johnson could determine a link with any of the prehistoric populations that date prior to 100 B.C. Since the peatlands of Minnesota have been developing over the past 5,000 years, however, the earliest Indian cultures (i.e., Paleo-Indian) may have been living in forested rather than peat-covered landscapes (Heinselman 1970).

Historic-Period Occupation

The subsistence economy of the historic Dakota emphasized the seasonal exploitation of key resources: deer, elk, moose, and other mammals; numerous species of fish, turtles, and birds; wild rice, maple sugar, and other wild plant foods; and limited horticulture, including corn and tobacco (Pond 1908; Howard 1984; Johnson 1985; Whelan 1987). Buffalo were also hunted, sometimes in joint expeditions with the Nakota and Lakota groups to the west (Landes 1968; Walker 1982). Villages acted autonomously for the most part, and the village is most properly thought of as the basic unit of production, distribution, and consumption.

The Dakota seasonal cycle consisted of two halves, winter and summer. Time was measured in terms of the lunar cycles, and seasonal activities were often reflected in the names they assigned the different moons (Riggs 1893). *Wozupiwi,* "the planting moon," translated as May; *Wipazokawastewi,* "the moon when the berries [June berries] are ripe," translated as June; and *Wasutonwi,* "the harvest moon," translated as August (Riggs 1890). Although these names are descriptive and not strictly a calendar of activities, they do fit the seasonal pattern of tasks described in historic and ethnographic sources. Winter would have been the time between November and March, while summer was the time between May and September. The moon names from May through October refer to horticultural and gathering activities, whereas the winter names refer primarily to animals. What is suggested linguistically, and also described ethnographically, is a seasonal round of animal harvest during winter and plant harvest during summer.

Pond's description of the Dakota economy in 1834 fits within the broad outline of this seasonal pattern. Deer were of importance for their meat and skin, and deer hunting usually took place in the "winter" season, between October and January (Pond 1908). During this time the entire group moved frequently, staying in temporary tent camps. The remainder of the winter season was spent in tipis near or in the summer village site, where stored food (especially corn and wild rice) could be added to the winter's meat supply. Occasional hunting, fur trapping, and fishing, as well as considerable

Fig. 15.1. *Spearing Muskrats in Winter.* Watercolor by Seth Eastman. Photo courtesy James Jerome Hill Reference Library, St. Paul, Minnesota.

socializing with friends and family, occupied the late winter (Fig. 15.1; Pond 1908).

Around April, the transitional moon between the winter and summer halves of the year, the village inhabitants splintered into smaller groups. Men went off to muskrating sites, for spring was the time when muskrat pelts were in prime condition; women went to their sugar-bush locations to process maple syrup (Pond 1908). April was also the month when the year's pelts were traded. May saw the group reunited in their summer village and living in bark lodges rather than tipis. Turtles, waterfowl, fish, and local mammals were hunted early in this season while the corn fields were prepared and the crop planted (Pond 1908).

As the summer progressed, domestic and wild crops became available to supply enough food to meet present demands, and gradually some surplus was accumulated and stored for the next winter. Travel was easy in the summer, when the waterways were open, and small groups frequently made forays to other villages or other parts of the *Oceti Sakowin* territory to visit and trade (Pond 1908).

October was the other transitional moon, linking the summer horticultural season to the winter hunting time. Pond reported that this was also a time when the group split into smaller units, with some going to the wild rice beds, others to cranberrying locations, and still others off on fall fur-trapping expeditions, in which muskrat figured prominently (Pond 1908; Fig. 15.2). Once these various tasks had been completed, the village reassembled and the annual cycle began again with another deer-hunting season.

Ojibwe seasonal movement and economic cycles are similar to Dakota seasonal activities (chapter 16). Maple sugaring and spring fishing took place at family "sugar bush" locations (groves of sugar maple trees, *Acer saccharum*) during the early spring. With the onset of consistently warmer weather, multifamily groups moved to semipermanent summer villages, where they engaged in horticultural and gathering activities (supplemented by fishing and hunting). Multifamily groups again assembled at wild rice camps in the late summer and early fall to harvest the wild rice crop, hunt birds, and fish. With the onset of winter, small family groups dispersed to protected winter camps, where they could live off

Fig. 15.2. *Rice Gatherers.* Oil by Seth Eastman. U.S. House of Representatives Collection. Photo courtesy Architect of the Capitol, Washington, D.C.

stored produce from the previous year, supplemented by winter hunting and ice fishing.

Peatland Resources

Four words for peat and related formations appear in the Dakota-language dictionaries compiled by 19th-century missionaries (Riggs 1890; Williamson 1886). Williamson's English-Dakota dictionary lists:

bog	*wiwi*; *maka coco* (p. 21)
marsh	*wiwi* (p. 21)
peat	*butkan aonpi* (p. 102)
swamp	*ptega*; *wiwi* (p. 129)

Riggs's 1890 dictionary lists the Dakota-to-English translation of these words as:

wiwi	a bog; a quagmire (p. 580)
maka coco	soft, muddy earth (pp. 305, 102)
butkan aonpi	a place where roots are layed down (pp. 40, 160)
ptega	a low, swampy place (p. 427)

These words suggest that the Dakota recognized and named an ecological community with the characteristics of peatland. This may be taken to indicate that the Dakota had some need for distinguishing this ecological community; the numerous peatland plant and animal taxa used by the Dakota support this notion. Furthermore, the Dakota made clear distinctions between wet, swampy areas (*wiwi* and *ptega*) and peatlands (*butkan aonpi*), suggesting that they had a rather precise taxonomy presumably built on different experiences in and uses for swamps and bogs.

The Ojibwe language also reveals distinctions between swampy areas and what might be interpreted as peat areas. Rhodes (1985), in his modern dictionary of Ojibwe language, lists the following terms:

marsh	*miishkooki*, or *waabshkoki* (p. 519)
swamp	*mshkiig* (p. 590)
swamp land	*mshkiigki* (p. 590)

The derivation of these words is informative. The first term for marsh (*miishkooki*) and the words for

Table 15.1. Minnesota Peatland Plant Taxa Commonly Used by Native Americans

Taxon	Use code	Taxon	Use code
Shrubs, Herbs, and Mosses		**Shrubs, Herbs, and Mosses (cont.)**	
Acorus Calamus (sweet flag)	M,R	*Potentilla palustris* (marsh cinquefoil)	M
Alnus rugosa (smooth alder)	M,O	*Ribes budsonianum* (swamp black currant)	M,F
Andromeda glaucophylla (bog rosemary)	F	*R. triste* (swamp red currant)	M,F
Asclepias incarnata (swamp milkweed)	R	*Rubus pubescens* (dwarf raspberry)	F
Aster simplex (paniculate aster)	R	*Sagittaria cuneata* (arrow-head)	M,F
Athyrium Filix-femina (lady fern)	M	*S. latifolia* (arrow-head)	M,F
Betula pumila (swamp birch)	M	*Salix spp.* (willow)	M,R,O
Caltha palustris (marsh marigold)	M,F	*Sarracenia purpurea* (pitcher plant)	M,O
Carex spp. (sedges)	O	*Scirpus validus* (larger bulrush)	F,R,O
Chamaedaphne calyculata (leatherleaf)	F,M	*Scutellaria galericulata* (marsh skullcap)	M
Cirsium arvense (thistle)	M	*Smilacina trifolia* (3-leaved Solomon's seal;	
Clintonia borealis (bluebead)	M,O	false Solomon seal)	M,O
Coptis trifolia (gold thread)	M,O	*Solidago graminifolia* (goldenrod)	M
Cornus canadensis (dwarf dogwood; bunchberry)	M,F	*Sphagnum spp.* (sphagnum moss)	O,X
Cornus stolonifera (red-osier dogwood)	M,O	*Spirea alba* (meadow sweet)	H
Cypripedium calceolus (yellow lady's slipper)	M	*Trientalis borealis* (starflower)	R
C. reginae (showy lady's slipper)	M	*Typha latifolia* (cattail)	M,O,X
Dicranum sp. (broom moss)	O,X	*Vaccinium angustifolium* (blueberry)	M,F,O
Dryopteris cristata (shield fern)	M	*V. macrocarpon* (cranberry)	M,F
Equisetum fluviatile (horsetail)	M,O	*V. Oxycoccos* (small cranberry)	M,F
Eupatorium maculatum (Joe Pye weed)	M,R	*Viburnum trilobum* (large-bush cranberry)	M,F
E. perfoliatum (boneset)	M	*Zizania aquatica* (wild rice)	F
E. purpureum (trumpetweed)	M,R,O	**Trees**	
Galium trifidum (small bedstraw)	M	*Abies balsamea* (balsam fir)	M,R,O
Gaultheria bispidula (creeping snowberry)	F	*Betula lutea* (yellow birch)	M,F
Iris versicolor (large blue flag; blue iris)	M,O	*Betula papyrifera* (paper birch)	M,F,O,X
Lathyrus palustris (marsh vetchling)	M,O	*Fraxinus nigra* (black ash)	M,O,X
Ledum groenlandicum (Labrador tea)	M,F	*Juniperus communis* (dwarf juniper)	M
Lonicera oblongifolia (swamp honeysuckle)	M	*Larix laricina* (tamarack)	M,O,X
Malaxis unifolia (adder's mouth)	M,R	*Picea glauca* (white spruce)	M,R
Mentha canadensis (wild mint)	M,F,R,O	*Picea mariana* (black spruce)	M,R,O,X
Nuphar advena (yellow water lily)	M,F	*Pinus resinosa* (red pine)	M,R,O,X
Nymphaea odorata (white water lily)	M,F	*Pinus Strobus* (white pine)	M,R,F,O,X
Phragmites communis (reed)	O,X	*Thuja occidentalis* (Northern white cedar)	M,R,O
Polygonum coccineum (swamp Persicaria)	M,O		

Note: M = medicinal; F = food; R = religious; O = other; X = utilized but unknown function(s).

Source: Plant taxa compiled from Soper 1917; Harper 1964; Heinselman 1970; Boelter and Verry 1977. Native American usages compiled from Gilmore 1919; Densmore 1928; Smith 1932; Moerman 1977, 1986.

swamp and swamp land (*mshkiig, mshkiigki*) contain a root word used to refer to grass or prairie (*miishkoons*, grass; *miishkoonskaag*, lots of grass; *mshkode*, prairie). This suggests that marshes, for the Ojibwe, are reedy or grassy fens. The word for cranberry, *mshkiigmin*, and the word for larch, *mshkiigwaatig*, are clearly associated with swamps. The second term for marsh, however, contains an entirely different root. *Waabshkoko* may be indicative of areas of muskeg or bog because its root is the word for the color white or gray (*waab*). This term does not refer to reeds, grass, or woody vegetation but probably implies low ground-cover plants that are characteristically white or gray in color (e.g., mosses?), perhaps seasonally.

The uses of individual peatland plants and animals by Dakota and Ojibwe people were many and

varied (Tables 15.1 and 15.2). Vegetation commonly found in Minnesota peatland was important as food and medicine, in religious and magical rituals, as raw material for mats and baskets, and in the construction of houses and canoes. Table 15.1 lists the more commonly occurring peatland plant taxa and the historically recorded usages in both Ojibwe and Dakota societies.

Historical knowledge of Dakota ethnobotany is decidedly limited. No systematic studies of Eastern Dakota plant usage have been published. Gilmore's 1919 study of the ethnobotany of the Missouri River cultures includes the Western Dakota subdivisions, and while they are culturally related to the Dakota people in northern Minnesota, their prairie habitat does not contain peatland areas similar to those found in northern Minnesota. Consequently, much of the infor-

Table 15.2. Minnesota Peatland Vertebrate Taxa Commonly Used by Native Americans

Taxon	Use code	Taxon	Use code
Amphibians and Reptiles		**Birds (cont.)**	
Chelydra serpentina (snapping turtle)	F, O–A, H	*Nyctea scandiaca* (snowy owl)	O–H
Chrysemys picta (painted turtle)	F, O–A, H	*Pandion haliaetus* (osprey)	O–A,H
Graptemys geographica (map turtle)	F, O–A, H	*Picoides arcticus* (black-backed three-toed	
Graptemys pseudogeographica (false map turtle)	F, O–A, H	woodpecker)	O–H
Trionyx spiniferus (spiny softshell)	F, O–A	*P. pubescens* (downy woodpecker)	O–H
		P. villosus (hairy woodpecker)	O–H
Birds		*Podiceps grisegena* (red-necked grebe)	F–A
Accipiter cooperii (Cooper's hawk)	X–A	*Podiceps podiceps* (pied-billed grebe)	F–A
Accipiter gentilis (goshawk)	X–A	*Strix nebulosa* (great gray owl)	O–H
Accipiter striatus (sharp-shinned hawk)	X–A	*Tympharuchus phasianellus* (sharp-tailed grouse)	F–A
Anas americana (American widgeon)	F,O–A		
Anas discors (blue-winged teal)	F,O–A	**Mammals**	
Anas platyrhynchos (mallard)	F,O–A	*Alces alces* (moose)	F,C,O–A,H
Ardea heodias (great blue heron)	X–A	*Bison bison* (buffalo)	F,C,O–A,H
Aythya collaris (ring-necked duck)	F–A	*Canis latrans* (coyote)	F,C,O–A,H
Bonasa umbellus (ruffed grouse)	F,O–A	*Canis lupus* (wolf)	F,C,O–A,H
Botaurus lentiginosus (American bittern)	X–A	*Castor canadensis* (beaver)	F,C,O–A,H
Branta canadensis (Canada goose)	F,O–A	*Cervus candensis* (elk)	F,C,O–A,H
Bucephala clangula (goldeneye)	F–A	*Erithizon dorsatum* (porcupine)	F,C,O–H
Buteo jamaicensis (red-tailed hawk)	X–A	*Glaucomys sabrinus* (northern flying squirrel)	X–A
B. lagapus (rough-legged hawk)	O–H	*Lepus americanus* (snowshoe hare)	F–H
B. paltypterus (broad-winged hawk)	O–H	*Lutra canadensis* (river otter)	F,C,O–A,H
Circus cyaneus (marsh hawk)	X–A	*Lynx canadensis* (lynx)	F,C,O–A,H
Dendragopus canadensis (spruce grouse)	X–A	*Lynx rufus* (bobcat)	F,C,O–A,H
Dryocopus pileatus (pileated woodpecker)	R,O–A,H	*Martes pennanti* (fisher)	F,C,O–A,H
Falco columbarius (pigeon hawk)	O–H	*Mephitis mephitis* (striped skunk)	F,C,O–A,H
F. rusticolus (gyrfalcon)	O–H	*Mustela erminea* (ermine)	F,C,O–A,H
F. sparverius (American kestrel)	O–H	*Mustela frenata* (long-tailed weasel)	F,C,O–A,H
Fulica americana (American coot)	F–A	*Mustela vison* (mink)	F,C,O–A,H
Gavia immer (common loon)	F, O–A	*Odocoileus virginianus* (white-tail deer)	F,C,O–A,H
Grus canadensis (sandhill crane)	F–A	*Ondatra zibethicus* (muskrat)	F,C,O–A,H
Haliaeetus leucocephalus (bald eagle)	R,O–A,H	*Procyon lotor* (raccoon)	F,C,O–A,H
Lophodytes cucullatus (hooded merganser)	X–A	*Rangifer tarandus* (caribou)	F,C,O–H
Melanerpes erythrocephalus (red-headed)		*Sylvilagus floridanus* (cottontail rabbit)	F,C–A,H
woodpecker)	O–A,H	*Ursus americanus* (black bear)	F,C,O–A,H
Mergus merganser (common merganser)	X–A	*Vulpes vulpes* (red fox)	F,C,O–A,H

Note: F = food; R = religious; C = clothing; O = other; X = utilized but unknown function(s); A = archaeological; H = historic/ethnographic reference.

Source: Animal taxa compiled from Eastman 1902; Pond 1908; Breckenridge 1944; Harper 1964; Green and Janssen 1975; Hazard 1982. Native American usages compiled from Eastman 1902; Pond 1908; Whelan 1987; Whelan n.d.

mation in Table 15.1 is derived from studies of Ojibwe ethnobotany.

Table 15.1. clearly illustrates the diversity of taxa collected from peatlands (more than 47% of all species found in those areas) and the variety of uses to which individual species were put. Many food plants were harvested from northern Minnesota marshes and swamplands, as described by 19th-century missionary Samuel Pond in 1908:

Of these native food plants the most important were the *psincha*, the *psinchincha*, the *mdo*, the wild turnip or pomme de terre, the water-lily, and wild rice. The *psinchincha* is a root, in shape resembling a hen's egg, and about half as large. The *psincha* is spherical and about an inch in diameter. They both grow at the bottom of shallow lakes, and the former

sometimes grows in marshy ground, where there is not much water. These roots and those of the water-lily were dug, some by the men but more by the women. They often gathered them where the water was waist-deep, feeling for them with their feet at the bottom of the lakes.

These plants are also listed as important elements in the Dakota diet by other authors (Schoolcraft 1854; Eastman 1902). Although the scientific designations of these species are not always certain, four of the six taxa listed are likely to have been found in marsh habitats: *psinchincha*, *psincha*, water lily, and wild rice.

Wild rice (*Zizania aquatica*) was a particularly important food source, and the prevalence of wild rice in northern Minnesota freshwater habitats was one factor that was considered by historic and prehistoric Indians in selecting settlement locations for the summer and fall (Jenks 1900). Charles Eastman, a Dakota man writing at the turn of the century, described the August wild rice harvest as a time of great importance in the subsistence cycle of the Dakota (1902). Wild berries (especially cranberries and swamp currants), *psinchincha*, and water fowl were also harvested at this time.

When our people were gathering the wild rice, they always watched for [*psinchincha*] that grows in the muddy bottom of lakes and ponds. (Eastman 1902)

Their eyes were on all kinds of fruit, watching the ripening process. Berries of all kinds were industriously gathered. In a word, they diligently sought out everything edible, whether it grew on bushes or trees, on the ground, or in the mud at the bottom of the lakes. While some were digging all day on the prairies for a peck of wild turnips, others were in the water up to their arms, exploring the bottom of the lakes in search of *psinchincha*. Nothing was so hidden that they did not find it, nor so hard to come at that they did not get it. (Pond 1908)

After the wild rice had been harvested and processed, the fall deer hunt began. Pond notes that "swamp grass" (*Carex* species?) was used to insulate the tipis occupied during the deer hunt, and it also provided warmth and comfort as mattress material on the hunt (1908). This may have been *Typha latifolia*, however, for the Dakota word for this plant is *wihuta-hu*, "the bottom of a tipi" (Gilmore 1919).

A number of peat plants provided medicines for the historic Dakota and Ojibwe, including *Acorus calamus* (sweet flag), used as an analgesic; *Sagittaria arifolia* (arrowleaf), boiled as a digestive aid; and *Typha latifolia* (cattail) down used as a burn dressing (Moerman 1977; Gilmore 1919). Another important taxon of the region was *Cornus stolonifera* (red-osier dogwood), which when mixed with tobacco was called *kinikinick* and was smoked for religious and ceremonial purposes, as well as for pleasure (Gilmore 1919). Tree species common in peatland areas were also used. Birchbark was a critical resource — used for making canoes, summer house coverings, and containers — although the Dakota living in southern Minnesota substituted other barks (such as elm and basswood) and began constructing dugout canoes (Gilmore 1919; Pond 1908; Schoolcraft 1854). Mats for flooring were woven from the fibers of a number of plant species, including bulrush, *Scirpus validus* (Gilmore 1919). Several kinds of trees were used in the construction of summer lodges and winter tipis, including the various *Pinus* species (Gilmore 1919). Pitch from the coniferous trees found in and around peatlands was used as glue and sealant for canoes and containers (Densmore 1928).

Animal resources common to peatland habitats were also important to the Indian way of life. Table 15.2 lists the most commonly occurring animal taxa and their uses (when known) by Dakota people. The ethnozoology of Native American people is poorly known, although some evidence is provided in historic writ-

ings. Pond noted that the deer (presumably *Odocoileus virginianus*) was the cornerstone of the Eastern Dakota hunting economy (1908); peatlands would not be ideal deer habitat (chapter 6), but the range of this species does include all of the counties of northern Minnesota (Hazard 1982). The importance of deer to the Dakota is indicated by the fact that two months of their calendar, November and December, are named for the deer (Riggs 1890). Riggs's dictionary lists November as *takiyuhawi*, "the deer rutting moon" and December as *tahecapsunwi*, "the moon when the deer shed their horns." Pond points out that a number of other animal species (most of whose ranges include the northern Minnesota peatlands) were relied upon as well:

Next in importance to deer as food were ducks and geese, and in some parts of the country they were perhaps of even greater importance. . . . Muskrats were esteemed good food in winter and early spring, but not in warm weather. Fish and turtles were consumed by [the Dakota] in great quantities, but they did not like to be confined to a diet of fish. . . . There was a well beaten path leading to each lake within reach far and near, and the fishermen did not suffer the grass to grow in them. They prowled around the margin of every lake and marsh, and no tortoise could venture on the land or turtle put his head above the water, without running the risk of being captured. (1908)

Muskrat were particularly important for their pelts during the historic fur trade period in Minnesota (roughly 1775-1850), and large numbers of these animals were taken in spring and again in the fall by the Dakota (Pond 1908; Whelan 1987). Beaver were also a significant resource both historically, as part of the fur trade, and prehistorically. Other small and medium-sized mammals found in peatlands or the surrounding uplands included both skunks and otters; these species were harvested by historic and prehistoric Siouan people for their pelts as well as their meat. Charles Eastman reported that otter skin, bear skin, and beaver skin were highly prized representations of wealth for display, exchange, and payment of debts (1902).

Northern Minnesota is decidedly richer in large game than many Midwestern locations: elk, two species of deer (mule and white-tail), moose, caribou, and bison are all found there. While peat is not the preferred habitat of most of these animals (chapter 6), evidence of the presence of all six species has been found in the region either historically or prehistorically (Hazard 1982; Lukens 1973; Whelan 1990). For historic and prehistoric Siouan people living primarily through hunting and gathering — supplemented to some degree with horticulture — the large game in northern Minnesota must have been attractive. Hazard suggests that large ungulates such as bison may have retreated into peatlands in order to escape from forest fires in the surrounding prairie and woodlands (1982). Archaeological evidence suggests that prehistoric peoples often drove large game animals into boggy areas in order to dispatch them more easily (Wedel 1978). The Itasca bison kill site in Clearwater County is an example

(Shay 1971). Here at least 16 bison were killed along a marshy stream near Lake Itasca around 7,000 to 8,000 years ago, perhaps after having been driven off an adjacent slope by early Archaic hunters. Thus the northern Minnesota peatlands may have been a source not only of plant and animal resources but also of opportunities for trapping large game animals.

Prehistoric Occupation

No archaeological or ethnohistoric evidence of the use of peat for fuel in the New World has yet been uncovered. In Western Europe and Great Britain, where peat has been burned for fuel for millennia, archaeological evidence of peat usage coincides with late Neolithic forest clearance (Fischer 1980; Glob 1971; Godwin 1981; Iversen 1941, 1949, 1960; Moore 1973). The Neolithic, beginning there as early as 8,000 years ago, is defined by the extensive use of domesticated plants and animals, the manufacture of pottery, and the adoption of settled village lifeways. With the coming of iron-plow technology in prehistoric Europe (about 3,000 years ago), land clearance for cultivation seems to have resulted in widespread deforestation (Godwin 1981; Iversen 1949, 1960). Larger quantities of wood were supporting pottery kilns and smelting ovens and heating the homes of a growing human population. These activities had the unintended result of encouraging peat growth in Europe through deforestation (Coles and Harding 1979).

As wood supplies dwindled, peat became more important as a source of fuel and as a building material (Coles and Harding 1979). Evidence of the prehistoric use of peat in Europe and Great Britain has increased significantly as a result of the continued modern use of peat for fuel (Glob 1971; Fischer 1980): modern peat cuttings have been responsible for the recovery of most of the prehistoric material found in bogs and fens. Since the New World doesn't use peat for fuel, it is not surprising that no evidence of prehistoric use of peat has been uncovered.

In the New World the Neolithic is referred to as the Woodland period; most of the defining characteristics — pottery, settled village life, and the use of domesticated plants and animals — are the same. Woodland people are further distinguished by the large earthen burial mounds they constructed. The way of life of New World Woodland people was significantly different from that in the Old World, however, and consequently peat fuel may not have been needed to the same extent as in Europe. Historically we know that large-scale land clearance was not practiced by the Dakota, in part because frost limited reliance on horticulture but also because human population densities were not large enough to necessitate intensive horticultural practices. The prehistoric Woodland period in northern Minnesota appears to be similar to the Neolithic of Europe; some horticulture was practiced in conjunction with the hunting and gathering of wild re-

Table 15.3. Frequency and Types of Sites Found in Minnesota Peatlands

County	# of sites	# of sites in peatlands	Site names	Site type	Cultural affiliation
Beltrami	35	2	21-BL-2 (Waskish)	village	Blackduck
			21-BL-33 (Battle River)	village	
Lake of the Woods	6	3	21-LW-1	unknown	
			21-LW-2	unknown	
			21-LW-3	village	
Koochiching	23	7	21-KC-3 (Smith Mound)	burial	Laurel Blackduck
			21-KC-4	burial	
			21-KC-5	burial	
			21-KC-6 (Houska Point)	village	Archaic Laurel
			21-KC-7 (Little Fork)	burial	Archaic
			21-KC-8 (Nett Lake Pictographs)	other	
			21-KC-9	burial and village	
Roseau	20	0			
Marshall	24	0			

sources, but extensive deforestation to clear land for settlements and horticultural plots was uncommon. Plows were not used prehistorically by Indian people in Minnesota (whereas the prehistoric European technology relied heavily on draft animals and iron plows), and floodplain and river terrace soils were consequently preferred for American-style hoe and digging-stick horticulture. In the absence of plows, much smaller fields were also cultivated. Still, in regions where timber is relatively scarce (as around the open bogs, fens, and marshes in Koochiching County, Minnesota) peat may have been burned for fuel in prehistoric times, though documentary or archaeological evidence must still be found to confirm this conjecture.

Some initial conclusions about the prehistoric use of peatlands can be reached by examining the distribution of archaeological sites (i.e., settlement pattern) in each county (Table 15.3). In settlement-pattern analysis, archaeologists assume that prehistoric people chose to occupy specific locations in order to exploit the resources nearby (Roper 1979; Vita-Finzi and Higgs 1970). Presumably, living sites in the peatlands were intended to be conveniently located near important peatland resources. The majority of the Minnesota peatlands are in Beltrami, Lake of the Woods, Koochiching, Roseau, and Marshall counties. In general, there are relatively few known archaeological sites in these counties (Anfinson 1987). Of the 35 recorded sites in Beltrami County, only 2 are located in the peatlands north of Red Lake. Lake of the Woods County has just 6 recorded sites, 3 of which are located in peat-covered areas. There are 23 known sites in Koochiching County, of which 6 are located on peatlands and 13 are on upland areas surrounded by peat. None of the 20 sites in Roseau County is located directly on peatlands, but any significant movement outside these villages would have necessitated at least crossing the peatlands. Finally, Marshall County has 24 recorded archaeological sites, none of which is located near peat deposits.

Even the sites located on high ground in these five counties would have been surrounded by peatlands, and the resources available there would have been familiar to the prehistoric people. The seasonal round of activities of the historic Dakota may be expected to hold in general terms for the prehistoric Woodland people as well, in which case periodic forays outside the village would have been common in the fall and spring. The location of prehistoric villages and burial-mound sites in or adjacent to peatlands thus implies that prehistoric Woodland people were intentionally living near peat resources they wished to exploit. The following section focuses on two of the prehistoric groups that are thought to be ancestral to the historic Dakota (Blackduck culture and Laurel culture) and describes the archaeological evidence of their ways of life and use of peat resources.

Woodland-Period Occupations

We have little archaeological evidence for occupation of northern Minnesota during the early Woodland period (1,000 B.C.-500 B.C.), but middle and late Woodland sites (500 B.C. to contact with Europeans) are more common. The most commonly identified middle Woodland expression in northern Minnesota is Laurel culture (100 B.C.-800 A.D.) and, according to Ossenberg (1974), the Laurel people were ancestral (along with the Blackduck culture) to the historically known Dakota Sioux. Mason (1981) argues that Laurel culture arose directly from Archaic antecedents; most significant changes included the introduction of Laurel pottery styles and the use of burial mounds for mortuary interment. Stoltman (1973) describes this middle Woodland culture:

What generally characterizes the Laurel culture, besides its Middle Woodland age . . . and northerly geographic position, is a hunting and gathering way of life focusing, no doubt in a seasonal rhythm, mainly upon fish, moose, and beaver. In addition most Laurel sites are characterized by a distinctive assemblage of artifacts that includes the earliest ceramics in the area.

Cooking with pottery has many technological advantages over cooking with nonceramic materials (Arnold 1985), and Laurel people presumably could have eaten a much broader range of plant and animal foods than their Archaic forebears. This may have included more green leafy vegetables available in the Minnesota peatlands (e.g., *Aster, Caltha*, or *Nymphaea*). Ceramic cooking technology also renders seeds more digestible, for they frequently require longer cooking times. Further, cooking vessels allow for the destruction of bacteria and parasites in meat and the breakdown of some plant toxins. Since Laurel culture sites are more numerous and larger than Archaic sites — evidence of a more successful adaptation to northern Minnesota, including the peatlands — ceramic technology may have played an important part in adaptation to this region.

Unfortunately, there is little direct evidence of the prehistoric use of plants from the Laurel period in Minnesota. Floral remains are rarely preserved and require exacting field-recovery techniques, which have been developed just since 1968. As a result, we can only hypothesize that Laurel groups used many of the same peatland plant taxa as the historic Dakota and Ojibwe cultures. Since the skeletal evidence suggests that Laurel populations in northern Minnesota are related to later Dakota people, it is reasonable to project historically known uses back in time. Though no evidence of burning peat fuel or using peat for construction in prehistoric times is currently available, this may be due to a lack of investigation into these questions.

Faunal evidence from this period is more widespread (Lukens 1973). As Stoltman (1973) describes the Laurel economy, moose, beaver, and fish were the faunal mainstays of the diet. All of these resources were abundant in the peatland habitats of northern Minnesota. Lukens reported on the faunal assemblages from three Laurel sites in Minnesota, analyzing a total of 8,094 bones and fragments (1973). Fish was the most numerous vertebrate — nearly 57% of the total. Mammal remains were next in frequency. Beaver, moose, and woodchuck — just 3 species out of 27 mammalian taxa identified — made up 75% of the mammal bones recovered. Birds (mainly aquatic) and turtles accounted for the remainder of the assemblage. Tools and other articles were also made from parts of peatland animals. Stoltman found evidence of worked moose and caribou phalanges, harpoon heads made of moose antlers, and engraving tools made of beaver incisors (1973). Clearly, the animal component of the Laurel economy emphasized species common to aquatic habitats, including peatlands.

Blackduck culture (800-1400 A.D.) was widespread across northern Minnesota, Manitoba, and southern Ontario during late Woodland times. Also ancestral to historic Dakota people, Blackduck populations replaced Laurel people in Minnesota. Again there is a decided lack of evidence for the use of plants by Blackduck groups in Minnesota. While it is reasonable to suggest that information on plant usage known from the historic Dakota and Ojibwe cultures can be projected back, confirming archaeological evidence is needed.

Faunal analysis undertaken by the author on animal bones from the Hill Point site (Blackduck culture) reveals a pattern of animal usage similar to that described by Lukens for Laurel sites (Whelan 1990). The Hill Point site (21-CE-2) is located in Clearwater County, Minnesota, along the north arm of Lake Itasca, and is not in the northern peatlands region. The Itasca moraine on which the site rests is dotted with wetlands, however. In keeping with the more ecologically moderate terrain, the mammalian fauna from the site is dominated by deer (roughly 60%). Beaver, muskrat, and moose are next in abundance; each is roughly 10% of the mammalian fauna. Turtle species (including snapping turtle and painted turtle) and various species of fish are also present in the assemblage. Thus, peatland resources were still valuable and exploited by these later Blackduck peoples, as they were during Laurel times.

Summary

Although neither historic nor prehistoric use of peat for fuel or construction has been demonstrated for the Indian people of northern Minnesota, peatland resources have been exploited as a valuable part of their economy for more than 2,500 years. Historic ethnobotanical evidence illustrates uses of peatland plants for food, medicine, and ceremonial purposes, as housing material, for basketry and fiber, and in the construction of utensils and containers. We have little evidence of prehistoric peat plant usage, but the prehistoric Indians probably used plants for many of the same purposes. Curiously, historic knowledge of the use of animals indigenous to Minnesota peatlands is scarce, and archaeological evidence must be substituted. The faunal data indicate the exploitation of a variety of animals, particularly moose, beaver, and fish. These species were utilized for food, skins, and the production of an array of different tool types.

Literature Cited

Anderson, G. C. 1980. Early Dakota migration and intertribal war: A revision. Western Historical Quarterly 11:17-36.

Anfinson, S. 1987. 1986 annual report, Minnesota municipal and county highway archaeological reconnaissance study. Minnesota Historical Society, St. Paul, Minnesota, USA.

Arnold, D. 1985. Ceramic theory and cultural process. Cambridge University Press, New York, New York, USA.

Boelter, D., and E. Verry. 1977. Peatland and water in the northern lake states. U.S. Department of Agriculture Forest Service General Technical Report NC-31.

Breckenridge, W. J. 1944. Reptiles and amphibians of Minnesota. University of Minnesota Press, Minneapolis, Minnesota, USA.

Coles, J. M., and A. F. Harding. 1979. The Bronze Age in Europe. Methuen, London, U.K..

Densmore, F. 1928. Uses of plants by the Chippewa Indians. Bureau of American Ethnology, Annual Report 44:275-397.

Eastman, C. 1902; reprint 1971. Indian boyhood. Dover, New York, New York, USA.

Fischer, C. 1980. Bog bodies of Denmark. Pages 177-193 in A. Cockburn and E. Cockburn, editors. Mummies, disease and ancient cultures. Cambridge University Press, New York, New York, USA.

Folwell, W. 1921. A history of Minnesota. Vol. 1. Minnesota Historical Society Press, St. Paul, Minnesota, USA.

Gilmore, M. R. 1919. Uses of plants by the Indians of the Missouri River region. Bureau of American Ethnology, Annual Report 33:43-154.

Glob, P. V. 1971. The bog people, Iron-Age man preserved. Faber and Faber, London, U.K.

Godwin, H. 1981. The archives of the peat bogs. Cambridge University Press, New York, New York, USA.

Green, J. C., and R. Janssen. 1975. Minnesota birds. University of Minnesota Press, Minneapolis, Minnesota, USA.

Harper, F. 1964. Plant and animal associations in the interior of the Ungava peninsula. Miscellaneous Publications of the University of Kansas Museum of Natural History, no. 38.

Hazard, E. B. 1982. The mammals of Minnesota. University of Minnesota Press, Minneapolis, Minnesota, USA.

Heinselman, M. L. 1970. Landscape evolution, peatland types, and the environment in the Lake Agassiz Peatlands Natural Area, Minnesota. Ecological Monographs 40:235-261.

Hickerson, H. 1962. The southwest Chippewa: An ethnohistorical study. American Anthropological Society Memoir no. 92.

———. 1974. Chippewa Indians IV: Ethnohistory of the Chippewa in central Minnesota. Garland, New York, New York, USA.

Hodge, F. W., editor. 1912. Handbook of American Indians north of Mexico. Smithsonian Institution, Bureau of American Ethnology, Bulletin no. 30, parts 1 and 2.

Howard, J. H. 1984. The Canadian Sioux. University of Nebraska Press, Lincoln, Nebraska, USA.

Iversen, J. 1941. Land occupation in Denmark's Stone Age. Danmarks Geologiske Undersogelse, R II, no. 66:1-66.

———. 1949. The influence of prehistoric man on vegetation. Danmarks Geologiske Undersogelse, Series IV, 3(6):1-25.

———. 1960. Problems of the early post-glacial forest development in Denmark. Danmarks Geologiske Undersogelse, Series IV, 4(3):1-32.

Jenks, A. E. 1900. The wild rice gatherers of the upper lakes: A study in American primitive economics. U.S. Bureau of American Ethnology, 19th Annual Report, 1897-1898 (2):1013-1137.

Johnson, E. 1985. The 17th century Mdewakanton Dakota subsistence mode. Pages 154-166 in J. Spector and E. Johnson, editors. Archaeology, ecology and ethnohistory of the prairie-forest border zone of Minnesota and Manitoba. J & L Reprints in Anthropology vol. 31. J & L Reprints, Lincoln, Nebraska, USA.

Karpinski, L. C. 1977. Maps of famous cartographers depicting North America. An historical atlas of the Great Lakes and Michigan, with bibliography of the printed maps of Michigan to 1880. 2nd edition. Meridian, Amsterdam, Netherlands.

Landes, R. 1968. The Mystic Lake Sioux. University of Wisconsin Press, Madison, Wisconsin, USA

Lukens, P. W., Jr. 1973. The vertebrate fauna from Pike Bay Mound, Smith Mound 4, and McKinstry Mound. Pages 37-45 in J. B. Stoltman, editor. The Laurel culture in Minnesota. Minnesota Prehistoric Archaeology Series no. 8. Minnesota Historical Society, St. Paul, Minnesota, USA.

Mason, R. 1981. Great Lakes archaeology. Academic Press, New York, New York, USA.

Moerman, D. 1977. American medical ethnobotany, a reference dictionary. Garland Press, New York, New York, USA.

———. 1986. Medicinal plants of Native America. University of Michigan Museum of Anthropology, Technical Reports, no. 19.

Moore, P. D. 1973. The influence of prehistoric cultures upon the initiation and spread of blanket bog in upland Wales. Nature 241:350-353.

Niemi, G. J., and J. Hanowski. 1983. Inter-continental comparisons of habitat structure as related to bird distribution in peatlands of eastern Finland and northern Minnesota, USA. Pages 59-73 in C. H.

Fuchsman and S. A. Spigarelli, editors. International Symposium on Peat Utilization. Bemidji State University, Bemidji, Minnesota, USA.

Ossenberg, N. S. 1974. Origins and relationships of woodland peoples: The evidence of cranial morphology. Pages 15-39 *in* E. Johnson, editor. Aspects of upper Great Lakes anthropology. Minnesota Historical Society Press, St. Paul, Minnesota, USA.

Pond, S. W. 1908. The Dakotas or Sioux in Minnesota as they were in 1834. Collections, Minnesota Historical Society 12:319-501.

Powers, W. 1975. Oglala religion. University of Nebraska Press, Lincoln, Nebraska, USA.

Rhodes, R. A. 1985. Eastern Ojibwa-Chippewa-Ottawa dictionary. Mouton Publications, New York, New York, USA.

Riggs, S. R. 1890. A Dakota-English dictionary. *In* J. O. Dorsey, editor. Contributions to North American ethnology no. 7. Government Printing Office, Washington D.C., USA

————. 1893; reprinted 1973. Dakota grammar, texts, and ethnography. Ross & Haines, Minneapolis, Minnesota, USA.

Roper, D. 1979. The method and theory of site catchment analysis: A review. Advances in archaeological method and theory 2:119-140.

Schoolcraft, H. R. 1854. Information respecting the history, condition and prospects of the Indian tribes of the United States. Lippincott, Grambo & Co., Philadelphia, Pennsylvania, USA.

Shay, C. T. 1971. The Itasca bison kill site: An ecological analysis. Minnesota prehistoric archaeology series no. 6. Minnesota Historical Society, St. Paul, Minnesota, USA.

Smith, H. H. 1932. Ethnobotany of the Ojibwe Indians. Bulletin, Public Museum of Milwaukee 4:327-525.

Soper, E. K. 1917. The origin, occurrence, and uses of Minnesota peat. Dissertation. University of Minnesota, Minneapolis, Minnesota, USA.

Stoltman, J. B. 1973. The Laurel culture in Minnesota. Minnesota prehistoric archaeology series no. 8. Minnesota Historical Society, St. Paul, Minnesota, USA.

Struever, S. 1968. Flotation techniques for the recovery of small-scale archaeological remains. American Antiquity 33:353-362.

Vita-Finzi, C., and E. S. Higgs. 1970. Prehistoric economy in the Mount Carmel area of Palestine: Site catchment analysis. Proceedings of the Prehistoric Society 36:1-37.

Walker, J. R. 1982. Lakota society. R. DeMallie, editor. University of Nebraska Press, Lincoln, Nebraska, USA.

————. 1983. Lakota myth. E. Jahner, editor. University of Nebraska Press, Lincoln, Nebraska, USA.

Warren, W. 1885. History of the Ojibway nation, based upon traditions and oral statements. Collections, Minnesota Historical Society 5:21-394.

Wedel, M. M. 1974. LeSueur and the Dakota Sioux. Pages 157-171 *in* E. Johnson, editor. Aspects of upper Great Lakes anthropology. Minnesota Historical Society Press, St. Paul, Minnesota, USA.

Wedel, W. 1978. The prehistoric plains. Pages 183-219 *in* J. D. Jennings, editor. Ancient Native Americans. Freeman, San Francisco, California, USA.

Whelan, M. No date. Faunal analysis from the Hill Point site (21-CE-2). Manuscript.

————. 1987. The archaeological analysis of a 19th century Dakota economy. Dissertation. University of Minnesota, Minneapolis, Minnesota, USA.

Williamson, J. P. 1886. An English-Dakota school dictionary. Yankton Agency, Dakota Territory.

Winchell, N. H. 1911. The aborigines of Minnesota. Pioneer Press Co., St. Paul, Minnesota, USA.

CHAPTER 16

The Red Lake Ojibwe

Melissa L. Meyer

Although archaeological evidence of the presence of humans in northwestern Minnesota dates back at least 12,000 years, the Red Lake Ojibwe are relative newcomers. Their predecessors had used available resources in a seasonal round of provisioning activities to supply their subsistence needs (chapter 15). Before the establishment of reservations, the Red Lake Ojibwe pursued a similar subsistence strategy. In fact, the Red Lake economy today continues to be based largely on the same resources that native inhabitants of the region have exploited for generations. In this sense, the Red Lake Ojibwe maintain a great deal of continuity with the past.

The peatlands to the north and east of the Red Lake Reservation are part of a regional water network and play an intrinsic role in filtering the waters entering Upper Red Lake. Over the course of the past century, reservation leaders have expressed an understanding of the area they inhabit as an ecosystem central to sustaining their way of life. Despite attempts to privatize their collectively held land base, to ditch and drain the peatlands, and to mine the peat for fuel, Red Lakers have steadfastly maintained their commitment to preserving the integrity of the ecosystem.

Ojibwe Migrations

By the late 18th century, Ojibwe bands had replaced Dakota villages at the lake and stream sites in northern Minnesota. Migrations that spanned many generations brought the Ojibwe people — driven by Iroquois raids from the east and lured by opportunities to the west — from the region of Sault Ste. Marie through the Great Lakes watershed into northern Minnesota. Hunting, trapping, and trading forays to the north and south of Lake Superior evolved into migrations and the eventual dispersion of the Ojibwe throughout the Hudson Bay drainage area, northern Wisconsin and Minnesota, and Michigan's upper peninsula. Separated geographically by Lake Superior, the northern and southern Ojibwe evolved culturally in different directions as a result of differences in both environment and trade relations. Ancestors of most of the Red Lake Ojibwe followed the northern route; they were one of very few groups in northern Minnesota to do so (Dunning 1958, 1969; Bishop 1974, 1976, 1978; Bishop and Smith 1970; Hickerson 1962, 1970).

In 1679 the Frenchman Daniel Grayson Duluth persuaded the Dakota to trade peacefully with the Ojibwe who had migrated southwestward by assuring both tribes of a reliable source of European trade goods. This alliance allowed the Ojibwe to establish their position as fur trade intermediaries and gave them access to game farther west, which was necessary because intensified hunting for the trade quickly depleted the limited game in boreal forest zones (Hickerson 1962, 1965, 1970).

The alliance between the Dakota and Ojibwe shattered when the French journeyed farther west in the 1720s and 1730s to establish direct trade relations with the Dakota, bypassing Ojibwe intermediaries. Faced with a reduction in game areas open to them, the southern Ojibwe moved into northern Minnesota to exploit as invaders the resources previously accessible to them as allies of the Dakota. Their migrations displaced Dakota communities that remained after the acquisition of the horse had enhanced bison hunting and trade opportunities, providing incentive for most of the Minnesota Siouan population to emigrate to the prairies of the Minnesota and Missouri rivers. Competition between the Dakota and Ojibwe settled into a stable pattern once the Ojibwe had successfully occupied the Minnesota coniferous woodland. The Red Lake band entered the scene from the north at a later time, thereby avoiding the most intense period of competition and profiting from the Dakota exodus (Hickerson 1962, 1965, 1970, 1974; Anderson 1980; Holzkamm 1983; Watrall 1968; White 1978).

The northern migration route had entailed less cultural reordering than that followed by the southern Ojibwe (Dunning 1958, 1969; Bishop 1974, 1976, 1978; Bishop and Smith 1970; Landes 1939; Densmore 1919, 1929). In addition, the isolation of the Red Lake area contributed to the reputation of Red Lake Indians as being among the most conservative Ojibwe bands in the state. The fact that the Red River flows north to Hudson Bay deterred settlement of the area by Euroamericans eager to develop market agriculture. Peatlands constituted a large portion of the habitat in which the Red Lake Indians practiced a subsistence strategy uniquely suited to the north country environment.

In the mid-19th century, the Red Lake Ojibwe followed an annual subsistence cycle of seasonal harvesting activities that closely resembled that of the early historic Dakota (chapter 15). Situated as they were on the shores of a bountiful fishery, the people relied on fish from Red Lake. Wild rice, the cereal grain staple of indigenous people of the western Great Lakes, also was an important part of their diet (Fig. 16.1). Their proximity to the peatlands insured that the variegated resources there would be as substantial a part of their subsistence as they had been for the Dakota (Tables 15.1, 15.2). Although seasonal migrations were part of the subsistence pattern, the stability of the timing and locations gave continuity to the annual cycle. Diverse resources provided a hedge against the failure of any particular resource. All in all, this annual subsistence cycle amply sustained the Ojibwe in the north country of Minnesota. Its flexibility allowed its long-term survival; the Ojibwe had even integrated the demands of the Euroamerican fur trade without excessive dislocation. This reliable seasonal circuit still serves as the cornerstone of autonomy for the Red Lake band (Densmore 1910, 1919, 1928, 1929; Mittelholtz 1957; Rogers 1974; James 1956; Cleland 1966; Yarnell 1964; Nelson and Dahl 1986; Rogosin 1958).

The Expanding Market Economy

The fact that the Red River flows north helped to isolate the Red Lake Ojibwe from the pressures generated by the expanding United States market economy. Full-scale incorporation of the area into the capitalistic economy awaited the arrival of railroads to bring larger urban markets within reach (Hall 1989, 1988; Wolf 1982). For generations the Ojibwe had interacted with Euroamericans, exchanging furs and foodstuffs for guns, traps, ammunition, metal goods, and other items. Intermarriage with French and British fur traders smoothed the cultural and economic exchange, and offspring of mixed descent found roles mediating between what became two ethnic groups (Peterson 1978, 1981, 1982; Brown 1975, 1980; Van Kirk 1980).

Intensive exploitation of fur-bearing animals for their pelts often led to overhunting, with all of its attendant environmental repercussions (Cronon 1983; White 1984). The more isolated Red Lake Ojibwe, however, were less centrally involved in the trade. Even though they intensified hunting, trapping, and hide processing, they did not deplete all furred animals, and they avoided one of the worst costs of overhunting — dependence on Euroamericans for significant contributions to their subsistence. When the fur trade began to decline, Ojibwe men found opportunities to work for wages both on nearby farms and in the lumber industry (Fig. 16.2). They continued to meet their need for manufactured goods by turning to seasonal wage labor, an alternative that fit easily into the traditional seasonal round. Opportunities to sell surplus produce also provided the cash necessary to pay for manufactured items. The Ojibwe minimized the disruptive impact of the developing market economy by incorporating wage labor

Fig. 16.1. Mr. Charles Anderson stands to pole the canoe through the wild rice fields while Mrs. Charles Anderson uses ricing sticks to pull stalks over and gently knock the ripened grains into the canoe. In time, they will fill the canoe and return to group camps on the shore where everyone cooperates to process the rice. (Photograph copyright by C. Brill; reprinted from *Red Lake Nation* [Minnesota, 1992].)

Fig. 16.2. A Red Lake man earned wages by driving a horse-drawn sleigh to transport logs to nearby sawmills. (Photograph copyright by C. Brill; reprinted from *Red Lake Nation* [Minnesota, 1992].)

and the continued sale of seasonal produce into their annual round of provisioning activities (Shifferd 1976; Meyer 1984, 1985, 1990, 1991; Kay 1985).

Lumbering and intensive agriculture, which represented the next wave of Euroamerican economic enterprise in the area, entailed far different relationships within the regional ecosystem. As railroads in the Red River Valley brought more lucrative markets within reach, the growing immigrant Euroamerican population came to covet "undeveloped" land set aside for Indians as reservations. Euroamerican methods of crop production demanded an expanding land base and displaced indigenous plant and animal species. The lumber industry pressed reservation timber resources from the east as it exhausted the forests in its path across northern Minnesota. Pressure from these two constituencies had eroded the land and resource base at the White Earth Reservation to the southwest of the Red Lake Reservation. As they confronted external pressures on their resources, Red Lake leaders benefited from observing the process of dispossession that unfolded at White Earth (Congressional Serial Set 2449, 2747; Meyer 1984, 1985, 1990, 1991).

The Campaign to Protect Red Lake Resources

In the face of mounting pressure from outsiders to gain access to their resources, the Red Lake Ojibwe articulated a defense of their reservation as an ecosystem in which they were human participants. The Red Lake Reservation was established by treaty in 1863 (Fig. 16.3; Kappler 1904-1941). Through the 1889 Nelson Act, the Red Lake band ceded a large acreage to the north and west, much of it peatland, in return for an established reservation. Red Lake leaders hoped that the provision of the Nelson Act establishing definite reservation boundaries would ultimately protect their claim to the lands surrounding Upper and Lower Red Lakes (Congressional Serial Set 2747).

Between 1887 and 1934, the U.S. government embarked on an experiment in social engineering designed to eradicate Indian cultures nationwide and replace them with a romantic version of the American agrarian ideal, the small-scale yeoman farmer. This policy emphasized allotment of the reservation land bases to individuals to farm. Policymakers aimed to sever tribal communal bonds and to instill in Indian people an appreciation of private property that would enable them to become full participants in the market economy. They hoped that education and Christianization would hasten this transformation. Federal officials envisioned that Indians and their cultures would meld into the dominant society, retaining little of their heritage. The 1889 Nelson Act was to achieve this goal among the Minnesota Ojibwe people (Carlson 1981; Fritz 1963; Hoxie 1984; Mardock 1971; Meyer 1985, 1991; Otis 1973; Prucha 1976, 1984; U.S. Statutes at Large 24:388-391, 25:642-646).

Red Lake band leaders selectively supported the Nelson Act negotiations. While they participated in an initial land cession, they adamantly refused to consider allotting the reservation land to individuals, selling the remainder, or allowing the proceeds to be deposited in a central fund to which all of the "Chippewa in Minnesota" shared rights. Leaders insisted on continuing to hold their land in common. They dismissed allotment provisions as a thinly veiled attempt to dispossess them of their lands and undermine their tribal relations. They would continue to maintain their opposition to these measures for the next hundred years, and they show no sign of reconsidering their stance today (Congressional Serial Set 2747).

Steadfast maintenance of their independent, collective rights formed the central facet of the Red Lake band's political ideology. The high proportion of swampland and forested land on the reservation aided them in resisting the proposed allotments. While the Ojibwe had many uses for swampland, policymakers intended that all Indians should instead become market-oriented farmers. U.S. legislators saw timber at Red Lake as a marketable resource that would provide the necessary funding to fuel their assimilation

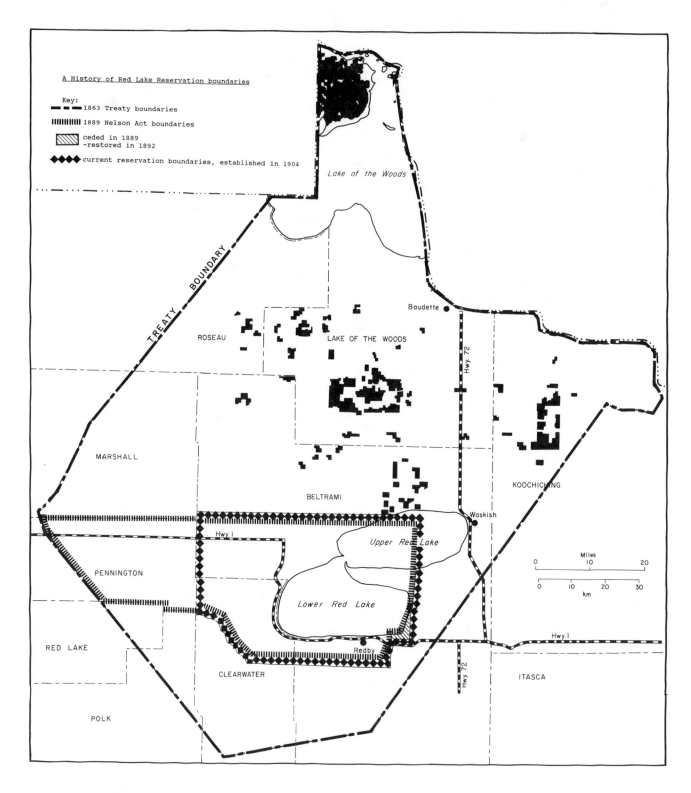

Fig. 16.3. A history of Red Lake Reservation boundaries.

programs. Allotting swampland and timberland to individuals would do little to enhance farming skills among the Red Lake Indians, so the nature of the land and the resistance of the Red Lake band continued to forestall allotment of the reservation land.

Fully aware of the threat posed by the Nelson Act, Red Lake leaders sought additional safeguards at the first opportunity. When government representatives showed interest in acquiring another land cession from the Red Lake band, Indian leaders attempted to manipulate the situation for their own ends. They remained adamant in their opposition to allotment of their land when Indian Office inspector James McLaughlin arrived to negotiate. Their leaders indicated, however, that they might be willing to agree to another land cession if they received legislative guarantees permitting them to retain communal ownership of their unallotted land base. McLaughlin agreed, and in 1904 the Red Lake Band secured the legislation on which their legal rights to what became known as the Red Lake Diminished Reservation are based. Government officials still hoped that the Red Lake band would eventually accept allotment of land, but the Red Lakers had reason to feel even more secure in their land tenure arrangements (Congressional Hearings 1920).

The chain of events that occurred at the White Earth Reservation demonstrated to Red Lake leaders how individual land rights could undermine tribal relations and threaten the very economic base of the reservation. Ojibwe leaders, many of them market-oriented entrepreneurs of mixed descent from White Earth, formed the General Council of the Chippewa in 1913 in response to mismanagement of their tribal resources by the U.S. government. Because the 1889 Nelson Act had created the Chippewa in Minnesota Fund as the depository for the proceeds of all land and timber sales on Ojibwe-ceded lands, exercising any control over the administration of these funds required the participation of all of the Ojibwe reservations in Minnesota. Initially, the Red Lake band joined the General Council of the Chippewa to secure control over their common tribal funds from the federal government (Meyer 1934, 1985, 1991).

In 1918, entrepreneurs of mixed descent from the White Earth Reservation attempted to use the General Council to divide the proceeds of all Ojibwe assets and distribute them among enrolled tribal members. As part of this plan, council leaders, very few of them from Red Lake, sought to have the Red Lake Diminished Reservation allotted and the surplus sold in accordance with the 1889 Nelson Act. The money from these land sales would contribute to the Chippewa in Minnesota Fund and would enrich the rest of the Ojibwe at the expense of the Red Lake band. Red Lake leaders severed their ties with the General Council of the Chippewa and formed their own independent General Council of the Red Lake Band to safeguard their interests (Bureau of Indian Affairs, White Earth Central Classified Files [BIA/WECCF] 054, 056; Bureau of Indians Affairs,

Red Lake Central Classified Files [BIA/RLCCF] 054, 056; Congressional Serial Set 8838, 10085, 10229).

As part of their move toward independent status, Red Lake band leaders cast their loyalties with the Indian Office to guard against fulfillment of the terms of the Nelson Act. They consistently referred to the 1904 McLaughlin Agreement, which established their separate reservation property rights, as the legal basis of their claims (BIA/RLCCF 054).

After 1918, in the midst of this crisis period, Red Lake leaders structured their independent council so that individualistic entrepreneurs could not easily gain control. They vested decision-making power in seven chiefs who, in turn, appointed five council members each. In the long run, such a structure would produce a tightly knit group, probably related by kinship, whose members shared a similar conservative outlook on reservation policy and cultural adaptation. This organizational structure provided for continuity and a degree of ethical exclusivity by adapting the traditional political band structure to allow for the concentration of power in the hands of these seven chiefs (BIA/RLCCF 054, 056).

As might have been expected, this form of political organization alienated younger, more commercially oriented members of the band. These young men, some of whose families had been associated with the formation of the threatening General Council of the Chippewa, charged the Red Lake council with operating undemocratically to freeze the younger element out of reservation politics. The discontented group formed the Red Lake Tribal Business Association in the 1920s. The council and the association operated simultaneously for a time, and the Indian Office even lent some legitimacy to the association by corresponding with it directly about administrative matters (BIA/RLCCF 054, 056).

Founders of the Red Lake council defended their right to determine the structure of reservation government by arguing that their tribal relations had never been severed by allotment of the reservation land base. In essence, they were using the terms of U.S. policy to preserve their own interests. They recognized that their separate, independent status made them unique among the Minnesota Ojibwe. Leaders argued that their successful perseverance through the harshest of the U.S. government's assimilation campaign entitled them to retain their council of chiefs as the governing body on the reservation (BIA/RLCCF 054, 056; Congressional Serial Set 8838, 10085, 10229).

The General Council of the Red Lake Band maintained its separate status in the face of the 1934 Indian Reorganization Act. Under authority of this act, leaders of the other Ojibwe reservations in Minnesota agreed to incorporate as the Minnesota Chippewa Tribe. After much deliberation and concern for their separate tribal assets, Red Lake leaders decided not to cooperate with the elective governmental structure of the Minnesota Chippewa Tribe, which would have jurisdiction over all the reservations. Fearing that restrictions on their political autonomy might follow any relationship with the Minnesota Chippewa Tribe, the Red Lake band refused to cooperate even on a temporary basis (BIA/RLCCF 054, 056; U.S. Statutes at Large 48:984; Congressional Serial Set 10085, 10229).

The entire governmental structure at Red Lake had evolved to protect land and resources from both external and internal threats. Today the Red Lake Reservation retains the only contiguous, unallotted reservation in Minnesota, and the band's accomplishment in retaining their land base is exceptional among Indian reservations nationwide. Only a few tribal groups have been able to retain a significant proportion of their land bases; the majority today possess only fragments of their original reservation holdings. The Red Lake band tenaciously persevered in retaining its landed reserve in all its diversity — peatlands, forest lands, prairie, lakes, and streams — understanding full well its importance to the survival of their culture.

Red Lake Leaders and Environmental Protection

In defending their resources, Red Lake leaders sought to protect the quality of their environment as well as their land base. Various state and federal programs throughout the years threatened to alter the water level of Red Lake — the backbone of the Red Lake Ojibwe subsistence economy. Plans to drain the peatlands or dam the outlet of Red Lake all failed to consider the wishes of Red Lake leaders, who favored maintaining the natural water level of Red Lake. When the water level rose too high it threatened low-lying gardens, meadows, and rice fields throughout the drainage areas of upstream tributaries. When the water level fell too low it exposed gravel and sand bars where critical fish species spawned, threatening the Red Lake fishing industry. Red Lake leaders understood that the long-term survival of their culture rested on their continued ability to adapt their subsistence base successfully to the changing conditions around them.

In 1908, Minnesota policymakers passed the Volstead Act aimed at draining the public swamplands in northern Minnesota (chapter 12). Peatlands that had been ceded by the Ojibwe were a sizable proportion of the wetlands targeted for this drainage project. Legislative terms had not vested titles to these lands in the U.S. government. Rather, the United States agreed to sell the land for the Ojibwe, depositing the proceeds in the commonly held Chippewa in Minnesota Fund. Settlers proved slow to make homestead entries on the wetlands, however, and often failed to sustain their efforts. The Red Lake band retained ownership of parcels of the ceded lands where homestead entry was not made or where would-be homesteaders failed to fulfill the terms of the legislation to receive title. This accounts for the contemporary dispersal of reservation

holdings throughout the peatlands. Unable to foresee this outcome, legislators reasoned that draining these lands would "reclaim" the swamplands, encourage successful homestead settlement, and afford some profit for the Ojibwe from otherwise "useless" lands. In their eyes, the project would produce revenue for the Indians, revenue for the state, and agricultural land for homesteaders. They believed that the bounty they envisioned justified earmarking Ojibwe tribal funds for a percentage of the costs of the land surveys and ditch construction (U.S. Statutes at Large 34:352-353, 1033; 35:82, 169-171; 38:88, 591-592; 39:978; 41:15; 42:1164-1165; 44:187; BIA/RLCCF 309; U.S. General Accounting Office 1950).

Red Lake leaders saw the situation differently. They expressed a more sophisticated understanding of ecological relationships as they protested against both the drainage project and the proposed use of their tribal funds. Ojibwe leaders explained that, because of its influence on the level and quality of the waters of Red Lake, the swampland supported plant and animal life central to their peoples' subsistence (BIA/RLCCF 309; U.S. General Accounting Office 1950).

As they articulated their opposition to the drainage project, Red Lake leaders described the different value systems on which they and their opponents based their arguments. The Ojibwe regarded swampland as central to their way of life. The wetlands environment nurtured many of their critical resources (see Tables 15.1 and 15.2). The network of lakes, peatlands, and forest supported the wild rice crop, cranberry bushes, fish, and various animals, especially beaver and muskrat. The entire Ojibwe subsistence cycle depended on the region's water system. Farmers, on the other hand, regarded swampland as useless. Prospective "developers" described in detail the difficulty of traveling through the muskeg while combating swarms of mosquitos and black flies. They could conceive of these wetlands as having no possible benefit without first "reclaiming" them to make them suitable for agricultural production. The difference of opinion, born of deeply held cultural values, could not have been more profound (BIA/RLCCF 309; Congressional Serial Set 5579).

The protests of Red Lake leaders made little difference in the long run. The paternalistic Indian policy of the early 20th century intended that U.S. legislators should make decisions for Indian people (Hoxie 1984). Both the surveys of the swampland and the proposed drainage project proceeded as planned, with Ojibwe tribal funds partially financing both of these operations. Although the Red Lake Indians retained property rights to their diminished reservation, without the vote they had no power to influence local or national political leaders. As wards of the federal government, the nation's only indigenous people were effectively denied any role in administering their affairs. Regardless of the staunch opposition to the drainage plans voiced by Red Lake leaders, local U.S. political leaders and administrators implemented the controversial drainage project.

Drainage of the countryside north of Red Lake raised the water level of the lake, as Indian leaders had predicted. Upriver tributaries also experienced higher water levels, which affected the delicate wild rice beds and flooded low-lying gardens and meadows. Lowering the water table in the peatlands may have been a factor in rendering the area more susceptible to fires during dry periods. Drainage eliminated the protective peatland buffers, and valuable stands of timber were destroyed by uncontrollable fires. Drainage of the peatlands may have temporarily created additional agricultural lands for homesteaders, but the effects rebounded within the ecosystem, causing problems for the Ojibwe in other areas (Congressional Serial Set 5579; BIA/RLCCF 309; BIA/WECCF 309).

In the 1920s and 1930s, Euroamericans who had settled downstream toward Thief River Falls favored building a dam at the outlet of Red Lake to regulate the flow of the Red Lake River and stem downstream flooding. Considering the problems generated by drainage of the peatlands, Red Lake leaders could support a dam that would also regulate the water level of Red Lake. Engineers worried that such a dam would raise the water level even further, endangering the drainage ditches that directed water away from the swamplands. Local Euroamerican settlers also complained of a water shortage and recommended dredging the river and straightening its sluggish, meandering channel to allow a more rapid, free-flowing current. Red Lake leaders reacted to these suggestions with alarm. Dredging and altering the river's channel would surely diminish the water level of Red Lake, exposing crucial fish spawning grounds. Leaders at Red Lake suspected that the proposal would benefit interested power companies more than the settlers, whom they advised to dig wells instead. Local Euroamerican interests tended to support projects that affected their immediate situation without considering the ramifications for the regional ecosystem (BIA/RLCCF 309; Congressional Serial Set 10795).

In the face of a national mood in the early 20th century that favored the administration of Indian affairs by local Euroamericans (Hoxie 1984), the Red Lake Indians battled weighty odds in their struggle to conserve and protect their resources. They objected to the peatland drainage projects on ecological grounds, yet U.S. policymakers assessed a proportion of the costs of the project against their tribal funds. To compound their financial woes, Congress repeatedly granted commutation rights to homestead settlers on Ojibwe-ceded lands, depriving the Chippewa in Minnesota Fund of much-needed revenue. Upon completion of the drainage project, authorities assessed Ojibwe tribal funds for "improvements" that their funds had partially financed in the first place. That the Red Lake band withstood the assault on its reservation resources during this period of massive external pressure is testimony to the band's

understanding of its role within the ecosystem (U.S. Statutes at Large 29:342; 30:595; 31:179, 241; 33:1005; 34:326; 36:265; Congressional Serial Set 8227, 8388; U.S. Government Accounting Office 1950).

In the 1970s, the Red Lake Tribal Council found itself confronted with yet another threat to the ecosystem. During an "energy crisis" in which concerns about the dependence of the United States on imported fuel escalated, the Minnesota Gas Company (Minnegasco) proposed to mine peat from the lands north of Red Lake as an alternative energy source. Such development would have a massive impact that would reverberate throughout the ecosystem, a fact that was not lost on Red Lake tribal leaders (see chapter 18).

Most importantly, mining the peatlands would have threatened the water quality of Red Lake, potentially harming the Red Lake fishing industry in particular. The introduction of dissolved and suspended solids would have affected temperature, light penetration, and the solubility of gases and minerals, and it may have clogged and abraded fish gills at higher levels. Heavy metals like mercury produced by mechanized equipment would have posed a threat to fish and to the entire food chain. Changes in pH levels would also have had deleterious effects on fish. At higher levels iron becomes scarce and at lower levels trace metals become soluble, both conditions that fish would have been unable to tolerate. Fertilizers necessary to convert mined peatland to farmland would have reduced the amount of oxygen available to fish. Anaerobic conditions that would have been produced by the blanketing of the lake bottom with suspended solids would have negatively affected all aquatic life (Walter Butler Company 1978).

Other components of the Red Lake economy also would have been affected. Peat mining consumes vegetative cover and would have changed the availability and quality of surface water by creating a series of ponds. Wildlife in the area would have felt the effects; waterfowl and water-dependent animals would have increased at the expense of species dependent on land. Mining operations, with their attendant noise and bustle of activity, might have interfered with migration routes and driven game away altogether. Swamp tree species that tolerate only a narrow range of water fluctuation would have suffered from changes in water flow and quality. At the time, this would have affected the plans of operators of the Red Lake sawmill to make greater use of swamp species. Finally, changes in water level and flow would have affected nearby wild rice stands and nascent plans for commercial paddy cultivation (Walter Butler Company 1978).

In effect, by threatening the existing water table and water quality of Red Lake, potential peat mining schemes put the entire nongovernmental economic base of the reservation at risk. Realizing this, Red Lake tribal leaders passed a resolution to preserve the peatlands untouched. Their action was in keeping with a decades-long commitment to the integrity of their ecosystem. Although a new emphasis on "self-determination" had altered the tenor of U.S. Indian policy since the early 20th century (Prucha 1984; Deloria and Lytle 1984), it is nonetheless unclear that the wishes of Red Lake leaders had anything to do with averting peat mining. After assessing the costs and benefits of the proposed project, Minnegasco abandoned its plans (Walter Butler Company 1978; see chapter 17).

Conclusion

Fish and other natural resources still serve as the basis for all of the nongovernmental employment on the Red Lake Reservation today. The components of the self-supporting element of the reservation economy remain the same as they have for generations — perhaps even centuries; only the techniques for managing and harvesting the resources have intensified. In a very real sense, the cultural and economic heritage of the Red Lake Ojibwe, one that very likely resembles the customs of all prior native inhabitants of the area, has carried them through the decades and continues to sustain them today. Their leaders clearly recognize the role that the peatlands have played in sustaining the overall system.

The Red Lake people have been commercial fishers for more than 70 years. Initially they made the transition from a subsistence economy simply by selling part of their harvest to traders or local merchants, allowing households to participate in the cash economy, if only peripherally (Fig. 16.4). By 1917, in response to a wartime scarcity of meat, the state began to operate a commercial fishery at Red Lake. The Minnesota State Fisheries operated as a wholesaler from 1920 to 1928, distributing Red Lake fish to Midwestern and eastern markets. By the terms of this arrangement, individual Red Lake Indians simply sold their catches to the state-operated fishery. In 1922, Red Lake leaders objected to the state-run operation, complaining that competition from the state's fishing crews and its marketing arrangements benefited the state of Minnesota rather than the Red Lakers. In 1929, the Red Lake Tribe organized the Red Lake Fisheries Association, which processed walleye and perch (50% of the production) and six other species of fish. The Red Lake fishery remains the major source of employment on the reservation today, generating two-thirds of the nonpublic employment, and stands out as the single most important component of the reservation economy. The peatlands, which serve as a natural filter for the waters of the lake, help to insure the health of this enterprise. (Congressional Serial Set; Annual Reports of the Commissioner of Indian Affairs 1917-1965; Walter Butler Company 1978).

Hunting represents much more than a cultural heritage to the Red Lake Indians; it is crucial to their economic subsistence. Fish, moose, deer, and wild fowl still make up more than 50% of the Red Lake diet. These game resources contribute to the delicate subsistence balance on the reservation, providing food that enables

Red Lake people to spend their meager cash income on other things (Walter Butler Company 1978).

Until the early 1980s, the Red Lake Indian Sawmills had served as a source of reservation income for more than 50 years, providing nearly one-third of the nonpublic employment on the reservation (Fig. 16.5). The first steam-operated sawmill was constructed in Redby in 1924, its prosperity resting on the profitability of processing white pine and Norway pine. Since then the mill has been rebuilt and updated, but the availability of highly profitable pine has decreased, and Red Lakers have had difficulty maintaining profitable returns. As early as the 1930s, sawmill managers recognized their need to obtain timber from the open market rather than the reservation to keep the mill in operation. In the past, there had been some discussion of using swamp species that would have been supported by a peatland environment. The mill closed in the early 1980s, but changes in the market might make this a profitable endeavor once again (Congressional Serial Set 10761, 11897; Annual Reports of the Commissioner of Indian Affairs 1924-1965; Walter Butler Company 1978; Irving 1989).

Wild rice is still harvested by the traditional canoe and flail method on the reservation, and tribal specialists have experimented with cultivated paddy operations using nonshattering strains of rice. Most of these enterprises have been located near peatlands. Until the early to mid-1980s, wild rice held the most promise for agricultural development at Red Lake. Since then, however, commercial paddy operations in California have eclipsed Minnesota as the major producer of wild rice in the world. The heavy capital outlays required to establish wild rice paddies will make it increasingly difficult for Red Lake people to break into what once seemed the most lucrative agricultural enterprise for the reservation (Vennum 1988; Berde 1980; Oelke et al. 1982; Winchell and Dahl 1984; Nelson and Dahl 1986; Walter Butler Company 1978).

All of these tribal enterprises have contributed to a slowly developing economic self-sufficiency, but Red Lake employment statistics still look bleak. Economic conditions at Red Lake fall below the national average for Indians. The average unemployment rate approaches and seasonally surpasses 50%, and the average per capita income falls far below the poverty level. At the same time, the population of Red Lake is increasing faster than the average Indian population, bolstered by the return migration of Indians from nearby urban areas. The marginal Red Lake economy depends centrally on the nongovernment employment afforded by fish and other natural resources. In this context the integrity and viability of the ecosystem, of which the peatlands are a central part, are crucial to the vitality of the Red Lake economy and society (Walter Butler Company 1978).

Archaeological and historical research dovetail to establish that the components of an ancient way of life evolved by indigenous people in Minnesota's north

Fig. 16.4. Nets and floats hang to dry on a scaffold. Subsistence-oriented fishers did not require much technological innovation to sell their modest surplus on the market. (Photo courtesy of Minnesota Historical Society.)

Fig. 16.5. The Red Lake Indian Sawmill. It may once again become a profitable economic enterprise. (Photo courtesy of Minnesota Historical Society.)

country (chapter 15) have persisted to leave a distinctive mark on contemporary Indian cultures. The resources of the Red Lake Reservation ecosystem, which include the peatlands, remain as central to the livelihood and well-being of the Indian people who live there today as they were to their ancestors. Prehistoric and early historic native peoples have seldom left the sort of records that reveal their attitudes toward the land and the resources that sustained their cultures and ways of life. Since the inception of the General Council of the Red Lake Band in the early 20th century, however, tribal leaders have actively sought to safeguard their resources against outside exploitation, consistently demonstrating an understanding of their role within the ecosystem as a whole. Today, tribal chairman Roger Jourdain continues in this tradition, saying, "The most important use of the peatlands is as a natural filter for the waters which maintain our lakes" (Irving 1989). "Peat provides habitat for our wildlife. Much of our forests grow on peat. The wildlife, timber and fish are our greatest resource, . . . (the) primary source of employment and income for the tribe" (Walter Butler Company 1978). For the native people who have lived on and near the peatlands, the importance of the ecosystem to their way of life has remained unchanged.

Literature Cited

Anderson, G. C. 1980. Early Dakota migration and intertribal war: A revision. Western Historical Quarterly 11:17-36.

Berde, S. 1980. Wild ricing: The transformation of an aboriginal subsistence pattern. Pages 101-125 in J. A. Paredes, editor. Anishinabe: Six studies of modern Chippewa. University Presses of Florida, Tallahassee, Florida, USA.

Bishop, C. A. 1974. The northern Ojibwa and the fur trade: An historical and ecological study. Holt, Rinehart, and Winston of Canada, Toronto, Ontario, Canada.

——. 1976. The emergence of the northern Ojibwa: Social and economic consequences. American Ethnologist 3:39-54.

——. 1978. Cultural and biological adaptations to deprivation: The northern Ojibwa case. In C. C. Laughlin, Jr., and I. A. Brady, editors. Extinction and survival in human populations. Columbia University Press, New York, New York, USA.

—— and M. E. Smith. 1970. The emergence of hunting territories among the northern Ojibwa. Ethnology 9:1-15.

Brown, J. S. H. 1975. Fur traders, racial categories, and kinship networks. In W. Cowan, editor. Papers of the Sixth Algonquian Conference, 1974. National Museum of Man, Mercury Series no. 23. Canadian Ethnology Service, Ottawa, Ontario, Canada.

——. 1980. Strangers in blood: Fur trade company families in Indian country. University of British Columbia Press, Vancouver, British Columbia, Canada.

Bureau of Indian Affairs. 1907-1950. National Archives and Records Service. Record Group 75. Red Lake Agency. Central Classified Files 054, 056, 309.

——. 1907-1950. National Archives and Records Service. Record Group 75. White Earth Agency. Central Classified Files 054, 056, 309.

Carlson, L. 1981. Indians, bureaucrats and land: The Dawes Act and the decline of Indian farming. Greenwood Press, Westport, Connecticut, USA.

Cleland, C. E. 1966. The prehistoric animal ecology and ethnozoology of the upper Great Lakes region. Museum of Anthropology, Anthropological Papers no. 29. University of Michigan Press, Ann Arbor, Michigan, USA.

Cronon, W. 1983. Changes in the land: Indians, colonists, and the ecology of New England. Hill and Wang, New York, New York, USA.

Deloria, V., and C. Lytle. 1984. The nations within: The past and future of American Indian sovereignty. Pantheon, New York, New York, USA.

Densmore, F. 1910-1913. Chippewa music. Bureau of American Ethnology, Smithsonian Institution 45, 53. U.S. Government Printing Office, Washington D.C., USA.

——. 1919. Material culture among the Chippewa. Smithsonian Miscellaneous Collections 70:114-118.

——. 1928. Uses of plants by the Chippewa Indians. Bureau of American Ethnology, Annual Report 44:275-397.

——. 1929. Chippewa customs. Bureau of American Ethnology, Smithsonian Institution 86. U.S. Government Printing Office, Washington D.C., USA.

Dunning, R. W. 1958. Some implications of economic change in northern Ojibway social structure. Canadian Journal of Economics and Political Science 24.

——. 1969. Social and economic change among the northern Ojibwa. University of Toronto Press, Toronto, Ontario, Canada.

Folwell, W. W. 1921. A history of Minnesota. Vols. 1 and 4. Minnesota Historical Society Press, St. Paul, Minnesota, USA.

Fritz, H. E. 1963. The movement for Indian assimilation, 1860-1890. University of Pennsylvania Press, Philadelphia, Pennsylvania, USA.

Hall, T. D. 1988. Patterns of Native American incorporation into state societies. Pages 23-38 in M. C. Snipp, editor. Public policy impacts on American Indian economic development. Native American Studies, Institute for Native American Development, University of New Mexico, Albuquerque, New Mexico, USA.

——. 1989. Social change in the Southwest, 1350-1880. University of Kansas Press, Lawrence, Kansas, USA.

Hickerson, H. 1962. The Southwestern Chippewa: An ethnohistorical study. American Anthropological Association Memoir no. 92.

——. 1965. The Virginia deer and intertribal buffer zones in the upper Mississippi valley. In A. Leeds and A. Vayda, editors. Man, culture and animals: The role of animals human ecological adjustments. American Association for the Advancement of Science, Publication no. 78. Washington, D.C., USA.

——. 1970. The Chippewa and their neighbors: A study in ethnohistory. Holt, Rinehart and Winston, New York, New York, USA.

——. 1974. Chippewa Indians IV: Ethnohistory of the Chippewa in central Minnesota. Garland, New York, New York, USA.

Holzkamm, T. E. 1983. Eastern Dakota population movements and the European fur trade: One more time. Plains Anthropologist 28:225-233.

Hoxie, F. E. 1984. A final promise: The campaign to assimilate the Indians, 1880-1920. University of Nebraska Press, Lincoln, Nebraska, USA.

Irving, E. 1989. Interviews with Bud Anderson, Roger Anderson, Joe Day, Roger Jourdain, and Anthony Wilson. Transcripts in the possession of Melissa L. Meyer.

James, E., editor. 1956. A narrative of the captivity and adventures of John Tanner during thirty years residence among the Indians in the interior of North America. Ross and Haines, Minneapolis, Minnesota, USA.

Jenks, A. E. 1900. The wild rice gatherers of the upper lakes: A study in American primitive economics. U.S. Bureau of American Ethnology, 19th Annual Report, 1897-1898 2:1013-1137.

Kappler, C. J., compiler. 1904-1941. Indian affairs: Laws and treaties. 5 vols. U.S. Government Printing Office, Washington, D.C., USA.

Kay, J. 1985. Native Americans in the fur trade and wildlife depletion. Environmental Review 9:118-130.

Landes, R. 1939. The Ojibwa woman. Columbia University Contributions to Anthropology no. 31. New York, New York, USA.

———. 1968. The Mystic Lake Sioux. University of Wisconsin Press, Madison, Wisconsin, USA.

Mardock, R. W. 1971. The reformers and the American Indian. University of Missouri Press, Columbia, Missouri, USA.

Meyer, M. L. 1984. Warehousers and sharks: Chippewa leadership and political factionalism on the White Earth Reservation, 1907-1920. Journal of the West 23:32-46.

———. 1985. Tradition and the market: The social relations of the White Earth Anishinaabeg, 1889-1920. Ph.D. dissertation. University of Minnesota, Minneapolis, Minnesota, USA.

———. 1990. Signatures and thumbprints: Ethnicity among the White Earth Anishinaabeg, 1889-1920. Social Science History 14: 305-345.

———. 1991. "We can not get a living as we used to": Dispossession and the White Earth Anishinaabeg, 1889-1920. The American Historical Review 96:368-394.

Mittelholtz, E. F. 1957. Historical review of the Red Lake Indian Reservation, Redlake, Minnesota: A history of its people and progress. Beltrami County Historical Society Collections no. 2.

Nelson, R. N., and R. P. Dahl. 1986. The wild rice industry: Economic analysis of rapid growth and implications for Minnesota. Institute of Agriculture, Forestry, and Home Economics, Department of Agricultural and Applied Economics Staff Paper P86-25. University of Minnesota, St. Paul, Minnesota, USA.

Oelke, E. A.,et al. 1982. Wild rice production in Minnesota. Agricultural Extension Service Bulletin 464. University of Minnesota, St. Paul, Minnesota, USA.

Otis, D. S. 1973. The Dawes Act and the allotment of Indian lands. E. P. Prucha, editor. University of Oklahoma Press, Norman, Oklahoma, USA.

Peterson, J. 1978. Prelude to Red River: A social portrait of the Great Lakes Metis. Ethnohistory 25:41-67.

———. 1981. The people in between: Indian-white marriage and the genesis of a Metis society and culture in the Great Lakes region, 1680-1830. Ph.D. dissertation. University of Illinois at Chicago Circle, Chicago, Illinois, USA.

———. 1982. Ethnogenesis: The settlement and growth of a "new people" in the Great Lakes region, 1702-1815. American Indian Culture and Research Journal 6:23-64.

Prucha, F. P. 1976. American Indian policy in crisis: Christian reformers and the Indian, 1865-1900. University of Oklahoma Press, Norman, Oklahoma, USA.

———. 1984. The Great Father: The United States government and the American Indians. University of Nebraska Press, Lincoln, Nebraska, USA.

Rogers, J. 1974. Red world and white: Memories of a Chippewa boyhood. University of Oklahoma Press, Norman, Oklahoma, USA. (First published in 1957 as A Chippewa Speaks by John Rogers.)

Rogosin, A. 1958. Wild rice (Zizania aquatica) in northern Minnesota, with special reference to the effects of various water levels and water level changes, seeding densities, and fertilizer. M.S. thesis. University of Minnesota, St. Paul, Minnesota, USA.

Shifferd, P. A. 1976. A study in economic change: The Chippewa of northern Wisconsin, 1854-1900. Western Canadian Journal of Anthropology 6:16-41.

Tanner, H. H., editor. Cartography by M. Pinther. 1987. The atlas of Great Lakes Indian history. Published for the Newberry Library by the University of Oklahoma Press, Norman, Oklahoma, USA.

U.S. Congress. 1890-1920. Congressional Hearings. U.S. Government Printing Office, Washington, D.C., USA.

———. 1890-1930. Congressional Record. U.S. Government Printing Office, Washington, D.C., USA.

———. 1890-1965. Serial Set. U.S. Government Printing Office, Washington, D.C., USA.

———. 1887-1950. Statutes at Large. U.S. Government Printing Office, Washington, D.C., USA.

U.S. General Accounting Office. 1950. United States General Accounting Office Report re. Petitions of the Minnesota Chippewa Tribe et al., and Red Lake Band, et al., August 21, 1950. Indian Claims Commission 188, 189, 189-Amended, 189-A 189-B, and 189-C.

Van Kirk, S. 1980. "Many tender ties": Women in fur trade society, 1670-1870. Watson and Dwyer, Winnipeg, Manitoba, Canada.

Vennum, T., Jr. 1988. Wild rice and the Ojibway people. Minnesota Historical Society Press, St. Paul, Minnesota, USA.

Walter Butler Company, Inc. 1978. Peat utilization and the Red Lake Indian Reservation. For the Minnesota Department of Natural Resources, Division of Minerals, requisition 13657. Walter Butler Company, Inc., St. Paul, Minnesota, USA.

Warren, W. 1885. History of the Ojibway Nation, based upon traditions and oral statements. Minnesota Historical Society Collections 5:21-394.

Watrall, C. 1968. Virginia deer and the buffer zone in the late Prehistoric-early Protohistoric periods in Minnesota. Plains Anthropologist 13: 81-86.

White, R. 1978. The winning of the West: The expansion of the western Sioux in the 18th and 19th centuries. Journal of American History 65:319-343.

———. 1984. The roots of dependency: Subsistence, environment, and social change among the Choctaw, Pawnee, and Navajo. University of Nebraska Press, Lincoln, Nebraska, USA.

Winchell, E. H., and R. P. Dahl. 1984. Wild rice: Production, prices, and marketing. Agricultural Experiment Station, Miscellaneous Publication 29. University of Minnesota, St. Paul, Minnesota, USA.

Wolf, E. R. 1982. Europe and the people without history. University of California Press, Berkeley, California, USA.

Yarnell, R. A. 1964. Aboriginal relationships between culture and plant life in the upper Great Lakes region. Anthropological Paper no. 23, Museum of Anthropology. University of Michigan Press, Ann Arbor, Michigan, USA.

A Note on the Sources

Documents from the records of the Bureau of Indian Affairs (Record Group 75) and its predecessor organization, the Office of Indian Affairs (or Indian Office), housed at the National Archives and the Kansas City Federal Regional Archives, served as the foundation for the discussion of the Red Lake Ojibwe and the peatlands. The individual documents from which this interpretation was constructed are too voluminous to be cited separately. Instead, numbers from the Central Classified Files of Record Group 75 were used to denote the general filing categories: tribal councils, delegations, petitions, swamplands, and dams and ditches. Each of these general categories contains a large amount of documentation pertinent to the topic. Sometimes a general chronological organization is discernible... sometimes not.

CHAPTER 17

Ditching of Red Lake Peatland During the Homestead Era

Kristine L. Bradof

Red Lake peatland in northern Minnesota is renowned for its complex vegetation patterns, which reveal the sensitivity of plants to water flow and water chemistry. Another pattern superimposed on the peatland, an extensive network of drainage ditches, has received little attention, yet it is the legacy of an important chapter in Minnesota's history.

These ditches failed to drain the peatland effectively for its intended conversion to agricultural uses. The high costs of ditch construction, however, did drain the financial resources of three northern Minnesota counties, leaving them on the brink of fiscal disaster by 1929. Ironically, the failed attempt to develop this vast wilderness nearly three quarters of a century ago ultimately led to its preservation as the Red Lake State Wildlife Management Area.

The homestead era came relatively late to much of northern Minnesota. Poorly drained, inaccessible, peat-covered lands were passed over in favor of more promising agricultural lands elsewhere. Newspapers published in Bemidji between 1907 and 1917 reveal a great public interest in the settlement and development of the region and in the construction of roads suitable for automobiles. Ditching to drain the "swamps" and "muck" soils was seen by many as a means to those ends. As a result, more than 1,500 miles of ditches were constructed in Beltrami County alone (*Bemidji Sentinel*, May 11, 1928).

The effects of this drainage were far-reaching: the near bankruptcy of three counties; the construction of roads where none had existed; the creation of Lake of the Woods County; and flood-control measures on Red Lake, including a proposal to construct a 100-mile canal to divert water to the Minnesota River. Drainage even became an election-campaign issue.

This chapter traces the drainage history of Red Lake

peatland and surrounding areas from the earliest drainage legislation in the 1850s to the establishment of the Red Lake Game Preserve in 1929. Included are sections on the petition and approval procedure used, methods of ditch construction, the influence of land developers, and the advice of drainage experts at the time.

Federal Involvement in Drainage

The drainage history of Red Lake peatland actually began more than 50 years before the first ditch was dug in Beltrami County. Interest in the reclamation of "swamp and overflowed lands" in the United States prompted Congress to transfer to the states ownership of all public lands unfit for cultivation because of poor drainage (U.S. Statutes at Large 1850, 1855). The legislation provided that all proceeds from these lands be used for the purpose of reclamation by means of levees and drains.

In 1858 the new state of Minnesota passed the first of a number of drainage laws that ultimately led to the procedure used in Red Lake peatland (Palmer 1915). None of these federal or state laws was applicable to the peatland, however, because it belonged to the Chippewa Indians until it was ceded to the United States along with other Chippewa lands in 1889 (U.S. Statutes at Large 1889).

A drainage survey conducted by the U.S. Geological Survey between 1906 and 1908 on 3,081,600 acres of ceded lands concluded that

these lands may be reclaimed at reasonable cost and that their reclamation will change vast areas of worthless swamps into fertile farms.... The entire area of the ceded lands north of Red Lake, as well as adjoining areas to the east, in Koochiching and Itasca Counties, now almost entirely given over to swamps and such wild animals as may exist there, is destined to support a large farming population when prop-

263

erly drained. . . . The area drained by the Rapid River [including most of Red Lake peatland] has a larger percentage of wet lands than any of the other drainage basins, and under present conditions the development of agriculture there, if any, will be very slow. A systematic drainage scheme must be provided, and just as soon as such reclamation becomes a reality every quarter section will have an occupant. (U.S. Department of the Interior 1909)

By this time the ceded Chippewa lands had been open to settlement for more than 18 years under the Homestead Act of 1862, but a million and a half acres remained unclaimed (U.S. Congress 1908). Congressmen Andrew Volstead and Halvor Steenerson of Minnesota introduced legislation to facilitate the drainage and settlement of this area. Steenerson was unable to persuade Congress to appropriate federal funds for drainage in Minnesota as it had for irrigation of arid lands in the West. Volstead's bill fared better, though it did face opposition. The Volstead Act, signed into law in 1908, allowed the state to apply its drainage laws to lands held in trust for the Chippewa Indians by the federal government. Drainage costs were to be assessed to the lands benefited and patents issued to those entrymen who met the qualifications under the homestead laws and paid a minimum of $1.25 per acre, that sum being partial compensation to the Indians for their lands. It was the Volstead Act that finally provided the legal mechanism for the large-scale drainage projects in northern Minnesota, including those in Red Lake peatland.

Petition and Approval Procedure

By 1905, when ditch construction was under way in Beltrami County, four agencies could establish public drainage ditches: (1) the town supervisors, (2) the county commissioners, (3) the district courts, and (4) the state drainage commission (Palmer 1915). The procedures to be followed before, during, and after ditch construction were much the same for the different agencies. Those followed by the district courts (Palmer 1915; *Bemidji Sentinel*, February 28, 1919) are outlined below because the drainage of Red Lake peatland involved only the so-called judicial ditches, some of which extended into Marshall, Koochiching, and what later became Lake of the Woods counties.

The first step toward the construction of a judicial ditch was the presentation to the district judge of a petition signed by one or more of the landowners whose lands would be affected by the ditch. This petition stated the reason for the ditch and the public benefit to be derived from it, as well as the proposed route. Township supervisors, city officials, and authorized agents of public institutions, corporations, and railroads could also petition the court to construct a ditch that would benefit their lands. The petitioners were required to put up a bond to cover preliminary expenses in case the petition should be denied.

Notices of the filing of the petition and of the time

and place of the public hearing regarding it were published in local newspapers and posted in public places in each township for at least three weeks. Copies were also mailed to nonresidents who owned lands within two miles of the proposed route. This was the responsibility of the district judge, as was the appointment of a qualified civil engineer to survey and make construction-cost estimates for the proposed route. Three "viewers" who were residents of the district not related to any of the petitioners and whose interests were not affected by the ditch were then selected by the judge. These viewers assessed the benefits or damages to each piece of property and the overall cost of construction relative to the hoped-for benefits.

At a public hearing, the petition and the reports of the engineer and the viewers were read and public comments heard. If the judge determined that construction of the ditch was warranted, he directed the county auditor (or auditors, if more than one county was involved) to publish in the newspapers for at least three weeks a notice of the letting of the work and requests for bids. The contract was awarded to the lowest bidder whose plans were approved by the engineer. Liens against the property benefited were then issued by the auditor and bonds sold by the county commissioners to finance the ditch construction. The liens were to be paid in 10 or 15 equal annual installments by the landowners, and the bonds were to be retired with this money.

Construction of Drainage Ditches and Roads

Information on the actual methods of ditch and peat-road construction is scarce. Letters written by Roy Bliler, Beltrami County surveyor and district highway engineer, mention two floating dredges and one gasoline dredge used during 1913 and 1914 for ditch and state road work near Kelliher and the Tamarack River. A roadbed constructed in 1910 consisted of "muck" piled between parallel ditches 6 to 8 feet deep and about 60 feet apart (Kelly 1978).

Some small ditches were dug by hand with the aid of teams of horses. Draglines, various steam shovels, and dry ditchers are mentioned in the literature (Soper 1919; Kelly 1978), but floating dredges were used in Red Lake peatland because of the depth of the peat and the wet conditions (Fig. 17.1). Peat depths recorded by the ditch survey for Judicial Ditches (J.D.) 30 and 36 in T.155-156 N., R.30-32 W. (Fig. 12.1) ranged from 1 to 15 feet; much of the peat was more than 10 feet deep (Soper 1919).

No written accounts of the operation of the dredging machines could be located, but interviews with Ernest W. Blanchard in 1983 and 1984 provided much firsthand information. Mr. Blanchard began to work on the floating dredges in Red Lake peatland in 1913, when he was 13 years old.

The winter before construction was to begin on a

Fig. 17.1. Cordwood-burning floating dredge of the type used in Red Lake peatland. Photo by E. J. Bourgeois, Beltrami County (Soper 1919).

Fig. 17.2. Floating dredge with cookshack and bunkhouse barges behind, Koochiching County. Photo by E. J. Bourgeois (Soper 1919).

ditch, a 66-foot-wide right-of-way was cleared, 33 feet on each side of the surveyed section line. Four-foot cordwood was hauled by ox team along the right-of-way and left there for the steam dredge to burn as it moved along. Later, oil-burning dredges that started on gasoline and switched over to kerosene after warming up became more common. The barrels of kerosene for those machines were trucked as close to the dredging site as possible during the preceding winter.

The dredge itself floated, and a cookshack and bunkhouse were located on separate barges behind the dredge (Fig. 17.2). Also at the site were a blacksmith forge and electric lights with reflectors, powered by a 6-horsepower generator, for working at night. Sometimes dams had to be built behind the barges to keep the water level up. "Roustabouts" loaded the cordwood or kerosene onto a little barge and pulled it along behind the dredge with ropes. They also had to pull the cookshack and bunkhouse barges along as the dredging proceeded.

The floating dredge consisted of a boom that could swing from side to side and a "cupstick," which pivoted in a vertical plane between the two beams that made up the boom, allowing the dipper, or cup, at its end to scoop up the peat. The boom then swung around to dump the peat on the bank. Dipper capacity was probably one-half to three-quarters of a cubic yard on these machines; correspondence from the division highway engineer in 1917 indicated that this was the capacity of several kerosene-powered dredges in use (Kelly 1978). The boom and cupstick were operated by cables and gears.

Stability of the dredge was maintained by two movable beams that extended outward like buttresses on either side. "Spudfeet" 6 or 8 feet long attached to the ends of these beams and apparently functioned like pontoons on the soft peat surface. In very wet conditions, tamarack or spruce poles 20 to 30 feet long were chained onto the spudfeet to keep the dredge from tipping. The dredge moved inchworm fashion as the dipper was thrown forward onto the bank ahead and,

while it remained stationary, the dredge was drawn forward. It might advance 200 to 500 feet per day.

Most of the ditches were excavated to depths of 6 to 12 feet, according to Blanchard. Averell and McGrew (1929) state that most of the ditches in the northern counties were 3 to 6 feet deep, with a top width of 12 to 20 feet and a bottom width of 4 to 10 feet. If the ditch base didn't reach the clay beneath, the peat would float up and have to be removed by hand with sod hooks (long-handled, pitchforklike tools with curved teeth). These sod hooks were also used for road grading on the peat where machines could not be used. "They got on there [the grade] and knocked the big piles of muskeg down so it looked level. That's all. They [these 'roads'] were never used anyway," according to Blanchard. On part of J.D. 30 two miles west of Hillman Lake, going north through the western water track, dredging was nearly impossible. "We just stirred it up and moved ahead.... This one runner, Art Weinkuch — the wind was blowing that day and he couldn't dig. He put on his report, 'High wind and a rough sea. Unable to dig.'"

The Drainage History of Red Lake Peatland and Beltrami County

My attempt to reconstruct the drainage history of Beltrami County and of the peatland in particular began with visits to state agencies and the Minnesota Historical Society in St. Paul, and to county agencies and libraries in Bemidji, the county seat. Unfortunately, very little useful information was to be found. The only means of obtaining a chronological record of drainage ditch and road construction and of public sentiment toward those projects was the tedious scanning of the *Bemidji Pioneer* and the *Bemidji Sentinel* weekly newspapers (hereafter referred to as *BP* and *BS*, respectively).

1907 to 1909

Relatively little ditch construction was undertaken during this period in Beltrami County, and most of that was done by the State Drainage Commission. More work appeared to be taking place in neighboring Koo-

BEMIDJI PIONEER.

BEMIDJI BELTRAMI COUNTY, MINNESOTA. THURSDAY, AUGUST 4. 1910　　　　$1.50 PF

A RE-CLAIMED SWAMP FARM SHOWS DRAINAGE BENEFIT

Island Farm, Near Floodwood, Produces Crop Values of $800 Per Acre on Former Moskeg Swamps.---Celery Grown Superior to Michigan Product.

(From the Big Falls Compass.)

For years the big swamps of Northern Minnesota have been scorned as worthless, or if one saw any possibilities in the muskeg, they regarded the expense of drainage an insurmountable obstacle.

Celery culture at a profit of from $800 to $3,200 per acre on the muskeg swamp lands of St. Louis county is the possibility that is held out to those who within the next few years are willing to grasp the opportunities offered to supply a constantly growing market at the Head of the lakes, says the Duluth Herald.

Through the efforts of a Duluth citizen, one of the worst muskeg swamps in northern Minnesota, a vast tract stretching for miles on either side of the Great Northern railway branch to Grand Forks, beginning about five or six miles northwest of Floodwood, is fast being transformed into a vast productive farm, and it will not be many years before the immense bog on which the railroad company spent thousands of dollars to keep its tracks above the water will be reclaimed, and Island Farm owned by G. G. Hartley of this city, will be considered one of the model farms of the state.

The secret of Mr. Hartley's success is drainage. His farm in the present condition, is a splendid example of what proper drainage will do for the big swamp districts of Northern Minnesota.

When one considers that practically all the effort that has been put forth to bring the Island Farm to its present stage of development dates not more than a year and a half, to two years back, the result must be regarded as almost wonderful.

An ardent believer in the possibilities of Minnesota's swamp lands, Mr. Hartley purchased what he considered one of the worst swamp tracts in this district, with the intention of proving his theory. By persistent effort he has succeeded in enlisting state aid in the project

of swamp drainage in northern Minnesota, and some of the state ditches already cut his land and the water is being carried off to the Floodwood river in one direction the Mississippi Lateral ditch mile apart col feed into the being of a widt water that woul sized boat.

Much of the only completed th results speak volume Where a few years from several inches water over the land spring, the ground w would not bear up the man, now four-horse team to a heavy bog plow are w shod, while the implement the the old bog turf and sticks and roots, which are pu and burned. Mowing machin run without trouble over meadows of grass that were u water a year ago.

The greatest success, howev that has been scored by the owne of Island Farm is the establishment of the facts that the muskeg swamps of this county are the greatest cel ery producers of the country, Michigan not excepted. The Island Farm is raising celery this year for which the commission houses in Duluth are paying a premium. It is of a quality that excels the celery shipped to the Head of the Lakes from Michigan or California, during the winter season.

That the present season was a backward one for market gardening of any sort, is a matter of gene knowledge. Owing to the prolonged cold weather last spring the gardeners were several weeks later than usual in setting out greenhouse grown plants of which celery is a notable example. On the Island Farm celery was planted in the field late in June, where in ordinary seasons it should have been set out in May, thus insuring two crops in a season.

In spite of the drawbacks, a late

growing season and the fact the ground wher out ~ ants were set broken, being four and one netted the cre, and, if inter mar nt plans, net re h higher made dozen price per

od L

FAIR TO HAVE ROAD BUILDING EXHIBIT

Visitors at the 1911 State Exposition Will be Given Opportunity to Witness Construction of One

ARRANGEMENTS NOW BEING MADE

This Year's Exhibition Will be of Greater Educational Value Then Any of its Predecessors.

From Wednesday's Daily.

Visitors at the 1911 Minnesota State Fair and Exposition will have the opportunity of watching the construction of a scientifically perfect road.

"Good Roads" is a topic of the most vital and general interest at the present time. Up to date progressive men, whether they live in the cities or country, are interested in improving the roads of their particular districts and of the state. Better roads mean better business and better general conditions. People all over Minnesota are awaking to this fact.

Members of the board of managers of the Minnesota State Agricultural Society appreciate the importance of and the growing interest in good roads and are planning to give the fair this fall

"Good Roads" is a topic of the

most vital and general interest at

the present time. Up to date pro

the present time. Up to date pro

gressive men, whether they live in

the cities or country, are interested

in improving the roads of their

particular districts and of the state.

Better roads mean better business

and better general conditions.

People all over Minnesota are

awaking to this fact.

SWAMP SETTLERS FIND ROAD-CONDITIONS HARD

ew Men Have to Pay Total Tax— State Does not Pay for Part Thru Its Own Land.

From Monday's Daily.

Mrs. May H. Bailey, of Happyland, Minnesota, four miles south of Little Fork, is spending a few days in the city. She is a homesteader there and is interested in the Northern Minnesota Development association.

She reports that the condition of some of the settlers in the upper part of Beltrami and Koochiching counties is pitiful. Much of the land is swamp land and belongs to the state. These settlers, she says, have gone in there and taken up small sections between swamps. After they have lived on them and improved them, the state has reclaimed them as swamp land, thus leaving the settler absolutely nothing for his work.

Mrs. Bailey says that not many roads have been cut there as the expense has to be bore by the few farmers, and the vast areas of state land do not contribute one cent. These people have a hard time getting to market and many of their lands are surrounded by swamps.

The people up there, she says, are vitally interested in Auditor Hayner's petition to have the state pay a part of the tax for building roads past state land.

Fig. 17.3. Articles from Bemidji newspapers during the period 1907 to 1912.

chiching County. Beltrami County's slower entry into the drainage business was probably due in part to the fact that the lands ceded by the Red Lake Chippewas were withdrawn from entry and settlement by Congress in 1906, at which time the survey by the U.S. Geological Survey was ordered. Congressman Steenerson and many northern Minnesotans hoped that the federal government would undertake the drainage of these lands, but the reclamation cost of $2 million estimated by the survey was more than Congress was willing to appropriate (*BP*, April 2, 1908; October 29, 1908). The compromise that resulted and that restored those lands to entry was the Volstead Act of 1908. By late 1907, there began to appear articles encouraging settlement of the swamplands that had only to be drained in order to be cheap, productive farmland (Fig. 17.3).

1910 to 1912

Large ditches more than 500 miles long were constructed northwest of Red Lake peatland in western Beltrami and eastern Marshall counties (*BP*, February 3, 1910; *BS*, August 20, 1920; *BP*, May 18, 1911).

With the construction of J.D. 13 from the Rapid River to Baudette (12 miles along what is now Minnesota Highway 72 in northern Lake of the Woods County) and J.D. 5 along the southern part of Red Lake peatland (T.154-155 N., R.30-31 W.), only 18 of the 98 miles between Bemidji and Baudette would be without a road (*BP*, April 13, 1911).

Road construction had become a major objective throughout the state by this time (Fig. 17.3). Settlers living on small parcels amid state-owned lands in northern Beltrami and Koochiching counties endorsed a petition by the Koochiching County auditor to have the state share in the cost of building roads through or adjacent to state land (*BP*, August 4, 1910). Beltrami County surveyor Roy Bliler noted the correlation between good roads and settlement, suggesting that prospective land buyers driven over good roads were more likely to buy land than those driven over poor roads (*BP*, August 18, 1910).

Passage of the Elwell Good Roads bill (Minnesota General Laws, 1911) authorized county boards to establish State Rural Highways on petition of six or more landowners, issuing bonds, letting contracts, and charging one-half of the costs against the state and one-quarter each against the county and the property benefited. This law was repealed by the legislature in 1915 (Minnesota State Highway Commission 1917).

1913 to 1915

The number of petitions for ditch construction greatly increased during this period. Two ditches of importance in the drainage of Red Lake peatland were completed: J.D. 14 north of Upper Red Lake (48 miles in T.155 N., R.31-32 W.) and J.D. 20, which provided

the 22-mile link between J.D. 13 and J.D. 5 for the proposed road from Bemidji to Baudette (today's Highway 72) along the range line between R.30 W. and R.31 W., T.156-158 N. These two ditches, which cost about $80,000 and $25,000, respectively, afforded access to the drainage projects that followed (*BP*, August 19, 1915; December 10, 1914).

In September 1914, ditch engineer Euclid J. (Ernie) Bourgeois walked from Kelliher to Baudette, about 57 miles, and stated that "all that remains to be done to have a road good enough for an auto road to travel between Bemidji and Baudette is eight and a half miles of work on ditch 20," which would be completed that month (*BP*, September 10, 1914).

Late in 1915, J.D. 30 and 36, two of the largest ditch systems within Red Lake peatland, were ordered established. J.D. 30, lying west of Highway 72 in T.155-158 N., R.31-32 W., was to be about 165 miles long (*BP*, September 2, 1915). J.D. 36, to the east of Highway 72 in T.155-158 N., R. 29-30 W. (partly in Koochiching County) with J.D. 20 on its western edge, was estimated to be 127 miles long (*BP*, November 18, 1915).

Ditches petitioned for, under construction, or already constructed as of April 1914 in Beltrami, Koochiching, Clearwater, and Marshall counties would require expenditure of $4 million over the next three years. About $2.8 million was spent as of July 1914 in Beltrami and Koochiching counties for the drainage of 1.8 million acres, creating 1,814 miles of ditch road at a cost of $1,650 per mile (*BP*, July 9, 1914).

As to the impact of the drainage ditches on settlement in the region, there was great interest in the purchase of tax-delinquent drained state lands under the provisions of the Volstead Act (Fig. 17.4). Beltrami County auditor J. L. George, speaking about the 1913 sale of tax-delinquent land, remarked that

it looks very much as if practically all the state lands will be sold at this sale and placed on the regular tax list next year. . . . This sale fully bears out my contention of the past two years that the ditch projects in the northern part of the county will be of untold benefit in opening up and develop-

Fig. 17.4. "A pioneer farm located in a large muskeg, northern Minnesota." Photo by E. J. Bourgeois (Soper 1919). Many settlers purchased tax-delinquent lands.

ing the county of Beltrami, one of the richest dairy and stock districts of the state. (*BP*, May 15, 1913)

Similarly, an article in the *Bemidji Pioneer* (Fig. 17.5) stated that the 1914 land sale demonstrated

not only the wisdom of ditch construction in the swamp acres, but the demand for and real value of swamp lands. It means, too, that the financial standing of Beltrami County cannot be justly questioned; that ditch construction, even with its consequent heavy bonded indebtedness, is an asset, not a liability. (*BP*, May 14, 1914).

Most of the tracts were purchased for an amount greater than the tax judgment, and that excess, about $16,000, would be applied toward the ditch liens for those tracts (*BP*, May 14, 1914; June 17, 1915). At the time of the 1915 sale of tax-delinquent lands, which netted about $17,000 in overbids, it was stated that the ditches had brought at least 300 new farmers to the county in the preceding year (*BP*, May 20, 1915; June 17, 1915).

The federal government, meanwhile, ruled that any excess collected above the tax due on a parcel of land was to be paid to the federal government rather than remain in the county to help pay the ditch lien. Auditor George protested that the land purchasers were burdened enough by having to pay $1.25 per acre to the federal government (as compensation to the Indians for their ceded lands) and the ditch tax, which might be as much as $2.50 per acre; it was unfair for the government to benefit by the land's increased value, which was the result of ditch construction paid for by the settlers, not the government. George called for amendment of the Volstead Act to allow these excess funds to be used by the counties to pay for and maintain the ditches (*BP*, June 17, 1915).

In another matter concerning the Volstead Act, it was decided that costs for roads constructed in conjunction with ditches could be assessed to the lands benefited, and that lands sold for delinquent road taxes could be patented under the provisions of the act (*BP*, November 25, 1915).

1916

Ditch construction continued into 1916, but fewer new projects were initiated, not because of declining interest in drainage, but because the drainage of the county was nearly completed. The year 1916 was marked by increased opposition to the issuance of ditch bonds for the funding of new drainage projects. Koochiching County was reported to be nearly bankrupt, bonded to the limit for the construction of ditches and roads on state land, which could not be assessed and taxed until it was purchased by private owners.

Residents of southern Beltrami County filed a protest with the county auditor against the construction of J.D. 25 (Fig. 17.5), urging that no further ditch construction should take place "under the present financial condition of the county" (*BP*, October 26, 1916;

November 30, 1916). The protest also stated that the bonded indebtedness of the county had "exceeded the safety limit," that "grave doubt exists as to whether the drained lands will be taken up by actual settlers to insure the maintenance of the ditches without taxing the whole county," that there was insufficient outlet for the existing ditches, and that the tax rate in the county was already the highest in the state (*BP*, November 2, 1916). Also indicative of the rising sentiment against the ditch projects was the support given to Charles S. Carter, a candidate for the state legislature. Carter, in his campaign to "stop ditching and blow more stumps" (that is, to develop logged-over lands), pledged his opposition

to further issuing of ditch bonds in this district until the present judicial ditch law is amended so that ditches will be constructed only where the quality of the soil, prospective settlement and general conditions warrant the same.

Ditch bonds at the time totaled $2,813,850, with an additional $278,000 in road bonds and $154,000 in refunded bonds (*BP*, October 26, 1916). One of Carter's supporters wrote in a letter to the editor:

The idea of permitting a few men to mortgage our homes in order that ditches might be built to satisfy a few people is a big joke. Bonds, bonds, and more bonds. That's all you hear when you go away from home. It's our duty to stop this sort of thing and to vote for men who pledge themselves to do so. Politics should never enter into this matter. We are all affected if we own property here. I have been through the ditch country where it has done much good, but there is so much hopeless ditching, where in a short time it must all be done over again that it's an absolute waste of money. (*BP*, November 2, 1916)

Though the vote was close, the drainage issue did not carry the election for Carter (*BP*, November 16, 1916).

Auditor George continued to deny that the ditches were responsible for the high taxes in the county, pointing out that there were "hundreds of thousands of acres placed on the tax list during the last few years that never would have been owned by private parties without the drainage which has been done" (Fig. 17.5). During the 1916 land sale, more than 150,000 acres of government lands were claimed by payment of delinquent interest on the drainage assessment; all or part of the ditch lien was in many cases paid as well (*BP*, June 15, 1916). An amendment to the Volstead Act passed in September 1916 turned the excess from such land sales over to the county for drainage purposes (U.S. Statutes at Large 1917).

An article in the *Bemidji Pioneer* (April 20, 1916) discussed another positive result of ditch construction: the creation of 1,600 miles of roads that would not otherwise exist (Fig. 17.5). Interest in roads continued to be high; many new roads were petitioned (*BP*, February 24, 1916). Beltrami County now had 294 of the estimated 95,000 automobiles in Minnesota (*BP*, February 10, 1916). An auto trip from Baudette to Bemidji via Waskish and Kelliher over ditch roads adjacent to J.D. 13 and J.D. 20, the route of modern Minnesota

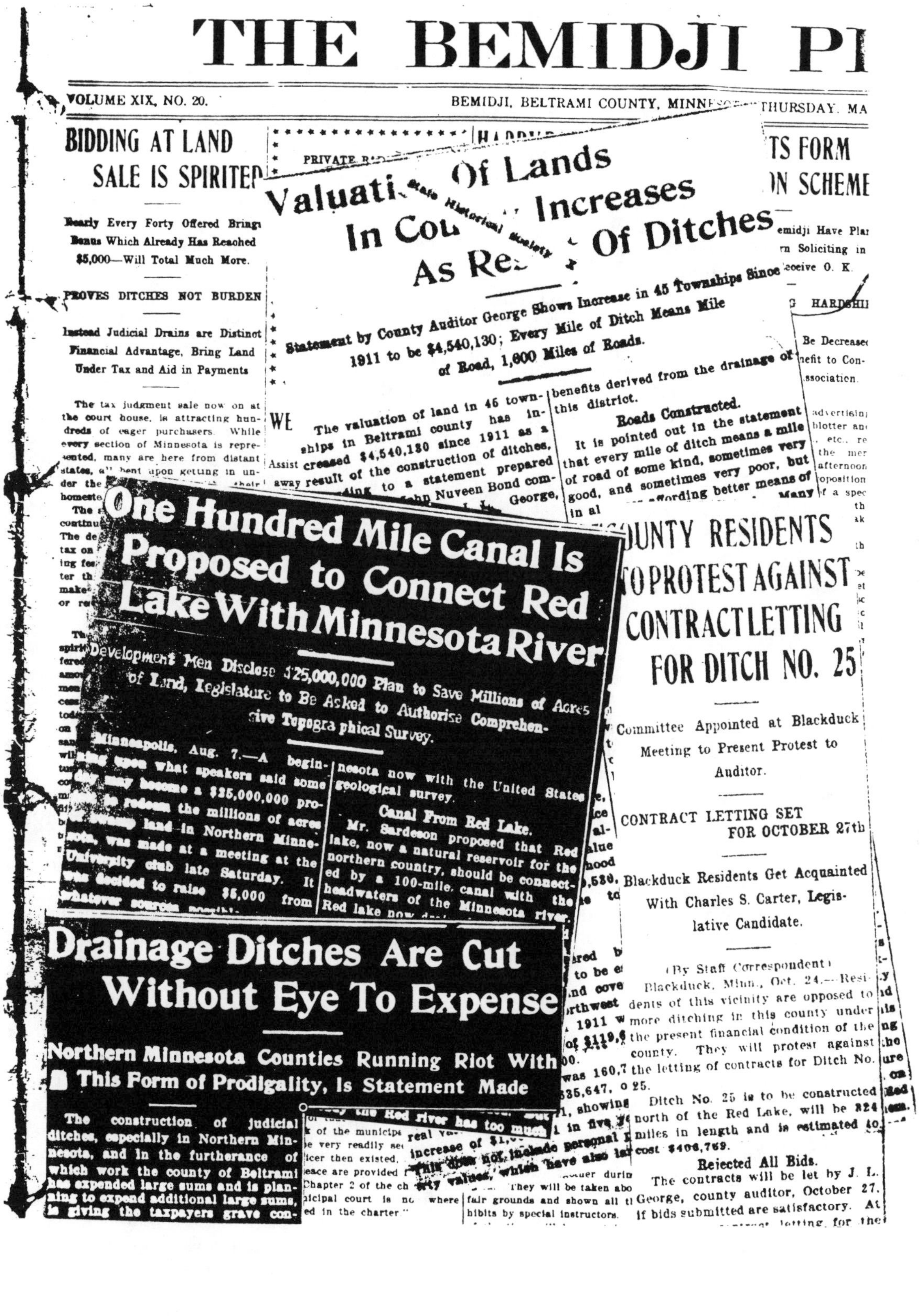

THE BEMIDJI PI

VOLUME XIX, NO. 20. BEMIDJI, BELTRAMI COUNTY, MINN... THURSDAY, MA

BIDDING AT LAND SALE IS SPIRITED

Nearly Every Forty Offered Brings Bonus Which Already Has Reached $5,000—Will Total Much More.

PROVES DITCHES NOT BURDEN

Instead Judicial Drains are Distinct Financial Advantage, Bring Land Under Tax and Aid in Payments

The tax judgment sale now on at the court house, is attracting hundreds of eager purchasers. While every section of Minnesota is represented, many are here from distant states, all bent upon getting in under the ...

Valuation Of Lands In County Increases As Result Of Ditches

Statement by County Auditor George Shows Increase in 45 Townships Since 1911 to be $4,540,130; Every Mile of Ditch Means Mile of Road, 1,600 Miles of Roads.

The valuation of land in 46 townships in Beltrami county has increased $4,540,130 since 1911 as a result of the construction of ditches, according to a statement prepared ... John Nuveen Bond com... George, ...

benefits derived from the drainage of this district.

Roads Constructed.

It is pointed out in the statement that every mile of ditch means a mile of road of some kind, sometimes very good, and sometimes very poor, but in all ... affording better means of ... Many ...

TS FORM ON SCHEME

...emidji Have Plan ...rn Soliciting in ...eceive O. K.

HARDSHIP

Be Decreased ...nefit to Con... ...ssociation.

One Hundred Mile Canal Is Proposed to Connect Red Lake With Minnesota River

Development Men Disclose $25,000,000 Plan to Save Millions of Acres of Land, Legislature to Be Asked to Authorize Comprehensive Topographical Survey.

Minneapolis, Aug. 7.—A beginning upon what speakers said some day may become a $25,000,000 project to redeem the millions of acres of swamp land in Northern Minnesota, was made at a meeting at the University club late Saturday. It was decided to raise $5,000 from whatever sources possible ... nesota now with the United States geological survey.

Canal From Red Lake.

Mr. Sardeson proposed that Red lake, now a natural reservoir for the northern country, should be connected by a 100-mile canal with the headwaters of the Minnesota river. Red lake now ...

COUNTY RESIDENTS TO PROTEST AGAINST CONTRACT LETTING FOR DITCH NO. 25

Committee Appointed at Blackduck Meeting to Present Protest to Auditor.

CONTRACT LETTING SET FOR OCTOBER 27th

Blackduck Residents Get Acquainted With Charles S. Carter, Legislative Candidate.

(By Staff Correspondent)

Blackduck, Minn., Oct. 24.—Residents of this vicinity are opposed to more ditching in this county under the present financial condition of the county. They will protest against the letting of contracts for Ditch No. 25.

Ditch No. 25 is to be constructed north of the Red Lake, will be 824 miles in length and is estimated to cost $406,769.

Rejected All Bids.

The contracts will be let by J. L. George, county auditor, October 27. If bids submitted are satisfactory. At ... letting for the ...

Drainage Ditches Are Cut Without Eye To Expense

Northern Minnesota Counties Running Riot With This Form of Prodigality, Is Statement Made

The construction of judicial ditches, especially in Northern Minnesota, and in the furtherance of which work the county of Beltrami has expended large sums and is planning to expend additional large sums, is giving the taxpayers grave con...

Fig. 17.5. Articles from Bemidji newspapers during the period 1913 to 1917.

Highway 72, was big news (*BP*, July 27, 1916). The 38-mile trip from Baudette to Waskish took four hours. The possibility of a major highway along that route was discussed by the Minnesota Highway Commission, but it was abandoned in favor of a route passing west of Red Lake, where permanent roads were more common (*BP*, February 13, 1916).

Flooding along the Red Lake River and in the Red River Valley brought renewed interest in controlling the level of Red Lake, which was four feet above its normal high-water stage in the spring of 1916 (*BP*, June 22, 1916). Local farmers, the Bemidji Commercial Club, and the Red River Development Association called for the federal and state governments to construct a dam at the outlet of Red Lake and to dredge and straighten 27 miles of the Red Lake River. The hope was that this would prevent flooding and provide an adequate outlet for runoff from ditches east and north of Red Lake, which were backed up because of the high lake level (*BP*, June 22, 1916; October 12, 1916; November 30, 1916).

A more ambitious approach to solving the drainage and flooding problem was proposed by F. W. Sardeson of the U.S. Geological Survey (*BP*, August 10, 1916). He pointed out that the Red River was already subject to flooding, so that improving the drainage of its tributary, the Red Lake River, would only magnify problems downstream. Drainage of Red Lake by means of ditches to Rainy Lake would flood the land north of Red Lake. Sardeson therefore recommended that Red Lake be connected to the headwaters of the Minnesota River by a 100-mile canal that would cost $25 million, according to an engineer's estimate (Fig. 17.5). Water that naturally flowed toward Hudson Bay would thus be diverted to the Gulf of Mexico.

1917 to 1918

Early in 1917, a grand jury urged that no more bonds be issued for ditch purposes until the value of existing ditches had been proven. It also said that state aid should be sought for future drainage projects (*BP*, February 22, 1917). New questions were raised about the value of ditches already constructed (Fig. 17.5). Even some of the early supporters of drainage and development objected to the way in which they were being accomplished. G. G. Hartley, who owned an experimental farm on drained peat land, commented that there was probably not "one bona fide settler for each 25,000 acres of land that we have paid for draining" and that areas drained but not settled ran a high risk of fire unless the ditches were dammed in order to maintain water levels. He recommended not draining any more swamps until settlers were secured for those lands already drained (*BP*, January 11, 1917).

Whatever the wisdom of further ditching may have been, the effect of U.S. involvement in World War I on the bond market and on the availability of labor and materials brought drainage projects to a virtual halt (Minnesota Department of Drainage and Waters 1921).

1919 to 1923

Much of the construction during these years involved projects initiated prior to the war, suspended, and then reopened after the war (Minnesota Department of Drainage and Waters 1923). Redigging of some ditches was deemed necessary. Averell and McGrew (1929) reported that "very few of the ditches have been maintained, with the result that many of them are partly filled and the channel obstructed by a growth of sedge and rushes in the bottom and brush along the banks." Some ditches had been dammed by beavers.

In February 1921 Congress authorized the Red Lake Drainage and Conservancy District (organized by district court in 1920) to deepen, widen, and straighten the Red Lake River and regulate the level of Red Lake (U.S. Statutes at Large 1921). Although 215 landowners signed a petition in 1919 calling for these improvements, the project was not without opposition (Fig. 17.6). Farmers and Red Lake Indians protested that they would be forced to pay a disproportionate share of the project costs compared to the water-power companies (Minnesota Department of Drainage and Waters 1921; *BS*, December 22, 1922). In addition, state fisheries experts cautioned against a significant lowering of the lake level that could cause the lake to freeze to the bottom, thus killing the whitefish spawn. A lower water level would also shift the lake's northern shore a mile southward (*BS*, November 4, 1921).

The county auditor's office continued to emphasize the benefits of drainage (Fig. 17.6), which added more than 913,000 acres with a valuation of nearly $8 million to the county tax rolls between 1911 and 1920 and created 1,500 miles of road, the $813,000 cost of which was borne by the lands benefited rather than by the county as a whole (*BS*, August 20, 1920).

Not everyone agreed that more drainage projects were advisable, however. Early in 1921, a Land Clearing Association formed in order to push for the development of 8 million acres of cut-over lands that were suitable for agriculture without requiring drainage (*BS*, January-November 1921). A proposal for Consolidated Ditch No. 1 in Beltrami and Marshall counties was rejected by the Beltrami County commissioners in 1921 because they felt that the financial condition of the county did not warrant its construction (*BS*, June 10, 1921). Along the same lines, the Minnesota Department of Drainage and Waters (1923) recommended "intensive reclamation of smaller areas rather than the superficial drainage of the larger virgin marshes." It also emphasized that it was more important to insure that people already farming make adequate profits than to induce more people to begin farming reclaimed areas.

Residents of Beltrami County sought state funding under the proposed Babcock road plan for the con-

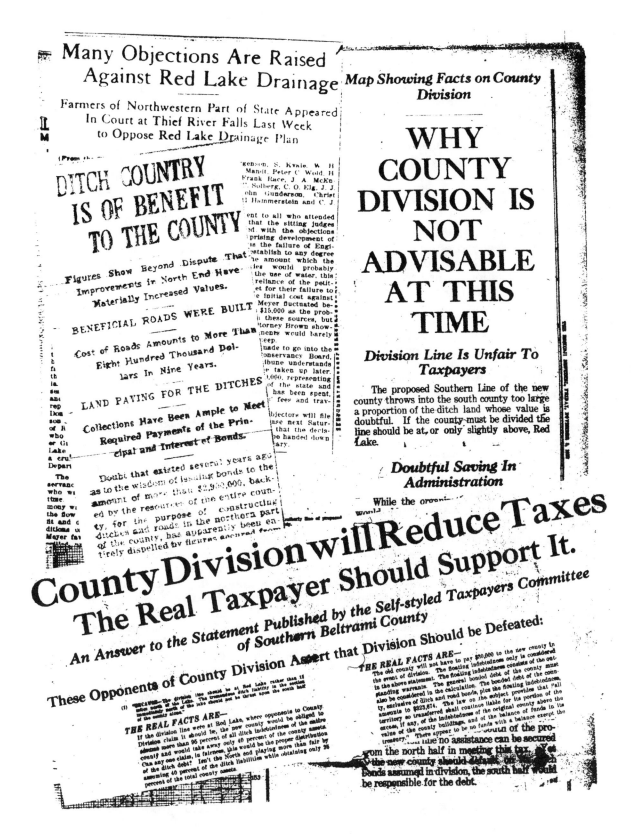

Fig. 17.6. Articles from Bemidji newspapers during the period 1921 to 1922.

struction of a trunk highway between Bemidji and Baudette (*BS*, January 10, 1919). The 16th amendment to the state constitution, known as the Babcock amendment, was adopted in 1920. It established a trunk highway system to connect county seats and other population centers (Kelly 1978). The Beltrami County commissioners issued $350,000 in bonds for road construction, including $100,000 for the proposed Bemidji-Baudette route, which was considered good for the development of the "north country" (*BS*, June 6, 1919; October 24, 1919).

This road, Trunk Highway 72, did not exist until 1927, but State Rural Highway No. 12, which was completed between Kelliher and Waskish by 1917, was extended northward across Red Lake peatland. The road was gravel-surfaced; part of it has since become Minnesota Highway 72. Gravel from Ludlow Island was placed on the road for the first time in the winter of 1921-22 by horses and sled and later by trucks (*BS*, February 24, 1922; March 24, 1922; E. Blanchard, interview).

By 1920 the idea of dividing the county (Fig. 17.6) had surfaced among residents of northern Beltrami County (*BS*, January 2, 1920). The general reason given for creating Lake of the Woods County was that the interests of northern Beltrami County were not being adequately served because travel to and communication with the county seat, Bemidji, was difficult. It was also argued that taxes would be lower if there were two counties instead of one.

When the issue was placed on the ballot in 1922, however, liability for ditch debts proved to be the divisive factor. Some "south-enders" maintained that the proposed division line 12 miles north of Upper Red Lake would throw "into the south county too large a proportion of the ditch land whose value is doubtful" and that the new county would be liable only for ditch liens within its own borders, whereas residents of the south county would share the responsibility if the new county defaulted on ditch bonds assumed as the result of division (*BS*, November 3, 1922).

The "north-enders" countered with the assertion that the new county would assume 40% of the ditch liability while obtaining only 25% of the total county assets. The suggestion by opponents to county division that the dividing line be at the north end of Upper Red Lake was rejected by supporters of the new county because that county would then assume 95% of the ditch liability with only 40% of total county assets (*BS*, November 3, 1922). County division was approved by a vote of 3,390 to 2,883 (*BS*, November 17, 1922).

1927 to 1929

During the 1920s, the delinquency rate on ditch taxes continued to rise steadily (Fig. 17.7), from 1.3% in 1916 to 77% in 1926 for Beltrami County and from 42.7% in 1922 to 70% in 1926 for Lake of the Woods County (*BS*, May 11, 1928). General tax delinquency was also high:

53% in Lake of the Woods (64% of taxable acreage) and 35% in Beltrami (60% of taxable acreage), as compared to the 4% rate in the remainder of the state outside the Red River Valley (*BS*, May 11, 1928; September 2, 1927). Much of this delinquency arose from lands that had been drained but on which settlers could not make a living. Only one ditch project in the two counties had a delinquency of less than 50%; the average rate was more than 70% (*BS*, April 20, 1928).

In neighboring Koochiching County, 35% of general taxes were delinquent in 1926; the tax rate had more than doubled since 1918 (*BS*, November 4, 1927). Drainage bonds worth $1,500,000 had been issued by the county; $1,045,000 was still outstanding, along with $375,000 in road bonds and other construction costs. A major factor in the financial problems of this county was that 800,000 acres, about 40% of the total area, were state lands subject to ditch taxes but not to general taxes.

In testimony before the Minnesota House Interim Committee on Local Indebtedness, Finance, and Taxation, Judge J. H. Brown noted that improvements near the state-owned tracts caused the value of those state lands to rise. "The state is holding this land for speculation just like a private individual," said Brown, who asked the state to assume some responsibility for the county's financial plight (*BS*, October 28, 1927). Some committee members agreed that the state should pay the ditch debts, but others, doubting that legislators from southern Minnesota could be persuaded of this, suggested a 50-year financing plan at a low rate of interest.

Beltrami County tried various means to meet its obligations. An opportunity for residents to pay back taxes at reduced rates brought in more than $50,000 (*BS*, November 4, 1927; November 17, 1927). The Board of County Commissioners levied a $50,000 general ditch tax in 1927 as it had for the three preceding years but later rescinded it, saying that no levies should be imposed until the state legislature had considered the whole ditch tax problem (*BS*, November 4, 1927; December 9, 1927). The board passed a resolution calling for a special session of the legislature early in 1928 (*BS*, October 21, 1927). The major points of the resolution were as follows:

1. When judicial ditches were constructed under state drainage laws, Beltrami County "through no act or will of its own and without the privilege of a day in court" was ordered by the district court to issue bonds, of which about $2.5 million remained unpaid.

2. Beltrami and Lake of the Woods counties had to issue refunding bonds in order to avoid defaulting on the original bonds; payments of ditch taxes assessed against benefited lands steadily decreased until they barely covered the interest on the bonded indebtedness.

BEMIDJI SENT

BEMIDJI, MINN., FRIDAY, MAY 11, 1928.

Ditch Burden Needs Aid Of Entire State

Empey Gives Graphic Account of Ditch Bond Situation at Association Meeting

STATE HO[L]... WER TO LIFT BU... HE CLAIMS

Effect... posed Railroad ... n Red Lake Line Also Discussed

Beltrami county is confronted wit an impossible situation Too much of the land area is in the hands of d linquent taxpayers. It would only folly to boost the tax rate with t idea of conquering the barrier with 10 years or so. It seems to me th the state holds the power to free from this burden. Let us not all this problem to go unsolved any lo er."

At a special meeting of the Bem Civic and Commerce associa Thursday evening at the city hall tiring Secretary George W. E made these startling statements advanced this one possible solutic the ensigma involved in the redu of the $1,500,000 debt of ditch against Beltrami county.

"To get a fair idea of the situa declared Mr. Empey, "we are no able to pay the $87,000 interest debt. This year it is estimate ... the interest will not

Tax Burden In Koochiching Is Ably Outlined

Data Compiled by Judge Brown Shows Problem is Deserving of State's Assistance

STATE-OWNED LAND IS CAUSING MAIN BURDEN

Drainage Debt Also Causing Concern There; Delinquency is Rapidly Increasing

Koochiching county's principal tax problem, differing considerably from the problems faced by Beltrami and Lake of the Woods counties in retiring an enormous ditch debt, is nevertheless a subject worthy of the wide consideration which is being given it by county leaders and the state legislative interim commissions.

For the benefit of persons in this territory who are not acquainted with the Koochiching county problem but who are interested in it, an assembly ... facts compiled by Judge J H

Ditch Relief Now Is Denied North Country

Brief Presented Before Interim Commission Shows Counties Need Outside Help

1507 MILES DITCHES COST THREE MILLIONS

High Rate of Delinquency Sure to Increase if Counties Even Try to Pay Debt

The State Executive Council is without authority to make a relief loan to aid Beltrami and Lake of the Woods counties which face a default of ditch bonds June 1, under emergency power given the council by the law, Attorney General O. A. Youngquist ruled Wednesday.

His ruling was made in reply to a query from Lieutenant Governor W. I. Nolan, chairman of the Minnesota Reforestation Commission, which then voted that it would be unable to recommend any relief measure at this time. The ruling and the Commission's decision, coming after a plea made by county officials to the forestry group for emergency aid spelled final defeat of the plan of the two counties.

A total of $200,000 is due on ditch bonds June 1 which are joint obligations of the two counties incurred when they were one county before 1922. Delinquent taxes ... cut over lan ...

Mass Meeting On Ditch Tax Friday Night

Able Speakers Are to Outline Procedure Designed to Bring About Relief

DELEGATION FROM LAKE OF WOODS TO ATTEND

... Attend Open

Tourist Camp is Already

Diamond Point Tourist ... its heavy summer service ... this week when a family ... children pitched their te... grounds. Other carloads of ... bound vacationers are daily ... the grounds en route to summe... on the lakes adjacent to Bemid... Board officials have renovate... grounds preparatory to the arr... the guests and will keep the pl... order for the rest of the season...

State Interim Group Is Told About Ditches

Beltrami and Lake of Woods County Delegation Appears At St. Paul Hearing

COMPREHENSIVE BRIEF OUTLINES CONDITIONS

Temporary Solution was Asked to Stave Off Default of Bonds Due June 1

Declaring that "the deplorable situation existing in Beltrami and Lake of the Woods counties, with the threatened disaster to the industrial and property interests of the citizens, which will be the inevitable consequence of default, presents a threatening calamity and disaster, which except for loss of life, will be as dire, annihilating and far-reaching in its consequences as were the results of the ...se Lake fire," members of the t tax committee are appearing to... before the state interim commis... with the hope of working out some diate solution that while even... rary would stave off the im... ng default of bonds due on June

members of the committee who emidji and Baudette Saturday nday were armed with copies retentious and complete brief g in detail the existing situa... presenting various plans of ief The brief, which repre... ts of intensive study in... ... compilation and numerous statistics, presents a com... e to the reader and will be e value in the arguments be made on behalf of the ected.

... brief opens with a comprehen-

Ditch Delinquency Nearly 79 Percent

County Auditor Gives Figures on Delinquent Tax Lists Published Recently

For the benefit of Beltrami county residents who have not been able to spend the time required to read over the twenty-page Delinquent Property Tax statement published in the Bemidji Daily Pioneer last Saturday, appearing in The Sentinel this week and in The Daily Pioneer again next Saturday, County Auditor A. D. John-son offered some figures this morning which reveal where the delinquencies exist and how great they were during 1926 in Beltrami county.

The county auditor's report shows that more than 22,000 parcels were delinquent in Beltrami county during 1926, excluding all previous years, and that 79 per cent of this delinquency has been in existence prior to 1926.

The report reveals.

That the total 1926 delinquency was $447,566.72. It is estimated that this amount is nearly one-fourth of the total delinquency for all years.

That the ditch tax of $177,000 levied in 1926 brought in only $97,819 in ac...

DITCH PROBLEM IS DEPLORABLE

State Interim Commission is Told at Hearing Today That Relief is Imperative

(Continued from Page 1) might result in a jail sentence for contempt of court. "It is noteworthy," declared the document, "that since this ultimatum no other Judicial Ditches have been established."

Section B of the brief is devoted to statistics of ditch construction and finance and the general financial condition of the two counties. This section presents some interesting statistics will be reprinted in part from time to time for the enlightenment of citizens. It points out among other that should default occur in the pal and interest and all out of bonds become due at once, proper legal procedure, a judg... evy for the entire amount of bonded indebtedness would be ... would mean that the levy

Fig. 17.7. Articles from Bemidji newspapers during the period 1927 to 1929.

3. Many drained lands placed on the tax rolls under the Volstead Act were later abandoned and therefore became tax delinquent.

4. Most of these drained lands were uninhabited and constituted a major fire hazard.

5. Lake of the Woods County was unable to pay the share of ditch bonds it assumed when it split from Beltrami County. Beltrami, having issued the bonds, was responsible for the Lake of the Woods payment. With both counties on the verge of default, irreparable harm would be done to the credit of northern Minnesota, thus discouraging further settlement in the region.

6. The board added a levy of nine mills to the county tax effective for four years in order to reduce the county's indebtedness, but taxpayers were taking legal action against the imposition of such general levies to pay ditch debts.

7. A judicial decision could "provide a remedy for only one of the two innocent classes of people, to-wit, the taxpayers and the bond holders."

Whereas the proximate cause of the deplorable condition that exists in northern Minnesota is the passage of the drainage legislation without provision for proper restriction, which has permitted the perpetration of this outrage, saddling with large obligations Beltrami and other counties, without granting the taxpayers now called upon for payment the right of representation in the judicial ditch proceedings, or otherwise, in violation of that first principle of the American Commonwealth forbidding taxation without representation, and it follows that the ultimate fair and equitable relief for all parties concerned must be legislative rather than judicial.

The House Interim Committee toured the ditch country between Bemidji and Baudette in October 1927. Lieutenant Governor W. I. Nolan affirmed that "northern Minnesota is indeed facing a serious situation because of the tremendous tax burden imposed upon the people through the unwise construction of ditches," a situation that the entire state would have to help resolve (*BS*, October 21, 1927).

Various solutions were suggested. At a Beltrami Farmer-Labor Party meeting, the state was urged to create a revolving fund for the purchase of county bonds issued to retire outstanding bonds; the fund would come from a small state tax (*BS*, March 23, 1928). Unfortunately, the state was facing its highest average tax rate in history at that time (*BS*, March 30, 1928).

The Bemidji superintendent of schools suggested that the peat in the district was valuable and might be used to repay some of the debt (*BS*, October 21, 1927). H. H. Chapman of the U.S. Forest Service proposed the development of northern Minnesota's 19 million acres of cut-over lands for forest crops as "the only solution for the delinquent tax problem" (*BS*, September 2, 1927). These lands were intended to be developed for agriculture, but, as of 1925, only 18.3% of this land was

owned by farmers and only 5.4% was producing crops. The State Drainage Commission studied the possible benefits of drainage on tree growth, but whether or not such benefits would exceed drainage costs remained uncertain (Averell and McGrew 1929).

The state legislature produced the ultimate solution to the problem in 1929, creating the Red Lake Game Preserve by an act that turned over tax-delinquent lands in Beltrami, Lake of the Woods, and adjoining Koochiching counties to the state in exchange for payment of the outstanding ditch debt (Minnesota Session Laws 1929). This preserve was made up of

all lands and waters in Lake of the Woods County lying south of Rainy River, and south of Lake of the Woods, and all full and fractional townships in Beltrami County lying north of the north line of Township 151, excluding, however, all of the lands and waters lying within the Red Lake Indian Reservation, and including also . . . part of Koochiching County.

Most of this area was encompassed by the Beltrami Island and Pine Island Conservation Projects (now state forests) in 1941 (Minnesota Session Laws 1941). The game preserve, now known as the Red Lake Wildlife Management Area, is managed by the Minnesota Department of Natural Resources.

In light of the dismal failure of much of the peatland drainage in northern Minnesota, one may well wonder why the situation was allowed to go unchecked for so long. The major factor appears to have been public sentiment in favor of settlement and development, which was encouraged by the optimistic predictions of land developers, a variety of public officials, and some experts on peat utilization.

A. A. Andrews and the Land Developers

In his 1923 biennial report, Minnesota commissioner of immigration Oscar H. Smith criticized the state for "following a hit and miss policy of land settlement," which encouraged people to locate on land not suited for agriculture on which it was impossible to make a living (*BS*, January 6, 1923). Sharing blame with the state for this situation, in Smith's view, were "unscrupulous land dealers who resort to dishonest and unfair tactics, both in the misrepresentation of lands and in gouging their victims to the limit."

The line between development boosters and these unscrupulous land dealers wasn't always clear. G. G. Hartley of Duluth was described as "one of the most vigorous 'boosters' of the north country" (*BP*, October 3, 1907). His Island Farm near Floodwood was considered a showpiece of what could be done with reclaimed swamp land (*BP*, October 17, 1907). State ditches and laterals about a quarter mile apart had lowered water levels sufficiently that a fine crop of celery was being produced on muskeg that had been burned, plowed, and enriched with manure. So impressed with Hartley's results were the secretary of state, the state

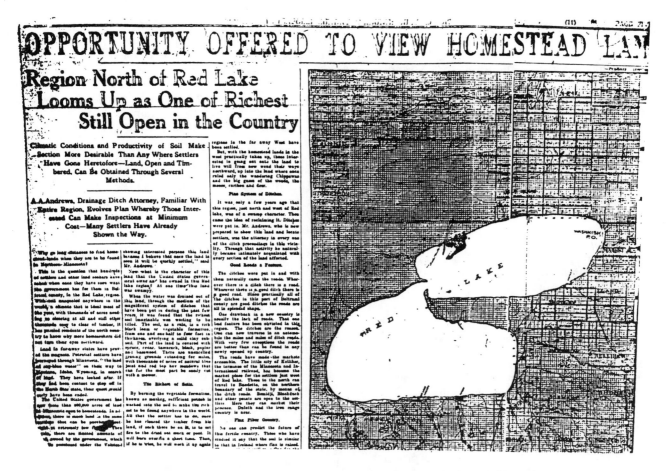

Fig. 17.8. "Publicity advertisement" in Minneapolis Sunday *Tribune*, July 29, 1917. See appendix for complete text.

immigration commissioner, the state drainage engineer, and a state senator that they bought 5,000 acres of swampland adjacent to Hartley's with the intention of raising hay and clover to demonstrate its stock-raising potential, then selling the land in parcels.

It may be that these demonstration farms were intended solely to encourage settlement by advertising the agricultural potential of the swamp lands proposed for drainage. This was not necessarily the case, however. Beltrami County attorney C. L. Pegelow, speaking at a public meeting to discuss the drainage debt situation in 1927, called attention to the acres of land that had become tax delinquent when speculators were unable to sell them at a profit (*BS*, October 27, 1927). These speculators had been encouraged to buy by "a scheming party of individuals that had found an especially fine piece of fertile soil near the bogs on which they had grown an attractive crop"; the impressive harvest was used as bait for unwary land seekers.

The farm near Waskish owned by A. A. Andrews was a prime example of one that induced prospective settlers to buy land unsuited for agriculture, including part of Red Lake peatland. Andrews was involved with drainage projects in Beltrami County from the very beginning as attorney for the petitioners in virtually every

ditch proceeding. In addition, he operated a real estate office out of Kelliher, taking prospective buyers across Red Lake to his farm on the northeast shore for a fee of $25. A full-page advertisement in the July 29, 1917, Minneapolis Sunday *Tribune* (Fig. 17.8; see appendix for full text), which describes Andrews's services, made the region north of Red Lake drained by "the magnificent system of ditches" seem like a homesteader's paradise.

Many settlers bought land sight unseen, having read such advertisements or having seen Andrews's farm and having been told that their land had as much potential (Vandersluis 1963; E. Blanchard, interview). According to Blanchard, who remembers Andrews's real estate activities well:

A lot of these fellows came up and wanted me to show them where their land was. They had been paying taxes on it for years. I would take them about halfway in and they would say they had seen enough!

Andrews was not the only land dealer interested in northern Minnesota's newly drained lands. Beltrami County surveyor Roy Bliler received inquiries in 1915 from land-investment firms in Minneapolis and St. Paul — and from out of state as well — about the area that includes Red Lake peatland (Kelly 1978). The

Minneapolis Real Estate Board contributed $500 to the Northern Minnesota Development Association (NMDA) in support of its work to encourage the reclamation of wetlands in northern Minnesota (BP, November 23, 1916). The NMDA reported a large number of inquiries in response to its advertising campaign (BP, December 7, 1916).

The development boosters' glowing descriptions of the opportunities offered by northern Minnesota to the land seeker undoubtedly were of considerable aid to the land dealers throughout the drainage era. Andrews himself was one of the foremost boosters, emphasizing that sales of the reclaimed lands added significantly to Beltrami County's assessed valuation (BP, October 2, 1913) and that the area had "the greatest opportunity of any section in the country" because of the productiveness of its peat and cut-over land, as shown by his demonstration farm (BS, November 17, 1922).

The NMDA annual report for 1916 had a similar message:

Here we have millions of acres of land of wonderful fertility where any man of energy may make a start with little capital. Our duty is to make this land accessible, to prepare it for the settler by roads and ditches, to advertise the wealth of our resources, to help the settler solve the problems that will confront him when he goes upon the land.

At the same time, the NMDA called on the next legislature to appropriate funds for "investigation and demonstration purposes to show what thousands of acres already reclaimed can be made to produce." It should be noted that although the association supported drainage and development, its members also passed a resolution stating that existing drainage laws should be amended so that ditches would be established "only where warranted" (BP, December 7, 1916).

Expert Opinion on Peatland Development During the Drainage Era

Articles published in the *Journal of the American Peat Society* and elsewhere during the drainage era in northern Minnesota indicated great interest in the uses and potential uses of peat in various forms. Among the potential uses mentioned most often in connection with Minnesota peat were the production of peat coke, machine peat for fuel, and producer-gas for generating electricity cheaply; peat could also serve as a raw material for nitrates, ammonia, dye stuff, paper, fabrics, artificial wool, moss litter, and packing material (Toltz 1912; BP, September 10, 1914; Soper 1916). The main purpose for which most of northern Minnesota's peatlands were drained, however, was agriculture.

If the articles summarized here are representative of the information available when the judicial ditch systems were being constructed, it appears that much of the expert advice was disregarded. The system by which ditch petitioners chose the location of the ditches, subject only to the approval of a local engineer and the

court, may have been at fault, or the experts may have failed to make their views known to the appropriate parties. Most likely, it was a combination of both factors. The situation could not have been helped by the fact that a mere 8.7% of Minnesota's area had been topographically mapped even as late as 1925, by which time $60 million worth of drainage projects had already been constructed within the state. Only Florida and Mississippi had lower percentages of their total areas mapped at that time (Minnesota Department of Drainage and Waters 1925).

That drainage was essential to the farming of peatlands was not disputed among the experts. F. J. Alway (1916), dean of the Division of Soils at the University of Minnesota, said that

in speaking of successful farming on peat it is always understood that a satisfactory drainage system has been installed — one that holds the water far enough below the surface to prevent waterlogging and makes it possible for men and horses to work upon it and yet does not lower it to such an extent that the crops suffer from drought.

H. M. Wilson (1908) of the U.S. Geological Survey and W. R. Beattie (1912) of the U.S. Department of Agriculture (USDA) emphasized that drainage methods should be tailored to local conditions, guided by careful engineering studies. Prior to installing a drainage system, the natural surface flow patterns and soil characteristics needed to be determined, according to Charles G. Elliott (1908), chief drainage engineer and chief of drainage investigations for the USDA. He added, "Too much emphasis cannot, however, be placed upon the need of guarding against errors in location which shall either impair the efficiency of the drains or add unnecessarily to their cost."

Presumably, the installation of drainage ditches at one- to two-mile intervals along section lines without regard to topography and surface hydrology, as was the case with most of the judicial ditches, would qualify as an error in location. Still, E. K. Soper, who worked for the Minnesota Geological Survey, was reportedly enthusiastic about the progress of drainage in Beltrami and Koochiching counties and the prospects for agriculture and dairy farming in the region (BP, September 10, 1914). When he was consulted by William Everts, the engineer for J.D. 30 west of Minnesota Highway 72 in Red Lake peatland, about the practicability of draining the swamp lands in the Red Lake area, he replied that the land could be drained into Red Lake and Rapid River (BP, April 4, 1915). "It would be easy to drain the upper four or five feet dry enough so that it could be cropped with hay," he said, recommending that Alway be consulted for more information about the agricultural possibilities.

Elliott (1908), Beattie (1912), and E. R. Jones (1921), the Wisconsin state drainage engineer, all recommended the use of tile drainage rather than open ditches, which were subject to slumping and filling in with weeds, in addition to taking up space that could otherwise be used for crops. Tile was usually more expensive initially, but maintenance costs were lower (Beattie 1912). Jones pointed out the high cost of excavating ditches 5 feet deep and 6 to 10 feet wide at the top. Elliott noted that in some muck soils, only open ditches could be used because the land would settle 25% to 50% percent after drainage, thus destroying the alignment of drain tile.

Ditches generally should be at least 4 feet deep and perhaps 500 feet apart, according to Elliott (1908). Ditches reaching sand or marl usually would not require additional tile drainage, but those reaching clay might need tile at spacings of 300 feet for hay or 150 feet for cultivated crops. Sluice gates placed along ditches would help maintain water levels during dry periods (Elliott 1908; Davis 1911; Beattie 1912). Jones (1921) warned against draining areas before they were ready to be cultivated because of fire hazard.

Wilson (1908) saw a number of advantages to drainage as compared with the irrigation projects that the federal government was undertaking. Drainage was simpler and cheaper, and part of its costs could be offset by using peat for fuel where coal was expensive and by selling the timber from peatlands cleared for agriculture. Wilson cautioned that land not suited for agriculture should not be drained.

The importance of assessing the specific characteristics of the peat in a given area was widely recognized. Charles A. Davis (1911) commented, "In those types of peats where the ash or mineral content is high and the vegetable matter well decomposed, drainage and clearing are often all that is necessary to prepare the land for cultivation." In other peats, lime, potash, manure, chemical fertilizer, legumes as green manure, or some combination of these was usually necessary (Davis 1911; Beattie 1912; Alway 1916). Beattie also mentioned the need for building up soil bacteria, which are typically deficient in waterlogged soils.

Limitations on the profitable cultivation of peatlands included the costs of labor, fertilizers and lime, transportation, and marketing (Alway 1916). A limited number of crops were suited to cultivation on peat, including celery, onions, lettuce, spinach, and potatoes (Davis 1911; Beattie 1912; Alway 1916). Alway added that because summer frosts occur more frequently on peat than on nearby mineral soils, frost-resistant crops and forage crops were most practical. John T. Stewart (1915), an agricultural engineer at the University of Minnesota, noted that problems arose because farmers wanted to grow certain crops whether or not they were adapted to peat. In his experience, not all reports about growth of crops on peat could be trusted because often the substrate referred to was not actually peat.

Carl G. Kleinstuck, described as a pioneer manufacturer of peat products and an expert on peat soils and suitable crops, considered the fuel value of Minnesota peat to be only a fraction of its value for crop production, especially for celery, onions, and peppermint (*BP*, August 27, 1914). Land bought in Michigan for a few dollars an acre was bringing $1,200 per acre for production of celery and about $550 for onions. Similarly, ceded Chippewa lands in Minnesota could be sold for under $5 per acre after drainage in a region where farmland was selling for $20 per acre (Wilson 1908). At least 3 million acres could be reclaimed at reasonable cost, according to Wilson.

Alway (1921) described all of the peatlands in Minnesota as potentially agricultural for hay and pasturage, with a small portion suited to potatoes, onions, celery, and lettuce. "The question as to whether any particular tract of peat land in Minnesota . . . can at present profitably be reclaimed for ordinary farming operations by proper drainage and fertilization, can not, in general, be definitely and reliably answered" (Alway 1916).

Peatland Agriculture in Northern Minnesota Today

The promise of peatland agriculture in Minnesota envisioned during the drainage era never materialized. The Minnesota Department of Natural Resources (DNR; 1981) reported that about 10% of Minnesota's peat acreage was being used for agriculture at that time. Very little agricultural use is made of northern Minnesota peatlands because of climatic factors, inaccessibility, and distance to markets. In addition to hay, pasture, and forage, the crops that grew best in experiments on drained peatlands in northern Minnesota were cabbage, broccoli, cauliflower, celery, potatoes, carrots, hybrid wheat, oats, barley, and turf grasses (Farnham and Levar 1980).

The drainage ditches in Red Lake peatland have, in general, not been maintained because the area was never settled except in scattered locations along the upland margins. Beaver dams on ditches have been removed where their impoundments threatened adjacent roads, according to the Beltrami County Highway Department and the Minnesota DNR at Waskish (personal communications). Other than that, maintenance or improvement of ditches would be done only at the request of landowners. The only petitions for ditch improvements in the area on file in the Beltrami County auditor's office, all from 1981, asked for permission to divert water for wild rice operations.

Epilogue: Drainage in the 1980s, Unresolved Conflicts

The controversy over drainage projects in Minnesota did not end in 1929, nor has it been limited to large projects in the northern counties. Two recent examples come to mind. Both involve ditches dug 40 to 75 years ago, now in need of repair, according to some landowners. Not all parties involved share that opinion, however.

Minnesota Public Radio aired a story in 1985 on its regional news program, *MPR Journal*, about a dispute

Outdoors

Star Tribune

Sunday
November 22/1987   **21C**

Ditch laws send a precious resource down drain

The county ditch sagas never end.

Ditch No. 5 in Scott County. It was going to be re-dug — "repair" is the catch word — and a beautiful marsh, Bradshaw Lake, was going to be destroyed.

For what purpose? To make more farm land.

There was Ditch 37 in Lincoln County that tried to drain the life water out of Anderson Lake.

For what purpose? To make more farm land.

And there's the saga of Ditch No. 4. Again in Scott County. Again a "repair" job on a ditch constructed more than 50 years ago when the only good country land was farm land.

Ron Schara

But times change. Values change. Repairing Ditch No. 4 today threatens a 235-acre marsh, although the involved landowners and Department of Natural Resources are discussing ways to save the wetland.

Another repair job. Another ditch

issue. Another controversy. Another close call for a chunk of wetlands, of which so many already have been drained.

And for what purpose? To make more farmland.

More farmland?

Excuse me, but haven't we paid about $27 billion to U.S. landowners this year to quit farming so much land, to quit producing surplus crops? Haven't we spent hundreds of millions of dollars in Minnesota for the same purpose? Doesn't the federal farm bill penalize farmers who accept cropland retirement payments and then bust swamps or unplowed prairies to create more crop land?

If this is so, why are there so many ditch projects?

Why is Pat Mulroy, 28, spending his nights knocking on doors in his Scott County farm neighborhood to stop the repair of Ditch No. 5?

He doesn't want more crop land.

"I'm farming ducks," Mulroy said the other day.

Mulroy created a slough on his property. But if the ditch repair goes through, Mulroy said his slough, too, would go down the ditch.

"When you ask if our state drainage-ditch laws reflect today's values," Tibor Gallo said, "it depends on whose values you're talking about."

Gallo is a special assistant in the state's attorney general office who

handles DNR's drainage-ditch issues.

Why the drainage ditch sagas in Minnesota? Because values are changing while the drainage-ditch laws have not.

"They were written in 1945 and haven't been changed much since," Gallo said.

In 1945, the only good slough was a drained one. In essence, the state's ditch laws still make the same claim.

"To justify a ditch project," Gallo said, "the county must show a benefit."

It's almost automatic the way the drainage laws are written. The assessed benefits are arbitrary and usually inflated to justify the assessments each landowner along

the ditch system must pay, Gallo said.

"Although there are many problems with the ditch law, it works the greatest injustice on the landowners themselves," he said. In other words, they usually pay for more ditch than they get in benefits.

In addition, the only "benefit" that counts is is converting "valueless" sloughs to "valuable" crop land. In other words, the more sloughs drained by a ditch, the "better" the ditch project, the more "benefits" are claimed.

What a wonderful ditch law.

If a ditch project is feasible, a county board is almost forced to complete it.

Schara continued on page 23C

Schara Continued from page 21C

regardless of local sentiments. One landowner along a ditch may demand a ditch repair, although the cost is assessed among all landowners along the ditch route based on alleged benefits.

"If you're a landowner with wet land, the drainage law is a friend because the neighbors have to help pay for the drainage ditch. But if you like wetlands, the drainage law will drive you nuts," Gallo said.

Indeed, the DNR once was assessed $34,000 in "benefits" because a ditch project would convert wildlife habitat into crop land.

"We claimed in court that the ditch was damaging, not benefiting, the DNR land," Gallo said. DNR won the case.

But the ditch law remains unchanged.

It's a law that requires farmers to pay a $10,000 bond for ditch plans, but the money is lost if the ditch never is built.

That's why there are so many ditches in Minnesota. And that's why people, such as Mulroy, are pounding doors in Scott County.

Not every Minnesota ditch is a wonderful idea. But those who protect the state's ditch laws won't hear of it. "That kind of thinking is foreign," Gallo said.

So the repair of Ditch 5 is a project still alive and the repair of Ditch 4 is coming up for a hearing before the Scott County Board.

What's there to decide?

Repair the drainage ditches, improve the crop land and let the taxpayers pay to keep the improved land out of crop production.

It made sense in 1945, right?

Fig. 17.9. Article on ditch laws by Ron Schara in the Minneapolis-based *Star Tribune*, November 22, 1987.

over the proposed enlargement of 4.5 miles of McLeod County Ditch No. 5 near Glencoe. Tom Meersman's report began with the premise that "the dispute raises a number of questions about whether drainage is always such a good idea and to what extent it may affect the environment."

In 1979 ten farmers signed a petition for ditch improvements, arguing that their lands were being flooded because many other farmers had drained their upland fields, and the subsequent runoff exceeded the ditch's capacity. A county engineer reviewed the situation and recommended that the 75-year-old ditch be deepened and widened substantially. The county commissioner approved the petition.

As has been the case since the late 1800s, however, the existing drainage law required ditch improvements to be paid for by all landowners whose lands drain into the ditch, even if those lands are several miles away. The cost estimate of nearly $900,000 for the project outraged many residents, who received assessments of $8,000 to $10,000, some up to $40,000, for "improvements" of little or no benefit to them. Not surprisingly, a petition opposing ditch improvements — this time signed by 133 residents — soon followed. The county commissioner was voted out of office, and letters of

protest were written to the Minnesota Department of Natural Resources (DNR), the Minnesota Pollution Control Agency, and the Army Corps of Engineers, all involved in issuing permits for the project.

Arguments against the ditch enlargement centered on the marginal agricultural value of the lands to be drained relative to the costs involved. In addition, it was recognized that the drainage would only shift flooding and erosion problems farther downstream. DNR area hydrologist Ken Stone observed that undrained wetlands were often more valuable for wildlife habitat and groundwater recharge than many people realized. The counties magnified the problems by approving ditching projects "on a piecemeal basis, usually with little or no consideration for long-term consequences," using guidelines established by antiquated drainage laws. A court decision was pending at the time of the broadcast.

An article by Ron Schara in the Minneapolis-based *Star Tribune* on November 22, 1987 (Fig. 17.9) sounded similar themes in the farmland versus wetland debate, noting the irony of creating farmland through ditch improvements at a time when billions of tax dollars were being paid to farmers to remove land from production.

Not all of the news about wetlands documents their disappearance. The Reinvest in Minnesota (RIM) Act of 1986 created new alternatives for land of marginal agricultural value. The RIM Reserve Wetlands Restoration Program of the Minnesota Department of Agriculture, in its first year, encouraged some 100 landowners in the state to plug ditches and break tile lines in an attempt to restore wetlands (Sierra Club North Star Chapter 1988).

Protection has been proposed for Red Lake peatland, the largest wetland area in Minnesota, along with 17 other "ecologically significant" peatlands (Minnesota DNR 1984). These areas are important, relatively undisturbed examples of the continental or forested raised-bog peatland type. Unique features distinguish Minnesota's peatlands from those occurring elsewhere in the world. They represent the southern limit of patterned peatlands and, as such, are valuable for comparative studies of peatland development and ecological processes. Permafrost, found in Canadian and northern European peatlands, does not occur in Minnesota.

Recent studies have only begun to reveal the hydrologic complexities of these extensive peatland systems. Despite the extreme living conditions of the peatland environment, it is home to rare and unusual animals and plants. The DNR report also cites the contributions of analysis of pollen and other fossil plant material in peat to our knowledge of past climatic and vegetative changes. Archaeological exploration of Minnesota peatlands may provide clues to cultural history as well.

The Minnesota DNR recommended to the state legislature that core areas be preserved as Scientific and Natural Areas or Peatland Scientific Protection Areas with Peatland Watershed Protection Areas surrounding them. For an update on peatland protection, see chapter 19.

In the more than 70 years since much of Red Lake peatland was ditched, the area has essentially been left alone despite periodic development threats. Economic studies by would-be developers and environmental studies by government agencies and independent researchers have thus far discouraged further development. The Upper Red Lake peatlands were designated a National Natural Landmark by the Secretary of the Interior on May 15, 1975 (Midwest Research Institute 1976), but the area still lacks a management plan to insure that activities detrimental to the integrity of the peatland ecosystem will not occur.

Ernie Blanchard has heard development schemes proposed for Red Lake peatland since he was a boy. As someone who came to know the peatland well, working on the dredging crews and later walking his district as a ranger for the Minnesota State Forest Service, he has a unique perspective from which to view the area and its fate. It seems somehow appropriate to allow him the last word here, just as he concluded our 1984 interview:

There was ditch all through this country. A lot of it did some good and a lot of it — just water over the dam, I guess. The peat was deep and there wasn't much you could do. They talked a lot about selling the peat off of there, and I don't believe in that. Because if they take that off, they're going to kill Red Lake. That peat acts as a sponge to feed the lake. That's all it does. Otherwise, if you took the peat off and got down to clay, you'd have these rains and it would all run down to the lake and down the Red [Lake] River, then to the Red River of the North, and that's the end of it. I think it's better to leave it the way it is. I think the good Lord wanted it this way.

APPENDIX

Text of "Publicity Advertisement" from Minneapolis Sunday *Tribune*, July 29, 1917

Why go long distances to find homestead lands when they are to be found in northern Minnesota?

This is the question that hundreds of settlers and other land seekers have asked when once they have seen what the government has for them in Beltrami County, in the Red Lake region. With soil unequaled anywhere in the world, a climate that is ideal most of the year, with thousands of acres needing no clearing at all and still other thousands easy to clear of timber, it has puzzled residents of the north country to know why more homeseekers did not turn their eyes northward.

Land in far-away states have proved the magnets. Potential settlers have journeyed through Minnesota, "the land of sky blue water," on their way to Montana, Idaho, Wyoming in search of land. They have looked afar. If they had been content to stop off in the North Star state, their quest would early have been ended.

The United States government has now more than 800,000 acres of land in Minnesota open to homesteads. In addition, there is much land in the same sections that can be purchased outright at extremely low figures. Then again, there are limited amounts of land, owned by the government, which can be purchased under the Volstead Act, that is by the payment of delinquent ditch taxes.

Land North of Red Lake

Much of the most desirable land is in the region north of Red Lake, the largest body of fresh water in any state in the union.

"Don't try to make an entry on any kind of government land without first zoning the land" — this is the advice which the United States gives continually. In order to follow that advice literally, hundreds of settlers, with their eyes on land in the Far West, have had to give up their dreams of carving out a farm home for themselves.

A. A. Andrews, a drainage attorney of Bemidji, familiar with every section of land in Beltrami County and particularly that land which has been drained and is open to entry or is now for sale either as patented land or under the Volstead Act, has evolved a method whereby these lands can be brought to the personal attention of those desiring to inspect it.

Headquarters at Kelliher

Mr. Andrews has opened headquarters at Kelliher, a station on the Minnesota and International rail road, a branch of the Northern Pacific. For a fee of just $25, he will take interested persons from Kelliher right into the region north of Red Lake, where the best land is located. He will transport them by automobile to Washkish, right on the shores of Red Lake. In his big launch, the "Storm King," he will take them across Red Lake to the land. His services as an attorney will be at the disposal of the prospective settlers. Upon payment of the one fee, Mr. Andrews will post prospective settlers on the methods of acquiring land. He will give them all of the legal advice necessary.

"I have decided upon this plan of showing interested persons this land because I believe that once the land is seen it will be quickly settled," said Mr. Andrews.

Now what is the character of this land that the United States government owns and has owned in this Red Lake Region? At one time this land was swampy.

When the water was drained out of this land, through the medium of the magnificent system of ditches that have been put in during the past few years, it was found that the richest soil imaginable was waiting to be tilled. The soil, as a rule, is a rich black loam or vegetable formation, from one and one-half to four feet in thickness, overlying a solid clay sub-soil. Part of the land is covered with spruce, cedar, tamarack, birch, poplar and basswood. There are unexcelled grazing grounds extending for miles, with thousands of acres of natural blue joint and red top hay meadows that can for the most part be easily cut with a mower.

The Richest of Soils

By burning the vegetable formation, known as muskeg, sufficient potash is worked into the soil to make the richest to be found anywhere in the world. All that the settler has to do, once he has cleared the timber from his land, if such there be on it, is to set fire in the dried out muck or peat. It will burn over in a short time. Then, if he is wise, he will work it up again by means of a harrow or disc and burn once more. The result will be soil that cannot be equaled anywhere for productive quality.

The climate of all of northern Minnesota is invigorating and healthful. During the summers the days are long, giving just that much more growing weather. In the winter, while the mercury goes far below zero, the cold is of the dry variety that simply puts "pep and punch" into the inhabitants. It is a cold that is far more comfortable than in the warmer climate where it is much damper.

Beltrami County has always been blessed with sufficient rainfall, the average annual precipitation being about 30 inches. There are no dry farming problems that must be solved here. Abundant water and pure is to be found everywhere.

With a climate that is practically ideal all of the year round, healthful in the extreme, it has been all the more strange that these homestead lands of the government have been passed by and those in the more arid regions in the far away West have been settled.

But, with the homestead lands in the West practically taken up, those interested in going out onto the land to live will from now on wend their ways northward, up into the land where once ruled only the wandering Chippewas and the big game of the woods, the moose, caribou and deer.

Fine System of Ditches

It was only a few years ago that this region, just north and west of Red Lake, was of a swamp character. Then came the idea of reclaiming it. Ditches were put in. Mr. Andrews, who is now prepared to show this land and locate settlers, was the attorney in every one of the ditch proceedings in this vicinity. Through that activity he naturally became intimately acquainted with every section of the land affected.

Good Roads a Feature

The ditches were put in and with them naturally came the roads. Wherever there is a ditch there is a road. Wherever there is a good ditch there is a good road. Since practically all of the ditches in this part of Beltrami County are good ditches the roads are all in splendid shape.

One drawback in a new country is usually the lack of roads. That one bad feature has been obviated in this region. The ditches are the reason. One can now traverse in an automobile the miles and miles of ditch roads. With very few exceptions the roads are better than can be found in any newly opened up country.

The roads have made the market accessible. The little city of Kelliher, the terminus of the Minnesota and International railroad, has become the market place for the settlers just west of Red Lake. Those to the north can travel to Baudette, on the northern boundary of the state, by means of the ditch roads. Bemidji, Blackduck and other points are open to the settlers. Here they can market their produce. Duluth and the iron range country is near.

Flax Fiber Country

No one can predict the future of this fertile country. Those who have studied it say that the soil is similar to that in Ireland where flax is raised. They say that it will raise flax for the fiber as well as for the flaxseed. Good flax fiber is in demand the world over. Linen makers in America have had to import much of it.

Experts who have analyzed the soil in this great region, which lacks now only the hand of man to transform it into a veritable paradise, say that some day the greatest flax fields in all the Americas will be found here. They believe that the time is not far distant when great linen factories will be located in the Twin Cities and Minnesota linen will take its place side by side with Irish and Dutch linen.

Mr. Andrews, in order to prove the flax raising qualities of the soil, planted a small tract of it this year. Within four weeks, he had as fine a stand of flax as could be found in any other portion of the state. His experiment has already proven that the soil is well fitted for flax raising.

Forage Crops Grow Well

Hay and forage crops grow to perfection in this new land. All varieties of clover, timothy hay, alfalfa grow with amazing results. Red clover has become a standard crop with the settlers who are already on the ground.

It would shock the Iowa and Missouri enthusiast to see how well corn grows in northern Minnesota. Those who have been so certain that Iowa, Southern Illinois and Missouri were the only places where corn can be grown successfully have something coming to them in the way of enlightenment. On account of the long hours of sunlight, corn has become one of the staple crops in northern Minnesota.

All of the small grains have done well on this land. Rye, barley and oats have been raised with much success. Although the farmers have not raised much wheat as yet, what has been grown has shown a good yield per acre.

It is in the root crops that the land has shown the most startling results. The reclaimed land, because of its heavy layer of decayed vegetation, is particularly adapted to the raising of potatoes, turnips and other roots. Celery raising has become one of the great industries of northern Minnesota. It is predicted that in a few years there will be much celery raising in this Red Lake region.

Dairying Sure to Come

Diversified farming, with emphasis on stock raising, is destined to characterize this section. In the timber land in its wild state many varieties of plants making good stock feed are grown. This is particularly true where the land had been burned over and most of the trees destroyed. Wild peas, wild clover and grasses and vines are to be found. There is enough of this to feed many herds of cattle. This wild growth comes out early in the spring. Many settlers run their cattle on the unfenced land.

Because of this natural forage and the adaptability of the country to the raising of hay and forage crops and the presence of fresh and pure water in abundance, the region will undoubtedly become a great stock raising section.

Large Scale Clearing Possible

To those who desire to take over large tracts of land to clear, the region in Beltrami County offers unusual opportunities. Experts who have analyzed carefully the situation say that, with the use of large tractors the land could be cleared and made ready for cultivation in a very short time. They say that a few heavy tractors that could draw large discs or harrows to tear up the land would do the business. By discing up the muskeg, or vegetable matter, and then burning it, a soil richer in quality than any known in the state is obtained.

That large interests ought to be attracted to this section is the opinion expressed by the experts.

"With some capital to purchase large tractors and other necessary machinery, hundreds, yes, thousands of acres of this land could be put into shape in a short time," said one man who went over the ground.

Summer Resort Possibilities

The summer resort possibilities of this region should not be overlooked by prospective buyers or settlers. Red Lake affords all of the advantages of the more settled lakes and has the added advantage of its shores being in their natural state.

Truly, here one finds the "land of sky blue water" that the Chippewa Indians sang about. Fine bathing beaches everywhere, lakes and streams teeming with fish, game of all varieties — these are the things that make Red Lake a Mecca for summer vacationists.

Once the wonders of this lake of lakes have been heralded to the world there is bound to be a rush for its shores. When that time comes — and it will come in the not far distant future — those who own land on its shores or in the vicinity will be the chief ones to benefit thereby.

The average settler could live by his fishing if he so desired. Muscalonge, that gamy fish of large variety, is to be found in all of the lakes. In the streams hundreds of trout leap. Here we find a fisherman's paradise practically untouched, save by the lazy Chippewa from the reservation between upper and lower Red Lake.

Many Light the Way

Many settlers have already lighted the way in this country. Those who go there now are not the first pioneers by any manner of means. Of course, those who are farming there now have not been there very long. But they have been there long enough to show the results. They have been there long enough to convince any one that the soil will raise any kind of a root crop, that the hay and forage crops are raised in abundance.

Take, for example, the cases of the Lastinec brothers, Austrian Slavs, who have carved out a farm home for themselves that is a paradise in itself. These men were never on a farm before they took up their homestead, not far from Washkish on Red Lake.

Glass makers by trade, clearing land and tilling soil were new tasks to them. But they possessed hope and persistence. They went up into the North country without funds of any kind. They didn't have any machinery. Lacking even the ordinary materials, they used their hands instead. They built a small log hut on their claim and then set to work. A drainage ditch had reclaimed the land.

Even working with their hands, it was an easy matter for those two intrepid workers from Europe to clear their land. The tamarack timber was soon cut down. Then they borrowed an ox from a neighbor and pulled the stumps. These stumps were then piled high and burned. Once the stumps and roots were out of the way, the so-called "peat," the soil covering, was set afire. It was burned over once, twice, thrice. The potash from the ashes was worked into the ground. Then the seed was planted. Lo and behold, the crop of clover and timothy that was produced on that land astounded everyone who saw it. This year there is a splendid crop of potatoes and of corn.

Can everyone carve out just such a paradise? Let the Slavs answer:

"We came here to take a homestead because we were told the soil was good once it was cleared. We have found it so. Never before did we do work outside. But we had to work hard here. But, then, that's all right. Only by hard work can anyone get anywhere. We are satisfied. We have a fine farm now and we are clearing more land every year. It isn't hard if one keeps right at it."

Far away Austria with the turmoil of revolution and world war, with vast armies advancing and retreating, have no places in the minds of these men. They are making things grow and have found true joy in the doing.

The farm of these men from central Europe is perhaps the finest example of what can be done on this type of land. But is far from being the only example. August Kosbau, for instance, has been on his tract a little more than three years. When he came it was a wilderness; now it is a garden spot such as would be found much farther south. He has large clearings of land that next year will be bringing him a good income. Even this year, he will harvest large crops of garden truck. He, too, has learned the secret of burning the peat and enriching the soil.

Volstead Act Explained

What are the main points of this Volstead Act pertaining to claims?

Here is the answer to the question as given in a government bulletin: "Under the Volstead Act the United States subjected certain of its Minnesota lands to assessment and taxation for drainage benefits, and permitted such lands to be sold for delinquent ditch taxes under the laws of Minnesota in the same manner as though the land were held in private ownership. No residence or improvements are necessary. The act provides, however, that only persons having the qualifications of homestead entrymen may acquire title to any lands of the United States so assessed, and limits the quantity to be purchased and held by any one person to 160 acres. The 160 acres need not be contiguous or in one body. If the purchaser has heretofore exhausted his homestead right by entering less than 160 acres, he may purchase and acquire title to the difference between the number of acres formerly entered and 160, provided this difference is not too small to fit the government subdivision which may be bid in by him.

"In order to protect his rights acquired at the tax sale, the purchaser should, within 90 days after the sale appear in person at the local land office or before some officer authorized to administer oaths in homestead cases in the district in which the land is situated, and execute a modified homestead affidavit and make the same payment that a commuting homestead entryman would have to make after 14 months residence and cultivation, less publication and witness fees. The minimum price to be paid the United States for the land within the deeded Red Lake reservation is about $1.28 per acre."

Expert Services Offered

To those who want to look at these homestead lands offered by Uncle Sam, those who want to buy up at low prices lands under the Volstead Act, or those who are simply interested as investors in either farm lands or summer resort spots will do well to get in touch with Mr. A. A. Andrews at Kelliher, Minnesota. The train that leaves the Twin Cities in the morning reaches Kelliher at 8 p.m. Mr. Andrews or one of his assistants will meet anyone who notifies him.

"We have made the fee for locating very low," said Mr. Andrews. "For the $25, we will show the land on the north shore of Red Lake. We'll locate homesteaders. They will have the benefit of our advice and experience in these matters.

"Instead of going far out into Montana, prospective buyers can for the price of the railroad fare from the cities to Kelliher, which is $10 round trip, and the $25, make the long automobile ride through the entire country, take the long boat trip across Red Lake and get all of the legal advice relative to locating."

Literature Cited

Alway, F. J. 1916. Some limitation on the cultivation of peat lands in Minnesota. Journal of the American Peat Society 9:65-73.

———. 1921. Availability of Minnesota peat land for agriculture. Report of Senate committee relative to iron ores and peat lands in the state of Minnesota. Senate Document no. 2.

Averell, J. L., and P. C. McGrew. 1929. The reaction of swamp forests to drainage in northern Minnesota. Minnesota State Department of Drainage and Waters, St. Paul, Minnesota, USA.

Beattie, W. R. 1912. The agricultural value of peat lands. Journal of the American Peat Society 5:193-204.

Davis, C. A. 1911. The agricultural side of peat bog utilization. Journal of the American Peat Society 4:97-100.

Elliott, C. G. 1908. Practical farm drainage. A manual for farmer and student. Second edition. Wiley, New York, New York, USA.

Farnham, R. S., and T. Levar. 1980. Agricultural reclamation of peatlands. Minnesota Department of Natural Resources, St. Paul, Minnesota, USA.

Glaser, P. H. 1987. The ecology of patterned boreal peatlands of northern Minnesota: A community profile. U.S. Fish and Wildlife Service Report 85 (7.14).

Jones, E. R. 1921. Tile drainage on peat lands. Journal of the American Peat Society 14:32-39.

Kelly, A. B. 1978. Pioneer roadbuilder: An historical biography. Published by the author, Redlake, Minnesota, USA.

Midwest Research Institute. 1976. Final report on peat program. Phase I. Environmental effects and preliminary technology assessment. Center for Peat Research. Midwest Research Institute, Minnetonka, Minnesota, USA.

Minnesota Department of Drainage and Waters. 1921. First biennial report of the Commissioner of Drainage and Waters. St. Paul, Minnesota, USA.

———. 1923. Second biennial report of the Commissioner of Drainage and Waters. St. Paul, Minnesota, USA

———. 1925. Third biennial report of the Commissioner of Drainage and Waters. St. Paul, Minnesota, USA.

Minnesota Department of Natural Resources. 1981. Minnesota Peat Program Final Report. St. Paul, Minnesota, USA.

———. 1984. Recommendations for the protection of ecologically significant peatlands in Minnesota. Minnesota Department of Natural Resources, St. Paul, Minnesota, USA.

Minnesota General Laws. 1911. General laws of the State of Minnesota, 1911. Chapter 254. Minnesota Secretary of State, St. Paul, Minnesota, USA.

Minnesota Session Laws. 1929. Session laws of the State of Minnesota 1929. Chapter 258. Minnesota Secretary of State, St. Paul, Minnesota, USA.

———. 1941. Session laws of the State of Minnesota, 1941. Chapter 215. Minnesota Secretary of State, St. Paul, Minnesota, USA.

Minnesota State Highway Commission. 1917. Report of the Minnesota State Highway Commission, 1915-1916. St. Paul, Minnesota, USA.

Palmer, B. 1915. Swamp land drainage with special reference to Minnesota. University of Minnesota Studies in the Social Sciences no. 5.

Siegel, D. I. 1981. Hydrogeologic setting of the Glacial Lake Agassiz peatlands, northern Minnesota. U.S. Geological Survey Water Resources Investigation 81-24.

Sierra Club North Star Chapter. 1988. New program restores wildlife habitat. Sierra North Star 35(3):4. Sierra Club North Star Chapter, Minneapolis, Minnesota, USA.

Soper, E. K. 1916. The peat deposits of Minnesota. Journal of the American Peat Society 9:81-88.

———. 1919. The peat deposits of Minnesota. Minnesota Geological Survey Bulletin 16.

Stewart, J. T. 1915. Peat lands in Minnesota and Wisconsin. Journal of the American Peat Society 8:16-22.

Toltz, M. 1912. The peat outlook in the northwest. Journal of the American Peat Society 5:37-38.

U.S. Congress. 1908. Congressional record, containing the proceedings and debates of the Sixtieth Congress, First Session. Vol. 42. Government Printing Office, Washington, D.C., USA.

U.S. Department of the Interior, 1909. A detailed report of a drainage survey of certain wet, overflowed, or swampy lands ceded by the

Chippewa Indians in Minnesota. 61st Congress, 1st Session. House of Representatives Document no. 27.

U.S. Statutes at Large. 1850. Public laws of the United States of America passed at the first session of the Thirty-first Congress, 1849-1850. Vol. 9. Chapter 84. Little, Brown, Boston, Massachusetts, USA.

———. 1855. Public laws of the United States of America passed at the second session of the Thirty-third Congress, 1854-1855. Vol. 10. Chapter 147. Little, Brown, Boston, Massachusetts, USA.

———. 1889. The statutes at large of the United States of America from December, 1887 to March, 1889. Vol. 25. Chapter 24. Part 1. Government Printing Office, Washington, D.C., USA.

———. 1917. The statutes at large of the United States of America from December, 1915, to March, 1917. Vol. 39. Part 1. Public acts and resolutions. Chapter 437. Government Printing Office, Washington, D.C., USA.

———. 1921. The statutes at large of the United States of America from May, 1919, to March, 1921. Vol. 41. Part 1. Public acts and resolutions. Chapter 64. Government Printing Office, Washington, D.C., USA.

Vandersluis, C. W. 1963. A brief history of Beltrami County. Beltrami County Historical Society, Bemidji, Minnesota, USA.

Wilson, H. M. 1908. Swamp lands and their reclamation. Part II. Journal of the American Peat Society 1:34-35.

CHAPTER 18

Management of Minnesota's Peatlands and Their Economic Uses

Mary E. Keirstead

Among the various perceptions of peatlands in Minnesota is the long-standing view that the peatlands are a vast, untapped economic resource. As early as 1870, the state legislature considered using peat as a locomotive fuel. In the early part of the present century, peatlands were thought to be both an important source of agricultural land and a fuel source that could free the state from dependence on other fuels. Since the 1940s, the Iron Range Resources and Rehabilitation Board (IRRRB) has conducted numerous studies in hope of bringing a viable peat-related industry to the iron-mining region of the state. In the midst of the "energy crisis" of the 1970s, Minnesota's peatlands were again thought to have promise as a source of fuel that could alleviate the problems of fuel shortages and rising energy costs.

The state's peatlands have in fact been exploited for a variety of economic uses. Peatlands have been cleared and drained for agriculture, and forested peatlands have been logged. Peat has been removed for use as a horticultural product and as fuel. But despite these uses, most of the state's peatlands remain undeveloped.

The responsibility for managing a substantial portion of the state's peatlands lies with the state of Minnesota, which owns or administers about half of the 6 million acres of peatland. Much of the peatland came to be state-owned because of failed attempts to clear and drain the land for farming. The Volstead Act of 1908 allowed large areas of peatland to be included in drainage districts in the northern counties, and miles of ditches were built in an unsuccessful attempt to drain the land (chapter 17). By 1929 several million acres had been forfeited for nonpayment of taxes because of ditch liens, and the state intervened. A legislative act (Laws of Minnesota, 1929, ch. 258) found the lands unsuitable for agriculture and established the state-owned Red Lake Game Preserve in Beltrami, Lake of the Woods, and Koochiching counties. Similar legislation (Laws of Minnesota., 1931, ch. 407; 1933, ch. 402) was later enacted for lands in Aitkin, Roseau, Mahnomen, and Marshall counties.

State legislators soon perceived other potential economic uses of peatlands, and in 1935 they enacted legislation (Minnesota Statute 92.461) that withdrew from sale all state lands "chiefly valuable by reason of deposits of peat in commercial quantities." Legislation (Minnesota Statute 92.50) was then enacted to allow peatlands to be leased for development.

Management of the peatlands was an easy task until 1975, when the state was suddenly faced with a Minnesota Gas Company (Minnegasco) request to lease 300,000 acres of peatland. Minnegasco was studying the possibility of producing high-Btu gas from peat, and the company's plans proposed the mining of large quantities of state-owned peatland in Koochiching, Lake of the Woods, and Beltrami counties.

The state responded to Minnegasco's request by declaring a moratorium on leasing and establishing the Minnesota Peat Program in the Department of Natural Resources. From 1975 to 1981, the Peat Program studied the environmental, economic, and social impacts of peatland development. The program's goal was to gather sufficient information to assess not only Minnegasco's proposal but also development of the peatlands in general, and to submit recommendations for the management of the state's peatlands to the legislature in 1981 (Minnesota Dept. of Natural Resources [MDNR] 1981).

These recommendations became the policies that guide the management of the state's peatlands. The basic premises of the policies are that peatlands have multiple uses and that the state should determine the most suitable uses of peatlands by evaluating environmental, geographic, and economic factors (MDNR 1981).

Much of the following discussion of the economic uses of peatlands is based on information gathered by the Minnesota Peat Program. The uses can be divided into three groups: (1) those that exploit the peatland as land on which to grow food crops, trees, and bio-energy crops; (2) those that require mining and processing of peat (horticultural products, fuel, and industrial chemicals); and (3) those that preserve peatlands for scientific, aesthetic, and environmental reasons.

Surveys of Minnesota's Peatlands

A key part of the Minnesota Peat Program is the Peat Inventory Project, which has completed the first comprehensive survey of the state's peatlands. Two previous surveys of peat deposits (Soper 1919; Office of IRRR 1964, 1965, 1966, 1970) were not comprehensive. Soil surveys such as the U.S. Department of Agriculture's Soil Atlas Project provide complete coverage of the state but not the detail needed for a peat inventory. Even the Peat Inventory Project's information must be supplemented by more detailed surveys when the use of a specific site is evaluated.

To appreciate the accomplishments of the Peat Inventory Project, one must consider the extent of the state's peatlands and the kind of information needed. The approximately 6 million acres of peatland in Minnesota are located in all but the southwestern and southeastern corners of the state (Fig. 18.1). Access to the interior of the very large peatlands in the northern counties is difficult. Assessment of the suitability for different uses requires information about the location and areal extent of the deposits, their depth, and the stratigraphy of peat types. To survey the entire state with any detail is clearly an enormous task.

The first survey of the state's peatlands was conducted by Soper during the summers of 1914 and 1915 as part of the U.S. Geological Survey's assessment of the potential of the peat resources in the United States as an alternative fuel.

In Soper's (1919) report, *The Peat Deposits of Minnesota*, he writes that

a detailed testing of some of the largest deposits would require years to complete, and in view of the constantly increasing demand in Minnesota for a report on the peat, or muskeg lands of the state, . . . it was decided to examine as many localities as possible and to publish a report of a more general nature which would include descriptions of most of the larger muskegs and peat deposits.

Soper surveyed every county containing peat deposits. His descriptions provide an assessment of the best use of the peat and information about peat type, average depth, vegetation, and chemical analyses. He

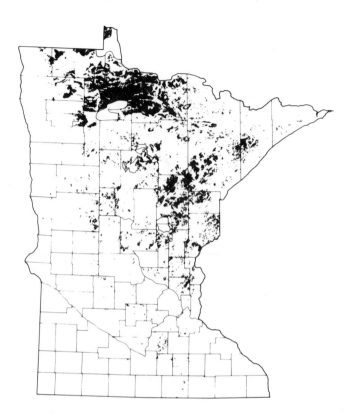

Fig. 18.1. Distribution of peat resources in Minnesota. (Courtesy of Minnesota Department of Natural Resources, Division of Minerals.)

estimated that Minnesota contains over 6 billion tons of "good peat fuel," a quantity probably greater than any other state's. Estimating the size and quality of Minnesota's peatlands was Soper's greatest accomplishment; limited time and restricted access prevented him from completing anything more than a general evaluation.

No further substantial survey work was conducted until the 1960s. After years of investigating the various uses of peat, the Iron Range Resources and Rehabilitation Commission (now Board) decided to support the horticultural peat industry by identifying possible mining sites (IRRRC 1958b). During the 1960s, the group completed surveys of 37 peatlands; all but four (Office of IRRR 1964, 1965, 1966, 1970) are unpublished. While these surveys provide detail about peat depth and type, they cover only a small portion of the state's peatlands.

In 1975, when the staff of the Peat Program began to address the use and management of peatlands, they soon realized that no existing survey provided the needed information. A peat survey project was then proposed by the Midwest Research Institute (1975). The planners of the survey concluded, as Soper had, that a general map was not adequate for site-specific planning, but a detailed inventory would be very expensive and take many years to complete. They therefore chose to limit the survey to the counties containing

Fig. 18.2. Peat Inventory Project staff obtain a peat core for analysis. (Courtesy of Minnesota Department of Natural Resources, Division of Minerals.)

the most peat and to conduct the survey at a reconnaissance level of detail; that is, boundaries were drawn between the mapping units by extrapolating from observation sites rather than by following the boundaries continuously on the ground as in detailed soil surveys (Johnson 1982).

The Peat Inventory Project staff began work in 1976. Their main objective was "to outline the dimensions of major peat areas in northern Minnesota and to determine the quantity, quality, type, and depth of the peat deposits" (MDNR 1979). In 1979 the U.S. Department of Energy and the Gas Research Institute, a private, not-for-profit organization representing the natural gas industry, provided funding to continue the project through 1986. Because the purpose of the federal program was to determine the quantity of fuel peat in the states containing substantial deposits, information was compiled in accordance with the Department of Energy's criteria for fuel-grade peat: (1) having a heating value of 8,000 Btu/pound or more in an oven-dried state, (2) containing less than 25% ash, (3) occurring in deposits at least 150 cm (about 5 feet) deep, and (4) covering a cumulative area of more that 30 hectares (80 acres) per 2.6 km² (square mile) (MDNR 1984a).

Survey work was carried out in nine counties: Koochiching, Lake of the Woods, Beltrami, St. Louis, Aitkin, Carlton, Itasca, Cass, and Lake. The staff began their work by locating the peatlands to be surveyed from aerial photos and U.S. Geological Survey maps. In the field, the surveyors selected observation sites along traverses and at random locations in the peatlands. At the observation sites, they sampled the peat with a Davis peat sampler at regular intervals in the vertical profile and examined it to determine the type of peat and the depth of the deposit. At some sites they obtained larger samples for laboratory analyses (Fig. 18.2).

The type of peat was described by its botanical content and the degree of decomposition. To describe de-

composition, the project staff used the von Post classification, a 10-point scale based on a visual examination of the physical properties of the peat when a small sample is squeezed in a clenched fist (MDNR 1984a). These results were later converted to the Soil Conservation Service's classification for decomposition: fibric peat, the least decomposed form, in which the botanical content can be readily identified; hemic peat, intermediate in decomposition, in which the plant fibers can usually be identified; and sapric peat, the most decomposed, in which the plant material's origin cannot be identified by visual inspection (MDNR 1984a).

To describe the botanical content, the project staff used the International Peat Society's classification: moss peat, composed mainly of plant remains derived from sphagnum and other mosses; herbaceous peat, composed mainly of plant remains derived from sedges, reeds, grasses, and related species; wood peat, composed of remains of trees and shrubs; and mixed groups (MDNR 1984a).

In the laboratory, peat samples were analyzed to determine ash content, moisture content, bulk density, pH, and heating value. Some samples were sent to the U.S. Department of Energy's laboratory to determine heating value, characteristics of the peat when it is burned, and constituents of the peat important in its use as a fuel.

The results of the survey can be used to locate sites suitable for development. Maps prepared for Aitkin and Koochiching counties and parts of Lake of the Woods, Beltrami, and St. Louis counties provide quick access to information about the areal extent and depth of peatlands (Fig. 18.3). Information about the depth and types of peat can be compiled to construct the stratigraphy of peat types in a deposit. A typical Minnesota peatland consists of a thin layer of sapric peat at the bottom of the deposit covered by a thicker layer of hemic peat, which may be overlain in parts of the peatland by fibric peat (Fig. 18.4).

Areas that meet general criteria for the size, depth, and peat type can then be examined further. These criteria differ according to the potential use of the peatland. A horticultural operation, for example, requires a peatland that contains a substantial amount of fibric peat, preferably sphagnum, which has the greatest water-holding capacity of the peat types. A fuel-peat operation, on the other hand, requires a peatland that contains primarily hemic peat, because it has a higher heating value than other types. Economic factors may necessitate a peat deposit of a certain size or depth.

The survey data are also available in a computerized data base. The use of the data with a geographic information system provides the means to consider additional factors that affect the potential uses of peat deposits (Hoshal and Johnson 1982). For example, the Peat Program staff can use the system to locate peatlands that are administered by the state and meet criteria set out in their management policies. Critical economic factors such as distance from cities and acces-

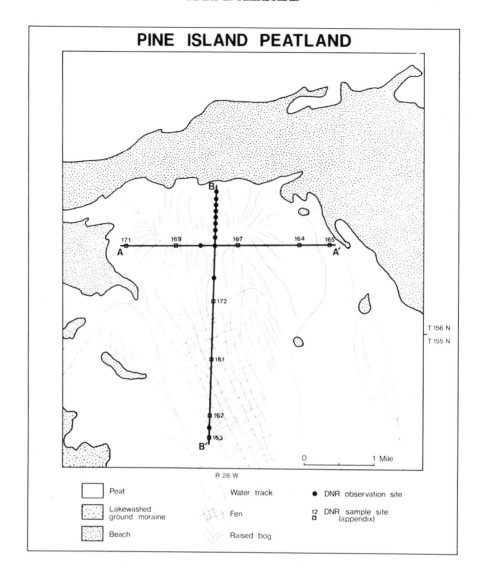

Fig. 18.3. Inventory map of Pine Island peatland in Koochiching County prepared from field survey information. (Reprinted, by permission, from L. Severson, H. Mooers, and T. Malterer. 1980. Inventory of peat resources: Koochiching County, Minnesota.)

sibility to transportation networks can be considered along with size, depth, and peat type.

Once peatlands that seem suitable for a particular use are located, more detailed information can be obtained from the data compiled from observation and sample sites. Botanical origin, decomposition, and pH are of greatest interest if the peat is to be a horticultural product. For fuel peat, decomposition, moisture content, ash content, and Btu value are most important.

For a final assessment of a site's suitability, additional survey work is necessary. Since 1983, the Peat Inventory Project's staff have been conducting more-detailed surveys of sites thought to have potential for fuel or horticultural peat. These surveys, conducted along grid lines spaced at intervals of 100 to 400 meters (109 to 437 yards), provide the same type of information as the reconnaissance-level surveys but considerably more. In addition, ease of access to the peatland is

determined, and an elevation survey is conducted to provide information about the ease of drainage.

The information collected by the Peat Inventory Project is critical to the success of the Peat Program's management of the state's peatlands. For the first time, the staff has the information and tools necessary both to locate peatlands suitable for particular uses and to evaluate requests to develop peatlands. These capabilities make possible the implementation of the program's policies, that is, to direct the development of peatlands.

Forest Resources

The largest use of Minnesota's peatlands in terms of acreage is the harvesting of trees from forested peatlands (Fig. 18.5). About 60% of Minnesota's 6 million acres of peatlands are forested, and about 28% — 1,623,000 acres — are classified as commercial forest

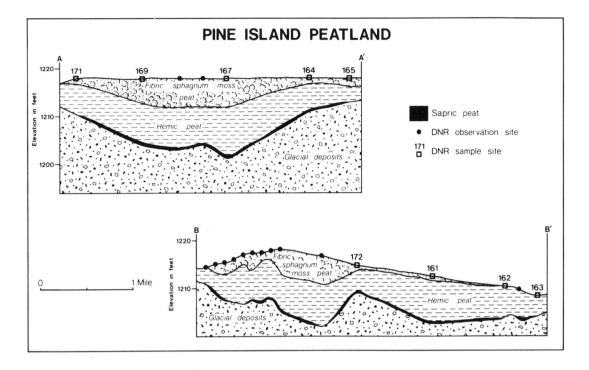

Fig. 18.4. Cross sections of Pine Island peatland in Koochiching County showing the stratigraphy of the peat deposits. (Reprinted, by permission, from L. Severson, H. Mooers, and T. Malterer. 1980. Inventory of peat resources: Koochiching County, Minnesota.)

land (MDNR 1981). The majority of these peatlands occur in the north-central and northeastern part of the state.

The major tree species that grow on peatlands are black spruce (*Picea mariana*), tamarack (*Larix laricina*), and northern white cedar (*Thuja occidentalis*). Lowland hardwoods such as American elm (*Ulmus americana*) and black ash (*Fraxinus nigra*) are also found on shallow peatlands but are not as common as the other peatland species. Black spruce, the most

Fig. 18.5. Black spruce forest in north-central Minnesota. (Courtesy of Minnesota Department of Natural Resources, Division of Minerals.)

abundant and valuable of the three major species, is cut primarily for pulpwood: its long fibers and bleachability are desirable characteristics for use in high-quality papers. It is also used for poles and Christmas trees. Tamarack is taken for pulpwood, fence posts, and poles. Northern white cedar is cut for food for wildlife in severe winters and is also used for lumber, posts, poles, and specialty products.

Kurmis and colleagues at the University of Minnesota conducted an assessment of the use of peatlands for their forest resource in 1978. They identified three factors that influence the degree to which peatland species are utilized: (1) accessibility of the timber, (2) location of markets, and (3) demand. Many of Minnesota's forested peatlands are inaccessible except in winter because of the absence of roads, and they have been too distant from markets to be cut economically. In the future, however, the increasing demand for timber will increase prices to the point that accessibility and distance from markets will no longer be economic barriers. The commercial sizes of spruce, tamarack, and cedar will become valuable resources for pulp and paper and lumber industries in Minnesota (John Krantz, personal communication).

The demand for black spruce is increasing, and the amount taken will continue to increase until 1995, when the recommended harvest level will be reached. The demand for tamarack is currently low; only 20% of the recommended harvest level is being cut (John Krantz, personal communication).

If demand for peatland tree species should increase

greatly, more-intensive management of forested peatlands may become desirable. Forest management practiced in Finland offers the best example of intensive use of peatlands. Finnish landowners began in the late 1950s to drain their forested peatlands to increase the growth of the trees. By 1980, 13.2 million acres (5.3 million ha) had been drained for forestry, and 65% of Finnish peatlands were expected to be drained by 1990. The productivity of some sites has been increased as much as 300% on some peat types. Finnish foresters have also planted trees on unforested peatlands with some success (Heikurainen 1982).

Revegetation with peatland tree species is one of the options for reclamation of mined peatlands. A study of management practices with several tree species conducted by White (1980) stressed the importance of adequate drainage and fertilization. Further studies of several species are being carried out by the reclamation staff of the Department of Natural Resources (MDNR 1987).

Agricultural Uses

Draining peatlands for agricultural use has long been a means of obtaining arable land, especially where good farmland is scarce. Irish farmers began draining and farming peatlands in the 18th century and in the early part of this century began farming mined peatlands (Cole 1984). In Finland, 30% of the country's arable land is drained peatland (Heikurainen and Laine 1980). Over 50% of the peatlands in Great Britain, Poland, and East Germany are used for agriculture (Kivinen 1980). Although peatlands are not as important a source of agricultural land in the United States as they are in some countries, thousands of acres have been drained for agriculture in some parts of the country — in California and Florida, for example.

In Minnesota, agriculture is the second largest use of the state's peatlands in terms of acreage: 10% of the state's peatlands, an estimated 670,000 acres, are used for agriculture (Farnham 1978). Only peatlands classified as commercial forest land constitute a larger use.

In the early part of the 20th century, it was hoped that peatlands would play a more important role in the state's agricultural development. Soper (1919) wrote in his report on the state's peatlands that "there is no problem more important to the agricultural development of the northern counties of Minnesota than that of the use of the peat swamps in that region."

Included in Soper's report is an article titled "Some Limitations on the Cultivation of Peat Lands in Minnesota" by F. J. Alway, dean of the Division of Soils at the University of Minnesota (1919). Alway explains the importance of suitable cultivation, fertilization, and favorable climatic conditions to successful farming on peat soils and emphasizes the need for a good drainage system to control the water level. He says that there are economic limitations in addition to climatic

ones — cost of labor, fertilizer, transportation, and marketing — and concludes that peat can be farmed profitably only where it offers some advantage over mineral soils.

Unfortunately, Alway's sound advice on the subject of agricultural development of peatlands was not heeded in the northern counties. Thousands of acres of peatland in Beltrami, Roseau, Lake of the Woods, and Koochiching counties were ditched in an unsuccessful attempt to drain the land for agricultural use (chapter 17). These efforts resulted in widespread forfeiture of the land by owners and legislation that declared these lands unsuitable for agriculture.

Much of what Alway said about successful farming on peatlands is still true today. Primarily because of climate and distance from markets, the northern peatlands continue to be largely unsuitable for agriculture. The majority of peatlands used for agriculture are in the southern half of the state, where the climate is more favorable for agriculture and the peatlands are smaller, shallower, and more easily drained. The largest acreages occur in Freeborn, Faribault, and Le Sueur counties. Hay and pasture land account for about 75% of the agricultural use of peatlands. Other uses include row crops, wild rice, turf grass, vegetables, and grass seed (Farnham 1978).

Experiments conducted by Farnham and Levar (1980) only proved further the importance of management practices such as control of the water level and fertilization to the success of farming on peatlands. This research, conducted in St. Louis County, also demonstrated that the problems created by the harsh climate of northern Minnesota can be overcome by choosing crops that grow well under these conditions. Best results were obtained with cabbage, broccoli, cauliflower, celery, potatoes, and carrots (Fig. 18.6).

Agricultural use of peatlands is not likely to increase greatly given the limitations created by the northern location of most peatlands. They will, however, con-

Fig. 18.6. Crops planted on peatland at Wilderness Valley Farms, Zim, Minnesota. (Courtesy of Minnesota Department of Natural Resources, Division of Minerals.)

tinue to provide a localized source of agricultural land, particularly for small or specialized farming operations.

Bio-energy Crops

Using Minnesota's peatlands to grow bio-energy crops — planted and harvested for use as fuel — was proposed in the 1970s. Because Minnesota contains large areas of wetlands, scientists studying bio-energy crops have focused their research on wetland plants, recognizing the advantage of being able to cultivate them on lands not considered prime land for traditional agricultural or silvicultural production.

Pratt and Andrews (1980) at the University of Minnesota began to evaluate the potential of native wetland species for use as bio-energy crops, focusing primarily on cattails (*Typha* spp.). Cattails were thought to be good candidates for bio-energy production because they grow naturally in monocultures, are easily propagated, and are very productive. Later research (Pratt *et al.* 1983), which measured the yields and nutrient requirements of eight wetland species, confirmed that three species of cattails (*Typha latifolia*, *Typha angustifolia*, and *Typha x glauca*) as well as the common reed (*Phragmites australis*) were promising bioenergy crops.

Field experiments with these four species provided further information about their productivity, adaptability to site conditions, resistance to pests, and planting and harvesting requirements (Fig. 18.7). The researchers concluded that two of the cattail species, *Typha angustifolia* and *Typha x glauca*, are the most promising of the four wetland plants in most situations (Pratt *et al.* 1985).

The search for suitable wetland species eventually became part of a comprehensive study of the feasibility of growing bio-energy crops in Minnesota, the University of Minnesota's Wetland Biomass Production Project. A

Fig. 18.7. Experimental plot of cattails planted on peatland at Wilderness Valley Farms, Zim, Minnesota. (Courtesy of Minnesota Department of Natural Resources, Division of Minerals.)

major part of the project was a study (Anderson and Craig 1984) that examined the availability of land, both peatlands and wet mineral soils, for growing energy crops. A computerized model was developed to consider land suitability (soil type, climate, hydrological setting, and current vegetative cover), land ownership, land use, and economic factors. Two areas of the state were analyzed to test the model. Extrapolating from the test areas, Anderson and Craig estimate that out of 8.8 million acres of wetland in the state, about 3 million acres could be suitable for growing bio-energy crops.

Woody plants that grow on wetlands have also been studied for bio-energy production in Minnesota. Larson and colleagues (1985) researched the intensive management of plantations of fast-growing, closely spaced trees. Research was conducted to select varieties of willows and poplars that have good yields, the ability to resprout after harvest, strong resistance to disease, and good survival and growth rates.

The results of field experiments with cattails and fast-growing trees have been encouraging. Along with peat, these crops represent the largest source of indigenous energy in the state. Many problems will have to be overcome before bio-energy crops can become a viable alternative to traditional fuels, however. Standardized fuel specifications for bio-energy crops are lacking, for example, and equipment that can use a variety of fuel types is hard to design (Minnesota Department of Energy and Economic Development 1985). Perhaps most important, bio-energy crops must be able to compete economically with other fuels.

Horticultural Products

The oldest continuing commercial use of mined peat in the United States is for horticultural purposes. While small amounts of peat have occasionally been mined for other uses such as fuel, almost all of the peat mined in the United States since the early part of the 20th century has been used in horticulture and agriculture (Singleton 1980).

Horticultural peat is familiar to most people as the peat moss that is used for potting soil and gardening, the predominant uses since the 1950s (Singleton 1980; Fig. 18.8). In 1988, 68% of the approximately 900,000 short tons of peat produced in the United States was used for soil improvement in gardens, lawns, and landscaping; 20% was used for potting soil. Nurseries consumed 5%, and the remaining 7% was used in a variety of ways, including in mixed fertilizers, for earthworm culture, in mushroom beds, and as seed inoculant (U.S. Department of Interior 1989).

The U.S. Bureau of Mines, which has collected data about peat production since 1908, classifies peat into five categories: sphagnum moss peat, hypnum moss peat, reed-sedge peat, peat humus, and other types. The moss peats are composed of poorly or moderately decomposed mosses; reed-sedge peat contains poorly decomposed reeds, sedges, and grasses; and

Fig. 18.8. Peat is put into bags at Michigan Peat Company, Cromwell, Minnesota. (Courtesy of Minnesota Department of Natural Resources, Division of Minerals.)

peat humus is so decomposed that the original plants cannot be identified (American Society for Testing and Materials 1969).

Sphagnum peat is particularly valued as a horticultural product because of its ability to retain water, its resistance to breakdown, and its acidity. Sphagnum peat is, however, the least plentiful peat type in the United States and accounted for only 10% of total production in 1988 (U.S. Department of Interior 1989).

Despite the fact that Minnesota contains the largest acreage of peatlands in the continental United States, the state's peat operations have contributed at the most only about 8% of the yearly production of peat in the United States from the mid-1950s to the present (Fig. 18.9; U.S. Department of Interior 1955-88). During these years, most peat has come from Michigan, Florida, and Indiana (Singleton 1980). Given that the market for horticultural peat has steadily increased since the 1930s, there has certainly been opportunity for growth in Minnesota's peat industry (Singleton

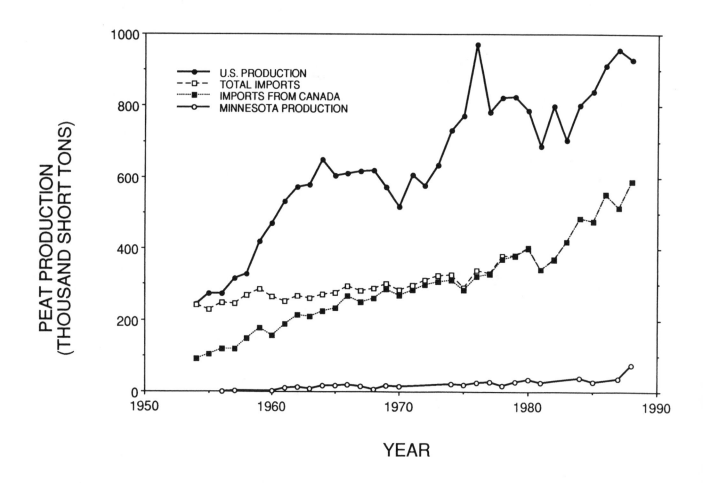

Fig. 18.9. Minnesota's share of yearly U.S. peat production, 1955-89. (*Source:* U.S. Department of Interior, Bureau of Mines, Minerals Yearbook: Vol. 1, Metals and Minerals, for the years cited.)

1980). Yet promoters of the industry, such as the Iron Range Resources and Rehabilitation Commission, have repeatedly been frustrated in their attempts to encourage increased production.

In 1945, the IRRRC started a horticultural-peat mining and processing operation at Floodwood, Minnesota, hoping to take advantage of the market created when importation of European peat halted during World War II (IRRRC 1945). The plant used a mechanical drying process that enabled greater quantities of peat to be produced than were possible at the few other production sites in Minnesota. Financial problems closed the plant in 1950.

By 1956, peat production in Minnesota had dropped so low that the state was no longer tabulated among the peat-producing states. Yet 300,000 short tons were produced by other states in 1956, and 250,000 short tons were imported from Canada and Germany (U.S. Department of Interior 1956).

Hoping to learn from the experience of European peat producers, members of the IRRRC visited several European countries in 1957. After their trip, commission members noted several problems with Minnesota's peat industry: small operations, inadequate equipment, and the lack of a uniform product (IRRRC 1958a). The commission nevertheless concluded that production of horticultural products was still the most promising use of Minnesota's peat. To help the industry, the IRRRC undertook a survey project to identify peat deposits of sufficient size, quality, and uniformity to compete with imported products (IRRRC 1958b).

Ten years later, five Minnesota companies were still producing only 3% of the yearly production in the United States, which had doubled in 10 years to 619,000 short tons (Minnesota Department of IRRR 1968, U.S. Department of Interior 1969). In 1976 the situation was much the same: four peat operations produced about 26,000 short tons, less than 3% of the total national production of 969,000 short tons (U.S. Department of Interior 1977).

Hoping to generate more interest in horticultural peat, the Peat Program conducted and published a survey of sphagnum-moss peat deposits (Malterer *et al.* 1979) and offered two peatlands containing sphagnum-moss peat at lease sales. Peat Program staff assumed, because other states were primarily producing other peat types, that Minnesota's producers could capture some of the market for moss peat, long filled by imported peat from Canada (Malterer *et al.* 1979; Singleton 1980).

The efforts of the Peat Program did generate some new interest in horticultural peat: more than a dozen companies produced horticultural peat in 1988, and production increased to about 74,000 short tons. In 1988 Minnesota operations produced about 8% of the national production, ranking third behind Florida and Michigan. Although sphagnum-moss peat imported from Canada still made up about 40% of the peat con-

sumed in the United States in 1988, U.S. sphagnum moss sales increased 142% primarily because of the development of new high-quality deposits in Minnesota (U.S. Department of Interior 1989).

A study by the Arrowhead Regional Development Commission (ARDC) (1985) clarified some of the problems faced by Minnesota peat producers. Peat distributors questioned about the market for horticultural peat responded that Canadian peat is considered a better quality product than domestic peat, especially by commercial growers, because it is more uniform in composition, less decomposed, and freer of wood. The ARDC concluded that Minnesota's producers would have to establish and adhere to strict standards of quality to compete with Canadian peat.

The ARDC (1985) recommended that Minnesota producers not try to compete directly with Canadian sphagnum-moss peat but rather try to market all types of Minnesota horticultural peat more aggressively, the goal being to create a greater awareness of Minnesota products and thus a bigger market. Support for this recommendation can be found in recent findings of the Peat Inventory Project. Sphagnum-moss peat deposits are an estimated 2% of Minnesota's peat resource, about 120,000 acres. But detailed surveys of potential horticultural-peat deposits have led to the conclusion that Minnesota's sphagnum resource is not as large as once thought. Furthermore, many of the sphagnum-moss deposits are too inaccessible to justify the expense of development, and variability of the decomposition of many of the deposits would make it difficult to compete with Canadian peat (David Olson, personal communication).

As of 1989 there were more than a dozen companies producing horticultural peat in Minnesota on sites ranging from as small as a few acres to as large as 1,500 acres. Two proposals to expand sites have been submitted to the Peat Program. Promoters of the horticultural-peat industry may continue to lament Minnesota's small share of the market, but economic conditions may simply keep the state's industry as a small but viable use of Minnesota's peat.

Fuel Peat

The history of the use of peat for fuel in Minnesota is notable for predictions about the great promise of the state's peat resources as a fuel supply. Since the beginning of the 20th century, peat enthusiasts have repeatedly extolled Minnesota's abundant peat deposits as an energy source, only to have their predictions fail to come true.

In 1910, Max Toltz, a peat promoter, wrote:

The time has come for the people of Minnesota to wake up to the fact that there are immense quantities of valuable materials lying idle and action should be commenced by our State government to at least locate and examine these peat deposits, seriously and systematically, as has been done in other states, and ascertain the supply of fuel.

Enthusiasm for the potential of Minnesota's peat was encouraged by activities in other areas of the country. In the eastern United States, some energy producers turned to peat after a coal miners' strike in 1903 caused a fuel shortage. Fuel peat was produced in small plants between 1908 and 1912 (Singleton 1980). In 1908, the U.S. Geological Survey began to survey the peat resources in states known to contain substantial deposits, primarily to ascertain fuel reserves.

Soper (1919), who surveyed the peat deposits in Minnesota, reported an avid interest among Minnesotans in the use of the state's peatlands. Soper's report did little to discourage the hopes of those interested in fuel peat. He wrote that "the presence of such large amounts of peat of excellent quality, situated[,] as many of the bogs are, immediately joining railroad tracks and near large and thriving towns, make[s] it certain that the machine peat industry will develop within a comparatively short time."

Soper (1919) also predicted that the manufacture of power in gas plants would soon be attempted and suggested using peat in the iron-ore industry for power and heat and as a binder for ore briquets and a source of coke. In the midst of these optimistic predictions, Soper inserted the cautionary note that the fuel-peat industry "may be slow to develop in the United States because of our enormous resources of excellent coal."

At first, peat promoters had reason to be optimistic. During the winter of 1919-20, pulverized peat produced by the Hennepin Atomized Fuel Company was successfully burned in the power plant of the Phoenix Building in Minneapolis. Several new companies began to produce fuel peat. In 1921, F. A. Wildes, superintendent of the state Bureau of Mines, conducted a demonstration of peat burning in the power plant of the state capitol for the members of the legislature, who later appropriated money for further experimentation. Henry Hindshaw, a staff member at the Bureau of Mines, is reported to have proclaimed that peat would free the state from dependence on coal from the eastern states (Osbon 1921).

By 1923 Soper's cautionary note about competition from coal seemed all too true, and enthusiasm for peat began to wane. Wildes (1924) wrote that there was considerable interest in using peat as fuel but no real progress toward that goal. Osbon (1923), editor of the *Journal of the American Peat Society,* pointed out that the test of peat in the Phoenix Building had been successful but costly. The final blow was dealt by a study published by the U.S. Bureau of Mines (Odell and Hood 1926) that concluded that peat could not compete with the more accessible and cheaper coal.

Not until the 1940s did the use of peat as a fuel again attract the interest of Minnesotans. In 1947 the recently formed Iron Range Resources and Rehabilitation Commission wrote in its biennial report that the many changes that had occurred in the 20 years since the Bureau of Mines' report provided sufficient reason to believe "that the time has now arrived to take advan-

tage of modern mechanical and scientific progress to convert the latent value of Minnesota peat into an industrial asset which will assure the economic welfare of the state for many years to come."

This optimism was again short-lived, however. In 1950, a bill introduced in the U.S. Congress by Senator Hubert H. Humphrey and Representative Fred Marshall of Minnesota to provide funding for research on fuel peat failed to pass. The IRRRC began to concentrate on other uses of peat, finally concluding that despite success in other countries, fuel peat was unlikely to be a profitable economic enterprise in Minnesota (IRRRC 1958b).

Almost 20 more years passed before energy shortages and rising fuel costs in the 1970s caused renewed interest in alternative fuels. Minnegasco's 1975 proposal to mine and gasify peat was symptomatic of the interest in peat. The state government, the federal government, other utilities, and private peat developers soon began to investigate the feasibility of using peat as a fuel.

While Minnegasco and the U.S. Department of Energy investigated gasification of peat, Minnesota's state government began to look at other options. Government representatives gathered information about the use of peat during visits to Europe and the Soviet Union in 1975. While attempts to use peat as a fuel were meeting with little success in the United States, several European countries and the Soviet Union were building peat-burning power plants. The Soviet Union's operations grew from one peat-fired power plant in 1931 to 76 power stations in 1975, consuming 70 million tons of peat and producing 4,000 megawatts of electricity (Midwest Research Institute 1976). Ireland began to use peat for energy on a commercial scale in 1946, when a national peat organization was founded. By 1951, peat supplied 7% of Ireland's electricity, 40% by 1965, and 25% in 1975 (Lang 1984). In Finland, peat was first used as a fuel in the 1940s, the first peat-fired district heating and power plant was built in 1972, and the fuel-peat industry was expected to keep growing (Midwest Research Institute 1976).

Having learned that the technology to use fuel peat was readily available, the Peat Program began to survey the state's peat deposits, to investigate how fuel peat might best be used in Minnesota, and to study the environmental and socioeconomic impacts of peat development. That economic factors might again determine the use of peat as a fuel was indicated by the results of a study sponsored by the Peat Program (Ekono, Inc. 1977), which found that neither retrofitting existing boilers to use peat nor building new energy facilities could compete with the current prices of coal. And in 1982 Minnegasco ended its study of peat gasification because it did not seem economically feasible (Pitzer 1982).

Realizing the importance of economic factors in the development of fuel peat, the Peat Program's staff designed a computer model to estimate the acres of peatland that could meet criteria that affect the cost of fuel peat. The staff used the geographic data in the Minnesota Land Management Information System with the Peat Inventory Project's survey information (Hoshal and Johnson 1982).

In choosing criteria, the staff made the following assumptions: a peatland must be at least a certain size to be worth the initial costs of development, the peatland should be close to a road to avoid the costs of building roads on the peatland, and the peatland should be within a certain distance of markets because of the high cost of transporting peat. Peatlands suitable for development were defined as (1) covering 1,000 contiguous acres, (2) within 50 miles of one of six cities in northern Minnesota, and (3) within one mile of a road. The peatlands that met these criteria were then grouped by ownership: those owned by the state and therefore available for leasing, those owned by the federal government or private parties, and those excluded from development.

This model was applied to an eight-county area in northern Minnesota. The analysis identified 950,000 acres of peatland that meet these criteria for fuel peat and are available for lease by the state. A map of the results provided a means of easily locating peatlands for more detailed evaluation.

These results also provided more realistic data about the quantity of peat that could be developed for fuel. For years, statements about the great potential of Minnesota peat were based on exploitation of all 6 million acres. This analysis demonstrated that factors other than the mere presence of peat deposits must be considered in predicting the potential of the state's peat as a fuel source.

Interest in fuel peat continued. In 1983, the legislature allocated funds to the Peat Program specifically to assist the development of fuel peat. The goals were to obtain fuel peat for use in industrial-scale boilers, to identify potential industrial consumers of peat with systems requiring little modification, and to conduct well-instrumented combustion tests according to strict engineering standards to demonstrate the feasibility of burning peat (MDNR 1987).

At the same time, Rasjo Torv, a Swedish company, asked the Peat Program for help in finding a site suitable for producing fuel peat. In 1984, the company signed a lease for 2,600 acres near Zim in St. Louis County and began to produce compressed cylinders of peat known as peat sods from a test area of 160 acres (Fig. 18.10).

The peat produced by Rasjo Torv was used in tests conducted at a variety of facilities in northern Minnesota between 1984 and 1986. The conclusion reached at the end of the testing was that peat can be successfully burned in combustion systems not designed for fuel peat. Most of the problems encountered during the testing involved the handling of the peat in the system rather than the combustion, and these problems could be solved easily (Great Lakes Peat Products Company 1986).

Rasjo Torv concluded, however, that producing sod peat for fuel was not an economically viable business in Minnesota given the current prices for coal (Great Lakes Peat Products Company 1986). The company stopped mining peat in 1986.

More promising economic results were obtained from a study by the Peat Program (Peatrex, Ltd. 1987). Boise Cascade Company in International Falls and the city of Hibbing's power plant were identified as potential peat consumers. The Peat Inventory Project surveyed peatlands close to these locations and conducted detailed surveys for two peatlands. Mining plans were designed for producing milled peat, which is shredded loose peat. Economic analysis of these proposed projects indicated that milled peat could be produced and delivered to these two sites at a price competitive with the price of coal. Neither of these projects was carried out, however.

As of 1990, no peat-energy operations were active in Minnesota. While it seems unlikely that any large-scale use of the peat for energy will be developed in the near future, small developments close to facilities that can use peat and other alternative fuels as a supplement to coal now seem the most realistic possibility for using Minnesota's peat as a fuel.

Industrial Chemicals

One of the less well known uses of peat is as a raw material for the production of industrial chemicals. While peat chemicals have been produced in the Soviet Union

Fig. 18.10. Tractor hauls peat sods off mining field at Great Lakes Peat Products Company near Cotton, Minnesota. (Photo by author.)

and several European countries for many years, there has been little interest in peat chemicals in the United States. Since most peat chemicals can also be produced from other raw materials, particularly petroleum and coal, the potential of peat as a chemical feedstock has been ignored in favor of abundant and less expensive fossil fuels.

This lack of interest in peat chemicals has confronted scientists in Minnesota more than once. From 1955 to 1958, the Iron Range Resources and Rehabilitation Commission provided funds for the Chemical Products from Peat Project at the University of Minnesota. Piret, Passer, and several colleagues researched preparation of three products: a peat-based fertilizer with a high level of nitrogen, peat reinforced with an alkali solution for use as a binder in pelletizing powdered taconite, and peat-derived humic acids for use as oil-well drilling fluids. Despite success in the laboratory, the project was discontinued in 1958 because of lack of interest in the commercial production of peat chemicals. Piret and his colleagues voiced doubts about the ability of peat chemicals to compete in the American chemical market (Piret *et al.* 1956, 1958).

In the mid-1970s, nationwide interest in alternatives to petroleum and coal encouraged Fuchsman (1978) at Bemidji State University to reexamine the potential of peat chemicals. Aware of the lack of interest encountered by his predecessors, Fuchsman pointed out that there was not only renewed interest in peat but also a substantial growth in the chemical industry resulting in an increased demand for raw materials.

Fuchsman (1978) reviewed the scientific literature about peat chemicals and visited several peat laboratories in the Soviet Union and Europe. From this research, Fuchsman identified four components of peat from which chemicals are derived — bitumens, carbohydrates, humic acids, and coke — and documented their uses.

Bitumens, the substances in peat that can be extracted by organic solvents, yield waxes and resins. Peat waxes have been produced in the Soviet Union for many years for use as industrial lubricants. Peat resins are primarily valuable for their steroid content, which is used by the pharmaceutical industry in the Soviet Union in the preparation of synthetic steroids.

Peat carbohydrates, primarily cellulose and hemicellulose, can be chemically treated to obtain sugars for use as a medium for growing protein-producing yeasts. In the Soviet Union these yeasts are used as a feed supplement for livestock.

Humic acids derived from peat have been used in fertilizers in European countries, and it is claimed that they promote the uptake of nitrogen and magnesium by plants and improve the root formation of seedlings and the resistance of crops to pests. Humic acids have also been used in the Soviet Union as viscosity modifiers in oil-well drilling muds. Although humic acids may have potential for use in the production of synthetic fibers and plastics, as components of paints,

and as flocculents and thickeners in water purification systems, other chemicals are now used for these purposes.

Peat coke is produced in Finland and Germany for use in the manufacture of high-purity silicon for the electronics industry and of special alloys in the metallurgical industry. In the Netherlands and Ireland, peat coke is used to produce activated carbon.

Almost all of the chemicals identified by Fuchsman are also produced from other raw materials. Peat waxes are similar to montan wax produced from brown coal. Yeasts are commonly grown on carbohydrates derived from agricultural products or obtained as a by-product from breweries and distilleries. Coke is produced from petroleum coke and coal coke.

After reviewing the scientific literature, Fuchsman analyzed samples of Minnesota peat to determine their suitability for producing peat chemicals. The results (Fuchsman *et al.* 1979) were promising, and he proceeded to contact American chemical companies. While these companies were willing to study or evaluate peat chemicals, they were not interested in producing them (Fuchsman 1981).

Fuchsman concluded that little financial incentive exists to produce any one peat chemical. A profitable operation would require sophisticated technology, production of large volumes, and the ability to enter markets already served by similar products. Only large companies have these capabilities, but they are not interested in such a speculative venture.

The alternative, Fuchsman (1981) suggests, is to adapt the production of peat chemicals to simpler technology when possible, emphasize unique characteristics of the chemicals, and market them close enough to the production site to cut transportation costs. The chemicals that best fit these criteria are peat waxes and peat carbohydrates.

Fuchsman's research was continued from 1983 to 1985 by Spigarelli and colleagues (1985) at Bemidji State University. Working from the premise that combining the production of peat waxes and peat fuel might reduce the costs of both enough to make the products competitive, Spigarelli and his colleagues developed a process by which peat waxes and peat fuel can be obtained from the same peat. They tested three methods of extracting the waxes and analyzed the remaining solids to determine their fuel value. Waste waters from one of the extraction methods, wet carbonization, were used as a growth medium for yeasts and fungi.

High-quality peat waxes and peat fuel were obtained from two of the extraction methods, and yeasts and fungi were successfully grown on the waste water. The successful production of several products still does not offer enough economic advantage to be competitive with similar products, however (MDNR 1987). While much has been learned about the production of peat chemicals, commercial production of peat chemicals remains only a potential use in Minnesota.

Protection of
Ecologically Significant Peatlands

Among the issues addressed by the Peat Program's policies for the management of the state's peatlands is the potential conflict between preservation of peatlands for their ecological value and development for other uses. How great the conflict could be was demonstrated when Minnegasco proposed in 1975 to lease and develop 300,000 acres in Red Lake peatland, also proposed to be preserved for its ecological significance. While this conflict was resolved in 1981 when Minnegasco dropped its proposal for economic reasons, it made evident the need to consider the preservation of peatlands in planning for the management of the resource.

In 1978, the Peat Program began working with an advisory group, the Task Force on Peatlands of Special Interest, to assess the ecological significance of peatlands larger than 3,000 acres. Preliminary recommendations for the preservation of 22 peatlands were included in the Peat Program's report to the legislature in 1981. Further study of these peatlands led to the recognition of the national and international significance of Minnesota's patterned peatlands and to a proposal in 1985 for legislation to protect 18 peatlands totaling 500,000 acres — about 8% of the state's peatlands (MDNR 1984b).

Preparing information in support of the legislation, the Peat Program's staff identified potential conflicts between protection of the 18 peatlands and other uses. Since many of the 18 peatlands are located in areas containing large, deep deposits of peat, conflict with development of these peatlands for fuel or horticultural products at first glance seems likely. Analysis of economic factors, however, shows that under current economic conditions little conflict exists between peat mining and protection of these peatlands.

This conclusion was reached with the help of a computer model designed to identify peatlands containing peat with potential for fuel or horticultural peat (MDNR 1984b). These "developable" peatlands were defined according to several criteria. First, the model singled out peatlands that are at least 160 acres in size, that are accessible within a mile of a paved road, and that are located within 50 miles, measured along paved highways, of 11 northern cities. These peatlands were then grouped into three categories: (1) those owned or administered by the state and therefore available for leasing, (2) those owned by the federal government or by private parties, and (3) those in management units that preclude development, including the 18 peatlands. Finally, in the counties for which information about the depth of peat deposits is available, the peatlands were divided into those deeper than five feet and those shallower.

The results of the analysis show the effect excluding the 18 peatlands would have on the quantity of peatlands available for development. In the five counties (Koochiching, St. Louis, Carlton, Beltrami, and Lake of the Woods) that contain half of the peatland acreage in the state and for which information about the depth of peat deposits is available, 360,000 acres are developable. When just access and ownership are considered, 1,800,000 acres in the entire state are developable.

The Peat Program's staff compared these results with development in Finland, where commercial-scale peat mining has been going on for many years. In Finland, fewer than 100,000 acres have been mined for fuel peat during the last 20 years. Furthermore, the Finnish government is committed to the preservation of ecologically significant peatlands and has already protected about 477,000 acres, about 5% of the country's peatlands (Ruuhijarvi 1978). The Peat Program staff concluded that exclusion of the 18 peatlands from development leaves sufficient acreage "to meet any imaginable demand for fuel peat in the foreseeable future" (MDNR 1984b).

By conducting further analyses and continuing to gather information, the Peat Program hopes to manage the state's peatlands to accommodate many uses with a minimum of conflict. This is made easier by the fact that few peatlands have so far been developed, a situation that makes it possible to assess the ecological significance and the economic potential of a substantially undeveloped resource and to direct development away from ecologically significant peatlands.

In 1991, the Minnesota legislature passed a wetland protection law that dedicates the state-owned portions of the 18 ecologically significant peatlands as scientific and natural areas (chapter 19). This legislation, along with the Peat Program policies, should ensure the management of the state's peatlands for their value as a significant ecological resource as well as for their economic potential.

Conclusion

Although Minnesota's peatlands have been used in a variety of ways, they remain for the most part undeveloped. Even the pressures for economic development of the last decade have not resulted in the exploitation of many additional acres. Economic conditions simply have not been favorable for extensive use of the peatlands, especially of the vast northern peatlands now known to be particularly significant ecologically.

The interest in the economic uses of peatlands has resulted in a beneficial increase in knowledge about many aspects of the state's peatlands: their environment, the quantity and quality of their peat deposits, the impact of development, and their potential uses. The state-owned peatlands are now managed under policies influenced by this knowledge. The development of both state-owned and privately owned peatlands is subject to laws and regulations that insure the control of environmental impacts and the reclamation of mined peatlands. While the future of the economic uses of peatlands is difficult to predict, recognition of their multifaceted value should guide their management in the years to come.

Literature Cited

Alway, F. J. 1919. Some limitations on the cultivation of peat lands in Minnesota. Pages 94-95 *in* E. K. Soper. The peat deposits of Minnesota. Minnesota Geological Survey Bulletin 16, Minneapolis, Minnesota, USA.

American Society for Testing and Materials. 1969. Standard classification of peats, mosses, humus, and related products. Philadelphia, Pennsylvania, USA.

Anderson, J. P., and W. J. Craig. 1984. Growing energy crops on Minnesota's wetlands: The land use perspective. Center for Urban and Regional Affairs, University of Minnesota, Minneapolis, Minnesota, USA.

Arrowhead Regional Development Commission. 1985. Recommendations for expansion of the Minnesota horticultural peat industry. Duluth, Minnesota, USA.

Cole, A. J. 1984. The history of agricultural development on peatland in Ireland. Pages 219-237 *in* Vol. III. Proceedings: 7th International Peat Congress, June 18-23, 1984, Dublin, Ireland. International Peat Society, Helsinki, Finland.

Ekono, Inc. 1977. Utilizing peat as a fuel. Minnesota Department of Natural Resources, St. Paul, Minnesota, USA.

———. 1981. Energy utilization potential of Minnesota peat. Minnesota Department of Natural Resources, St. Paul, Minnesota, USA.

Farnham, R. S. 1978. Status of present peatland uses for agricultural and horticultural peat production. Minnesota Department of Natural Resources, St. Paul, Minnesota, USA.

——— and T. Levar. 1980. Agricultural reclamation of mined peatlands. Minnesota Department of Natural Resources, St. Paul, Minnesota, USA.

Fuchsman, C. H. 1978. The industrial chemical technology of peat. Minnesota Department of Natural Resources, St. Paul, Minnesota, USA.

———. 1981. Potential industrial chemical utilization of Minnesota peat. Minnesota Department of Natural Resources, St. Paul, Minnesota, USA.

———, K. R. Lundberg, and K. A. Dreyer. 1979. Preliminary analytical survey of Minnesota peats for possible industrial chemical utilization. Minnesota Department of Natural Resources, St. Paul, Minnesota, USA.

Great Lakes Peat Products Company. 1986. Evaluation report on the Fens bog pilot project. Cotton, Minnesota, USA.

Heikurainen, L. 1982. Peatland forestry. Pages 53-62 *in* Peatlands and their utilization in Finland. Finnish Peat Society and Finnish National Committee of the International Peat Society, Helsinki, Finland.

Hoshal, J. C., and R. L. Johnson. 1982. An integration of a GIS with peatland management. Pages 319-327 *in* Vol. II. National Aeronautics and Space Administration Conference Publication 2261, National Conference on Energy Resource Management, September 9-12, 1982, Baltimore, Maryland, USA.

Johnson, R. L. 1982. Mapping a resource: Peat. Plan B paper. University of Minnesota, Minneapolis, Minnesota, USA.

Iron Range Resources and Rehabilitation Commission. 1945. Biennial Report 1943-1945. St. Paul, Minnesota, USA.

———. 1947. Developing the resources of Minnesota: Biennial report of Robert E. Wilson, Commissioner of Iron Range Resources and Rehabilitation. St. Paul, Minnesota, USA.

———. 1958a. Forward in developing natural resources: Biennial report on the work of the Office of Iron Range Resources and Rehabilitation 1956-1958. St. Paul, Minnesota, USA.

———. 1958b. Minnesota peat mission to Europe. St. Paul, Minnesota, USA.

Kivinen, E. 1980. New statistics on the utilization of peatlands in different countries. Pages 48-51 *in* Proceedings: 6th International Peat Congress, August 17-23, 1980, Duluth, Minnesota, USA. International Peat Society, Helsinki, Finland.

Kurmis, V., H. L. Hansen, J. J. Olson, and A. R. Aho. 1978. Vegetation types, species and areas of concern and forest resources utilization of northern Minnesota peatlands. Minnesota Department of Natural Resources, St. Paul, Minnesota, USA.

Lang, J. F. 1984. A review of the role of peat as a fuel for generation of electricity in Ireland. Pages 349-370 *in* Vol. II. Proceedings: 7th International Peat Congress, June 18-23, 1984, Dublin, Ireland. International Peat Society, Helsinki, Finland.

Larson, W. E., D. F. Grigal, and W. E. Berguson. 1985. Agroforestry research on Minnesota peatlands, final report to Legislative Commission on Minnesota Resources. University of Minnesota, St. Paul, Minnesota, USA.

Malterer, T. J., D. J. Olson, D. R. Mellem, B. Leulling, and E. J. Tome. 1979. Sphagnum moss peat deposits in Minnesota. Minnesota Department of Natural Resources, St. Paul, Minnesota, USA.

Midwest Research Institute. 1975. Planning document for a peat inventory program (modified version). Minnesota Department of Natural Resources, St. Paul, Minnesota, USA.

———. 1976. A report on European peat technology. Minnesota Department of Natural Resources, St. Paul, Minnesota, USA.

Minnesota Department of Energy and Economic Development. 1985. Ten years after the oil crisis: Lessons for the coming decade. 1984 Energy Policy and Conservation Biennial Report. St. Paul, Minnesota, USA.

Minnesota Department of Iron Range Resources and Rehabilitation. 1968. Biennial report 1966-68. St. Paul, Minnesota, USA.

Minnesota Department of Natural Resources. 1979. Minnesota Peat Program: Legislative status report. St. Paul, Minnesota, USA.

———. 1981. Minnesota Peat Program: Final report. St. Paul, Minnesota, USA.

———. 1984a. Inventory of peat resources: An area of Beltrami and Lake of the Woods counties, Minnesota. St. Paul, Minnesota, USA.

———. 1984b. Recommendations for the protection of ecologically significant peatlands in Minnesota. St. Paul, Minnesota, USA.

———. 1987. Minnesota Peat Program: Summary report, 1981-86. St. Paul, Minnesota, USA.

Odell, W. W., and O. P. Hood. 1926. Possibilities for the commercial utilization of peat. Bulletin 253. United States Department of Commerce, Bureau of Mines, Government Printing Office, Washington, D.C., USA.

Office of Iron Range Resources and Rehabilitation. 1964. Peat resources of Minnesota — report of inventory no. 1 — W. Central Lakes Bog, St. Louis County, Minnesota. St. Paul, Minnesota, USA.

———. 1965. Peat resources of Minnesota — report of inventory no. 2 — Cook Bog, St. Louis County, Minnesota. St. Paul, Minnesota, USA.

———. 1966. Peat resources of Minnesota — report of inventory no. 3 — Red Lake Bog, Beltrami County, Minnesota. St. Paul, Minnesota, USA.

———. 1970. Peat resources of Minnesota, potentiality report, Fens Bog area, St. Louis County, Minnesota. St. Paul, Minnesota, USA.

Osbon, C. C. 1921. News of the domestic industry. Journal of the American Peat Society 14(1):21.

———. 1923. Peat fuel resources of United States. Journal of the American Peat Society 16(1):10-21.

Peatrex, Ltd. 1987. Planning of milled peat production systems at two sites in northern Minnesota. Minnesota Department of Natural Resources, St. Paul, Minnesota, USA.

Piret, E. L., M. Passer, W. P. Martin, and A. J. Madden, editors. 1956. University of Minnesota Chemical Products from Peat Project: Progress report, 1955-56. University of Minnesota, Minneapolis, Minnesota, USA.

————. 1958. University of Minnesota Chemical Products from Peat Project: Progress report, 1957-58. University of Minnesota, Minneapolis, Minnesota, USA.

Pitzer, Mary J. 1982. Perpich: Peat can solve fuel, unemployment problems. (University of) Minnesota *Daily*, August 16:1, 3.

Pratt, D. C., and N. J. Andrews. 1980. Peatland energy crops: The productive potential of cattails and other wetland plants on Minnesota peatlands. Pages 444-450 *in* Proceedings: 6th International Peat Congress, August 17-23, 1980, Duluth, Minnesota, USA. International Peat Society, Helsinki, Finland.

————, D. R. Dubbe, E. G. Garver, and P. J. Linton. 1983. Wetland biomass production: Emergent aquatic management options and evaluations. Bio-energy Coordinating Office, University of Minnesota, St. Paul, Minnesota, USA.

————, D. R. Dubbe, and E. G. Garver. 1985. Energy from biomass in Minnesota — Part I; Final report Jan. 1, 1984-June 30, 1985, Minnesota Department of Energy and Economic Development and Legislative Commission of Minnesota Resources. University of Minnesota, St. Paul, Minnesota, USA.

Ruuhijarvi, R. 1978. Basic plan for peatland preservation in Finland. Suo 29:1-10.

Singleton, R. H. 1980. Peat in the United States. Pages 96-102 *in* Proceedings: 6th International Peat Congress, August 17-23, 1980, Duluth, Minnesota, USA. International Peat Society, Helsinki, Finland.

Soper, E. K. 1919. The peat deposits of Minnesota. Minnesota Geological Survey Bulletin 16, Minneapolis, Minnesota, USA.

Spigarelli, S. A., F. H. Chang, K. R. Lundberg, and D. Kumari. 1985. Industrial chemicals and other nonfuel products from Minnesota peat. Minnesota Department of Natural Resources, St. Paul, Minnesota, USA.

Toltz, M. 1910. The peat resources of Minnesota. Journal of the American Peat Society 3(1):1-11.

United States Department of Interior, Bureau of Mines. 1954-89. Minerals yearbook: Vol. I, metals and minerals. United States Government Printing Office, Washington, D.C., USA.

White, E. H. 1980. Forestry reclamation of peatlands in northern Minnesota. Minnesota Department of Natural Resources, St. Paul, Minnesota, USA.

Wildes, F. A. 1924. Peat developments in Minnesota in 1923. Journal of the American Peat Society 17(1):39-40.

CHAPTER 19

Peatland Protection

*Norman E. Aaseng and
Robert I. Djupstrom*

The Red Lake peatland, the "big bog" of northern Minnesota, stretches for 45 miles from east to west and 13 miles from north to south. Unbroken tracts of ribbed fen, ovoid islands, forested bogs, and teardrop islands are crisscrossed only by the trails of moose, eastern timber wolves, and the woodland caribou that disappeared in the 1940s. With the exception of some failed ditches, the landscape appears as it did in the 1850s — like an ocean of grass in which an occasional teardrop island looks like a ship at sea. This one-half-million acre ecosystem is one of the last true wilderness areas in the United States. Long considered a wasteland, this peatland is now internationally recognized for its diverse landform patterns of incomparable size and complexity. After a century of disregard, the Red Lake peatland and 17 other patterned peatlands are now protected by state law.

The preservation of Minnesota's patterned peatlands is the culmination of a gradual increase in appreciation of the peatland ecosystem. In 1892, U.S. government land surveyors, probably the first nonnative Americans to set foot in what is now called the Red Lake peatland, declared the area to be "practically unfit for any purpose" (U.S. Public Land Survey 1892). Later the peatland was subjected to an ambitious drainage project, split by a state highway, and bombed by aircraft during World War II training missions. During the environmental awakening of the early 1970s, the significance of this vast peatland began to be recognized, and portions of the Red Lake peatland were nominated as a National Natural Landmark. Like most peatlands, however, it was viewed by most people as an unimportant wasteland. Finally, the threat to the Red Lake peatland from a 1975 proposal to mine 300,000 acres for peat gasification led to an acceleration of research and renewed efforts by conservationists, sci-

entists, and resource managers to protect the patterned peatlands.

The research that was conducted in response to the peat gasification proposal, much of which is summarized in earlier chapters, has been crucial to the understanding and appreciation of the patterned peatlands and their ecological, scientific, educational, and aesthetic values. Peatlands lack the species diversity and richness found in many upland habitats, but their environmental conditions, ranging from extremely acid and low in nutrients to highly calcareous, have created unique habitats for specially adapted plant and animal species. More than two dozen species found in the state's patterned peatlands are officially listed as endangered, threatened, or of special concern (Minnesota Department of Natural Resources 1984a), and more have been recently proposed for inclusion on the list. One species of moss, *Calliergon aftoninum*, discovered growing in Minnesota's peatlands, had previously been reported to occur only as a fossil in peat cores.

Some of the peatland species — such as the insectivorous pitcher plant, sundew, and bladder wort — exhibit unusual adaptations to the austere peatland environment. Numerous species of orchids and ericaceous plants, such as the lingenberry, occur only in peatland habitats. Peatlands are also an important habitat for certain animal species. The palm warbler, the northern bog lemming, and the Disa alpine butterfly are highly dependent on peatland habitats.

In addition to the unusual biota of the peatland ecosystem, the remarkable landforms of Minnesota's patterned peatlands are renowned worldwide. The characteristic landform patterns of these peatlands result from the intricate relationship among vegetation, hydrology, topography, and climate (chapter 1). Elsewhere in the United States, peatland patterns are found only in Maine and Alaska and, to a very limited extent, in Michigan, Wisconsin, and New York. Outside of Minnesota, however, these patterns are found in comparable size or complexity only in the Hudson Bay lowland of Canada (and possibly in the western Siberian plain).

In a worldwide context, Minnesota's patterned peatlands are extremely valuable for the study of ecological and developmental processes in peatlands. The continental climate of Minnesota's patterned peatlands, in contrast to the maritime climate of peatlands in Europe and eastern Canada, provides a unique environmental setting for the comparative study of peatland processes. Furthermore, Minnesota's peatlands are free of permafrost, which is a complicating factor in the study of many Canadian and northern European peatlands. Most importantly, Minnesota's patterned peatlands are valuable for research because they are relatively undisturbed, unlike most peatlands in Europe, and they are more accessible for study than similarly pristine peatlands of Canada and Siberia.

Beyond the ecological values, peatlands offer an unusual opportunity for research on cultural and natural history. Because the peatland environment inhibits decomposition, artifacts of prehistoric significance are often found preserved in peat. Pollen and other plant fossils that have been deposited over thousands of years provide a wealth of information on past climate and vegetation.

At a time when the need to preserve entire ecosystems is becoming increasingly clear, protection of Minnesota's peatlands offers the rare opportunity to preserve a functioning ecosystem intact. Too often it is left to our imaginations to deduce what a vast ecosystem was like when only fragments are left. Minnesota's peatlands provide an important laboratory for ecological research and an opportunity for wilderness appreciation in a world where most ecosystems have been irreversibly altered.

Past Peatland Protection Efforts in Minnesota

The earliest protection of Minnesota's peatlands occurred in the late 1920s as a result of the state's efforts to save local governments from economic crisis. Extensive peatland drainage projects, conducted to encourage farming in these areas, had been funded through the sale of county bonds. When the farms failed and settlers left, the counties were faced with bankruptcy (chapter 17). A report by Dan Rose, cruiser for the Watab Paper and Pulp Company, cites the case of 29 townships in the "big bog" that contained but 80 settlers and 868,160 acres of swamp (Official Boards of Beltrami et al. 1928). Rose said, "I do not hesitate in stating that this swamp is of no value for farming as can be readily seen by the abandoned homes now in the swamp and also by watching the progress of those who

tried every way in their power to make a living out of it and moved away." Some 282 miles of public ditches were constructed in just six of these townships, to serve but 32 settlers.

Given the prospect of massive tax forfeiture and bankruptcy, local governments sought help from the state legislature. Turning over the tax-delinquent land to the state for the formation of a state wildlife game preserve and sanctuary in exchange for release from ditching debt was one solution that gained acceptance. To support this plan, the boards of Beltrami and Lake of the Woods counties, the Bemidji Civic and Commerce Association, and the Baudette Chamber of Commerce submitted a report to the legislature in 1928 calling for the establishment of a major northern Minnesota game refuge (Official Boards of Beltrami et al. 1928). The report extols the virtues of the area's vast wildlife resources and notes its lack of timber resources. This "solid tract of more than a million acres suitable and even excelling as a natural game refuge and at the same time at present uninhabited and available for purchase, presents a situation that does not have its equal in the United States," the report says. Economic benefits derived from the creation of Yellowstone and Yosemite National Parks and other game refuges are also mentioned.

With the support of sportsman organizations and the northern Minnesota counties, the legislature established the Red Lake Game Preserve in 1929 (Fig. 19.1). Some 1.3 million acres of tax-delinquent land were transferred to the Department of Conservation (known today as the Department of Natural Resources), and in return the state paid off the cost of financing the ditching effort (Ahrens 1987). The preserve has since been replaced in part by the Red Lake Wildlife Management Area and the Pine Island and Beltrami Island State Forests. Much of the Red Lake peatland has been eliminated from its boundaries. Thus this first action protecting peatlands in Minnesota provided only temporary protection to the state's foremost peatland; the long-term value was the return of considerable acreage of important peatlands to public ownership.

It was not until the early 1960s that ecological considerations came to bear on the protection of peatlands. At that time, a U.S. Forest Service researcher, Miron Heinselman, began to notice changes taking place in the areas he had worked for many years. Scars from tracked vehicles were still visible many years after a single foray into a bog. Winter roads cleared of trees to gain access to cut Christmas trees remained open years after use. Dense willow, alder, and tamarack stands appeared on the upslope of ditches and roads that interrupted the flow of the bogs.

Heinselman recognized the value of a pristine peatland and proposed to the Department of Conservation that the Myrtle Lake peatland, an area of some 22,000 acres, be set aside for scientific research. With the support of Department of Conservation regional forester Art Keenan and moral support from local foresters

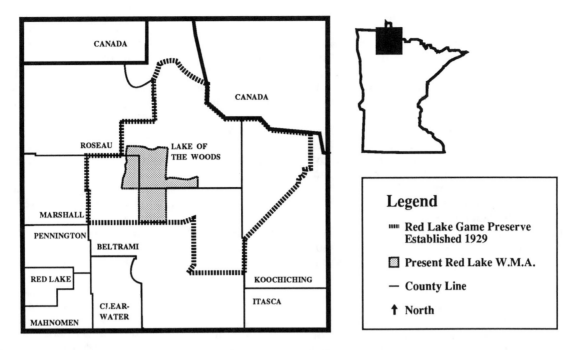

Fig. 19.1. Map of the Red Lake Game Refuge and current Red Lake Wildlife Management Area. The original boundaries of the Red Lake Game Preserve included all or part of 7 of the 18 peatlands that would later be identified as ecologically significant.

Eugene Jamrock and Clarence Buckman and Division of Forestry director Edward Lawson, commissioner Clarence Prout issued a policy statement in 1962 that officially established the Lake Agassiz (Myrtle Lake) Natural Area (Prout 1962). This first official action taken in Minnesota to protect a peatland for its ecological value decreed that the area would be preserved in its natural state for scientific study and observation. The Department of Conservation, within the constraints of limited staff and budget, actively enforced the policy, removing trespass hunting camps built on state land and shifting the harvest of the area's stunted black spruce Christmas trees to bogs of less significance.

Recognition of the Myrtle Lake peatland as a scientific and research site resulted in its being "designated" as a National Natural Landmark in 1964 by the National Park Service. Since the state agreed to maintain the site in its natural condition, it also received "registered" National Natural Landmark status. Although the agreement is not binding and offers little in the way of actual protection, this status further reaffirmed the national significance of one of Minnesota's finest peatlands.

With a greater understanding of Minnesota's peatlands beginning to develop (primarily through work in the Myrtle Lake peatland), ecologists focused some of their attention on other patterned peatlands. In a natural-history theme study for the National Park Service entitled *Inland Wetlands of the United States* (Goodwin and Niering 1971), two other peatlands were recommended for National Natural Landmark status. Minnesota's premier peatland, the Red Lake peatland, was designated in 1974 for its great diversity of peat

landform features and large size (over 200,000 acres). Due to local opposition, however, the Department of Natural Resources declined to enter into an agreement with the National Park Service that would have given the peatland registered status. The National Park Service has not acted on the other recommended site, the North Black River peatland.

Up to this point, alteration of the Minnesota patterned peatland ecosystem had been relatively benign. The peatlands seemed safe from the widespread alteration that had affected other ecosystems.

Recent Peatland Protection Efforts in Minnesota

A request by the Minnesota Gas Company (Minnegasco) in 1975 to lease 300,000 acres of peatland from the state for peat gasification set in motion new efforts to protect some of Minnesota's most ecologically significant peatlands. The prospect of large-scale development highlighted the need for a comprehensive policy to manage the poorly understood peatland resource. The responsibility for developing a management policy belonged to the Minnesota Department of Natural Resources (DNR), which established the Peat Program in the DNR's Minerals Division.

One important component of this policy was to be the preservation of selected peatland areas in their natural state. Plans for preservation often are not initiated until development has left only isolated and less-than-natural remnants to be protected. Such is the case with the tall-grass prairie and the "big woods" deciduous forest of Minnesota; both are less than 1% of their original area. In contrast, development in patterned peatlands has been minimal in Minnesota. By instituting a management policy to balance prudent development of the resource with protection, the state could take advantage of a rare opportunity for preservation before pressures of development restricted the options and before needless conflicts arose.

The Peat Program staff soon realized that needed information about Minnesota's peatlands was lacking. As is often the case, basic ecological research and inventory of an ecosystem, such as the patterned peatlands, begin only after it is considered to be of economic importance. The perception of Minnesota's peatlands as wastelands and difficult places in which to work was apparent from the paucity of basic biological information. To provide baseline data, the Peat Program sponsored research on the fauna, flora, vegetation, ecology, and hydrology of peatlands.

Efforts to draw on outside experience in the identification, evaluation, and management of ecologically significant peatlands met with limited success. Although much information was available about the economic development of peatlands in Europe, where peat has been exploited for centuries, little published information was available on protecting them for their ecological value. Preservation efforts have begun in many countries, but too late in those parts of continental Europe where much of the peat has been drained (Goodwillie 1980; Kivinen and Pakarinen 1980). Many countries are still conducting ecological surveys of their peatlands.

Finland's efforts to balance development and protection of peatlands provided a valuable model. Despite Finland's economic dependence on peatlands for energy and forest products (one-third of the country is peatland), a government policy calls for the protection of 7% of all peatlands in nature reserves (Heikurainen and Laine 1980). Finland has an advantage over most countries in its knowledge of the peatland ecosystem, as evidenced by a detailed peatland classification system and a list of peatlands identified as ecologically significant. A tradition of ecological classification dates back to the early 1900s and the pioneering work of Cajander (1913) on forest site types. This comprehensive body of knowledge has enabled Finland to develop protection plans. The difficulties of land ownership and the rapid increase in drainage has prevented conservation goals from being completely realized, however (Ruuhijarvi 1983).

In the United States, only Maine was wrestling in a significant way with the issue of protecting representative examples of peatland. Responding to the same development pressures initiated by the energy crisis that hit Minnesota in the late 1970s, Maine undertook a similar effort to evaluate the ecological value of its peatlands (Davis *et al.* 1983).

As Minnesota's fledgling Peat Program was getting off the ground, the state's academic community, working with the DNR Scientific and Natural Areas Program, officially called for the establishment of a portion of the Red Lake peatland as a Scientific and Natural Area (SNA), the highest degree of protection afforded by the state. The proposal was prompted by the threat posed by peat mining. Establishment as an SNA would have prohibited all development within its boundaries. Given the lack of a comprehensive inventory of Minnesota's peatlands, the lack of crucial internal DNR support, and the lack of a broad, well-organized base of external support, the DNR opted not to establish the SNA.

The Peat Program staff did see, however, an immediate need to determine the conflicts between preservation and peat development. To evaluate the information that was available at that time, the Peat Program formed a Task Force on Peatlands of Special Interest in 1978 that brought together peatland experts from various fields, including botany, ecology, hydrology, geology, wildlife, and forestry. By 1981, 22 peatlands had been identified as candidates for protection by the task force (DNR 1981). Efforts by Minnesota's environmental community culminated in legislation in 1983 that directed the DNR to assess by mid-1986 all 22 peatlands of special interest for their suitability as units of the Outdoor Recreation System under Minnesota Statutes 86A (as potential state parks, SNAs, and wilderness areas).

By 1984, ecology and hydrology studies were complete, a workable peatland classification system existed, and field evaluations were accomplished. Evaluation of the candidate peatlands was carried out (see appendix for a description of the process). Five peatlands were dropped and one was added to the list proposed for protection. These final 18 peatlands contain the best examples of the full range of patterned peatland complexes found in Minnesota (Fig. 19.2).

Two types of management areas were defined for each of the 18 peatlands: a core area and a watershed protection area. The core area includes the features of greatest ecological significance. The watershed protection area is the peatland buffer required to maintain the ecological integrity of the core area (Fig. 19.3). The total acreage of the 18 proposed peatland protection areas is just under half a million acres, 172,200 acres in the core areas and 323,800 acres in the watershed protection areas. An assessment of the land ownership revealed that the state owns about 85% (419,300 acres) of both core and watershed protection areas (Fig. 19.4). As major landowner, the state was in an excellent position to ensure that these areas were protected.

It was inevitable that a proposal to protect the core and watershed areas of 18 peatlands would generate controversy even though it concerned only 7% of Minnesota's total peatland area. Because compromises may be needed to gain broad-based acceptance, ranking of the 18 peatland areas identified where concessions could be made with the least ecological impact (Fig. 19.5).

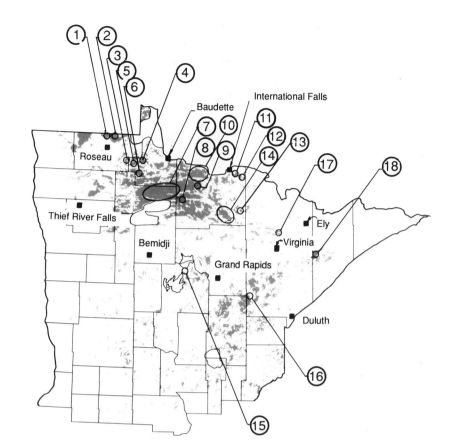

1 Pine Creek
2 Sprague Creek
3 Luxemberg
4 Winter Road Lake
5 Norris Camp
6 Mulligan Lake
7 Red Lake
8 Lost River
9 South Black River
10 North Black River
11 West Rat Root River
12 East Rat Root River
13 Nett Lake
14 Myrtle Lake
15 Hole-in-the-Bog
16 Wawina
17 Lost Lake
18 Sand Lake

Fig. 19.2. Areal extent of peatland protection areas in northern Minnesota.

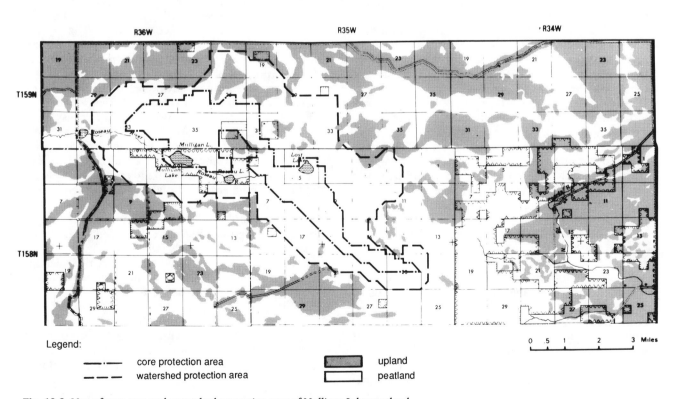

Legend:

——·— core protection area

——— watershed protection area

upland

peatland

Fig. 19.3. Map of core area and watershed protection area of Mulligan Lake peatland.

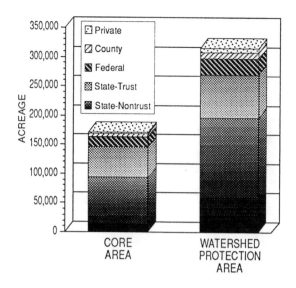

Fig. 19.4. Acreage and ownership of core and watershed protection area for the 18 ecologically significant peatlands.

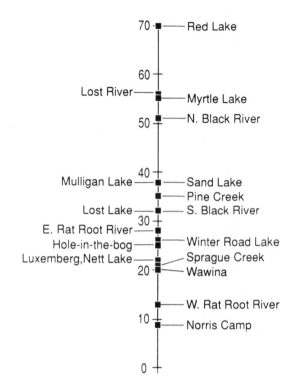

Fig. 19.5 Ranking of the 18 ecologically significant peatlands.

During this evaluation process, an internal DNR Peatland Protection Task Force was established with membership from the major resource management divisions: Forestry, Minerals, Parks, Fish and Wildlife, and Lands. The major function of this new task force was to develop departmental agreement on management options and restrictions that would support a balance of development and protection for Minnesota's peatlands.

Meanwhile, the economics of energy had changed. By 1984 the oil embargo was over, and the costs of converting peat to energy were higher than conventional energy sources. Minnegasco withdrew its request to lease 300,000 acres of peatland.

Identification and establishment of peat for horticultural uses and other peat-utilization ventures had, however, prompted publicly funded economic incentives for peat development. At the same time, interest in Minnesota's nonferrous metallic minerals was increasing, and vast areas of Minnesota's forest and peatland acres were viewed as a potential source.

Believing that 1985 might be the year to pass a peatland protection act, the environmental community returned to the legislature in 1984 to advance new legislation directing the DNR to develop final protection recommendations by November 15, 1984. Moving up the deadline complicated matters for the DNR, which had envisioned a lengthy public discussion and consensus-building process as essential to ultimately passing legislation. Suddenly, the time frame was shortened by two years. The agency was close to internal agreement but had no public consensus-building process.

The publication "Protection of Ecologically Significant Peatlands: Preliminary Report" (DNR 1984b) began the formulation of final recommendations. To assist in this process, the advisory Peatland Protection Area

Review Committee (PARC) was formed to represent the diverse groups with an interest in the DNR's recommendations: county officials; environmental groups; forest, mineral, and peat-mining industries; and the Native American and university communities. The primary areas of conflict were mineral exploration and mining, timber harvesting, peat mining, the use of motorized vehicles, and school trust-fund compensation.

Minor modification of proposed boundaries to exclude some commercially valuable timber and modifications to allow continued use of winter access roads through some of the peatland protection areas alleviated forest-industry concerns. Similarly, consideration for continued snowmobile trails on already disturbed corridors resolved this potentially controversial issue. Peat-mining concerns were laid to rest when assessments of statewide peat resources for fuel and horticulture indicated substantial supplies after the 18 peatland protection areas were excluded far in excess of foreseeable energy needs (DNR 1984c).

About 26% of the state-owned land in the peatland protection areas is school trust land. Income from this land is apportioned to the Permanent School Fund. Since activities that could have provided income to the fund were restricted or prohibited, it was agreed that the fund would be compensated for foregone income. Differences of opinion surfaced, however, about compensation for land that has no economic value at present but may have value in the future as economic conditions change.

Resolution of the major issue, mineral exploration

Fig. 19.6. The circular islands of the Red Lake peatland are one of the many unusual peat landform features occurring in the state's most highly ranked peatland. (Don Luce photo from 1984 report; courtesy Don Luce, Bell Museum of Natural History, Minneapolis.)

and mineral extraction, proved elusive. A major breakthrough was the proposal to create a two-tiered level of protection for peatlands. Under this proposal, the core areas in Minnesota's four most significant peatlands (Red Lake — Fig. 19.6, Myrtle Lake, and portions of North Black River[1] and Lost River) would be recommended as Scientific and Natural Areas to protect them from mineral development or exploration. The remaining cores, although qualified in every respect as SNAs, would be open to mineral exploration and development with certain restrictions; they would be called Peatland Scientific Protection Areas (PSPA).

PARC reaction to the proposal was relatively positive. Environmental groups at first rejected anything less than full SNA protection for all 18 peatland areas,

arguing that if they qualified they should be fully protected. Mineral interests, on the other hand, wanted all areas open for possible mineral finds, arguing that if more areas were left open for exploration, the chance of a find would be greater. A review of the known mineral potential and the state and federal land ownership in the four proposed SNAs, however, reduced mineral-exploration interest in these sites. Environmental groups gave their support to the proposal, with agreement on mineral exploration and mining guidelines, mitigation standards for potential mineral development, and a maximum of 1,500 acres for development within the proposed PSPAs. In addition, mining companies would be required to purchase for the state two acres of core area for every acre of core area affected. The blueprint for peatland protection legislation had been created.

The DNR report to the legislature — *Recommendation for Protection of Ecologically Significant Peatlands in Minnesota* (DNR 1984c) — proposed legislation protecting the 18 peatland complexes (Table 19.1) and addressed the issues of land ownership, economic opportunities lost or gained, School Trust Fund com-

---

[1] The U.S. Justice Department has been reviewing the land ownership in the North Black River peatland. Based on preliminary findings, it appears that portions of the core area proposed for SNA designation in the North Black River peatland belong to the Red Lake Indian Reservation. If this proves to be true, only core areas of Red Lake, Myrtle Lake, and portions of Lost River would be proposed for SNA designation.

Table 19.1. Summary of DNR findings and recommendations on the protection of ecologically significant peatlands (to apply to state-administered land only)

- Identified 18 peatlands for protection
- Proposed two types of core areas and a watershed protection area:

  Scientific and Natural Areas (110,800 acres)
  - includes at least some or all cores areas from the four highest-ranked peatlands
  - restricted activities include
      motorized vehicles
      forestry practices
      peat mining
      mineral exploration and mining

  Peatland Scientific Protection Area (34,000 acres)
  - includes remaining peatland core areas
  - restricted activities include
      motorized vehicles
      forestry practices (except in certain areas)
      peat mining
  - special provisions
      metallic mineral mining and exploration under certain
        conditions
      area that could be affected by mining activities
        restricted to 1,500 acres

  Watershed Protection Area (272,000 acres)
  - activities restricted from both core and watershed protection area
      new or improved ditches
      peat mining
      industrial mineral mining (gravel pit operations)

pensation, and management of activities in the protected areas.

Draft legislation was written by key members of the House and Senate and received the support of the governor's office. When counterlegislation that the environmental community felt drastically weakened the peatland protection objectives was introduced, the principal authors withdrew the proposed legislation rather than risk the passage of a weakened law.

From 1984 to 1991, the legislature took no action on the DNR's recommendations. Although county land would not have been affected by the proposed legislation, local opponents said that too much land in northern Minnesota is already "locked up." Bitter disputes over the Boundary Waters Canoe Area Wilderness and Voyageurs National Park left many northern Minnesotans opposed to any protection plan, whether or not there is a significant hindrance to economic development.

In 1985, the DNR provided some protection of the 18 ecologically significant peatlands in the Minnesota Rules for Peatland Reclamation, Chapter 6131. In the section pertaining to siting of peat mining operations, the rules state that the 18 peatland protection areas are "avoidance areas" where peat mining is allowed only "if no reasonable or prudent alternative exists and, in the case of state-owned land, the affected areas [are] replaced by an area of equal or greater public value serving the same purpose as the affected area." State metallic mineral leasing also follows the recommendations outlined in the DNR report.

ADDENDUM: MINNESOTA PEATLAND PROTECTION BILL PASSED

Seven years after the 1984 DNR report, broad-based public support for the protection of wetlands finally resulted in peatland protection. The legislature included language in a wetlands bill to protect all 18 ecologically significant peatlands. The original bill, introduced by Representative Willard Munger, would have protected the entire watershed and core areas of the 18 sites, over 500,000 acres. Once again, various concerns were raised about mining, exploration, forestry, recreation, local development, and environmental safeguards. The DNR convened a meeting of the various stakeholders and brought the issues to the table for discussion. Realizing that some form of peatland protection would probably become law, each worked to develop a bill acceptable to its clientele. The final legislation is a compromise.

On June 4, 1991, Governor Arne Carlson signed into law the Wetlands Conservation Act of 1991 that dedicates all 146,239 acres of state-owned lands in the core areas of the 18 peatlands as scientific and natural areas. The language is very similar to that proposed in the 1984 report. The concept of watershed protection areas is not contained in the legislation. Instead, peatlands outside the core areas, as well as all other wetlands, are exempt from draining and filling until 1993, when new regulations governing draining, filling, and mitigation for wetland losses will take effect.

Minnesota's ecologically significant peatlands, those vast inaccessible wilderness areas once viewed as everything from wastelands to fuel sources, are now protected. With the increasing understanding of the need to protect complete ecosystems, Minnesota has taken the lead and preserved an ecosystem unlike any other. The heart of the peatland ecosystem, with its rare species and complex patterns, is preserved for future generations to study and to appreciate.

APPENDIX

Evaluation Process for Peatland Protection in Minnesota

The process ultimately used by the Minnesota Department of Natural Resources (DNR) in the mid-1980s to develop a peatland protection plan evolved by trial and error over several years. It can be divided into three main tasks: (1) establishment of criteria for peatland significance, (2) evaluation of peatlands according to the criteria, and (3) development of management guidelines.

Development of Criteria

To systematically identify and evaluate significant peatlands, it first must be determined what factors are to be included in the evaluation of significance. Various attributes can be used as a basis for designating peatlands for protection. Minnesota's approach has been guided by the State Outdoor Recreation Act, which gave the DNR responsibility for "insuring adequate preservation of Minnesota's natural and historical heritage for public understanding and enjoyment" (Minnesota Statutes sec. 86A.02, 1980). Specifically, this act calls for the protection of a representative cross section of Minnesota's "landforms, fossil remains, plant and animal communities, rare and endangered species, and other biotic features and geological formations for scientific study and public edification."

With this directive in mind, seven attributes were identified as crucial in the assessment of ecological significance, and an attempt was made to incorporate them into the evaluation process. The attributes are as follows:

Representativity. Emphasis was given to protecting the full range of peatland features, such as peatland types, landforms, and plant and animal communities existing in the patterned peatland ecosystem.

Quality. Importance was given to the extent to which a peatland complex type, plant community, or landform corresponds to an idealized concept of that element as exhibited by its degree of definition or distinctiveness, completeness of features, and lack of human impact.

Rarity. Greater value was given to peatlands that provide habitat to rare and endangered plants and animals or that exhibit unusual or unique landforms or vegetation types.

Diversity. Within each representative peatland type, higher value was given to peatlands that have a greater diversity of vegetation or landform types.

Viability and defensibility. Preference was given to peatlands that have better prospects for the long-term survival of their significant features. Protection efforts are better spent on features that are relatively stable than on those that are merely transitional or ephemeral. In addition, peatlands that are less susceptible to human disturbance are more desirable than ones that are vulnerable.

Scientific value. Additional importance was given to peatlands that are valuable for current or future scientific research.

Geographical representativity. It is desirable that peatland features be protected across the range of their occurrence. Peatlands located in different parts of the state are subject to subtle differences in their environment. These differences can influence the type or degree of ecological development. Also, the occurrence of the full range of environmental conditions ensures that genetic diversity in plant and animal species is maintained.

Evaluation Process

Research and Inventory

Applying these criteria to Minnesota's peatlands requires an extensive knowledge of peatland flora, fauna, and community types as well as an understanding of the process of peatland development. To provide baseline data, the DNR sponsored research on the fauna, flora, vegetation, ecology, and hydrology of peatlands. These studies, many of them summarized in earlier chapters, provided the basis for the evaluation of peatlands. This research has been supplemented by field surveys as additional information was needed and as time and resources allowed.

The magnitude of the task of evaluating the state's peatlands required the DNR to reduce the area it was evaluating. Because the immediate need was to assess the impact of large-scale peatland development, the ecological evaluation was limited to peatlands greater than 3,000 acres. This restriction includes the major patterned peatlands of northern Minnesota. Evaluation of the smaller peatlands, including those in the southern half of the state, is a task for the future.

Even with this reduction, the study area included over 4 million acres of peatland, much of it largely inaccessible except by helicopter. Added to the problem of size was the relatively short period of time in which the project had to be completed.

The correlation of peatland vegetation habitat with patterns and landforms and an understanding of their ecology and evolution provided a valuable approach that enabled researchers to use aerial photographs as a major tool in the evaluation process. A preliminary survey of candidate peatlands over a large area meant that the limited time and resources could be spent most efficiently on sites targeted for detailed field analysis.

Fig. 19.7. Rare plant and animal species, the extensive ribbed fen pattern, and the lack of disturbance make the Mulligan Lake peatland the state's best example of a complex type 6 peatland. (Black-and-white infrared aerial photo)

Peatland Classification System

A peatland classification system that covers the broad spectrum of peatland types is the key to ensuring that the full range of Minnesota's patterned peatlands is adequately represented in the protection plan. The task of classifying peatland types in a useful system was successfully accomplished only after research produced a better understanding of peatland formation and evolution. The ecological classification adopted is based on peatland genesis and incorporates the major factors interacting in the development of peatland patterned complexes — vegetation, hydrology, and topography.

A total of 11 peatland complex types were identified in Minnesota. These types were differentiated on the basis of (1) the relative areas of bog and water track and (2) the path of runoff flowing across a peatland complex (see chapter 1 for a more detailed discussion).

Evaluation of Peatland Complex Types

With the adoption of a classification system, the process of identifying and ranking the best examples of each of the 11 types could begin. Peatland complex types were evaluated on the basis of the following attributes: the rarity of plants and animals, quality and completeness of bog and fen landform features, viability, extent of disturbance, and scientific research value. Points were awarded on established criteria and then normalized to a scale of 10, except in the case of wildlife, which was normalized to a scale of 5 (e.g., complex type 6; Fig. 19.7).

RARITY

The state list of endangered, threatened, and special-concern plant and animal species was used to assess rarity (Minnesota DNR 1984a). Rare species also served

Table 19.2. Rare vascular plant species and their relative values used in the peatland evaluation process

Scientific name	Common name	Value	State status
Carex exilis	A species of sedge	5	Special concern
Drosera linearis	Linear-leaved sundew	5	Threatened
Juncus stygius	Bog rush	5	Special concern
Drosera anglica	English sundew	4	Threatened
Eleocharis rostellata	Beaked spike-rush	4	Threatened
Rhynchospora fusca	Sooty-colored beak-rush	4	Special concern
Xyris montana	Yellow-eyed grass	4	Special concern
Nymphaea tetragona	Small white water lily	3	Threatened
Cladium mariscoides	Twig-rush	2	Special concern
Cypripedium arientium	Ram's head lady's slipper	2	Threatened
Rhynchospora capillacea	Hairlike beak rush	1	Threatened
Tofielda glutinosa	False asphodel	1	Special concern
Triglochin palustris	Marsh arrow-grass	1	Special concern
Carex sterlis	A species of sedge	1	Threatened
Carex capillaris	A species of sedge	1	Proposed special concern
Arethusa bulbosa	Dragon's mouth	0	Special concern

indirectly as indicators of unique water chemistry or hydrological conditions.

Vascular Plant Species. Sixteen rare vascular plant species were found in the patterned peatlands (Table 19.2). Because of the extensive knowledge of vascular plants in peatlands, they could be dealt with at a more sophisticated level than animal species. Field data were being collected on bryophytes but were not completed in time to be used in the evaluation. The weight or value given for each species in the evaluation was based on (1) the extent to which the species is found in Minnesota, (2) the degree to which the species depends upon patterned peatlands as habitat, and (3) the extent to which the species is already protected.

The highest value (5) was assigned to those species, such as *Juncus stygius*, that occur in very few sites in the state, are confined to patterned peatland habitats, and do not occur in any sites that are currently legally protected. Species given a low value, such as *Triglochin palustris*, are less dependent on patterned peatlands, since they occur in other peatland types such as calcareous fens in the prairie region. *Triglochin* is also found on more numerous sites, some of which are designated protected sites.

Although *Arethusa bulbosa* is listed as a species of special concern, it was not used in the evaluation. Because *Arethusa* was found in nearly every peatland that was field checked, the possibility that the absence of *Arethusa* at a particular site was a result of insufficient field checking could not be ruled out.

Animal Species. Eight rare animal species were found in patterned peatlands (Table 19.3). Habitat requirements for most of these wildlife species, unlike those for plant species, are uncertain, and peat landforms are not as useful in identifying where they are likely to be found. Surveys of wildlife in the large patterned peatlands have been minimal. The documentation of the occurrence of some species, such as the northern bog lemming and bog copper butterfly, is of-

ten the result of a chance encounter during the few scattered surveys that have been carried out. Because of this lack of data, less weight was given to wildlife values than to other attributes. Points given for each species, except the wolf, ranged from 1 to 2, depending on whether it occurred in or adjacent to the peatland type. Wolf Management Zones were used to award up to 3 points for the timber wolf.

QUALITY AND COMPLETENESS OF LANDFORM FEATURES

Each peatland complex type was evaluated through aerial photographs and ground surveys of both bog and fen landform features. An assessment of the presence and extent of fen and bog landform features was based on (1) areal extent, (2) pattern distinction, and (3) completeness of the range of features.

Evaluation of landforms was initially based on the assessment of features observed on black-and-white (1:90,000 and occasionally 1:15,840) and color infrared (1:60,000) aerial photographs. This information was supplemented by field observations.

Table 19.3. Rare animal species and their relative values used in the peatland evaluation process

Common name	Value	State status
Mammals		
Eastern timber wolf	1-3	Threatened
Northern bog lemming	1-2	Special concern
Birds		
Greater sandhill crane	1-2	Special concern
Sharp-tailed sparrow	1-2	Special concern
Short-eared owl	1-2	Special concern
Wilson's phalarope	1-2	Special concern
Yellow rail	1-2	Special concern
Butterfly		
Bog copper butterfly	1-2	Special concern

SCIENTIFIC VALUE

The scientific value assigned to each candidate peatland is based on (1) past research activity, (2) current or proposed research, (3) potential of an area for answering future research questions, and (4) accessibility and proximity to research institutions.

LACK OF DISTURBANCE

This category indicates the extent to which significant pristine areas still exist in each peatland and the degree to which vegetation and landform features have been modified by human disturbance. Aerial photos were primarily used to judge the extent of disturbance. Effects from roads and ditching were the most severe, whereas power lines and winter roads had a less significant impact.

GEOGRAPHICAL REPRESENTATIVITY

Because the evaluation resulted in the selection of examples of peatland types across their geographical range, it was not necessary to incorporate geographical representativity in the evaluation.

The results of the evaluation produced a rating of the top examples for each of the 11 peatland complex types. The number of candidate peatlands evaluated varied with the complex type. Two complex types are represented by a single example. Others, such as complex type 1, a simple raised bog, are more common. The top examples of the 11 complex types were found to occur in 18 peatlands. The number of complex types per peatland ranged from one to three.

Although the evaluation process attempted to evaluate the significance of the candidate peatlands objectively, it required many subjective judgments and was limited by the data available. The evaluation process was neither arbitrary nor rigid; it was a flexible tool that could incorporate feedback and fine tuning from peatland experts. The result is a process that can document the factors that were used, uniformly apply a given criterion to all peatlands, and produce findings that can be examined, critiqued, and supported by experts.

Rating of Ecological Significance

If it was feasible to preserve completely the best examples of each of the 11 peatland complex types, there would be no need for a comparative rating of ecological significance. The inevitable conflict between the need for preservation and the potential loss of economic opportunities, however, dictates that compromises may ultimately be necessary to gain broad-based political support for any protection plan. The ranking of peatlands provided a basis during negotiations for determining which compromises would have the lesser impact on peatland protection.

Because it is more practical from a resource-management point of view to manage entire peatlands rather than individual complexes, this evaluation combined data from all complex types occurring within each peatland. The same criteria that were used to evaluate individual peatland-complex types — rarity, quality, and completeness of landform features, lack of disturbance, and scientific value — were used to evaluate entire peatlands. In addition, the following factors were included:

PEATLAND COMPLEX TYPES

Peatlands were judged on the basis of the number and quality of complex types present. The relative quality of each complex type had been determined by its normalized score in the previous evaluation.

VIABILITY AND DEFENSIBILITY

This category gave an indication of the long-term stability of the peatland and how conducive its physical setting is to successful management for protection. Although peatland succession is only beginning to be understood, the following factors were considered influential in providing a secure habitat or environment for the peatland features of interest.

(1) *Size of peatland.* It was assumed that larger peatlands are more stable and less likely to undergo significant changes over a short time. A larger peatland is less vulnerable to climatic oscillations and human impacts such as air pollution.

(2) *Complexity of diversity of habitats.* It is less likely that rare species or plant communities will be displaced if a variety of habitat types exists for them to migrate to as peatland succession takes place.

(3) *Hydrologic isolation of peatland area.* Peatlands that are completely bordered by mineral soil are much more assured protection than those situated in the middle of an expanse of peatland.

(4) *Location in relationship to peatland watershed.* Peatland features downstream of disturbance are much more vulnerable to hydrologic impacts than those upstream. As a result, features in patterned peatlands that occur near the source of the watershed are more secure than those downslope.

After the evaluation of the 18 peatlands was completed, they were arranged according to their total points (Table 19.4). Caution should be used in interpreting this table: it is both difficult and artificial to characterize the ecological significance of a peatland with a single value. The total point values are shown only to provide a relative ranking. The subjective nature of the evaluation makes insignificant differences of only a few points. These disclaimers notwithstanding, a few generalizations are apparent from this analysis. First, the Red Lake peatland stands out clearly as the most significant peatland in the state. In fact, it is one of the largest and most complex and diverse peatlands of the continental peatland ecosystem. Comparable peatlands are not known in Europe or Asia. Only Hudson Bay lowlands may have peatlands of similar stature, and

Table 19.4. Summary of evaluation of 18 ecologically significant peatlands (updated 1989)

	Rarity		Quality and completeness		Complex type	Viability	Disturbance	Scientific research value	Total score
	Plant	Animal	Fen	Bog					
Red Lake	10	5	10	10	8	10	7	10	70
Lost River	7	1	7	7	9	8	9	8	56
Myrtle Lake	6	2	6	9	4	8	10	10	55
N. Black River	4		4	8	10	8	9	8	51
Sand Lake	3	2	4	3	3	7	9	7	38
Mulligan Lake	4	3	5		4	7	10	5	38
Pine Creek	4	1	4		4	7	10	5	35
Lost Lake	3		2	2	4	5	10	6	32
S. Black River	1	1	2	5	2	7	10	4	32
E. Rat Root River			3	3	3	6	10	3	28
Winter Road Lake		1	2		2	6	10	5	26
Hole-in-the-Bog				4	4	3	9	5	25
Nett Lake	2		1		3	1	10	5	22
Luxemberg	1	1	2		1	3	10	4	22
Sprague Creek	2	1	1		2	7	5	3	21
Wawina			1	2	2	5	5	5	20
W. Rat Root River			1	2	1	3	2	4	13
Norris Camp		1	1			4	1	2	9

these are of the boreal nonforested type. This peatland is regarded as internationally significant.

Three peatlands — Myrtle Lake, Lost River, and North Black River — are considered to be of national significance. Along with Red Lake peatland, they have been nominated as National Natural Landmarks, although only Myrtle Lake and portions of the Red Lake peatland have been so designated.

The remainder can be divided into three groups: peatlands with scores clustered in the thirties, those clustered in the twenties, and those close to 10 or less. These peatlands are considered of state significance. The vast majority of the state peatlands would not show up on this scale at all.

The results of the two phases of the evaluation provided the basis for determining which peatlands need protection. The degree to which the peatlands actually will be protected, however, will depend on the extent to which competition for the peatland resources can be resolved.

Development of Management Guidelines

Management Area Concept

Long-term protection of peatland features, unlike protection of terrestrial ecosystems, cannot be achieved merely by restricting disturbance in the immediate area. Protection of peatland ecosystems is complicated by the intimate interdependence between the features and the surrounding hydrology. The processes that perpetuate the peatland ecosystem, as well as plant communities and rare species, are extremely sensitive to changes in water levels and water chem-

istry. Even slight alteration in the surrounding water quality and quantity can cause significant change in vegetation. Patterned peatlands resemble riverine systems in that water is constantly flowing across a very gently sloping landscape. Consequently, features in the center of a peatland are particularly vulnerable to hydrological disruption occurring upstream or laterally.

Preservation of significant peatland features requires two types of protection. First, the peatland features must be protected from direct, on-site physical disturbance. Secondly, the hydrology of the surrounding peatland area must be sufficiently protected to maintain the ecological integrity of the features.

To accommodate this need, a two-level management approach was developed. First, the area that contains the features of greatest ecological importance was defined as the core area, where management guidelines should be concerned with direct on-site disturbance. A peatland could have one or more core areas. Activities that would have to be evaluated and possibly excluded or restricted include logging, ditching, peat and mineral mining, using motorized vehicles, and chemical applications.

The area surrounding the core area, the buffer required to maintain its ecological integrity, was defined as the watershed protection area. Management guidelines in this area are confined to those activities such as ditching and peat and mineral mining that could have a significant hydrological impact.

Compromises worked out between protection and development interests resulted in the splitting of the core areas into two types: Scientific and Natural Areas (SNAs) and Peatland Scientific Protection Areas (PSPAs).

Table 19.5. Prohibited and permitted activities
in the peatland protection areas

Activities	Management categories		Watershed protection areas
	Core areas		
	SNA	PSPA	
New Ditches	No	No	No
Improve Ditches	No[a]	No[a]	No
Repair ditches	No[a]	No[a]	No
Peat mining	No	No	No
Industrial minerals mining	No	No	No
Mineral exploration	No	Yes[a]	Yes
Timber harvesting	No	No	Yes
New corridors of disturbance	No	Yes[b]	Yes
Nonmotor recreation activities	Yes	Yes	Yes
Scientific & educational work	Yes	Yes	Yes
Maintenance & use of disturbance corridors	Yes[a]	Yes	Yes
Motorized uses on corridors of disturbance	Yes[c]	Yes	Yes
Disease, fire control	Yes[a]	Yes[a]	Yes
New winter roads	No	Yes[b]	Yes
Metallic mining	No	Yes[a]	Yes[a]
Any other adverse action	No	No	No

[a] with conditions or exceptions
[b] if permitted by management plan
[c] existing roads only

The SNAs would receive the greater protection. A summary of proposed restrictions is shown in Table 19.5.

The boundaries of the core and watershed protection areas of candidate peatlands have been based on the available data and the current understanding of peatland hydrology. An example of these boundaries is shown in a map for the Mulligan Lake peatland (Fig. 19.4). Much is still unknown, however, about the hydrology of large contiguous peatlands or the extent to which impacts would extend there. Therefore, boundaries of watershed protection areas in which natural divisions are marked by mineral-soil uplands, shallow peat, or river could be drawn with more confidence than boundaries of those that occur in a great expanse of peatland.

The total area of the 18 ecologically significant peatlands is just under 500,000 acres (Table 19.6). Approximately one-third of this acreage (172,239 acres) is core area, and the remaining two-thirds (323,835 acres) is watershed protection area. Just under half of the total acreage is in one peatland, the Red Lake peatland.

The 18 ecologically significant peatlands amount to about 7% of the state's peatland resource. Less than 3% of the state's peatlands are included in the more restrictive core area.

Table 19.6. Acreage of core and watershed protection areas for the 18 ecologically significant peatlands

	Core areas			Watershed protection areas		Total core and watershed protection areas	Total state owned
	State owned		Nonstate lands[a]	State owned	Nonstate lands[a]	All lands	(in %)
	SNA	PSPA					
Red Lake	82,783		4,796	137,682	8,245	233,506	94
Myrtle Lake	22,630		320	12,194	420	35,564	98
Lost River	6,198	5,650	40	46,045	3,244	61,177	95
N. Black River		1,220	9,573	12,744	18,815	42,352	33
Sand Lake		4,545	379	5,518	2,929	13,371	75
Mulligan Lake		5,236	909	12,205	2,386	20,736	84
Pine Creek		944	0	1,652	0	2,596	100
Lost Lake		200	2,460	1,780	2,542	6,982	28
S. Black River		5,992	0	8,499	78	14,569	99
Winter Road Lake		2,469	1,832	11,209	3,475	18,985	72
E. Rat Root River		2,732	160	3,375	1,648	7,915	77
Hole-in-the-Bog		1,482	140	882	660	3,164	75
Wawina		0	4,092	305	4,285	8,682	4
Nett Lake		0	400	0	820	1,220	0
Luxemberg		592	540	1,630	360	3,122	71
Sprague Creek		820	0	10,710	80	11,610	99
W. Rat Root River		1,430	20	2,450	100	4,000	97
Norris Camp		1,316	340	4,226	640	6,522	85
Total acres	111,611	34,628	26,001	273,106	50,727	496,073	85

[a] Lands not affected by proposed legislation.

Land Ownership

The degree to which the state can influence management of these ecologically significant peatlands is directly related to the extent of the state's ownership within them. Fortunately, the high percentage of public ownership of peatlands gives Minnesota a distinct advantage in managing the peat resource. In Europe and probably in most other peatland states, peatlands are dominated by private ownership. Minnesota manages 85% of the ecologically significant areas that have been identified, and 9% are under federal ownership (Table 19.6).

Of the 18 peatland protection areas, all but four are more than 70% owned by the state, and seven have state ownership exceeding 90%.

Of the remaining four peatlands, the county controls most of Wawina (83%) and Lost Lake (67%), whereas Indian reservation lands predominate in North Black River (48%) and Nett Lake (89%).

Literature Cited

Ahrens, M. H. 1987. A contribution to the history of land administration in Minnesota: The origins of the Red Lake Game Preserve. M.A. thesis. University of Minnesota, Minneapolis, Minnesota, USA.

Cajander, A. K. 1913. Studien uber die Moore Finnlands. Acta Forestalia Fennica. 2(3):1-208.

Davis, A. K., G. L. Jacobson, Jr., L. S. Widoff, and A. Zlotsky. 1983. Evaluation of Maine peatlands for their unique and exemplary qualities.

Goodwillie, R. 1980. European peatlands. European Committee for the Conservation of Nature and Natural Resources. Council of Europe Nature and Environment Series no. 19.

Goodwin, R. H., and W. A. Niering. 1975. Inland wetlands of the United States: Evaluated as potential registered landmarks. National Park Service. Superintendent of Documents, Government Printing Office.

Heikurainen, L., and J. Laine. 1980. Finnish peatlands and their utilization. Unpublished manuscript.

Kivinen, E., and P. Pakarinen. 1980. Peatland areas and the proportion of virgin peatlands in different countries. Pages 52-54 *in* 6th Proceedings of the International Peat Congress, Duluth, Minnesota, USA.

Minnesota Department of Natural Resources. 1981. Minnesota Peat Program: Final report.

———. 1984a. Checklist of endangered and threatened animal and plant species of Minnesota.

———. 1984b. Protection of ecologically significant peatlands in Minnesota.

———. 1984c. Recommendations for the protection of ecologically significant peatlands: Preliminary report.

Official Boards of Beltrami and Lake of the Woods Counties, Bemidji Civic and Commerce Association, and Baudette Chamber of Commerce. 1928. Proposed northern Minnesota game refuge.

Prout, C. 1962. Policy statement: Lake Agassiz Peatlands Natural Area. Minnesota Department of Conservation.

Ruuhijarvi, A. K. 1983. The Finnish mire types and their regional distribution. Pages 46-47 *in* A. J. P. Gore, editor. Mires: Swamps, bogs, fen and moor. Vol. 4B, Regional Studies. Elsevier, Amsterdam, Netherlands.

U.S. Public Land Survey. 1892.

Contributors

Norman E. Aaseng is a plant ecologist with the Natural Heritage Program of the Minnesota Department of Natural Resources (DNR). He was peatland ecologist for the Peat Program of the DNR Minerals Division from 1979 to 1989. He has a M.S. (1976) in forest ecology and a B.S. (1974) in forest science from the University of Minnesota College of Forestry.

William E. Berg received his B.S. in wildlife management and M.S. in wildlife ecology from the University of Minnesota. His master's research focused on the ecology of moose in the northwestern Minnesota peatlands. A wildlife biologist with the Minnesota Department of Natural Resources since 1971, Berg has been based at the Red Lake Wildlife Management Area in the Red Lake peatlands and with the Forest Wildlife Populations and Research Group at Grand Rapids. Much of his research has focused on prescribed burning and on the ecology and populations of moose, sharp-tailed grouse, and large furbearers including timber wolf, coyote, bobcat, beaver, and otter.

Kristine L. Bradof has a B.S. in geology and environmental geology from Beloit College and an M.S. in geology from the University of Minnesota. Her research on Red Lake peatland was supported by University of Minnesota and National Science Foundation Graduate Fellowships. Drawn to Lake Superior and the North Woods, she has worked as an interpretive naturalist for the Environmental Learning Center (now Wolf Ridge ELC), the Gunflint Ranger District, and the Minnesota Department of Natural Resources, and as a researcher for the National Park Service at Apostle Islands National Lakeshore. A former employee of Chicago's Field Museum of Natural History and the Science Museum

of Minnesota, she is now coordinator of the Regional Groundwater Education in Michigan (GEM) Center at Michigan Technological University in Houghton.

Kenneth Brooks is a professor in the faculty of the University of Minnesota College of Natural Resources. After receiving a Ph.D. in watershed management and hydrology at the University of Arizona in 1970, he worked for the Corps of Engineers from 1970 to 1975. Since he joined the University of Minnesota in 1975, his research and teaching have focused on the hydrology of peatland and upland forested watersheds and the development of a hydrologic model to predict streamflow response from such watersheds. Brooks is the author or co-author of more than sixty technical papers as well as a textbook, *Hydrology and the Management of Watersheds* (1991).

Barbara A. Coffin received an M.S. from the University of Minnesota Department of Botany in 1977. She was plant ecologist and then director of the Minnesota Department of Natural Resources Peat Program, which sponsored much of the research presented in this book. From 1979 until 1990, Coffin was director of the Minnesota Natural Heritage Program, a project sponsored by The Nature Conservancy and the Minnesota DNR and dedicated to the conservation of endangered species and threatened habitats. She is now natural science acquisition editor for the University of Minnesota Press.

Robert Djupstrom has been supervisor of the Scientific and Natural Areas Program of the Minnesota Department of Natural Resources since 1982. Before joining MDNR in 1975, Djupstrom was employed by the U.S. Department of the Interior in the Bureau of Outdoor Recreation. A graduate of the University of Wisconsin, Madison, he holds a degree in natural resources management with a major in wildlife ecology.

Paul H. Glaser received his A.B. degree from Rutgers — The State University and his M.S. and Ph.D. degrees from the University of Minnesota. He has conducted research on the ecology, paleoecology, and hydrogeology of peatlands across the boreal region of North America and has also studied blanket bogs in western Ireland. He is currently a research associate at the Limnological Research Center of the University of Minnesota.

JoAnn M. Hanowski is an avian ecologist in the Center for Water and the Environment of the Natural Resources Research Institute at the University of Minnesota, Duluth. Over the past twelve years she has coordinated several projects in which birds were used to assess the effects of environmental perturbations in the Great Lakes region.

Barbara Hansen is a fossil-pollen analyst who has a palynological consulting business, the Pollen Collection, in Stillwater, Minnesota. She is an affiliate of the University of Minnesota Limnological Research Center, where she participates in multidisciplinary research projects.

Miron L. Heinselman was educated at the University of Minnesota, where he received a Ph.D. in forestry and peatlands ecology in 1961. During a research career with the U.S. Forest Service from 1948 to 1974, he studied the relations among peatland topography, stratigraphy, water movements, water chemistry, and vegetation patterns; in 1961 Heinselman identified and described the patterned peatlands of Minnesota's Glacial Lake Agassiz region. For many years he studied the ecological role of fire in the virgin forests of the Boundary Waters Canoe Area wilderness, and since retirement he has continued research and writing on the natural role of fire in forest ecosystems.

C. R. Janssen is a professor of historical plant geography at the University of Utrecht, the Netherlands, with special interest in paleoecology and vegetation history of the past 15,000 years. He has published studies pertaining to the reconstruction of past plant communities and environments in the Netherlands, in northern Minnesota, and in mountain ranges in France, Spain, and Portugal.

Jan A. Janssens is a research associate in the University of Minnesota Department of Ecology, Evolution, and Behavior. Most of his work on the Red Lake peatland was done while he was affiliated with the Limnological Research Center at the University of Minnesota. He

is currently compiling bryophyte baseline data for the state of Minnesota and studying the effects of climatic change on the wetlands. Long-term research goals are an annotated moss checklist for the state, a wetland moss flora, and a North American quantitative reference data set with edaphic and climatic constraints on moss species for use in paleoenvironmental reconstruction.

Daryl R. Karns is an associate professor of biology at Hanover College in southeastern Indiana. A vertebrate biologist with research interests in the ecology and evolution of amphibians and reptiles, he received a Ph.D from the University of Minnesota in 1984. His work has involved studies of amphibians and reptiles in the dry savanna of West Africa, the peatlands of northern Minnesota, the temperate forests of the Midwest and the Appalachians, and, most recently, the tropical rain forests of Borneo.

Mary Keirstead has worked as an editor, writer, and environmental planner for the Minnesota Department of Natural Resources since 1979.

Melissa L. Meyer is an assistant professor of history at the University of California at Los Angeles. Among her publications are " 'We Can No Longer Get a Living as We Used to': Dispossession and the White Earth Anishinaabeg, 1889-1920," *The American Historical Review* 96 (1991); "Signatures and Thumbprints: Ethnicity among the White Earth Anishinaabeg, 1889-1920," *Social Science History* 14 (1990); and *The White Earth Tragedy: Ethnicity and Dispossession at a Minnesota Anishinaabe Reservation, 1889-1920*, forthcoming from the University of Nebraska Press.

Gerald J. Niemi is director of the Center for Water and the Environment of the Natural Resources Research Institute and an associate professor of biology at the University of Minnesota, Duluth. Niemi has been studying northern Minnesota birds for over fifteen years and has written more than twenty-five publications on the subject.

Gerda E. Nordquist coordinates animal surveys for the County Biological Survey of the Minnesota Department of Natural Resources. Her graduate research focused on small mammals in Minnesota peatlands and their relationship to structural and compositional attributes of peatland habitats.

Donald I. Siegel, associate professor of geology at Syracuse (New York) University, is a University of Minnesota-trained hydrogeologist and geochemist who specializes in research that integrates wetland ecology, hydrology, and chemistry. He is currently conducting studies on wetland solute and gas transport and paleo-hydrogeology.

Mary Whelan is an assistant professor of anthropology at the University of Iowa. Her interests include North American archaeology, zooarchaeology, and environmental archaeology.

Cathy (Barnosky) Whitlock is an associate professor of geography at the University of Oregon. Her research interests focus on the Quaternary vegetation and climate history of the western United States, Minnesota, Pennsylvania, Ireland, and China. She received a Ph.D. in geological sciences at the University of Washington in 1983 and was a NATO Postdoctoral Fellow at Trinity College in Dublin in 1984. Whitlock was a curator of paleobotany at Carnegie Museum of Natural History from 1984 to 1990 and assistant professor of geology at the University of Pittsburgh from 1989 to 1990. Together with H. E. Wright Jr., she co-edited *Late Quaternary Environments of the Soviet Union* (English edition, University of Minnesota Press).

Herbert E. Wright, Jr. a graduate of Harvard University, has been on the geology faculty at the University of Minnesota since 1947. He retired in 1988 as Regents' Professor of Geology, Ecology, and Botany and director of the Limnological Research Center. His research interests have focused on the history of landscapes — landforms, vegetation, and lakes — as they have been affected by climatic change and cultural developments.

Index